Foundation GCSE Mathematics: Revision and Practice

A. Ledsham and M. E. Wardle

Oxford University Press

Oxford University Press, Great Clarendon Street, Oxford OX2 6DP

Oxford New York
Athens Auckland Bangkok Bogota Buenos Aires
Calcutta Cape Town Chennai Dar es Salaam
Delhi Florence Hong Kong Istanbul Karachi
Kuala Lumpur Madrid Melbourne Mexico City
Mumbai Nairobi Paris São Paulo Singapore
Taipei Tokyo Toronto

and associated companies in
Berlin Ibadan

Oxford is a trade mark of Oxford University Press

First published 1996
Reprinted 1996, 1997, 1998

ISBN 0 19 914576 8 School edition
ISBN 0 19 914671 3 Bookshop edition

Typeset and illustrated by Tech-Set, Gateshead, Tyne and Wear.
Printed in Great Britain by
Butler & Tanner Ltd, Frome and London

Preface

This book covers all the material required for Levels 4, 5 and 6 in the Mathematics National Curriculum. It is particularly suitable for **Foundation Level** candidates working at Key Stage 4 towards GCSE in Mathematics, but it has also been designed so that it can be used by any student working at Levels 4–6.

Most of the material is divided into four sections: Number, Algebra, Shape and Space, and Data Handling, corresponding to Attainment Targets 2, 3 and 4. Each section is divided into chapters, which are sub-divided into separate topics. Within each topic the exercises provide progress, where appropriate, from Level 4 to Level 6.

For each topic there are hints and reminders of key ideas and language, together with worked examples followed by graded exercises. These are specifically designed to enable the student to develop confidence and gain a clear idea of what will be required. Questions which require the broader range of abilities and work associated with Attainment Target 1 (Using and Applying Mathematics) are included in a separate section towards the end of the book. This material contains fifty starting points for investigations, which, together with the examples of longer questions, give an indication of what might be required in coursework or examinations. However, it is strongly recommended that the particular syllabus is consulted for the specific style of questions required by the board. This book does not attempt to cover the related use of the computer, given the wide diversity of equipment and facilities now available in schools and colleges.

While primarily designed as a revision and practice book for GCSE, the material is equally suitable for use as a general resource of practice exercises for students at any stage from 11 to 16, supplementing other material or following on from normal class teaching.

The essence of this book is simplicity and directness.

A. Ledsham and M.E. Wardle
September 1995

CONTENTS

SECTION A: NUMBER

1 Understanding place value and the decimal system

1.1 Reading, writing and ordering whole numbers

All numbers are made up of figures. When numbers are written in words, the place value of each figure is mentioned unless the figure is nought. For example:

235 has three figures. It is made up of two hundreds, three tens and five units
205 is written as two hundred and five
230 is written as two hundred and thirty

Exercise 1.1

Write these numbers in words:

1 **a** 25 **b** 29 **c** 32 **d** 37

2 **a** 43 **b** 56 **c** 74 **d** 91

3 **a** 120 **b** 130 **c** 136 **d** 220

4 **a** 400 **b** 405 **c** 905 **d** 900

5 Salisbury Cathedral is 123 metres high.
Write this height in words.

6 The Dead Sea shoreline is 394 metres below sea level.
Write this number in words.

7 The River Thames flows 336 kilometres from the Father Thames statue to Southend.
Write this number in words.

8 The distance from London to Glasgow is 640 kilometres.
Write this number in words.

Exercise 1.2

Write these numbers in figures:

1 Eighteen
2 Thirteen
3 Twenty eight
4 Thirty four
5 Ninety two
6 Ninety
7 Three hundred and seventy nine
8 Five hundred and twenty four
9 Seven hundred and ninety five
10 Eight hundred and twenty five

In Questions 11–15, a number is written in a sentence.
Rewrite the number in figures.

11 The distance from London to Aberdeen is eight hundred and thirty eight kilometres.

12 The planet Mars takes six hundred and eighty seven days to orbit the sun.

13 The weight of a ferry boat is nine hundred and five tonnes.

The place value of any figure in a number can easily be found.

The place value of each figure in the number 453 is worked out as:

Hundreds	Tens	Units
4	5	3

The place value of the 4 is four hundreds or 400.
The place value of the 5 is five tens or 50.
The place value of the 3 is three units or 3.

Exercise 1.3

Give the place value of the underlined figure:

1 $1\underline{3}$ **2** $1\underline{7}$ **3** $24\underline{6}$ **4** $3\underline{9}1$ **5** $\underline{7}80$
6 $\underline{6}02$ **7** $47\underline{2}$ **8** $1\underline{9}0$ **9** $3\underline{5}2$ **10** $\underline{3}01$

The place value of figures can be used to arrange numbers in order of size. The example below shows how this is done.

Example 1

Arrange these numbers in order of size, starting with the largest:

302 203 230 320

The order is:

320 (three hundreds, two tens and no units)
302 (three hundreds, no tens and two units)
230 (two hundreds, three tens and no units)
203 (two hundreds, no tens and three units)

Exercise 1.4

Arrange these numbers in order of size, starting with the largest:

1 504 540 405 450
2 120 201 102 210
3 213 312 132 321 123 231
4 798 987 897 789 978 879
5 302 320 203 230 330 303 220 202

Arrange these numbers in order of size starting with the smallest:

6 132 123 133 122
7 599 589 588 598
8 934 944 933 943
9 122 212 112 221 211 121
10 656 565 655 665 556 566

It is well known that more than one number can be made up from the same set of figures.

For example, the figures 5, 3 and 2 can be used to make up these six numbers:

532 five hundred and thirty two
352 three hundred and fifty two
253 two hundred and fifty three
523 five hundred and twenty three
325 three hundred and twenty five
235 two hundred and thirty five

The six numbers have different values because the place value of at least one figure is changed when the figures are arranged in a different order.

Exercise 1.5

For each question find all the three-figure numbers that can be made using each of the figures once.
Arrange them in descending order.

1 3, 4 and 7 **2** 2, 5 and 9 **3** 4, 6 and 8 **4** 1, 3 and 5

5 7, 2 and 0 **6** 9, 6 and 0 **7** 5, 5 and 2 **8** 8, 8 and 4

9 3, 3 and 9 **10** 6, 6 and 0

For each question, find all the three-figure numbers that can be made using each of the figures once.
Arrange them in ascending order.

11 3, 6 and 8 **12** 4, 5 and 7 **13** 2, 4 and 6 **14** 1, 6 and 9

15 8, 3 and 0 **16** 5, 4 and 0 **17** 7, 7 and 3 **18** 2, 2 and 6

19 1, 1 and 8 **20** 9, 9 and 0

1.2 Multiplying whole numbers by 10 or 100

Multiplying a number by 10 changes the place value of each of the figures.
For example:

$7 \times 10 = 70$	seven units become seven tens
$51 \times 10 = 510$	five tens and one unit become five hundreds and one ten

It can easily be seen that if we multiply a number by 10, each number appears one place to the left.

Multiplying a number by 100 also changes the place value of each of the figures.
For example:

$5 \times 100 = 500$	five units become five hundreds
$10 \times 100 = 1000$	one ten becomes one thousand

We can see that if we multiply a number by 100, each number appears two places to the left.

Exercise 1.6

Multiply each of these numbers by 10:

1 2 **2** 8 **3** 25 **4** 43 **5** 70 **6** 99

Multiply each of these numbers by 100:

7 4 **8** 7 **9** 6 **10** 9 **11** 3 **12** 2

Given that 1 centimetre = 10 millimetres, change each of these distances to millimetres:

13 15 centimetres **14** 27 centimetres **15** 60 centimetres
16 90 centimetres **17** 9 centimetres **18** 4 centimetres

Given that 1 metre = 100 centimetres, change each of these distances to centimetres:

19 8 metres **20** 3 metres **21** 6 metres

22 Jodie's cat weighs 2 kilograms. Her dog is 10 times heavier. How heavy is her dog?

23 Sean has a toy soldier which is 16 centimetres tall. If he is 10 times taller than the toy soldier, how tall is he?

24 The distance from London to Wood Green is 9 kilometres.
  a If Huntingdon is 10 times further away from London than Wood Green, how far is it from London to Huntingdon?
  b If Inverness is 100 times further away from London than Wood Green, how far is it from London to Inverness?

25 Copy and complete the following:

a $21 \times 10 = ?$	**b** $35 \times 10 = ?$	**c** $90 \times 10 = ?$
d $8 \times 100 = ?$	**e** $2 \times 100 = ?$	**f** $7 \times 100 = ?$
g $5 \times ? = 500$	**h** $9 \times ? = 900$	**i** $4 \times ? = 400$
j $36 \times ? = 360$	**k** $45 \times ? = 450$	**l** $40 \times ? = 400$
m $80 \times ? = 800$	**n** $100 \times ? = 1000$	**o** $? \times 10 = 750$
p $? \times 10 = 420$	**q** $? \times 10 = 100$	**r** $? \times 100 = 500$
s $? \times 100 = 600$	**t** $? \times 100 = 1000$	

1.3 The relationship between place values in whole numbers

The place value of a figure usually means more than one thing. The figure 5 in the number 500 can mean:

500 is FIVE hundreds
or 500 is FIFTY tens
or 500 is FIVE HUNDRED units

Example 1

State the number of **a** thousands, **b** hundreds, **c** tens and **d** units in the number 3000.

a The number of thousands is THREE $(3000 = 3 \times 1000)$
b The number of hundreds is THIRTY $(3000 = 30 \times 100)$
c The number of tens is THREE HUNDRED $(3000 = 300 \times 10)$
d The number of units is THREE THOUSAND $(3000 = 3000 \times 1)$

Exercise 1.7

For Questions 1 to 4 state the number of **a** thousands, **b** hundreds, **c** tens and **d** units in the number:

1 9000 **2** 4000 **3** 2000 **4** 6000

For Questions 5 to 8 state the number of **a** hundreds, **b** tens and **c** units in the number:

5 800 **6** 300 **7** 200 **8** 900

For Questions 9 to 12 state the number of **a** tens and **b** units in the number:

9 70 **10** 50 **11** 30 **12** 80

Example 2

9000 centimetres of dressmaking tape are wound on a reel.
If the tape is cut up into 100 centimetre lengths, how many lengths are made?

9000 is NINETY hundreds $(9000 = 90 \times 100)$, therefore 90 lengths are made.

Exercise 1.8

1 A grocer has 3000 grams of salt. She pours it into 100 gram packets.
How many packets will she fill?

2 A coal merchant has 8000 kilograms of coal. He packs it into 100 kilogram bags.
How many bags will he fill?

3 A farmer has 2000 kilograms of grain. He packs it into 100 kilogram cartons.
a How many cartons will he fill?
b If, instead, he packs the grain into 10 kilogram cartons, how many of these cartons will he fill?

4 7000 centimetres of string are wound in a ball.
a If the string is cut up into 100 centimetre lengths, how many lengths will be cut?
b If, instead, the string is cut up into 10 centimetre lengths, how many lengths will be cut?

5 A baker has 900 loaves. She packs them on to trays which hold 10 loaves each.
How many trays will she require?

6 I have 5 one pound coins (or 500 pence) in my pocket.
If I take them to a bank and change them for 10 p pieces, how many 10 p pieces will I receive?

7 I have three 1 kilogram weights (or 3000 grams).
a If I try to balance them on a pair of scales with 10 gram weights, how many 10 gram weights will I need?
b If I then try to balance them on a pair of scales with 100 gram weights, how many 100 gram weights will I need?

8 Rosie has a flask which contains 2 litres (or 2000 millilitres) of tea.
How many cups of capacity 100 millilitres will it fill?

Example 3

A hardware shop manager has 2000 screws. He distributes them equally into 10 boxes.
a How many screws will there be in each box?
b If, instead, he distributes them equally into 100 small packets, how many screws will there be in each packet?

a 2000 is TWO HUNDRED tens ($2000 = 200 \times 10$), therefore there will be 200 screws in each box.
b 2000 is TWENTY hundreds ($2000 = 20 \times 100$), therefore there will be 20 screws in each packet.

Exercise 1.9

1 A grocer has 5000 grams of rice.
If he pours it out equally into 10 packets, how many grams will there be in each packet?

2 A school orders 200 books, one for each pupil in the school.
If the school has 10 classes of equal size, how many pupils are there in each class?

3 A gardener has 4000 grams of seeds.
a If she pours them out equally into 10 large packets, how many grams will there be in each packet?
b If, instead, she pours them out into 100 small packets, how many grams will there be in each small packet?

4 A chemist has 2000 sleeping pills.
a If she distributes them equally into 10 large bottles, how many pills will there be in each bottle?
b If, instead, she distributes them into 100 small packets, how many pills will there be in each packet?

5 500 people are waiting for an excursion train. When the empty train arrives they have to sit in reserved seats which are equally distributed between 10 coaches.
How many people will sit in each coach?

1.4 Decimal notation

Measurements involving mixed units can often be written out more simply by using a decimal point.
Look at this screwdriver:

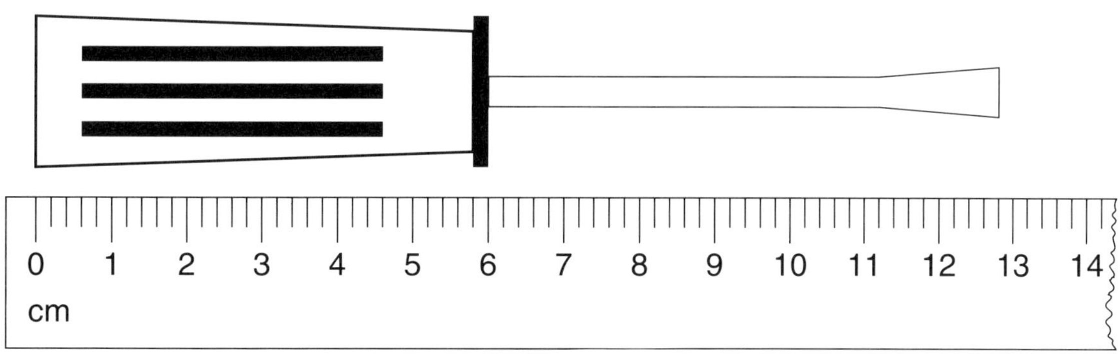

The small divisions on the ruler are 2 millimetres each, so the length of the screwdriver is 12 centimetres 8 millimetres.

This can be written as 12.8 centimetres (12.8 cm), or 128 millimetres (128 mm).

Exercise 1.10

Give the lengths of these in **a** centimetres and millimetres, **b** centimetres, **c** millimetres.

1

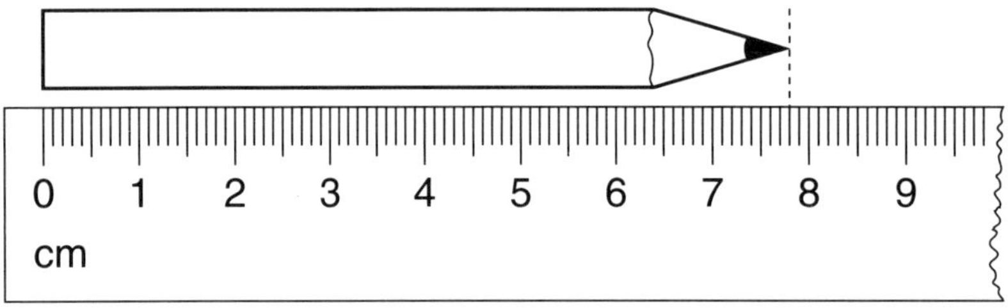

2

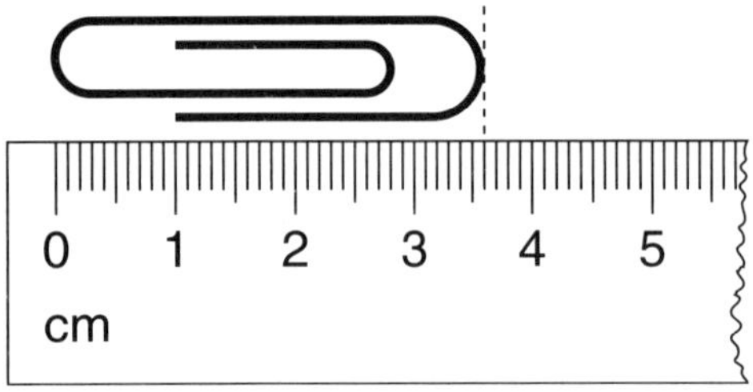

3

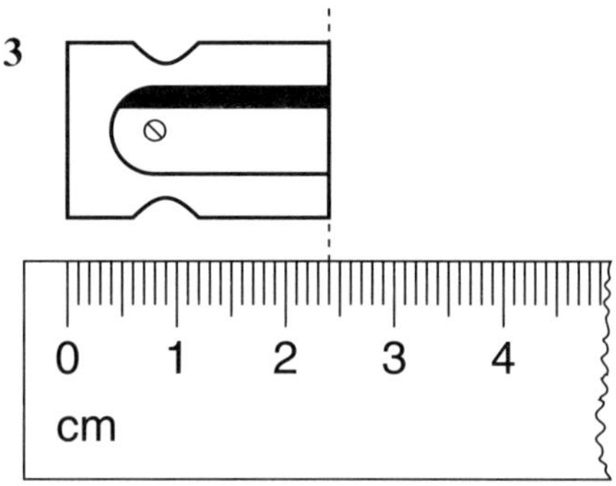

4

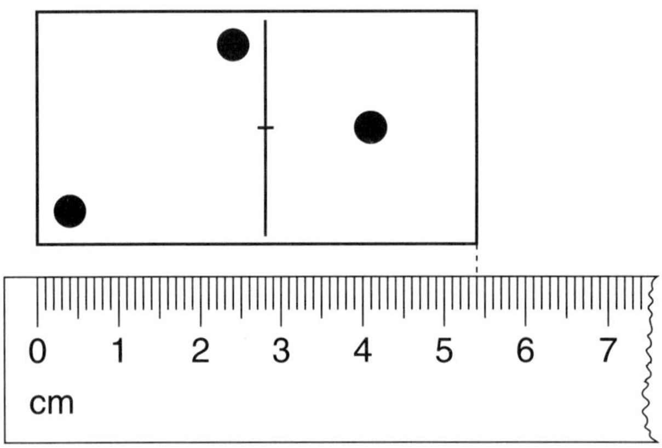

5

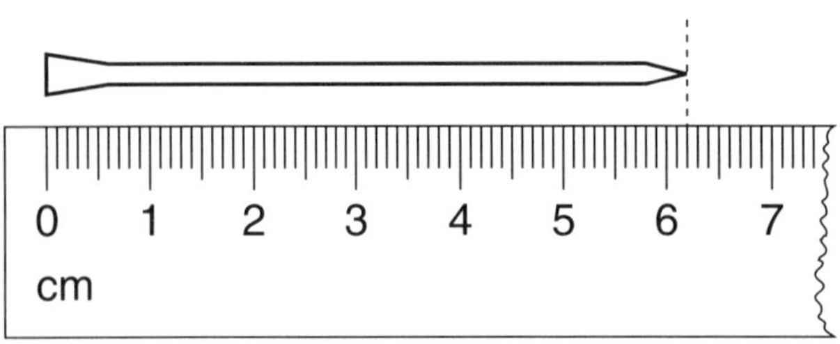

Example 1

A sewing needle is 5 cm 7 mm in length.
What is its length in **a** cm and **b** mm?

a 5 cm 7 mm = 5.7 cm **b** 5 cm 7 mm = 57 mm

Exercise 1.11

Copy and complete this table:

Object	Length (cm and mm)	Length (cm)	Length (mm)
1 Matchbox	6 cm 2 mm		
2 Toy car	8 cm 6 mm		
3 Rubber	3 cm 7 mm		
4 Pen	14 cm 3 mm		
5 Screwdriver	16 cm 2 mm		
6 Spanner		18.5 cm	
7 Penknife		13.1 cm	
8 Teaspoon		12.7 cm	
9 Toy lorry		25.2 cm	
10 Pencil case			243 mm
11 Knitting needle			284 mm
12 Pair of pliers			225 mm

Look carefully at the illustration below. A measuring tape, which is marked in metres and centimetres, is being used to measure the length of a piece of wood.
The length is:

1 metre 5 centimetres or 1.05 metres or 105 centimetres.

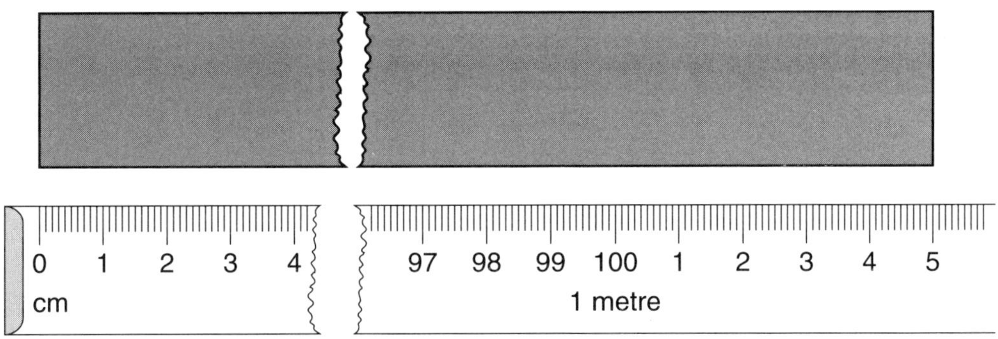

Exercise 1.12

The table below gives the height of each pupil in the 'Normans' house at Littletots Junior School.
Copy and complete the table:

Pupil	Height (m and cm)	Height (m)	Height (cm)
1 Joshua	1 m 36 cm		
2 Tom	1 m 28 cm		
3 Kate	1 m 16 cm		
4 Sally	1 m 8 cm		
5 Mandeep	1 m 6 cm		

Exercise 1.13

The table below gives the height of each pupil in the 'Vikings' house at Littletots Junior School.
Copy and complete the table:

Pupil	Height (m and cm)	Height (m)	Height (cm)
1 Jotinder		1.38 m	
2 Mark		1.29 m	
3 Lucy		1.3 m (or 1.30 m)	
4 Kelly		1.07 m	
5 Carl		1.05 m	

Exercise 1.14

The table below shows the height of each pupil in the 'Saxons' house at Littletots Junior School.
Copy and complete the table:

Pupil	Height (m and cm)	Height (m)	Height (cm)
1 Amy			134 cm
2 Drashma			125 cm
3 Liam			113 cm
4 Robert			109 cm
5 Andrew			101 cm

Exercise 1.15

The table below shows several dimensions.
Copy and complete the table.

Dimension to be measured	Measurement made (m and cm)	Measurement made (m)	Measurement made (cm)
1 Length of a table	1 m 55 cm		
2 Width of a table	1 m 10 cm		
3 Height of a ceiling		2.15 m	
4 Height of a chair		1.05 m	
5 Width of a front door			115 cm
6 Length of a garden fence			350 cm

1.5 Ordering decimals and appreciating place values

In Section 1.3 we studied the relationship between place values in whole numbers. We will now extend this to consider the place values of decimal numbers.

Example 1

Write down the place value of each figure in: 32.756

The figure 3 means THREE tens.
The figure 2 means TWO units.
The figure 7 in the first decimal place means SEVEN tenths.
The figure 5 in the second decimal place means FIVE hundredths.
The figure 6 in the third decimal place means SIX thousandths.

Note the following:

0.75 can be regarded as either $\frac{7}{10} + \frac{5}{100}$ or $\frac{75}{100}$

0.756 can be regarded as either $\frac{7}{10} + \frac{5}{100} + \frac{6}{1000}$ or $\frac{756}{1000}$

Exercise 1.16

Write down the place value of each figure in:

1 64.257 **2** 92.346 **3** 25.783 **4** 35.29
5 5.647 **6** 3.852 **7** 96.1 **8** 9.7

Example 2

Write in descending order of size:

0.7 0.57 0.399 0.81

The four terms can be written as follows:

0.700 0.570 0.399 0.810

Therefore the order of descending size is:

0.810 0.700 0.570 0.399

or 0.81 0.7 0.57 0.399

Exercise 1.17

Write in descending order of size:

1 0.46 0.65 0.37 0.58

2 0.276 0.458 0.367 0.549

3 0.732 0.55 0.823 0.641

4 0.458 0.279 0.54 0.36

5 0.6 0.54 0.48 0.721

6 0.173 0.4 0.2 0.36

Example 3

Write in ascending order of size:

3.042 3.42 3.024 3.24

The order of ascending size is:

3.024 3.042 3.24 3.42

Exercise 1.18

Write in ascending order of size:

1 5.243 5.234 5.324 5.342 5.432 5.423

2 6.453 6.435 6.245 6.543 6.354 6.534

3 7.564 7.654 7.645 7.465 7.546 7.456

4 7.809 7.098 7.089 7.89 7.98 7.908

5 1.2 1.22 1.002 1.02 1.202 1.022

2 Solving numerical problems

2.1 Simple multiplication and division problems

The multiplication table can be used for solving simple multiplication problems.
An example is given below.

Example 1

Jonathan starts with 9 marbles. After winning a game he has 3 times as many.
How many has he now?

From the table $9 \times 3 = 27$
Therefore Jonathan now has 27 marbles

×	1	2	3	4	5	6	7	8	9	10
1	1	2	3	4	5	6	7	8	9	10
2	2	4	6	8	10	12	14	16	18	20
3	3	6	9	12	15	18	21	24	27	30
4	4	8	12	16	20	24	28	32	36	40
5	5	10	15	20	25	30	35	40	45	50
6	6	12	18	24	30	36	42	48	54	60
7	7	14	21	28	35	42	49	56	63	70
8	8	16	24	32	40	48	56	64	72	80
9	9	18	(27)	36	45	54	63	72	81	90
10	10	20	30	40	50	60	70	80	90	100

Exercise 2.1

Use the multiplication table to answer these questions.

1 Peter has 8 sweets. His brother James has 4 times as many.
How many does James have?

2 Abbie has 7 coloured pencils. Her sister Samantha has 5 times as many.
How many does Samantha have?

3 In Short Street there are 9 houses. In Long Street there are 7 times as many.
How many houses are there in Long Street?

4 On a small farm there are 6 cows. On a larger farm there are 4 times as many.
How many cows are there on the larger farm?

5 Pencils cost 7 p each.
What is the cost of 6 pencils?

6 Erasers cost 8 p each.
What is the cost of 3 erasers?

7 Coloured pencils cost 9 p each.
What is the cost of 5 coloured pencils?

8 A toy car is 5 cm long. A toy lorry beside it is 4 times as long.
How long is the toy lorry?

The multiplication table can also be used for solving simple division problems.
An example is given below:

Example 2

Mrs Chavda has 24 chocolates. She shares them out equally between her 3 children.
How many does each child receive?

From the table $24 \div 3 = 8$.
Therefore each child receives 8 chocolates.

×	1	2	3	4	5	6	7	8	9	10
1	1	2	3	4	5	6	7	8	9	10
2	2	4	6	8	10	12	14	16	18	20
3	3	6	9	12	15	18	21	24	27	30
4	4	8	12	16	20	24	28	32	36	40
5	5	10	15	20	25	30	35	40	45	50
6	6	12	18	24	30	36	42	48	54	60
7	7	14	21	28	35	42	49	56	63	70
(8)	8	16	(24)	32	40	48	56	64	72	80
9	9	18	27	36	45	54	63	72	81	90
10	10	20	30	40	50	60	70	80	90	100

Exercise 2.2

1 Mr Smith has 32 sweets. He shares them out equally between his 4 children.
How many does each child receive?

2 Mrs Jones has 27 toffees. She shares them out equally between her 3 children.
How many does each child receive?

3 Tyrone has 45 nuts. He does not like nuts so he shares them out between 5 other boys.
How many does each boy receive?

4 Kirsty has 24 marshmallows. She does not like marshmallows so she shares them out between 6 other girls.
How many does each girl receive?

5 Dean has a piece of wood which is 63 cm long. He has to cut it up into 7 equal pieces.
How long will each piece be?

6 Annie has a piece of string which is 72 cm long. She has to cut it up into 9 equal pieces.
How long will each piece be?

2.2 Adding and subtracting mentally

Very often it is possible to add or subtract a pair of simple numbers mentally. For such simple cases there is no need to arrange the numbers with the figures in columns.

Example 1

Add together mentally: 25 and 7.
You could split 7 into 5 and 2 and add the 7 on in two stages:

$25 + 5 = 30$
$30 + 2 = 32$

So $25 + 7 = 32$

There are other ways you might think of.

Exercise 2.3

Add together mentally:

1 52 and 5 **2** 31 and 7 **3** 13 and 6 **4** 90 and 9
5 24 and 13 **6** 51 and 18 **7** 32 and 26 **8** 51 and 33

Solve these problems mentally:

9 Jack pays 30 p for a cup of tea and 16 p for a chocolate biscuit. How much does he pay altogether?

10 Natalie pays 25 p for a pen and 12 p for a pencil. How much does she spend altogether?

11 Marcus pays 60 p for a compass and 13 p for an eraser. How much does he pay altogether?

12 A knife has a handle of length 11 cm and a blade of length 16 cm. How long is the knife altogether?

13 A dog lead has a chain of length 55 cm and a handle of length 12 cm. How long is the lead altogether?

14 Sandeep arrives at a railway station 25 minutes before her train arrives. She then sits for 7 minutes in the train before it departs. How many minutes pass after she arrives at the station until the train departs?

Example 2

Find the difference between this pair of numbers mentally: 45 and 8.

You could split the 8 into 5 and 3 and take the 8 off in two stages:

$45 - 5 = 40$
$40 - 3 = 37$

So $45 - 8 = 37$

Exercise 2.4

Find mentally the difference between:

1 37 and 5	**2** 99 and 8	**3** 78 and 6	**4** 75 and 7
5 69 and 24	**6** 86 and 33	**7** 55 and 16	**8** 42 and 15

Solve these problems mentally:

9 David's brother is 25 years old and he is 8 years older than David. How old is David?

10 Chantelle's mother is 39 years old and she is 23 years older than Chantelle. How old is Chantelle?

11 Rani has a piece of string which is 80 cm long. She cuts off a piece of length 25 cm. How long is the remaining piece?

12 Marie buys an ice cream which costs 24 p. If she pays with a 50 p piece, how much change does she receive?

2.3 Mental addition of single-digit numbers

If numbers are very simple it is possible to add several of them together mentally. An example is shown below:

Example 1

Add together mentally: 9, 2, 3 and 6
Add one number at a time:

$9 + 2 = 11$
$11 + 3 = 14$
$14 + 6 = 20$

So $9 + 2 + 3 + 6 = 20$

Exercise 2.5

Add these numbers together mentally:

1 1, 2 and 6 **2** 1, 2 and 4 **3** 4, 3 and 1
4 4, 1 and 4 **5** 6, 2 and 4 **6** 1, 2, 3 and 5
7 5, 1, 2 and 7 **8** 3, 7, 2 and 1

Solve these problems mentally:

9 Fill-yoo-ful Stores own 6 supermarkets in London, 3 in Birmingham, 2 in Manchester and 1 in Liverpool.
How many do they own altogether?

10 Aisha's school consists of four buildings; North South, East and West. There are 4 classrooms in North block, 5 in South block, 3 in East block and 2 in West block.
How many classrooms does her school have altogether?

11 In Simon's house there are 3 chairs in the kitchen, 4 in the living room, 2 in his bedroom and 3 in his parents' bedroom.
How many chairs are there in his house altogether?

12 A local bus in a town runs from the bus station and stops only in the High Street, at the church and at the park before reaching the airport.
One day a bus leaves the bus station and passengers join as follows:

Bus Station: 6 High Street: 4 Church: 5 Park: 2

If all the passengers stay on the bus to the airport, how many get off there?

2.4 Addition and subtraction

When whole numbers are added together they must be arranged with the units in the right hand column, the tens in the next column, the hundreds in the next and so on.

Example 1

Work out this addition: 425 + 187

$$\begin{array}{r} 425 \\ +187 \\ \hline 612 \\ \hline {\scriptstyle 1\,1} \end{array}$$

Exercise 2.6

Work out these sums.
For each question state which part has a different answer from the other two.

1 **a** 53 + 25 **b** 37 + 42 **c** 12 + 66

2 **a** 56 + 31 **b** 74 + 12 **c** 63 + 23

3 **a** 38 + 27 **b** 46 + 19 **c** 29 + 35

4 **a** 535 + 144 **b** 252 + 427 **c** 316 + 362

5 **a** 453 + 514 **b** 352 + 616 **c** 237 + 730

6 **a** 548 + 334 **b** 223 + 669 **c** 775 + 117

7 A wheelbarrow weighs 29 kilograms and 65 kilograms of garden soil is loaded into it.
What is the weight of the loaded barrow?

8 Nathan is 137 centimetres tall.
If he is standing on a stool which is 56 centimetres tall, how high is he standing above the floor?

9 Swindon is 123 kilometres from London, and Newport is 89 kilometres from Swindon.
How far is it from London to Newport?

10 A car is towing a horse box which weighs 455 kilograms. A horse of weight 372 kilograms is travelling in the box.
What is the total weight that the car is towing?

When two whole numbers are subtracted they must be arranged in columns in exactly the same way as they are for addition.

Example 2

Work out this subtraction: 525 − 119

$$\begin{array}{r} 5\not{2}^{1}\not{5}^{15} \\ -119 \\ \hline 406 \\ \hline \end{array}$$

Exercise 2.7

Work out these subtractions.
For each question state which part has a different answer from the other two.

1 **a** 69 − 45
b 75 − 52
c 95 − 71

2 **a** 58 − 13
b 77 − 31
c 89 − 43

3 **a** 72 − 34
b 94 − 55
c 66 − 27

4 **a** 672 − 348
b 453 − 128
c 891 − 567

5 **a** 785 − 219
b 992 − 425
c 673 − 107

6 **a** 517 − 232
b 969 − 694
c 326 − 51

7 **a** 649 − 86
b 625 − 62
c 667 − 94

8 **a** 655 − 279
b 843 − 457
c 560 − 184

9 **a** 343 − 259
b 561 − 476
c 702 − 617

10 Pritesh is cycling to his uncle's house which is 85 kilometres from his own. After he has cycled for 37 kilometres he stops to buy some sweets.
How much further does he have to go?

11 Jane's mother is 160 centimetres tall.
a If Jane is 29 centimetres shorter, how tall is Jane?
b If Jane's sister Mary is 41 centimetres shorter than their mother, how tall is Mary?
c How much taller than Mary is Jane?

12 Jake's father weighs 75 kilograms.
If Jake is 38 kilograms lighter, how heavy is Jake?
If Jake's brother Daniel is 46 kilograms lighter than their father, how heavy is Daniel?
How much heavier than Daniel is Jake?
If their mother is 16 kilograms lighter than their father, how heavy is she?
How much heavier is the mother than each of the two boys?

2.5 Multiplication and division

The result of multiplying any single figure number by another is shown in the table opposite.

For example:

3×5 (3 times 5) $= 15$
6×9 (6 times 9) $= 54$
8×3 (8 times 3) $= 24$

×	0	1	2	3	4	5	6	7	8	9
0	0	0	0	0	0	0	0	0	0	0
1	0	1	2	3	4	5	6	7	8	9
2	0	2	4	6	8	10	12	14	16	18
3	0	3	6	9	12	15	18	21	24	27
4	0	4	8	12	16	20	24	28	32	36
5	0	5	10	15	20	25	30	35	40	45
6	0	6	12	18	24	30	36	42	48	54
7	0	7	14	21	28	35	42	49	56	63
8	0	8	16	24	32	40	48	56	64	72
9	0	9	18	27	36	45	54	63	72	81

When a two-figure number is multiplied by a single-figure number, the unit figure of the two-figure number is multiplied by the single-figure number first.

Example 1

Work out: 23×5

$$\begin{array}{r} 23 \\ \times\ 5 \\ \hline 115 \\ {\scriptstyle 1} \end{array}$$

Exercise 2.8

Work out these multiplications.
For each question state which part has a different answer from the other two.

1 **a** 34×2 **b** 22×3 **c** 17×4

2 **a** 14×7 **b** 16×6 **c** 24×4

3 **a** 21×4 **b** 28×3 **c** 17×5

4 **a** 39×2 **b** 16×5 **c** 13×6

5 **a** 30×3 **b** 13×7 **c** 15×6

6 **a** 19×5 **b** 48×2 **c** 12×8

7 **a** 14×8 **b** 16×7 **c** 29×4

8 **a** 12×9 **b** 13×8 **c** 27×4

9 Watford is 26 kilometres from London.
Find the distance from London to: **a** Milton Keynes **b** Rugby **c** Coventry and **d** Birmingham, if they are respectively 3, 5, 6 and 7 times further away from London than Watford.

10 Vicky's younger sister Josie weighs only 19 kilograms.
If Vicky is three times heavier than Josie, how heavy is Vicky?
If their father is five times heavier than Josie, how heavy is he?

Division is the opposite process to multiplication, and some simple divisions can be done by using the multiplication table on page 20.

For example:

$72 \div 8$ (72 divided by 8) $= 9$
$54 \div 9$ (54 divided by 9) $= 6$
$35 \div 5$ (35 divided by 5) $= 7$

When a two-figure number is divided by a single-figure number, the single-figure number is divided into the tens figure of the two-figure number first.

Example 2

Work out: $72 \div 6$

$$\begin{array}{r} 12 \\ 6\overline{)7\,^{1}2} \end{array}$$

(7 divided by 6 is 1 with a remainder of 1, 6 then goes into 12 twice.)

Exercise 2.9

Work out these divisions.
For each question state which part has a different answer from the other two.

1 **a** $39 \div 3$ **b** $28 \div 2$ **c** $65 \div 5$

2 **a** $68 \div 4$ **b** $54 \div 3$ **c** $90 \div 5$

3 **a** $91 \div 7$ **b** $96 \div 8$ **c** $60 \div 5$

4 **a** $52 \div 4$ **b** $78 \div 6$ **c** $99 \div 9$

5 **a** $75 \div 5$ **b** $98 \div 7$ **c** $90 \div 6$

6 **a** $51 \div 3$ **b** $96 \div 6$ **c** $64 \div 4$

7 **a** $96 \div 4$ **b** $50 \div 2$ **c** $75 \div 3$

8 **a** $45 \div 3$ **b** $60 \div 4$ **c** $80 \div 5$

9 Five identical sacks of flour are placed on a weighing machine and the dial reads 85 kilograms.
What is the weight of each sack?

10 A small school has 84 pupils. There are 7 classes with the same number of pupils in each class.
How many pupils are there in each class?

2.6 Further problems

Very often quantities can be written out more simply by using a decimal point.
For example:

Two pounds and thirty five pence is usually written as £2.35
Three metres and twenty four centimetres is 3.24 metres.

Example 1

If I go to a shop and buy a shirt which costs £8.45 and a tie which costs £2.75, find:

a how much I spend
b the change I receive if I pay with a £20 note.

a The amount I spend = £8.45 + £2.75

The figures and decimal points are correctly arranged in columns.

$$\begin{array}{r} £8.45 \\ +\ £2.75 \\ \hline £11.20 \\ \hline \end{array}$$

I therefore spend £11.20

b My change from a £20 note = £20 − £11.20

The figures and decimal points are again correctly arranged in columns.

$$\begin{array}{r} £20.00 \\ -\ £11.20 \\ \hline £8.80 \\ \hline \end{array}$$

My change is therefore £8.80

Exercise 2.10

1 Harjeet goes into a shop and buys a jumper which costs £7.55 and a pair of shoes which costs £8.65.
Find:
a how much she spends **b** her change out of a £20 note.

2 Mrs Patel goes to a sale and buys a table which costs £11.25 and a chair which costs £4.15.
Find:
a how much she spends **b** her change out of a £20 note.

3 Tim goes to a market and buys a radio which costs £6.35 and a spare plug which costs 85 p (£0.85).
Find:
a how much he spends **b** his change out of a £10 note.

4 Sophie goes into a shop and buys a box of crayons which costs £3.95 and a pen which costs £1.35.
Find:
a how much she spends **b** her change out of a £10 note.

5 Robert requires two pieces of wood, one of length 1 metre 55 centimetres and the other of length 1 metre 25 centimetres.
If the wood is only sold in 2 metre or 3 metre lengths, find each of the following:
a the total length he requires
b the length of the piece he must buy
c the length of wood he will have left over.

6 Mrs Brown requires two lengths of curtain rail, one of length 1 metre 65 centimetres and the other of length 85 centimetres (0.85 metres).
If the curtain rail is only sold in 2 metre or 3 metre lengths, find each of the following:
a the total length she requires
b the length of the piece she must buy
c the length of curtain rail she will have left over.

Example 2

Becky buys 4 pens which cost £1.35 each.
Find:

a the total cost
b the change from a £10 note.

a Total cost = £1.35 × 4

There are two figures after the decimal point, but the multiplication is done without the decimal point.

$$\begin{array}{r} 135 \\ \times 4 \\ \hline 540 \end{array}$$

The answer must include two figures after a decimal point in order to match the information.

So the Total cost is £5.40

b Her change from a £10 note = £10 − £5.40

The figures and decimal points are correctly arranged in columns.

$$\begin{array}{r} £10.00 \\ -£5.40 \\ \hline £\ \ 4.60 \end{array}$$

So her change is £4.60

Exercise 2.11

1 Paul buys 4 toy cars which cost £2.15 each.
Find:
a the total cost **b** the change from a £10 note.

2 Saleem buys 5 radio batteries which cost £1.18 each.
Find:
a the total cost **b** the change from a £10 note.

3 Polly buys 3 magazines which cost £1.30 each.
Find:
a the total cost **b** the change from a £5 note.

4 Keeley buys 8 hair slides which cost 55 p (£0.55) each.
Find:
a the total cost **b** the change from a £5 note.

5 Leroy requires 5 pieces of wood which are each 54 cm (0.54 metres) long.
If the wood is only sold in 2 metre or 3 metre lengths, find each of the following:
a the total length he requires
b the length of the piece he must buy
c the length of wood he will have left over.

6 Mr Bates requires 6 pieces of wire which are each 25 cm (0.25 metres) long.
If the wire is only sold 2 metre or 3 metre lengths, find each of the following:
a the total length he requires
b the length of the piece he must buy
c the length of wire he will have left over.

Example 3

A milkman has 99 milk bottles and he has to put them into crates which hold 8 bottles each.
How many crates will he fill, and how many bottles will be left over?

99 must be divided by 8 $8\overline{)9^{1}9}$
12 remainder 3

He will fill 12 crates and have 3 bottles left over.

Example 4

A carpenter requires 43 screws for making a large cabinet.
If he can only buy the screws in packets of 4, how many packets must he buy and how many screws will he not use?

43 must be divided by 4 $4\overline{)43}$
10 remainder 3

10 packets will not be enough, so he must buy 11 packets and have 1 screw left over.

Exercise 2.12

1 A greengrocer has 22 oranges and he puts them in packets of 4 in order to sell them.
How many packets can he fill and how many oranges will he have left over?

2 A builders' merchant has 66 floor boards and he packs them in bundles of 5 in order to sell them.
How many bundles can he make and how many boards will be left over?

3 A tree trunk of length 34 metres is to be sawn up to make telegraph poles of length 8 metres.
How many poles can be made and what length of trunk will be left over?

4 If buttons cost 6 p each, how many can I buy with a 50 p piece and how much change will I receive?

5 I require 100 tiles for my bathroom wall but I can only buy the tiles in packets of 6.
How many packets will I need and how many tiles will I not use?

6 Mary needs 62 candles for her grandmother's birthday cake but she can only buy them in packets of 4.
How many packets will she need? How many candles will she not use?

2.7 Long multiplication and division

Example 1

Work out: 43×14

$43 \times 14 = 43 \times 10$ add 43×4 　 but 　 $43 \times 10 = 430$
and 　 $43 \times 4 = 172$
therefore 　 $43 \times 14 = 430 + 172 = 602$

It is probably more convenient to set out the working as:

$$\begin{array}{rl} 43 & \\ \times\underline{14} & \\ 430 & \leftarrow 43 \times 10 \\ +\underline{172} & \leftarrow 43 \times 4 \\ \underline{602} & \leftarrow 43 \times 14 \end{array}$$

Therefore $43 \times 14 = 602$

Exercise 2.13

1 Work out each of these multiplications:

a 52 × 13	**b** 62 × 14	**c** 73 × 13	**d** 64 × 12
e 41 × 16	**f** 31 × 18	**g** 53 × 12	**h** 21 × 17

2 Work out each of these multiplications (clearly show any carried figures)

a 56 × 13	**b** 76 × 12	**c** 42 × 17	**d** 23 × 18
e 47 × 14	**f** 43 × 19	**g** 57 × 15	**h** 69 × 14

In Exercise 2.13 we multiplied by numbers which had values between 10 and 20. The same method can be used for multiplying by any two-figure number.

Example 2

Find the answer which is different.

a 88 × 23 **b** 56 × 36 **c** 96 × 21

a		**b**		**c**	
88		56		96	
× 23		× 36		× 21	
1760	← 88 × 20	1680	← 56 × 30	1920	← 56 × 20
+ 265	← 88 × 3	+ 336	← 56 × 6	+ 96	← 56 × 1
2024	← 88 × 23	2016	← 56 × 36	2016	← 96 × 21

The answer which is different is **a**.

Exercise 2.14

Find the answer which is different.

1 a 36 × 26	**b** 39 × 24	**c** 43 × 22
2 a 34 × 26	**b** 36 × 24	**c** 48 × 18

So far only numbers which themselves have two figures have been multiplied by two-figure numbers. We will now extend the method to multiply three-figure numbers by two-figure numbers.

Example 3

Find the answer which is different.

a 184 × 38 **b** 152 × 46 **c** 249 × 28

a		**b**		**c**	
184		152		249	
× 38		× 46		× 28	
5520	← 184 × 30	6080	← 152 × 40	4980	← 249 × 20
+ 1472	← 184 × 8	+ 912	← 152 × 6	+ 1992	← 249 × 8
6992	← 184 × 38	6992	← 152 × 46	6972	← 249 × 28

The answer which is different is **c**.

Exercise 2.15

Find the answer which is different.

1	**a**	126×32	**2**	**a**	117×28	**3**	**a**	142×38
	b	112×36		**b**	136×24		**b**	128×42
	c	184×22		**c**	192×17		**c**	112×48

Example 4

A tall cabinet has shelves which each hold 45 books.
If the cabinet has 12 shelves how many books can it hold?

Each shelf holds 45 books.
So 12 shelves hold 45×12 books.
$45 \times 12 = 540$
So the cabinet can hold 540 books.

Exercise 2.16

1 At Highfield School there are 25 classes which each have 18 pupils. How many pupils attend the school altogether?

2 At a railway depot 15 locomotives are awaiting wheel inspections. If each locomotive has 12 wheels, how many wheels are inspected altogether?

3 The distance from London to Slough is 26 kilometres. What is the distance from London to
a Plymouth if it is 14 times greater
b Penzance if it is 19 times greater?

4 A man is digging a trench and his wheelbarrow can hold 50 kilograms of soil. If he fills his wheelbarrow 18 times, how much soil does he remove?

Example 5

Find the answer which is different.

a $273 \div 13$ **b** $391 \div 17$ **c** $322 \div 14$

a
$$\begin{array}{r l} 21 & \\ 13\overline{)273} & \\ -260 & \leftarrow 13 \times 20 \\ \hline 13 & \\ -\ 13 & \leftarrow 13 \times 1 \\ \hline 00 & \end{array}$$

b
$$\begin{array}{r l} 23 & \\ 17\overline{)391} & \\ -340 & \leftarrow 17 \times 20 \\ \hline 51 & \\ -\ 51 & \leftarrow 17 \times 3 \\ \hline 00 & \end{array}$$

c
$$\begin{array}{r l} 23 & \\ 14\overline{)322} & \\ -280 & \leftarrow 14 \times 20 \\ \hline 42 & \\ -\ 42 & \leftarrow 14 \times 3 \\ \hline 00 & \end{array}$$

The answer which is different is **a**.

Exercise 2.17

Find the answer which is different.

1 a 299 ÷ 13
b 384 ÷ 16
c 414 ÷ 18

2 a 325 ÷ 13
b 364 ÷ 14
c 312 ÷ 12

3 a 576 ÷ 18
c 544 ÷ 17
c 627 ÷ 19

4 a 645 ÷ 15
b 672 ÷ 16
c 588 ÷ 14

5 a 676 ÷ 13
b 848 ÷ 16
c 954 ÷ 18

6 a 736 ÷ 23
b 792 ÷ 24
c 832 ÷ 26

Exercise 2.18

1 At Daniel's school 192 pupils stay to lunch.
If the dining hall has 16 tables, how many pupils sit at each table?

2 Some pipes are to be laid in a trench which is 385 metres long.
If each pipe is 11 metres in length, how many can be laid?

3 Nadim's new video tape takes 240 seconds to rewind from one end to the other, and it has 15 cartoon films on it of equal length.
How long will it take to rewind only one film?

4 A train is to carry 540 tonnes of coal from a colliery to a power station.
If each wagon holds 45 tonnes, how many wagons are required?

5 A shelf has a length of 54 centimetres (540 millimetres).
How many books of thickness 18 millimetres can be stacked on to it?

2.8 Multiplying and dividing single-digit powers of 10 mentally

Example 1

Work out mentally: 500 × 60

$500 \times 60 = 30\,000$
($5 \times 6 = 30$ and then three noughts must be added)

Exercise 2.19

Work these out mentally.

1 80 × 60 **2** 60 × 20 **3** 40 × 30 **4** 80 × 70
5 200 × 30 **6** 500 × 30 **7** 800 × 30 **8** 900 × 40

Example 2

Work out mentally: $350 \div 70$

$350 \div 70 = 35\not{0} \div 7\not{0} = 35 \div 7 = 5$
(dividing the top and bottom each by 10)

Exercise 2.20

Work these out mentally.

1 $210 \div 30$	**2** $320 \div 40$	**3** $180 \div 30$	**4** $240 \div 60$
5 $280 \div 40$	**6** $360 \div 90$	**7** $240 \div 80$	**8** $420 \div 70$
9 $120 \div 40$	**10** $450 \div 50$	**11** $120 \div 20$	**12** $180 \div 20$

3 Estimating, approximating and checking

3.1 Simple estimations

It is very useful to make a rough estimate in order to check whether an addition or subtraction has been worked out correctly. Look at the examples below.

Example 1

Find the sum of 49 and 32 and check your answer with a rough estimate.

49	Adding 50 and	50	(49 to the nearest ten)
$+32$	30 is a rough	$+30$	(32 to the nearest ten)
81	estimate.	80	

It can be seen that the estimate agrees roughly with the correct answer.

Example 2

Find the difference between 315 and 197 and check your answer with a rough estimate.

315	Subtracting 200	300	(315 to the nearest hundred)
-197	from 300 is a	-200	(197 to the nearest hundred)
118	rough estimate.	100	

It can be seen that the estimate agrees roughly with the correct answer.

To make rough estimates, two-figure numbers are corrected to the nearest ten, whereas three-figure numbers are corrected to the nearest hundred.

Exercise 3.1

Find the sum and check your answer with a rough estimate.

1 41 and 28	**2** 93 and 36	**3** 52 and 19	**4** 67 and 42
5 41 and 53	**6** 73 and 22	**7** 304 and 199	**8** 213 and 389
9 176 and 294	**10** 97 and 189	**11** 329 and 207	**12** 530 and 95

Find the difference and check your answer with a rough estimate.

13 51 and 29	**14** 82 and 37	**15** 93 and 56	**16** 78 and 39
17 80 and 62	**18** 311 and 124	**19** 603 and 329	**20** 521 and 230
21 677 and 288	**22** 591 and 97	**23** 721 and 287	**24** 509 and 379

3.2 Simple uses of a calculator (I)

Any multiplication of whole numbers will always result in a whole number answer.

When a whole number is divided by another whole number the answer is not always a whole number.

Example 1

Use your calculator to find: $38 \div 5$
Give your answer correct to the nearest whole number.

Press your calculator keys as shown below:

[AC] [3] [8] [÷] [5] [=] 7.6

The answer to the nearest whole number is 8 because the figure after the decimal point is more than 5.

Exercise 3.2

Use your calculator to work out these divisions.
Give each answer correct to the nearest whole number.

1 $34 \div 5$	**2** $49 \div 5$	**3** $24 \div 5$	**4** $39 \div 5$
5 $37 \div 10$	**6** $97 \div 10$	**7** $38 \div 5$	**8** $23 \div 5$
9 $316 \div 20$	**10** $396 \div 20$	**11** $31 \div 5$	**12** $26 \div 5$

Example 2

Use your calculator to find: $1254 \div 40$
Give your answer correct to the nearest whole number.

Press your calculator keys as shown below:

[AC] [1] [2] [5] [4] [÷] [4] [0] [=] 31.35

The answer to the nearest whole number is 31 because the first figure after the decimal point is less than 5.

Exercise 3.3

Use your calculator to work out these divisions.
Give each answer correct to the nearest whole number.

1 $102 \div 8$ **2** $126 \div 8$
3 $794 \div 40$ **4** $1278 \div 40$
5 $410 \div 16$ **6** $350 \div 16$
7 $157 \div 4$ **8** $181 \div 4$
9 $2258 \div 40$ **10** $1687 \div 20$
11 $265 \div 8$ **12** $950 \div 16$

Example 3

If 5 pens cost 44 p, find the cost of:

a 1 pen **b** 3 pens

Give each answer correct to the nearest penny.

a $44\,p \div 5 = 8.8\,p$ or 9 p to the nearest penny.
b $8.8\,p \times 3 = 26.4\,p$ or 26 p to the nearest penny.

Exercise 3.4

1 If 5 lollipops cost 39 p, find the cost of:
a 1 lollipop **b** 6 lollipops **c** 7 lollipops **d** 2 lollipops
e 4 lollipops **f** 9 lollipops **g** 3 lollipops **h** 8 lollipops
Give each answer correct to the nearest penny.

2 If 5 erasers cost 41 p, find the cost of:
a 1 eraser **b** 6 erasers **c** 7 erasers **d** 2 erasers
e 4 erasers **f** 9 erasers **g** 3 erasers **h** 8 erasers
Give each answer correct to the nearest penny.

3 There are 5 notebooks on a table and they form a pile of height 7 cm.
a Find the thickness of one notebook.
Find the height of a pile of:
b 6 notebooks **c** 3 notebooks **d** 8 notebooks **e** 4 notebooks
f 9 notebooks **g** 2 notebooks **h** 7 notebooks
Give each answer to the nearest centimetre.

Sometimes, for a practical reason, the answer to a division calculation has to be rounded up to the next higher whole number, whatever the decimal figures may be. Example 4 shows a calculation of this kind.

Example 4

A lift cannot carry more than 15 people at once.
If 51 people want to use the lift, find:

a The number of ascents that the lift will have to make.

b The number of people in the lift during the last ascent if it is to be full for all the others.

a $51 \div 15 = 3.4$
The cage must make 4 ascents because 3 would not be enough.

b $15 \times 3 = 45$
The number of people in the cage for the last ascent is:
$51 - 45 = 6$.

Exercise 3.5

1 The taxis which work from a certain rank cannot carry more than 5 passengers.
If 32 people are standing in the taxi rank queue, find:
a the number of taxis required to transport them
b the number of people in the last taxis if that is the only one which is not to be full.

2 A coach operator is taking some children to the seaside. He has a fleet of coaches which have 50 seats each.
If 210 children are going on the trip, find:
a the number of coaches required
b the number of children in the last coach, if that is the only one which is not full.

3 A milkman has some crates which can each hold 20 bottles.
If he wants to put 50 bottles into crates, find:
a the number of crates he will require
b the number of bottles he will put in the last crate, if that is the only crate which he is not going to fill.

4 A special train is to be hired for an excursion. The coaches can seat 80 passengers each.
If 500 people are going on the excursion, find:
a the number of coaches required
b the number of passengers in the last coach if all of the others are to be full.

5 A ferryboat can only carry 12 people.
If 39 people are waiting to cross the river, find:
a the number of crossings the boat must make
b the number of people in the boat during the last crossing if it is full for all the other crossings.

Example 5

A carpenter requires 93 lengths of floorboard for a new house.
If the boards are only sold in bundles of 6 find:
a the number of bundles he must buy
b the number of lengths that he will not use.

a $93 \div 6 = 15.5$
Therefore he must buy 16 bundles because 15 would not be enough.
b $16 \times 6 = 96$
The number of boards not used is: $96 - 93 = 3$.

Exercise 3.6

1 Mr Khan needs 54 screws for making a cabinet.
If the screws he needs are only sold in packets of 12, find:
a the number of packets he must buy
b the number of screws he will not use.

2 Mrs Bennett needs 42 curtain hooks.
If the hooks are only sold in packets of 10, find:
a the number of packets she must buy
b the number of hooks she will not use.

3 A secretary needs 51 address labels.
If they are only sold in sheets of 12 labels, find:
a the number of sheets she must buy
b the number of labels she will not use.

4 A woman needs 22 clothes pegs for her washing line.
If the pegs are only sold in packets of 8, find:
a the number of packets she must buy
b the number of pegs she will not use.

5 A builder has built a house with 13 doors. Each door needs 3 hinges.
a Find the total number of hinges needed.
If the hinges are only sold in packets of 4, find:
b the number of packets the builder must buy
c the number of hinges he will not use.

6 Barbara requires 15 picture hooks in order to hang up her framed pictures.
If the hooks are only sold in packets of 4, find:
a the number of packets she must buy
b the number of hooks she will not use.

3.3 Simple uses of a calculator (II)

Using a calculator makes most arithmetic processes easier.

Several examples are shown below.

Press the [AC] button to clear your calculator before each new calculation.

Example 1

Use your calculator to find the 'odd answer out'.
a $85 + 56 + 5$ **b** $98.3 + 11.1 + 35.6$ **c** $103 + 36.1 + 6.9$

Using a calculator gives the answers:
a 146 **b** 145 **c** 146
Therefore the 'odd answer out' is **b**.

Exercise 3.7

Use your calculator to find the 'odd answer out'.

1 **a** $18.65 + 14.72 + 15.43$
b $21.43 + 11.31 + 15.96$
c $20.94 + 16.51 + 11.25$

2 **a** $35.34 + 17.23 + 5.93$
b $35.28 + 19.81 + 3.51$
c $40.57 + 11.21 + 6.72$

3 **a** $35.6 + 24.7 + 13.2$
b $41.5 + 20.3 + 11.8$
c $29.8 + 18.4 + 25.3$

4 **a** $32.6 + 27.3 + 6.5$
b $41.7 + 19.2 + 5.6$
c $51.2 + 11.9 + 3.4$

5 **a** $41.4 + 25.2 + 15$
b $36.7 + 12.9 + 33$
c $51.5 + 17.1 + 13$

6 **a** $9.8 + 5.7 + 8$
b $7.6 + 8.9 + 7$
c $6.2 + 9.3 + 9$

Example 2

Use your calculator to find the 'odd answer out'.
a $31.5 - 25.9$ **b** $15.2 - 9.6$ **c** $24 - 18.3$

Using a calculator gives the answers:
a 5.6 **b** 5.6 **c** 5.7
The odd answer out is **c**.

Exercise 3.8

Use your calculator to find the 'odd answer out'.

1 **a** $35.27 - 16.12$
b $53.84 - 34.59$
c $46.73 - 27.58$

2 **a** $31.75 - 8.37$
b $29.46 - 5.98$
c $30.63 - 7.15$

3 **a** 99.2 – 34.8
b 94.1 – 29.7
c 81.7 – 18.3

4 **a** 53.2 – 9.7
b 51.4 – 8.9
c 50.1 – 7.6

5 **a** 65 – 21.46
b 73.8 – 30.16
c 82.05 – 38.41

6 **a** 9 – 4.39
b 8.2 – 3.69
c 7.25 – 2.64

Example 3

Use your calculator to find the 'odd answer out'.
a 45 × 16 **b** 50 × 14.4 **c** 60 × 12.5

Using a calculator gives the answers:
a 720 **b** 720 **c** 750
The 'odd answer out' is **c**.

Exercise 3.9

Use your calculator to find the 'odd answer out'.

1 **a** 35.2 × 1.5
b 33 × 1.6
c 31 × 1.7

2 **a** 8.96 × 2.5
b 9.75 × 2.4
c 13 × 1.8

3 **a** 9.2 × 0.75
b 8 × 0.85
c 15 × 0.46

4 **a** 415 × 24
b 356 × 28
c 623 × 16

5 **a** 28 × 2.5
b 12.5 × 5.6
c 6.25 × 9.6

6 **a** 80 × 0.15
b 60 × 0.25
c 25 × 0.48

Example 4

Use your calculator to find the 'odd answer out'.
a 153 ÷ 3.6 **b** 170.1 ÷ 4.2 **c** 27.2 ÷ 0.64

Using a calculator gives the answers:
a 42.5 **b** 40.5 **c** 42.5
The 'odd answer out' is **b**.

Exercise 3.10

Use your calculator to find the 'odd answer out'.

1 **a** 59.4 ÷ 1.8
b 83.2 ÷ 2.6
c 95.7 ÷ 2.9

2 **a** 95.4 ÷ 3.6
b 60.5 ÷ 2.2
c 38.5 ÷ 1.4

3 **a** 80.29 ÷ 1.55
b 71.28 ÷ 1.35
c 97.68 ÷ 1.85

4 **a** 55.8 ÷ 36
b 46.2 ÷ 28
c 49.6 ÷ 32

5 **a** 48 ÷ 1.28
b 72 ÷ 1.92
c 47 ÷ 1.25

6 **a** 32.5 ÷ 0.52
b 43.2 ÷ 0.64
c 51.3 ÷ 0.76

Example 5

A carpenter has a 17 metre length of skirting board.
He cuts off six pieces whose lengths are:
1 m 70 cm, 2 m 60 cm, 1 m 30 cm, 3 m 80 cm, 2 m 90 cm and 4 m 30 cm.
Find:
a the total length he cuts off.
b the length of skirting board he has left.

a $1.70 + 2.60 + 1.30 + 3.80 + 2.90 + 4.30 = 16.60$
The length cut off is 16 m 60 cm
b $17 - 16.60 = 0.40$
The length left is 40 cm.

Exercise 3.11

For Questions 1–3 find:
a the total amount paid
b the change from a £10 note.

1 Joseph goes to the supermarket where he spends £1.36 on biscuits, £1.05 on butter, £2.19 on cheese, £2.05 on coffee and £1.20 on tea.

2 Laura goes to the baker's where she spends £1.23 on bread, £1.25 on cakes, £1.05 on jam tarts, £1.10 on doughnuts and £1.12 on croissants.

3 Annie goes to the newsagents where she spends £1.25 on magazines, £1.15 on sweets, 60 p on newspapers, 80 p on comics and 75 p on ice cream.

4 Sian has a 10 metre length of string. In order to tie up some parcels she cuts off six pieces whose lengths are:
1 m 20 cm, 1 m 40 cm, 1 m 15 cm, 1 m 25 cm, 1 m 30 cm and 1 m 35 cm.
Find:
a the total length she cuts off.
b the length of string she has left.

5 Joanne has a 5 metre length of lace. In order to decorate some dresses she cuts off four pieces whose lengths are:
1 m 18 cm, 1 m 10 cm, 78 cm and 65 cm.
Find:
a the total length she cuts off.
b the length of lace she has left.

Example 6

Find:

a the cost of 3 shirts at £7.99 each together with 2 ties at £3.99 each, and
b the change from a £50 note.

a 3 shirts cost £7.99 × 3 = £23.97
2 ties cost £3.99 × 2 = £ 7.98
Therefore the total cost is: £23.97 + £7.98 = £31.95
b The change from a £50 note is: £50 − £31.95 = £18.05.

Exercise 3.12

For Questions 1–4 find:
a the total cost
b the change from a £50 note.

1 3 pairs of shoes at £5.95 a pair and 2 pairs of slippers at £2.95 a pair.

2 3 screwdrivers at £4.45 each and 2 spanners at £2.65 each.

3 4 tablecloths at £3.15 each and 3 tea towels at £2.25 each.

4 4 adults' tickets for the theatre at £5.50 each and 3 children's tickets at £2.75 each.

5 A carpenter has a 6 metre length of wood. In order to make a window frame he cuts off 2 pieces of length 1 m 40 cm and 2 pieces of length 1 m 20 cm.
Find:
a the total length he cuts off
b the length he has left.

6 A picture framer has a 6 metre length of framing wood. She cuts off 2 pieces of length 80 cm and 2 pieces of length 60 cm in order to make a frame.
Find:
a the total lengths he cuts off
b the length remaining.

Example 7

On one market stall 3 kilograms of apples cost £4.86. On a second stall 2 kilograms of the same apples cost £3.26.
Which is the better buy?

1 kilogram at the first stall costs: £4.86 ÷ 3 = £1.62
1 kilogram at the second stall costs: £3.26 ÷ 2 = £1.63

The apples on the first stall are the better buy (because they are cheaper).

Exercise 3.13

For all questions find which is the better buy.

1 3 kilograms of grapes costing £9.87 or 2 kilograms of the same grapes costing £6.56.

2 3 kilograms of strawberries costing £5.82 or 2 kilograms of the same strawberries costing £3.80.

3 4 kilograms of pears costing £5.28 or 3 kilograms of the same pears costing £3.99.

4 4 kilograms of apricots costing £5.80 or 3 kilograms of the same apricots costing £4.32.

5 6 oranges costing £1.26 or 5 similar oranges costing £1.10.

Example 8

Which is the best buy?
a 3 litres of squash for £2.01
b 2 litres of squash for £1.30
c 1.5 litres of squash for 99 p

a If 3 litres cost £2.01, then the price per litre is:
$£2.01 \div 3 = £0.67 = 67\,p$
b If 2 litres cost £1.30, then the price per litre is:
$£1.30 \div 2 = £0.65 = 65\,p$
c If 1.5 litres cost 99 p, then the price per litre is: $99\,p \div 1.5 = 66\,p$

Therefore **b** is the best buy.

Exercise 3.14

For Questions 1–3 find which is the better buy.

1 A 3 kg packet of raisins costing £3.63 or a 2.5 kg packet of the same raisins costing £3.05.

2 A 3 kg packet of lentils costing £3.30 or a 2.5 kg packet of the same lentils costing £2.70.

3 A 2 kg packet of sugar costing £1.70 or 1.5 kg packet of the same sugar costing £1.26.

For Questions 4 and 5 find which is the best buy.

4 **a** A 2.5 litre bottle of lemonade costing £1.70
b A 2 litre bottle of lemonade costing £1.38
c A 1.5 litre bottle of lemonade costing £1.05

5 **a** A 1.2 litre bottle of vinegar costing £1.14
b A 1.25 litre bottle of vinegar costing £1.15
c A 1.5 litre bottle of vinegar costing £1.35

3.4 Using 'trial and improvement' methods

It is often possible to get the answer to a problem by making a guess, seeing what its effect is, and then modifying the guess in order to get a better answer.

Example 1

Which whole number, multiplied by 27 gives the number nearest to 1000?

First guess 30: $30 \times 27 = 810$, but this is too small.
Second guess 40: $40 \times 27 = 1080$, but this is too big.
1080 is closer to 1000 than 810, so 40 is a better guess than 30.
Third guess 36: $36 \times 27 = 972$, but this is too small.
Fourth guess, 37: $37 \times 27 = 999$, and this is very close.

37 is therefore the nearest whole number.

Exercise 3.15

For Questions 1–5 find the nearest whole number which when multiplied by the given number gives 1000 as the answer.

1 17 **2** 28 **3** 19 **4** 13 **5** 24

6 Crates of eggs weigh 23 kg each. How many crates can be loaded on to a large van for which the maximum load is about 1 tonne? (1 tonne = 1000 kg).

7 Sacks of potatoes weigh 29 kg each. How many sacks can be loaded on to a trailer for which the maximum load is about 1 tonne?

8 Cherries weigh 18 grams each. How many can be put in a bag which has written on it 'approximate weight of contents 1 kg'?

For Questions 9–13 find the whole number which when multiplied by the given number gives the answer nearest 500.

9 16 **10** 13 **11** 11 **12** 19 **13** 15

14 A builder is building a swimming pool of area 500 square metres. If the width is 14 metres, find the length to the nearest whole number of metres (area = length $\times$ width).

15 Pencil sharpeners cost 17 p each. How many can be bought for £5? (£5 = 500 p).

16 Mr Green is making a paddling pool of area 200 square metres. If the width is 12 metres, find the length to the nearest whole number of metres.

Example 2

The playground at Diane's school is square and has an area of 450 square metres. Find the side length to the nearest metre.

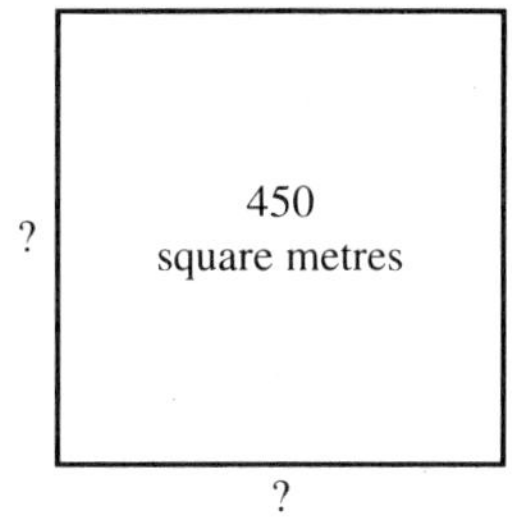

First guess 20: $20 \times 20 = 400$, but this is too small.
Second guess 21: $21 \times 21 = 441$, but this is still too small.
Third guess 22: $22 \times 22 = 484$, but this is too large.
441 is closer to 450 than 484, so 21 is a better guess than 22.

The answer correct to the nearest metre is 21 metres.

Exercise 3.16

For Questions 1–3 find the side length of the square correct to the nearest metre.

1 The playground of Paul's school is square and has an area of 730 square metres.

2 In a town there is a square car park of area 620 square metres.

3 Mr Brown's orchard is square and has an area of 780 square metres.

For Questions 4–6 find the side length of the square correct to the nearest centimetre.

4 A square clock has an area of 580 square centimetres.

5 The window in Christine's front door is square and has an area of 2200 square centimetres.

6 The door on a rabbit hutch is square and has an area of 2800 square centimetres.

For Questions 7–9 find the side length of the square correct to the nearest millimetre.

7 The small squares on a chessboard each have an area of 23 square centimetres (which is 2300 square millimetres).

8 A small square table mat has an area of 61 square centimetres.

9 Some square wall tiles each have an area of 94 square centimetres.

3.5 Approximating by using significant figures or decimal places

Significant figures

The number 1734 contains four figures.
The most significant is 1 because this is the number of thousands.
The least significant is 4 because this is the number of units.

We can write 1734 correct to one significant figure as 2000 since it is closer to 2000 than 1000.

We can write 1734 correct to three significant figures as 1730 since it is closer to 1730 than 1740.

Example 1

Write correct to three significant figures:
a 5438 **b** 8213 **c** 7345 **d** 6197

a 5438 is 5440 correct to three significant figures.
(The third figure is increased by one because the fourth figure is greater than 5.)
b 8213 is 8210 correct to three significant figures.
(The fourth figure is less than 5 so the units are ignored.)
c 7345 is 7350 correct to three significant figures.
(The third figure is also increased by one if the fourth figure is exactly 5.)
d 6197 is 6200 correct to three significant figures.
(The fourth figure is greater than 5 and in this case 19 has to be increased to 20.)

Exercise 3.17

Write correct to three significant figures:

1 3257 **2** 7369 **3** 3766 **4** 5942 **5** 7241 **6** 2481
7 3975 **8** 7206 **9** 8408 **10** 6705 **11** 3795 **12** 8599

Example 2

Write correct to two significant figures:
a 243 **b** 3961

a 243 is 240 correct to two significant figures.
(The third figure is less than 5 so the units are ignored.)
b 3961 is 4000 correct to two significant figures.
(The third figure is greater than 5, so in this case the 39 has to be increased to 40.)

Exercise 3.18

Write correct to two significant figures:

1 382	**2** 361	**3** 964	**4** 476	**5** 849	**6** 755
7 709	**8** 504	**9** 301	**10** 903	**11** 395	**12** 799

Exercise 3.19

Write correct to two significant figures:

1 6385	**2** 9386	**3** 9275	**4** 6437	**5** 4636	**6** 9754
7 4082	**8** 5976	**9** 4993	**10** 6956	**11** 5602	**12** 4206

Example 3

Write the number 56.92 correct to:
a three significant figures
b two significant figures.

a 56.92 is 56.9 correct to three significant figures.
(The fourth figure is less than 5 so it is ignored.)
b 56.92 is 57 correct to two significant figures.
(The second figure is increased by 1 because the third figure is greater than 5.)

Exercise 3.20

Write correct to three significant figures:

1 27.39	**2** 51.25	**3** 34.62	**4** 53.07	**5** 46.09	**6** 79.05
7 37.93	**8** 56.94	**9** 42.99	**10** 54.96	**11** 35.02	**12** 43.03

Exercise 3.21

Write correct to two significant figures:

1 64.74	**2** 82.94	**3** 32.53	**4** 57.25	**5** 67.43	**6** 76.04
7 49.73	**8** 59.91	**9** 89.56	**10** 70.83	**11** 30.74	**12** 20.06

Example 4

Write correct to two significant figures:
a 0.314 **b** 0.0569

a 0.314 is 0.31 correct to two significant figures.
(Note that the nought at the beginning is not a significant figure.)
b 0.0569 is 0.057 correct to two significant figures.
(Note that neither of the two noughts at the beginning are significant figures.)

Exercise 3.22

Write correct to two significant figures:

1 0.327	**2** 0.747	**3** 0.582	**4** 0.107
5 0.0938	**6** 0.0735	**7** 0.0563	**8** 0.0691

Decimal places

The number 29.346 contains three decimal places. If we want to approximate this number we can write it correct to two or to one decimal place(s).

Correct to two decimal places it is 29.35 (because the third decimal figure is a 6).
Correct to one decimal place it is 29.3 (because the second decimal figure is a four).

Example 5

Write correct to two decimal places:
a 4.286 **b** 2.132 **c** 5.197

a 4.286 is 4.29 correct to two decimal places.
(The second decimal figure is increased by 1 because the third decimal figure is greater than 5.)
b 2.132 is 2.13 correct to two decimal places.
(The third decimal figure is less than 5, so it is ignored.)
c 5.197 is 5.20 correct to two decimal places.
(The third decimal figure is greater than 5 and in this case 19 has to be increased to 20.)

Exercise 3.23

Write correct to two decimal places:

1 5.249	**2** 7.837	**3** 3.682	**4** 6.923	**5** 4.625	**6** 7.465
7 8.207	**8** 5.706	**9** 7.901	**10** 7.496	**11** 8.599	**12** 1.695

Example 6

Write correct to one decimal place:
a 5.26 **b** 6.024

a 5.26 is 5.3 correct to one decimal place.
(The first decimal figure is increased by 1 because the second decimal figure is greater than 5.)
b 6.024 is 6.0 correct to one decimal place.
(The second decimal figure is less than 5 so it is ignored.)

Exercise 3.24

Write correct to one decimal place:

1 4.72 **2** 7.51 **3** 6.38 **4** 1.68 **5** 2.85 **6** 6.06
7 4.09 **8** 7.02 **9** 2.98 **10** 5.99 **11** 3.95 **12** 4.94

Exercise 3.25

Write correct to one decimal place:

1 4.874 **2** 2.431 **3** 3.645 **4** 7.638 **5** 6.352 **6** 5.756
7 4.973 **8** 8.992 **9** 9.305 **10** 3.084 **11** 2.068 **12** 6.032

Exercise 3.26

Copy and complete this table.

		Number correct to four decimal places	Number correct to three decimal places	Number correct to two decimal places	Number correct to one decimal place
1	3.138 27				
2	5.247 39				
3	6.426 18				
4	17.915 83				
5	19.807 52				
6	12.476 32				

Example 8

Write correct to two decimal places:
a 0.846 **b** 0.073

a 0.846 is 0.85 correct to two decimal places.
(The second decimal figure is increased by 1 because the third decimal figure is greater than 5.)
b 0.073 is 0.07 correct to two decimal places.
(The third decimal figure is less than 5 so it is ignored. Note that the nought after the decimal point does count as a decimal place even though it does not count as a significant figure.)

Exercise 3.27

Write correct to two decimal places:
1 0.239 **2** 0.427 **3** 0.548 **4** 0.376 **5** 0.285 **6** 0.642
7 0.029 **8** 0.092 **9** 0.063 **10** 0.081 **11** 0.054 **12** 0.072

3.6 The use of estimations to check calculations

It is very easy to press the wrong key when using a calculator, so we must be able to check the result.

One way is to repeat the calculation and see if we get the same answer, but this could mean repeating the mistake as well.

Another way is to do the calculation a different way on the calculator. For example, does 163×52 give the same result as 52×163, namely 8476?
This method, however, is not always possible.

The best way is to do a rough estimate on paper. This will at least indicate whether the size of the answer is about right.

Example 1

Use your calculator to find these multiplications.
Check each answer with a rough estimate.

a 392×61 **b** 79×41

a $392 \times 61 = 23\,912$
Now 392 is 400 to one significant figure and 61 is 60 to one significant figure.
Therefore 392×61 is approximately 400×60 or 24 000.
The rough estimate (24 000) is therefore in good agreement with the exact answer (23 912.)

b $79 \times 41 = 3239$
79×41 is approximately $80 \times 40 = 3200$.
The rough estimate (3200) is therefore in good agreement with the exact answer (3239).

Exercise 3.28

Use your calculator to find these multiplications.
Check each answer with a rough estimate.

1 47×34	**2** 86×43	**3** 92×39	**4** 74×47
5 194×31	**6** 463×11	**7** 231×26	**8** 314×19
9 1017×8	**10** 2109×19	**11** 2987×31	**12** 1963×38

Example 2

Use your calculator to find these divisions.
Check each answer with a rough estimate.

a $819 \div 39$ **b** $1792 \div 512$

a $819 \div 39 = 21$
$819 \div 39$ is approximately $800 \div 40$ or 20.
The rough estimate (20) is therefore in good agreement with the exact answer (21).

b $1792 \div 512 = 3.5$
$1792 \div 512$ is approximately $2000 \div 500$ or 4.
The rough estimate (4) is therefore in good agreement with the exact answer (3.5).

Exercise 3.29

Use your calculator to find these divisions.
Check each answer with a rough estimate.

1 $910 \div 52$ **2** $608 \div 64$ **3** $729 \div 54$ **4** $798 \div 38$
5 $588 \div 56$ **6** $1991 \div 362$ **7** $1708 \div 488$ **8** $2706 \div 492$

Example 3

Use your calculator to find: $\dfrac{91 \times 78}{42}$

Check your answer with a rough estimate.

$$\frac{91 \times 78}{42} = 91 \times 78 \div 42 = 169$$

$91 \times 78 \div 42$ is approximately $90 \times 80 \div 40$ or 180.
The rough estimate (180) is therefore in good agreement with the exact answer (169).

Exercise 3.30

Use your calculator for these sums. Check each answer with a rough estimate.

1 $\dfrac{57 \times 18}{76}$ **2** $\dfrac{87 \times 28}{48}$ **3** $\dfrac{69 \times 57}{36}$

4 $\dfrac{58 \times 57}{87}$ **5** $\dfrac{56 \times 77}{49}$ **6** $\dfrac{91 \times 44}{52}$

4 Fractions and percentages

4.1 Simple uses of fractions

The simplest kind of fraction refers to one given part of a whole quantity.

Look at the circles below:

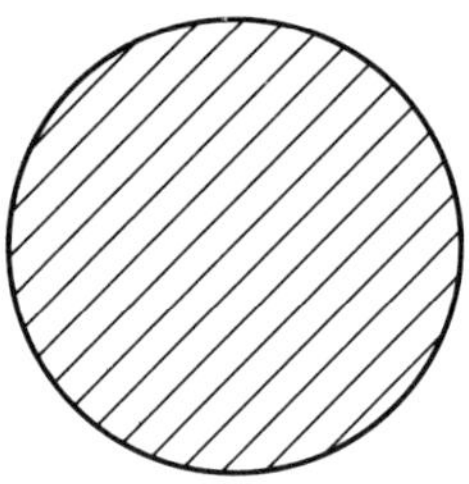

One whole circle

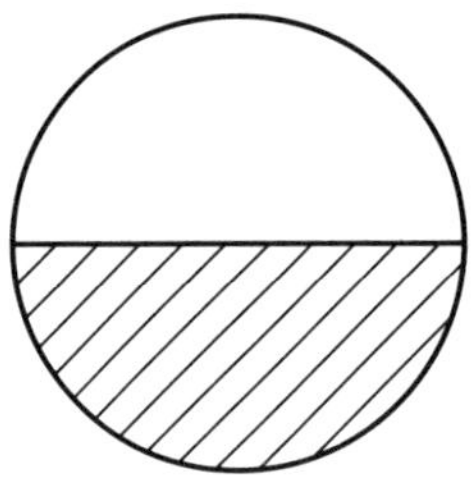

One half (or $\frac{1}{2}$) of the circle

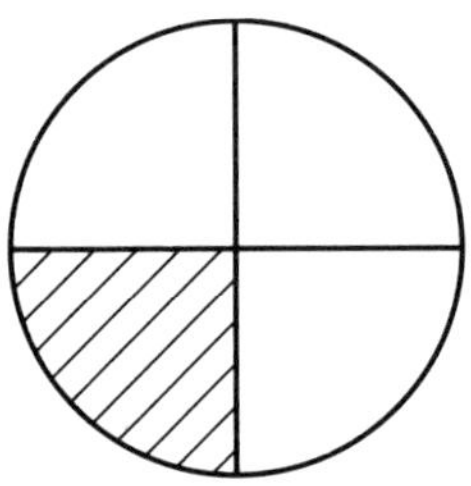

One quarter (or $\frac{1}{4}$) of the circle

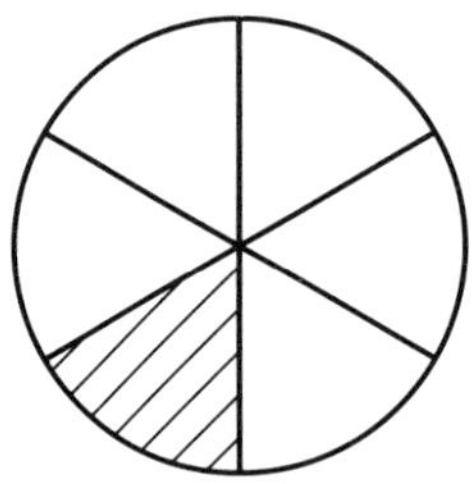

One sixth (or $\frac{1}{6}$) of the circle

Example 1

Hitesh has saved £36. He spends $\frac{1}{2}$ of it on a small radio, $\frac{1}{3}$ of it on an electronics kit, and $\frac{1}{6}$ of it on a new encyclopaedia.
How much does he spend on each?

His radio costs $\frac{1}{2}$ of £36 $= £36 \div 2 = £18$
His electronics kit costs $\frac{1}{3}$ of £36 $= £36 \div 3 = £12$
His encyclopaedia costs $\frac{1}{6}$ of £36 $= £36 \div 6 = £6$

Exercise 4.1

1 The distance from London to Glasgow is 648 kilometres.
Find the distance from London to:
a Preston, which is $\frac{1}{2}$ of the way to Glasgow
b Stafford, which is $\frac{1}{3}$ of the way to Glasgow
c Nuneaton, which is $\frac{1}{4}$ of the way to Glasgow
d Northampton, which is $\frac{1}{6}$ of the way to Glasgow
e Milton Keynes, which is $\frac{1}{8}$ of the way to Glasgow
f Bletchley, which is $\frac{1}{9}$ of the way to Glasgow.

2 Kelly's school is 900 metres along the road from her house.
Find the distance from Kelly's house to:
a the post office if it is $\frac{1}{2}$ of the distance to her school
b the supermarket if it is $\frac{1}{3}$ of the distance to her school
c the village hall if it is $\frac{1}{4}$ of the distance to her school
d the letter box if it is $\frac{1}{5}$ of the distance to her school
e the telephone box if it is $\frac{1}{6}$ of the distance to her school
f the police station if it is $\frac{1}{9}$ of the distance to her school
g the church if it is $\frac{1}{10}$ of the distance to her school.

3 Sally has gathered 120 nuts.
Find how many these girls have gathered:
a Jane if she has gathered $\frac{1}{2}$ as many as Sally
b Carly if she has gathered $\frac{1}{3}$ as many as Sally
c Sandeep if she has gathered $\frac{1}{4}$ as many as Sally
d Amy if she has gathered $\frac{1}{5}$ as many as Sally
e Nicola if she has gathered $\frac{1}{6}$ as many as Sally
f Kate if she has gathered $\frac{1}{8}$ as many as Sally
g Lucy if she has gathered $\frac{1}{10}$ as many as Sally.

Fractions can also refer to more than one part of a whole quantity.
Look at the rectangles below:

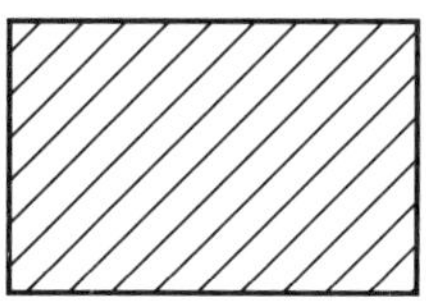
One whole rectangle

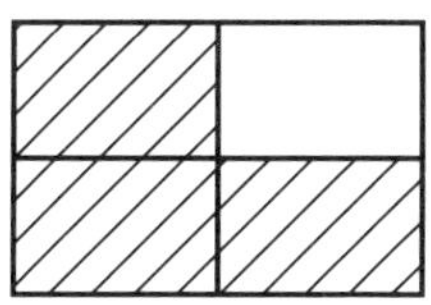
Three quarters (or $\frac{3}{4}$) of the rectangle

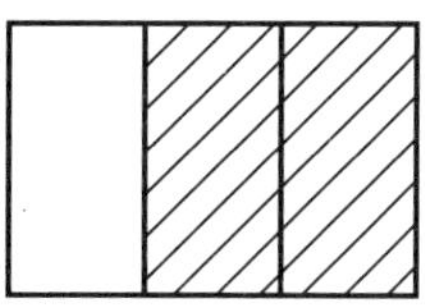
Two thirds (or $\frac{2}{3}$) of the rectangle

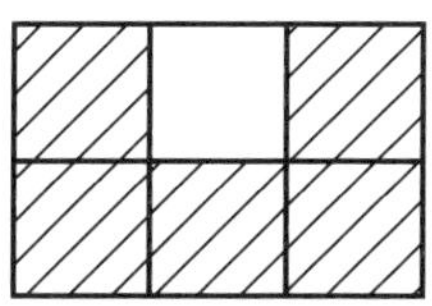
Five sixths (or $\frac{5}{6}$) of the rectangle

Example 2

Stickee's sugar is sold in three different sized packets – giant, large and small.
The giant pack contains 660 grams.

Find the weight of sugar in:
a the large sized packet if it contains $\frac{3}{4}$ as much as the giant packet.
b the small sized packet if it contains $\frac{2}{5}$ as much as the giant packet.

a $\frac{1}{4}$ of 660 grams $= 660 \div 4 = 165$ grams
so $\frac{3}{4}$ of 660 grams $= 165 \times 3$ grams $= 495$ grams
Or the weight of $\frac{3}{4}$ of 660 grams $= 660 \div 4 \times 3$ grams
$= 165 \times 3$ grams
$= 495$ grams

b $\frac{1}{5}$ of 660 grams $= 660 \div 5$ grams $= 132$ grams
so $\frac{2}{5}$ of 660 grams $= 132 \times 2$ grams $= 264$ grams
Or the weight is $\frac{2}{5}$ of 660 grams $= 660 \div 5 \times 2$ grams
$= 132 \times 2$ grams
$= 264$ grams

Exercise 4.2

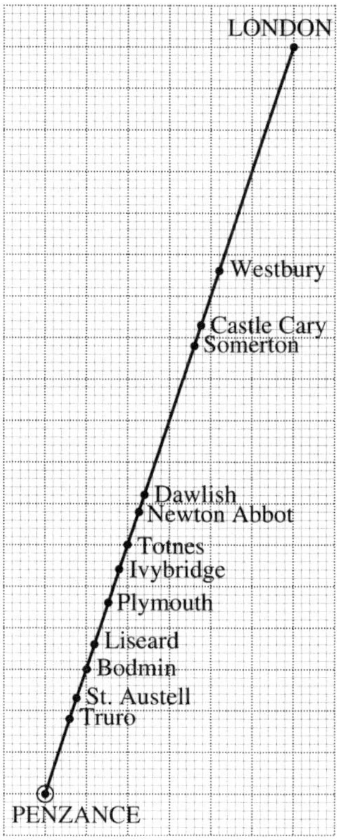

1 The distance from London to Penzance is 480 kilometres.
Find the distance from London to:
a Truro, which is $\frac{9}{10}$ of the way to Penzance
b Ivybridge, which is $\frac{7}{10}$ of the way to Penzance
c Westbury, which is $\frac{3}{10}$ of the way to Penzance
d Liskeard, which is $\frac{4}{5}$ of the way to Penzance
e Dawlish, which is $\frac{3}{5}$ of the way to Penzance
f Somerton, which is $\frac{2}{5}$ of the way to Penzance
g Bodmin, which is $\frac{5}{6}$ of the way to Penzance
h Plymouth, which is $\frac{3}{4}$ of the way to Penzance
i Totnes, which is $\frac{2}{3}$ of the way to Penzance
j St Austell, which is $\frac{7}{8}$ of the way to Penzance
k Newton Abbot, which is $\frac{5}{8}$ of the way to Penzance
l Castle Cary, which is $\frac{3}{8}$ of the way to Penzance.

2 A supermarket sells seven different kinds of soap powder. A packet of Washwell powder is the largest and contains 540 grams. Find the weight of powder in these other packets:
a the Soapswell packet if it contains $\frac{8}{9}$ as much as a Washwell packet
b the Rinsewell packet if it contains $\frac{7}{9}$ as much as a Washwell packet
c the Foamswell packet if it contains $\frac{5}{9}$ as much as a Washwell packet

d the Sudswell packet if it contains $\frac{5}{6}$ as much as a Washwell packet
e the Cleanswell packet if it contains $\frac{3}{4}$ as much as a Washwell packet
f the Whiteswell packet if it contains $\frac{2}{3}$ as much as a Washwell packet.

4.2 Simple percentages

At Willow Bank school there are 100 pupils: 53 boys and 47 girls.

$\frac{53}{100}$ of the pupils are boys and $\frac{47}{100}$ of the pupils are girls.

We can also say that:
53 per cent (53%) of the pupils are boys, and
47 per cent (47%) of the pupils are girls.

The term 'per cent' means 'out of one hundred'.

Example 1

Write this fraction as a percentage: $\frac{9}{100}$

$\frac{9}{100} = 9\%$

Example 2

Write this percentage as a fraction: 31%

$31\% = \frac{31}{100}$

Exercise 4.3

1 Write these fractions as percentages:

a $\frac{91}{100}$	**b** $\frac{73}{100}$	**c** $\frac{69}{100}$	**d** $\frac{83}{100}$
e $\frac{99}{100}$	**f** $\frac{77}{100}$	**g** $\frac{39}{100}$	**h** $\frac{21}{100}$

2 Write these percentages as fractions:

a 97%	**b** 51%	**c** 49%	**d** 89%
e 93%	**f** 61%	**g** 27%	**h** 33%

3 A farmer has 100 sheep. 57 of them are white and 43 of them are black.
a What fraction of the sheep are white?
b Write your answer to **a** as a percentage.
c What fraction of them are black?
d Write your answer to **c** as a percentage.

4 A baker has 100 loaves in his delivery van.
59 of them are white and 41 of them are brown.
a What fraction of the loaves are white?
b Write your answer to **a** as a percentage.
c What fraction of the loaves are brown?
d Write your answer to **c** as a percentage.

5 There are 100 apples in a box.
63 of them are red and 37 of them are green.
a What fraction of the apples are red?
b Write your answer to **a** as a percentage.
c What fraction of the apples are green?
d Write your answer to **c** as a percentage.

Although the term 'per cent' means 'out of one hundred', a percentage does not have to refer to a quantity of one hundred.

Example 3

There are 30 chocolates in a box and 20% of them are wrapped.
Find the number of wrapped chocolates.

$$\begin{aligned}\text{The number of wrapped chocolates} &= 20\% \text{ of } 30\\ &= \tfrac{20}{100} \text{ of } 30\\ &= \tfrac{2}{10} \text{ of } 30\\ &= 30 \div 10 \times 2\\ &= 3 \times 2\\ &= 6\end{aligned}$$

Exercise 4.4

1 There are 20 boys in Class 2A and 30% of them have red hair.
Find the number of boys with red hair.

2 There are 40 chocolates in a box and 60% of them have soft centres.
Find the number of chocolates with soft centres.

3 There are 50 commuters waiting at a railway station for their train and 80% of them are wearing an overcoat.
Find the number of commuters wearing an overcoat.

4 There are 20 people employed at a small factory and 80% of them travel to work by car.
Find the number who travel to work by car.

5 A large school has 70 teachers and 60% of them are men.
Find the number of men teachers.

6 There are 40 vehicles in a car park. 20% of them are vans.
Find the number of parked vans.

7 A box contains 40 packets of crisps. 50% of the packets contain plain crisps, 30% of the packets contain cheese and onion flavoured crisps, and 20% of the packets contain salt and vinegar flavoured crisps.
Find the number of packets of each kind.

8 Only 60 people live in Short Street. At an election 60% of them voted Conservative, 30% of them voted Labour, and 10% of them voted Liberal Democrat.
Find the number who voted for each of the three parties.

Very often a percentage of a sum of money or a measured quantity has to be calculated.
Look at Example 4.

Example 4

Find 20% of £40

$$\begin{aligned} 20\% \text{ of } £40 &= \tfrac{20}{100} \text{ of } £40 \\ &= \tfrac{2}{10} \text{ of } £40 \\ &= £40 \div 10 \times 2 \\ &= £4 \times 2 \\ &= £8 \end{aligned}$$

Exercise 4.5

Find:

1 20% of £80	**2** 30% of £60	**3** 60% of £90	**4** 40% of £70
5 40% of 60 m	**6** 30% of 40 m	**7** 20% of 30 m	**8** 60% of 20 m

Sometimes when a percentage of a quantity has to be calculated it is more convenient to change this quantity to a smaller unit. Two such cases are shown in Example 5.

Example 5

Find 50% of 3 metres

$$\begin{aligned} 50\% \text{ of } 3\,\text{m} &= 50\% \text{ of } 300\,\text{cm} \\ &= \tfrac{50}{100} \text{ of } 300\,\text{cm} \\ &= \tfrac{5}{10} \text{ of } 300\,\text{cm} \\ &= 300 \div 10 \times 5\,\text{cm} \\ &= 30 \times 5\,\text{cm} \\ &= 150\,\text{cm, or } 1.50\,\text{m, or } 1\,\text{m } 50\,\text{cm} \end{aligned}$$

Exercise 4.6

Find:

1 80% of £2 **2** 60% of £3 **3** 70% of £4 **4** 90% of £6
5 50% of 5 m **6** 70% of 9 m **7** 40% of 6 m **8** 30% of 8 m

One quantity can be expressed as a percentage of another.
Example 6 below shows a very simple case of this.

Example 6

A Labrador bitch has 10 puppies in her litter and 7 of them are yellow.
What percentage of the litter are yellow?

The percentage of puppies which are yellow

$= \frac{7}{10}$
$= \frac{70}{100}$
$= 70\%$

Exercise 4.7

1 There are 10 boys in Class 1B and one day 8 of them are present.
What percentage of the boys are present?

2 There are 5 books on a library shelf and Samantha borrows 3 of them.
What percentage of the books does she borrow?

3 Four children are waiting at the dentist's and three of them are girls.
What percentage of the children are girls?

4 Linford had a 5-day holiday from school and it rained on one of the days.
What percentage of the days were wet?

5 There are 10 students in Class 4A. 5 of them walk to school, 3 of them travel by bus to school and 2 of them cycle to school.
Find the percentage for each case.

4.3 Calculating fractions and percentages of quantities using a calculator where necessary

In Section 4.1 we saw how to calculate fractions and percentages of a quantity, but a calculator can make the task much easier.
Look at the example below:

Example 1

Calculate $\frac{2}{5}$ of £80.

Press the calculator keys as shown:

[AC] [8] [0] [×] [2] [÷] [5] [=] 32

The answer is £32.

Exercise 4.8

Find the 'odd answer out':

1 a $\frac{2}{5}$ of £110
b $\frac{3}{5}$ of £75
c $\frac{4}{5}$ of £55

2 a $\frac{7}{8}$ of £40
b $\frac{5}{8}$ of £56
c $\frac{3}{8}$ of £96

3 a $\frac{7}{12}$ of £96
b $\frac{5}{12}$ of £132
c $\frac{11}{12}$ of £60

4 a $\frac{7}{15}$ of £90
b $\frac{4}{15}$ of £165
c $\frac{11}{15}$ of £60

5 a $\frac{7}{10}$ of £120
b $\frac{9}{10}$ of £90
c $\frac{3}{10}$ of £280

6 a $\frac{3}{20}$ of £65
b $\frac{7}{20}$ of £27
c $\frac{9}{20}$ of £21

7 a $\frac{5}{24}$ of 744 m
b $\frac{7}{24}$ of 528 m
c $\frac{11}{24}$ of 336 m

8 a $\frac{1}{3}$ of 102 m
b $\frac{1}{4}$ of 136 m
c $\frac{1}{7}$ of 252 m

9 a $\frac{1}{6}$ of 156 m
b $\frac{1}{8}$ of 224 m
c $\frac{1}{15}$ of 420 m

10 a $\frac{1}{40}$ of 250 m
b $\frac{1}{24}$ of 162 m
c $\frac{1}{16}$ of 100 m

11 a $\frac{1}{12}$ of 99 m
b $\frac{1}{25}$ of 206 m
c $\frac{1}{20}$ of 165 m

12 a $\frac{11}{40}$ of 34 m
b $\frac{21}{40}$ of 18 m
c $\frac{27}{40}$ of 14 m

Example 2

There are 300 boys at Westmead School. $\frac{1}{3}$ of them are in North House, $\frac{2}{5}$ of them are in Town House and $\frac{4}{15}$ of them are in South House.
Calculate the number of boys in each of the three houses.

The number in North House = $\frac{1}{3}$ of 300 = 100
The number in Town House = $\frac{2}{5}$ of 300 = 120
The number in South House = $\frac{4}{15}$ of 300 = 80

Exercise 4.9

1 There are 240 boys at Eastgate School and they are given three sports options. $\frac{2}{5}$ choose football, $\frac{7}{20}$ choose rugby and $\frac{1}{4}$ choose cross-country.
Find the number who chose each of the three sports.

2 There are 360 girls at High Lane School. $\frac{3}{4}$ of them walk to school, $\frac{1}{6}$ cycle to school and $\frac{1}{12}$ travel to school by bus.
Find the number of girls for each of the three cases.

3 There are 480 people living in a small village. $\frac{2}{5}$ of them are men, $\frac{1}{3}$ of them are women, $\frac{1}{10}$ of them are boys and $\frac{1}{6}$ of them are girls.
Find the number in each category.

4 There are 120 vehicles on a cross-channel ferry. $\frac{1}{2}$ of them are cars, $\frac{3}{8}$ of them are vans, $\frac{1}{10}$ of them are lorries and $\frac{1}{40}$ of them are coaches.
Find the number of vehicles of each kind.

5 One day a vending machine dispensed 180 drinks. $\frac{3}{20}$ of them were orange, $\frac{1}{4}$ of them were lemon, $\frac{1}{3}$ of them were grapefruit, $\frac{1}{6}$ of them were lime and $\frac{1}{10}$ of them were raspberry.
Find the number of drinks of each flavour.

Example 3

Calculate 56% of 15 centimetres.

Press the calculator keys as shown:

[AC] [1] [5] [×] [5] [6] [÷] [1] [0] [0] [=] 8.4

The answer is 8.4 cm or 8 cm 4 mm.

Exercise 4.10

Find the 'odd answer out':

1 a 75% of £124	**2 a** 50% of £136	**3 a** 60% of £160
b 70% of £130	**b** 85% of £80	**b** 55% of £180
c 65% of £140	**c** 90% of £70	**c** 50% of £198
4 a 36% of £34	**5 a** 24% of £64	**6 a** 32% of £45
b 17% of £72	**b** 28% of £57	**b** 35% of £44
c 33% of £37	**c** 32% of £48	**c** 36% of £40
7 a 72% of 75 cm	**8 a** 35% of 240 cm	**9 a** 38% of 250 cm
b 70% of 80 cm	**b** 48% of 175 cm	**b** 32% of 300 cm
c 90% of 60 cm	**c** 68% of 125 cm	**c** 40% of 240 cm
10 a 70% of 4 cm	**11 a** 45% of 8 cm	**12 a** 26% of 25 cm
b 48% of 5 cm	**b** 14% of 25 cm	**b** 35% of 18 cm
c 15% of 16 cm	**c** 40% of 9 cm	**c** 21% of 30 cm

> **Example 4**
>
> On a bank holiday weekend an ice cream man sold 750 cornets. He sold 36% of them on Saturday, 24% of them on Sunday and 40% of them on Monday.
> Find the number of cornets he sold on each of the three days.
>
> The number he sold on Saturday = 36% of 750 = 270
> The number he sold on Sunday = 24% of 750 = 180
> The number he sold on Monday = 40% of 750 = 300

Exercise 4.11

1 On a certain day a vending machine sold 125 hot drinks. 32% of them were cups of tea, 48% were cups of coffee and 20% were cups of soup.
Find the number of drinks of each kind that were sold.

2 75 people are travelling on an over-crowded bus. 40% of them are sitting downstairs, 48% of them are sitting upstairs, and 12% of them are standing.
Find the number of people in each group.

3 300 people are travelling in a train. 65% of them are sitting in standard class seats, 20% of them are sitting in first class seats and 15% of them are sitting in the buffet.
Find the number of people in each group.

4 There are 50 doughnuts on a baker's tray. 40% of them are plain, 32% of them are jam-filled, 16% of them are apple-filled and 12% of them are custard-filled.
Find the number of each kind of doughnut.

5 250 boys are watching a school football match. 24% of them are watching from the north side of the pitch, 28% of them are watching from the south side of the pitch, 26% of them are watching from the east side of the pitch and 22% are watching from the west wide of the pitch.
Find the number of boys who are watching from each of the four sides.

6 There are 75 bottles of pop on a supermarket shelf. 24% of them contain orange pop, 28% of them contain lemon pop, 12% of them contain raspberry pop, 16% of them contain lime pop and 20% of them contain cherry pop.
Find the number of bottles of each flavour.

7 One day 80 trains left Euston station in London. 35% of them went to Birmingham, 25% of them went to Manchester, 20% of them went to Liverpool, 15% of them went to Glasgow, and 5% of them went to North Wales.
Find the number of trains that went to each destination.

4.4 Understanding and using equivalent fractions and ratios and relating these to decimals and percentages

Equivalent fractions:

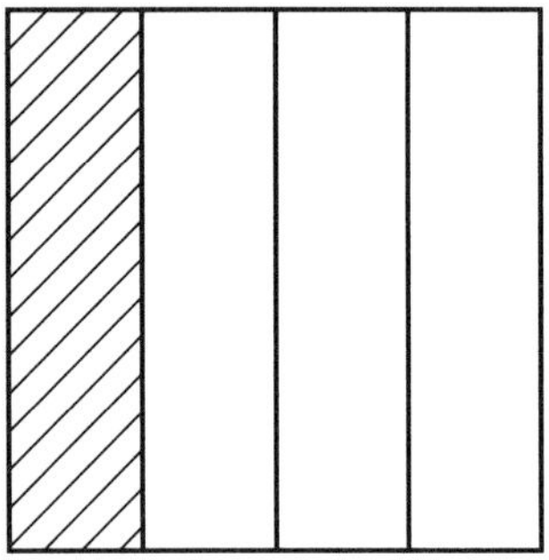

$\frac{1}{2}$

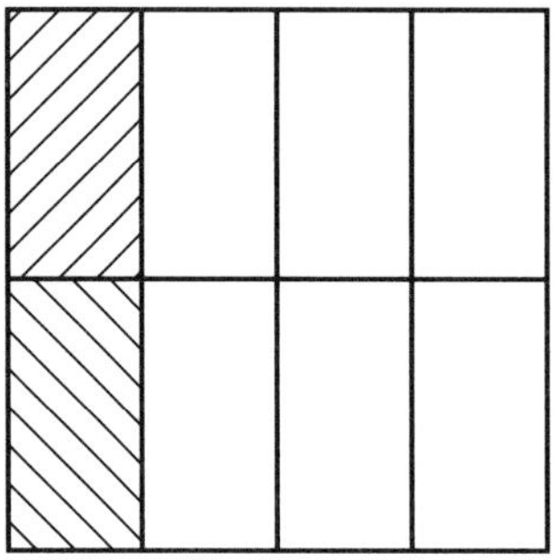

$\frac{2}{8}$

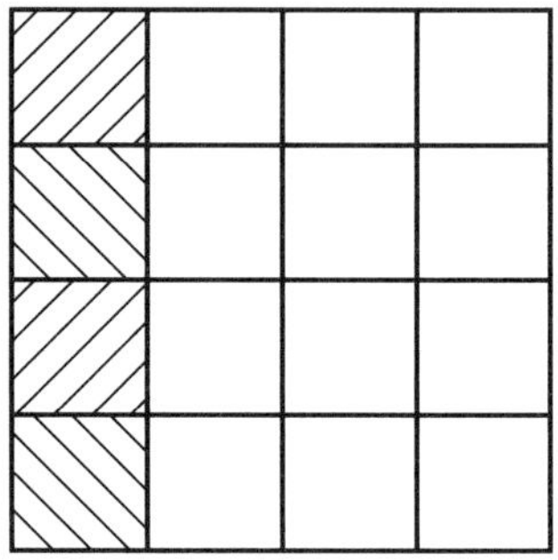

$\frac{4}{16}$

In each of the squares above **one quarter** has been shaded in.
In the first square this is shown as $\frac{1}{4}$.
In the second square it is also shown as $\frac{2}{8}$.
In the third square it is also shown as $\frac{4}{16}$.

$\frac{1}{4}$, $\frac{2}{8}$ and $\frac{4}{16}$ are therefore equivalent fractions.
They are different ways of writing the same fraction.

Example 1

Copy and complete:

a $\frac{1}{4} = \frac{?}{20} = \frac{?}{24}$ **b** $\frac{3}{4} = \frac{6}{?} = \frac{30}{?}$

a $\frac{1}{4} = \frac{5}{20} = \frac{6}{24}$ **b** $\frac{3}{4} = \frac{6}{8} = \frac{30}{40}$

Note that in each example the numerator and denominator (the upper and lower figures) are both multiplied by the same number.

Exercise 4.12

Copy and complete:

1 $\frac{1}{3} = \frac{?}{12} = \frac{?}{15} = \frac{?}{18} = \frac{?}{21} = \frac{?}{27}$

2 $\frac{1}{5} = \frac{?}{20} = \frac{?}{30} = \frac{?}{35} = \frac{?}{40} = \frac{?}{50}$

3 $\frac{1}{6} = \frac{?}{12} = \frac{?}{30} = \frac{?}{36} = \frac{?}{60} = \frac{?}{72}$

4 $\frac{1}{8} = \frac{?}{24} = \frac{?}{32} = \frac{?}{40} = \frac{?}{56} = \frac{?}{80}$

5 $\frac{1}{9} = \frac{?}{27} = \frac{?}{45} = \frac{?}{63} = \frac{?}{72} = \frac{?}{90}$

6 $\frac{1}{10} = \frac{?}{30} = \frac{?}{40} = \frac{?}{70} = \frac{?}{80} = \frac{?}{100}$

7 $\frac{3}{4} = \frac{?}{12} = \frac{?}{16} = \frac{?}{20} = \frac{?}{32} = \frac{?}{40}$

8 $\frac{2}{5} = \frac{?}{10} = \frac{?}{20} = \frac{?}{30} = \frac{?}{45} = \frac{?}{60}$

9 $\frac{3}{5} = \frac{?}{10} = \frac{?}{30} = \frac{?}{35} = \frac{?}{40} = \frac{?}{50}$

10 $\frac{4}{5} = \frac{?}{10} = \frac{?}{20} = \frac{?}{30} = \frac{?}{45} = \frac{?}{50}$

11 $\frac{5}{6} = \frac{?}{24} = \frac{?}{30} = \frac{?}{36} = \frac{?}{54} = \frac{?}{60}$

12 $\frac{3}{8} = \frac{?}{24} = \frac{?}{40} = \frac{?}{48} = \frac{?}{72} = \frac{?}{96}$

13 $\frac{1}{3} = \frac{2}{?} = \frac{3}{?} = \frac{5}{?} = \frac{9}{?} = \frac{12}{?}$

14 $\frac{1}{5} = \frac{2}{?} = \frac{3}{?} = \frac{5}{?} = \frac{9}{?} = \frac{12}{?}$

15 $\frac{1}{6} = \frac{3}{?} = \frac{4}{?} = \frac{7}{?} = \frac{8}{?} = \frac{9}{?}$

16 $\frac{1}{8} = \frac{2}{?} = \frac{6}{?} = \frac{8}{?} = \frac{9}{?} = \frac{12}{?}$

17 $\frac{4}{5} = \frac{12}{?} = \frac{20}{?} = \frac{28}{?} = \frac{32}{?} = \frac{48}{?}$

18 $\frac{5}{6} = \frac{10}{?} = \frac{15}{?} = \frac{35}{?} = \frac{40}{?} = \frac{60}{?}$

Example 2

Copy and complete:

a $\frac{5}{10} = \frac{?}{2}$ **b** $\frac{14}{21} = \frac{?}{3}$

a $\frac{5}{10} = \frac{1}{2}$ **b** $\frac{14}{21} = \frac{2}{3}$

Note that in each example both the numerator and the denominator are divided by the same figure.

Exercise 4.13

Copy and complete:

1 $\frac{8}{10} = \frac{?}{5}$	**2** $\frac{6}{10} = \frac{?}{5}$	**3** $\frac{4}{10} = \frac{?}{5}$	**4** $\frac{2}{10} = \frac{?}{5}$
5 $\frac{2}{12} = \frac{?}{6}$	**6** $\frac{6}{8} = \frac{?}{4}$	**7** $\frac{9}{15} = \frac{?}{5}$	**8** $\frac{6}{15} = \frac{?}{5}$
9 $\frac{35}{50} = \frac{?}{10}$	**10** $\frac{15}{20} = \frac{?}{4}$	**11** $\frac{5}{20} = \frac{?}{4}$	**12** $\frac{5}{25} = \frac{?}{5}$
13 $\frac{18}{24} = \frac{?}{4}$	**14** $\frac{16}{40} = \frac{?}{5}$	**15** $\frac{32}{40} = \frac{?}{5}$	**16** $\frac{16}{24} = \frac{?}{3}$

Example 3

Copy and complete:

a $\frac{6}{24} = \frac{1}{?}$ **b** $\frac{24}{27} = \frac{8}{?}$

a $\frac{6}{24} = \frac{6 \div 6}{24 \div 6} = \frac{1}{4}$ **b** $\frac{24}{27} = \frac{24 \div 3}{27 \div 3} = \frac{8}{9}$

Note again that in each example the numerator and denominator are divided by the same figure.

Exercise 4.14

1 $\frac{10}{16} = \frac{5}{?}$	**2** $\frac{6}{16} = \frac{3}{?}$	**3** $\frac{14}{16} = \frac{7}{?}$	**4** $\frac{2}{16} = \frac{1}{?}$
5 $\frac{9}{30} = \frac{3}{?}$	**6** $\frac{9}{12} = \frac{3}{?}$	**7** $\frac{12}{28} = \frac{3}{?}$	**8** $\frac{20}{28} = \frac{5}{?}$
9 $\frac{5}{15} = \frac{1}{?}$	**10** $\frac{5}{10} = \frac{1}{?}$	**11** $\frac{5}{20} = \frac{1}{?}$	**12** $\frac{15}{20} = \frac{3}{?}$
13 $\frac{24}{40} = \frac{3}{?}$	**14** $\frac{8}{40} = \frac{1}{?}$	**15** $\frac{8}{16} = \frac{1}{?}$	**16** $\frac{8}{80} = \frac{1}{?}$

Example 4

'Cancel' these fractions to their lowest terms and find the 'odd answer out'.

a $\frac{48}{60}$ **b** $\frac{35}{40}$ **c** $\frac{36}{45}$

a $\frac{48}{60} = \frac{4}{5}$ (The numerator and denominator are both divided by 12.)
b $\frac{35}{40} = \frac{7}{8}$ (The numerator and denominator are both divided by 5.)
c $\frac{36}{45} = \frac{4}{5}$ (The numerator and denominator are both divided by 9.)
The 'odd answer out' is **b**.

Exercise 4.15

Cancel the fractions and find the 'odd answer out'.

1 a $\frac{18}{30}$ **b** $\frac{27}{45}$ **c** $\frac{25}{40}$
2 a $\frac{21}{28}$ **b** $\frac{18}{24}$ **c** $\frac{14}{21}$
3 a $\frac{20}{25}$ **b** $\frac{16}{20}$ **c** $\frac{25}{30}$
4 a $\frac{24}{60}$ **b** $\frac{35}{70}$ **c** $\frac{16}{40}$
5 a $\frac{9}{36}$ **b** $\frac{21}{56}$ **c** $\frac{15}{60}$

Example 5

Change these fractions to decimals and find the 'odd answer out'.
a $\frac{15}{20}$ **b** $\frac{18}{24}$ **c** $\frac{24}{30}$

a $\frac{15}{20} = 15 \div 20 = 0.75$
b $\frac{18}{24} = 18 \div 24 = 0.75$
c $\frac{24}{30} = 24 \div 30 = 0.8$
The 'odd answer out' is **c**.

Exercise 4.16

Change the fractions to decimals and find the 'odd answer out'.

1 a $\frac{42}{50}$ **b** $\frac{68}{80}$ **c** $\frac{63}{75}$
2 a $\frac{102}{150}$ **b** $\frac{78}{120}$ **c** $\frac{52}{80}$
3 a $\frac{36}{60}$ **b** $\frac{27}{45}$ **c** $\frac{15}{24}$
4 a $\frac{22}{50}$ **b** $\frac{36}{80}$ **c** $\frac{27}{60}$
5 a $\frac{12}{75}$ **b** $\frac{28}{160}$ **c** $\frac{35}{200}$

We can use equivalent fractions to enable us to add or subtract fractions of different types.

Example 6

Evaluate: **a** $\frac{1}{2} + \frac{1}{4}$ **b** $\frac{1}{3} + \frac{8}{15}$

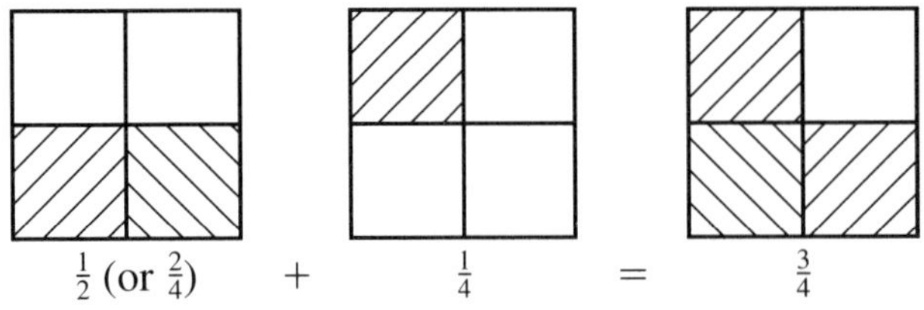

a $\frac{1}{2} + \frac{1}{4} = \frac{2}{4} + \frac{1}{4} = \frac{3}{4}$
b $\frac{1}{3} + \frac{8}{15} = \frac{5}{15} + \frac{8}{15} = \frac{13}{15}$

Exercise 4.17

Copy and complete:

1 $\frac{1}{5}+\frac{3}{20}=\frac{?}{20}+\frac{3}{20}=\frac{?}{20}$ **2** $\frac{1}{5}+\frac{7}{20}=\frac{?}{20}+\frac{7}{20}=\frac{?}{20}$
3 $\frac{2}{5}+\frac{1}{20}=\frac{?}{20}+\frac{1}{20}=\frac{?}{20}$ **4** $\frac{2}{5}+\frac{3}{20}=\frac{?}{20}+\frac{3}{20}=\frac{?}{20}$
5 $\frac{3}{5}+\frac{1}{20}=\frac{?}{20}+\frac{1}{20}=\frac{?}{20}$ **6** $\frac{3}{5}+\frac{7}{20}=\frac{?}{20}+\frac{7}{20}=\frac{?}{20}$

Exercise 4.18

Evaluate:

1 $\frac{1}{5}+\frac{8}{15}$ **2** $\frac{1}{5}+\frac{11}{15}$ **3** $\frac{1}{5}+\frac{4}{15}$ **4** $\frac{1}{5}+\frac{1}{15}$
5 $\frac{2}{5}+\frac{2}{15}$ **6** $\frac{2}{5}+\frac{8}{15}$ **7** $\frac{2}{5}+\frac{7}{15}$ **8** $\frac{3}{5}+\frac{4}{15}$
9 $\frac{2}{3}+\frac{2}{9}$ **10** $\frac{2}{3}+\frac{1}{9}$ **11** $\frac{1}{10}+\frac{7}{20}$ **12** $\frac{1}{10}+\frac{17}{20}$
13 $\frac{3}{10}+\frac{1}{20}$ **14** $\frac{3}{10}+\frac{7}{20}$ **15** $\frac{3}{10}+\frac{11}{20}$ **16** $\frac{3}{10}+\frac{13}{20}$
17 $\frac{1}{6}+\frac{5}{12}$ **18** $\frac{5}{6}+\frac{1}{12}$ **19** $\frac{1}{2}+\frac{3}{8}$ **20** $\frac{1}{2}+\frac{1}{8}$

Example 7

Evaluate: **a** $\frac{1}{2}-\frac{1}{4}$ **b** $\frac{2}{3}-\frac{7}{12}$

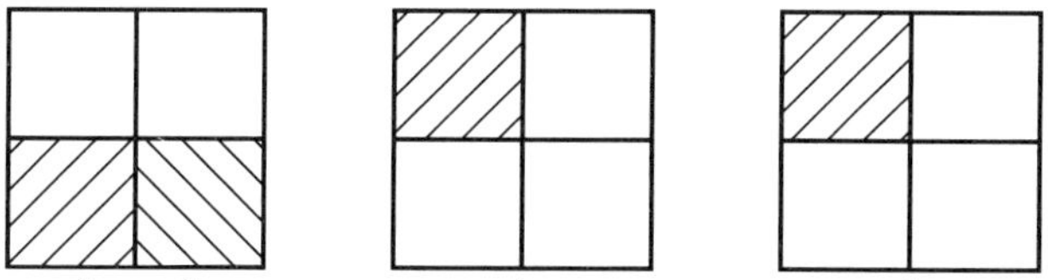

$\frac{1}{2}$ (or $\frac{2}{4}$) $-$ $\frac{1}{4}$ $=$ $\frac{1}{4}$

a $\frac{1}{2}-\frac{1}{4}=\frac{2}{4}-\frac{1}{4}=\frac{1}{4}$
b $\frac{2}{3}-\frac{7}{12}=\frac{8}{12}-\frac{7}{12}=\frac{1}{12}$

Exercise 4.19

Copy and complete:

1 $\frac{4}{5}-\frac{3}{20}=\frac{?}{20}-\frac{3}{20}=\frac{?}{20}$ **2** $\frac{4}{5}-\frac{7}{20}=\frac{?}{20}-\frac{7}{20}=\frac{?}{20}$
3 $\frac{3}{5}-\frac{1}{20}=\frac{?}{20}-\frac{1}{20}=\frac{?}{20}$ **4** $\frac{3}{5}-\frac{3}{20}=\frac{?}{20}-\frac{3}{20}=\frac{?}{20}$
5 $\frac{2}{5}-\frac{1}{20}=\frac{?}{20}-\frac{1}{20}=\frac{?}{20}$ **6** $\frac{2}{5}-\frac{7}{20}=\frac{?}{20}-\frac{7}{20}=\frac{?}{20}$

Exercise 4.20

Evaluate:

1 $\frac{9}{10}-\frac{7}{20}$ **2** $\frac{9}{10}-\frac{11}{20}$ **3** $\frac{9}{10}-\frac{17}{20}$ **4** $\frac{9}{10}-\frac{9}{20}$
5 $\frac{4}{5}-\frac{7}{10}$ **6** $\frac{4}{5}-\frac{1}{10}$ **7** $\frac{2}{5}-\frac{1}{10}$ **8** $\frac{2}{5}-\frac{3}{10}$
9 $\frac{3}{4}-\frac{1}{8}$ **10** $\frac{1}{4}-\frac{1}{8}$ **11** $\frac{1}{3}-\frac{1}{6}$ **12** $\frac{5}{6}-\frac{5}{12}$
13 $\frac{2}{3}-\frac{1}{9}$ **14** $\frac{1}{3}-\frac{1}{9}$ **15** $\frac{1}{3}-\frac{2}{9}$ **16** $\frac{1}{2}-\frac{3}{8}$
17 $\frac{1}{2}-\frac{1}{8}$ **18** $\frac{1}{2}-\frac{5}{12}$ **19** $\frac{1}{2}-\frac{1}{12}$ **20** $\frac{1}{3}-\frac{1}{15}$

Example 8

Evaluate: **a** $\frac{1}{3}+\frac{1}{6}$ **b** $\frac{1}{6}+\frac{5}{18}$
and in each case simplify your answer.

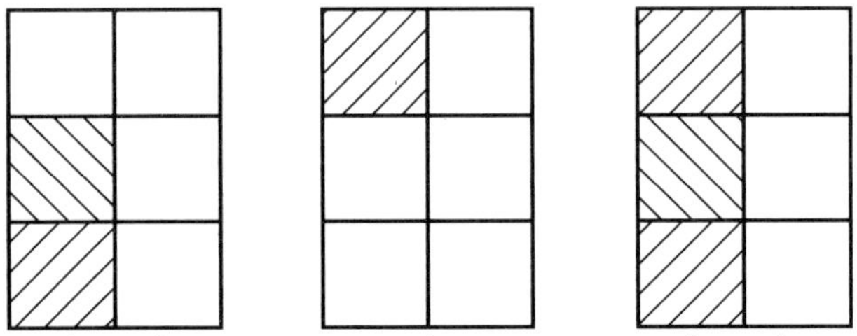

$\frac{1}{3}$ (or $\frac{2}{6}$) $+$ $\frac{1}{6}$ $=$ $\frac{3}{6}$ (or $\frac{1}{2}$)

a $\frac{1}{3}+\frac{1}{6}=\frac{2}{6}+\frac{1}{6}=\frac{3}{6}=\frac{1}{2}$
b $\frac{1}{6}+\frac{5}{18}=\frac{3}{18}+\frac{5}{18}=\frac{8}{18}=\frac{4}{9}$

Exercise 4.21

Evaluate:

1 $\frac{1}{4}+\frac{9}{20}$ **2** $\frac{1}{4}+\frac{13}{20}$ **3** $\frac{1}{4}+\frac{1}{20}$ **4** $\frac{3}{4}+\frac{3}{20}$
5 $\frac{1}{2}+\frac{3}{10}$ **6** $\frac{1}{2}+\frac{1}{10}$ **7** $\frac{1}{2}+\frac{1}{6}$ **8** $\frac{1}{3}+\frac{7}{15}$
9 $\frac{2}{3}+\frac{1}{12}$ **10** $\frac{1}{6}+\frac{1}{12}$ **11** $\frac{1}{6}+\frac{7}{12}$ **12** $\frac{1}{9}+\frac{1}{18}$
13 $\frac{1}{4}+\frac{5}{12}$ **14** $\frac{1}{4}+\frac{1}{12}$ **15** $\frac{1}{5}+\frac{11}{20}$ **16** $\frac{1}{5}+\frac{1}{20}$
17 $\frac{3}{10}+\frac{9}{20}$ **18** $\frac{7}{10}+\frac{1}{20}$ **19** $\frac{1}{5}+\frac{7}{15}$ **20** $\frac{1}{5}+\frac{2}{15}$

Example 9

Evaluate: **a** $\frac{1}{2}-\frac{1}{10}$ **b** $\frac{5}{6}-\frac{5}{18}$
and simplify your answer.

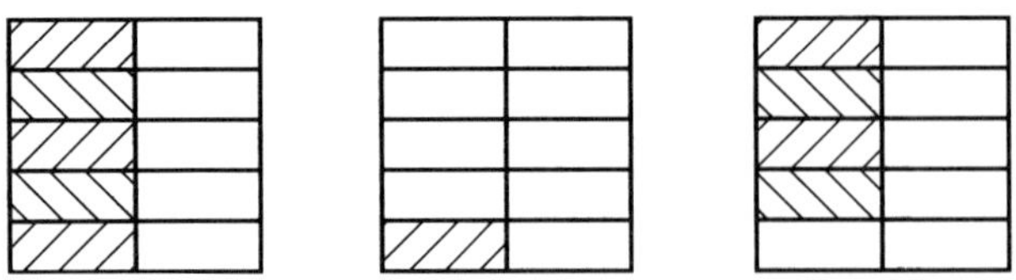

$\frac{1}{2}$ (or $\frac{5}{10}$) $-$ $\frac{1}{10}$ $=$ $\frac{4}{10}$ (or $\frac{2}{5}$)

a $\frac{1}{2}-\frac{1}{10}=\frac{5}{10}-\frac{1}{10}=\frac{4}{10}=\frac{2}{5}$
b $\frac{5}{6}-\frac{5}{18}=\frac{15}{18}-\frac{5}{18}=\frac{10}{18}=\frac{5}{9}$

Exercise 4.22

Evaluate:

1 $\frac{3}{4} - \frac{9}{20}$ **2** $\frac{3}{4} - \frac{13}{20}$ **3** $\frac{3}{4} - \frac{1}{20}$ **4** $\frac{1}{4} - \frac{3}{20}$
5 $\frac{1}{6} - \frac{1}{18}$ **6** $\frac{5}{6} - \frac{1}{18}$ **7** $\frac{2}{3} - \frac{4}{15}$ **8** $\frac{2}{3} - \frac{7}{15}$
9 $\frac{5}{6} - \frac{1}{12}$ **10** $\frac{5}{6} - \frac{7}{12}$ **11** $\frac{2}{3} - \frac{1}{6}$ **12** $\frac{2}{9} - \frac{1}{18}$
13 $\frac{3}{4} - \frac{5}{12}$ **14** $\frac{3}{4} - \frac{1}{12}$ **15** $\frac{9}{10} - \frac{13}{20}$ **16** $\frac{9}{10} - \frac{3}{20}$
17 $\frac{3}{5} - \frac{7}{20}$ **18** $\frac{2}{5} - \frac{3}{20}$ **19** $\frac{4}{5} - \frac{7}{15}$ **20** $\frac{4}{5} - \frac{2}{15}$

Example 10

Evaluate: **a** $\frac{3}{4} + \frac{1}{6}$ **b** $\frac{1}{4} - \frac{1}{6}$

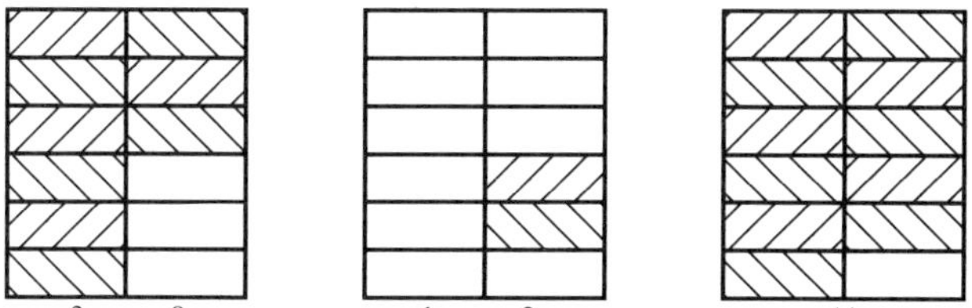

$\frac{3}{4}$ (or $\frac{9}{12}$) $+$ $\frac{1}{6}$ (or $\frac{2}{12}$) $=$ $\frac{11}{12}$

a $\frac{3}{4} + \frac{1}{6} = \frac{9}{12} + \frac{2}{12} = \frac{11}{12}$
b $\frac{1}{4} - \frac{1}{6} = \frac{3}{12} - \frac{2}{12} = \frac{1}{12}$

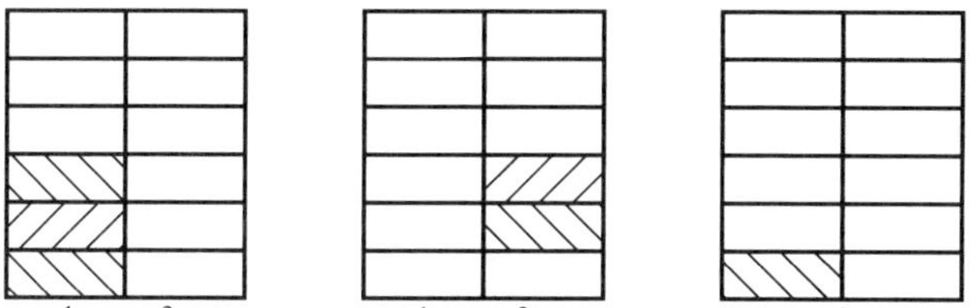

$\frac{1}{4}$ (or $\frac{3}{12}$) $-$ $\frac{1}{6}$ (or $\frac{2}{12}$) $=$ $\frac{1}{12}$

Exercise 4.23

Evaluate these sums simplifying your answers where possible.

1 $\frac{1}{8} + \frac{7}{10}$ **2** $\frac{1}{8} + \frac{3}{10}$ **3** $\frac{1}{8} + \frac{1}{10}$ **4** $\frac{3}{8} + \frac{1}{10}$
5 $\frac{1}{8} + \frac{1}{12}$ **6** $\frac{3}{8} + \frac{1}{12}$ **7** $\frac{5}{8} + \frac{1}{12}$ **8** $\frac{7}{8} + \frac{1}{12}$
9 $\frac{1}{6} + \frac{3}{8}$ **10** $\frac{1}{6} + \frac{5}{8}$ **11** $\frac{1}{6} + \frac{1}{8}$ **12** $\frac{1}{6} + \frac{1}{10}$
13 $\frac{5}{6} + \frac{1}{10}$ **14** $\frac{5}{6} - \frac{1}{10}$ **15** $\frac{1}{6} - \frac{1}{10}$ **16** $\frac{1}{6} - \frac{1}{15}$
17 $\frac{4}{15} - \frac{1}{6}$ **18** $\frac{3}{4} - \frac{1}{10}$ **19** $\frac{1}{4} - \frac{1}{10}$ **20** $\frac{5}{6} - \frac{1}{8}$

A percentage can be regarded as a special kind of ratio. For example, 92% could be written as 92 : 100.
It is therefore possible to see if ratios are equivalent by converting them to percentages.

Look at Example 11 below:

Example 11

Express each ratio as a percentage and find the 'odd answer out'.
a 20 : 25 **b** 45 : 60 **c** 21 : 28

a 20 : 25 as a percentage $= \frac{20}{25} \times 100$
$= 20 \times 100 \div 25$ i.e. $\frac{4}{5}$ or $\frac{80}{100}$
$= 80\%$
b 45 : 60 as a percentage $= \frac{45}{60} \times 100$
$= 45 \times 100 \div 60$ i.e. $\frac{3}{4}$ or $\frac{75}{100}$
$= 75\%$
c 21 : 28 as a percentage $= \frac{21}{28} \times 100$
$= 21 \times 100 \div 28$ i.e. $\frac{3}{4}$ or $\frac{75}{100}$
$= 75\%$
The 'odd answer out' is **a**.

Exercise 4.24

Express each ratio as a percentage and find the 'odd answer out'.

1	**a** 19 : 20	**b** 72 : 75	**c** 48 : 50
2	**a** 18 : 40	**b** 24 : 50	**c** 27 : 60
3	**a** 32 : 40	**b** 42 : 50	**c** 105 : 125
4	**a** 12 : 50	**b** 15 : 60	**c** 9 : 36
5	**a** 9 : 60	**b** 8 : 50	**c** 6 : 40

4.5 Recognising patterns in equivalent fractions

Example 1 below will remind you how to reduce a fraction to its lowest terms.

Example 1

Reduce to its lowest terms: **a** $\frac{5}{25}$ **b** $\frac{6}{21}$

a $\frac{5}{25} = \dfrac{5 \div 5}{25 \div 5} = \frac{1}{5}$ **b** $\frac{6}{21} = \dfrac{6 \div 3}{21 \div 3} = \frac{2}{7}$

Exercise 4.25

Reduce to its lowest terms:

1 $\frac{6}{8}$	**2** $\frac{10}{12}$	**3** $\frac{10}{16}$	**4** $\frac{6}{16}$	**5** $\frac{6}{10}$	**6** $\frac{4}{10}$
7 $\frac{6}{9}$	**8** $\frac{15}{18}$	**9** $\frac{15}{24}$	**10** $\frac{9}{24}$	**11** $\frac{27}{30}$	**12** $\frac{21}{30}$
13 $\frac{4}{16}$	**14** $\frac{4}{12}$	**15** $\frac{4}{8}$	**16** $\frac{4}{24}$	**17** $\frac{20}{24}$	**18** $\frac{12}{40}$
19 $\frac{5}{20}$	**20** $\frac{5}{25}$	**21** $\frac{12}{30}$	**22** $\frac{18}{30}$	**23** $\frac{24}{30}$	**24** $\frac{18}{24}$

Example 2

A set of equivalent fractions is shown below.
Find the next five fractions in the sequence.

$\frac{3}{4}$ $\frac{6}{8}$ $\frac{9}{12}$ $\frac{12}{16}$

The numerators (the upper figures) are the multiples of 3, so the next five numerators are:

15 18 21 24 and 27

The denominators (the lower figures) are the multiples of 4, so the next five denominators are:

20 24 28 32 and 36

The next five fractions are therefore:

$\frac{15}{20}$ $\frac{18}{24}$ $\frac{21}{28}$ $\frac{24}{32}$ and $\frac{27}{36}$

Exercise 4.26

Find the next five fractions in the sequence:

1 $\frac{2}{5}, \frac{4}{10}, \frac{6}{15}, \frac{8}{20} \ldots$
2 $\frac{3}{5}, \frac{6}{10}, \frac{9}{15}, \frac{12}{20} \ldots$
3 $\frac{5}{8}, \frac{10}{16}, \frac{15}{24}, \frac{20}{32} \ldots$
4 $\frac{7}{8}, \frac{14}{16}, \frac{21}{24}, \frac{28}{32} \ldots$
5 $\frac{1}{6}, \frac{2}{12}, \frac{3}{18}, \frac{4}{24} \ldots$
6 $\frac{1}{8}, \frac{2}{16}, \frac{3}{24}, \frac{4}{32} \ldots$

Example 3

A set of equivalent fractions is shown below.
Find the remaining terms in the sequence until the simplest of the equivalent fractions is found.

$\frac{20}{30}$ $\frac{18}{27}$ $\frac{16}{24}$ $\frac{14}{21}$

The numerators are descending multiples of 2, so the remaining numerators are:

12 10 8 6 4 and 2

The denominators are descending multiples of 3, so the remaining denominators are:

18 15 12 9 6 and 3

$\frac{2}{3}$ is therefore the simplest of the equivalent fractions.

Exercise 4.27

Find the remaining terms in the sequence until the simplest of the equivalent fractions is found:

1 $\frac{40}{50}, \frac{36}{45}, \frac{32}{40}, \frac{28}{35}, \ldots$
2 $\frac{50}{60}, \frac{45}{54}, \frac{40}{48}, \frac{35}{42}, \ldots$
3 $\frac{90}{100}, \frac{81}{90}, \frac{72}{80}, \frac{63}{70}, \ldots$
4 $\frac{30}{200}, \frac{27}{180}, \frac{24}{160}, \frac{21}{140}, \ldots$
5 $\frac{10}{50}, \frac{9}{45}, \frac{8}{40}, \frac{7}{35}, \ldots$
6 $\frac{10}{20}, \frac{9}{18}, \frac{8}{16}, \frac{7}{14}, \ldots$

Example 4

Find the simplest fraction which is equivalent to $\frac{18}{30}$ and write down two other equivalent fractions.

$$\frac{18}{30} = \frac{18 \div 6}{30 \div 6} = \frac{3}{5}$$

For two other equivalent fractions any two (except $\frac{3}{5}$ and $\frac{18}{30}$) may be chosen from the sequence:

$\frac{3}{5}$ $\frac{6}{10}$ $\frac{9}{15}$ $\frac{12}{20}$ $\frac{15}{25}$

(Note that the numerators are the multiples of 3 and the denominators are the multiples of 5.)

Exercise 4.28

Reduce the fraction to its lowest terms and write down two other equivalent fractions.

1 $\frac{8}{10}$ **2** $\frac{14}{20}$ **3** $\frac{10}{16}$ **4** $\frac{10}{24}$ **5** $\frac{6}{15}$ **6** $\frac{12}{21}$

7 $\frac{15}{50}$ **8** $\frac{15}{40}$ **9** $\frac{35}{100}$ **10** $\frac{5}{35}$ **11** $\frac{12}{42}$ **12** $\frac{12}{54}$

Example 5

Draw a graph of numerator against denominator for the following set of equivalent fractions:

$\frac{2}{3}$ $\frac{4}{6}$ $\frac{6}{9}$ $\frac{10}{15}$ $\frac{16}{24}$

From your graph find three more equivalent fractions.

The above details can be tabulated as follows:

Denominator	3	6	9	15	24
Numerator	2	4	6	10	16

The above five pairs of coordinates are plotted and the points are joined with a straight line.

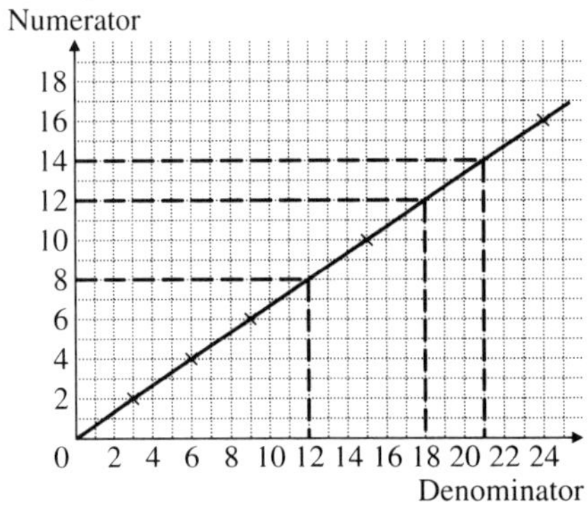

It can be seen from the broken lines that three further equivalent fractions are:

$\frac{8}{12}$ $\frac{12}{18}$ and $\frac{14}{21}$

Exercise 4.29

Three copies of the grid illustrated below are required, one for each question.

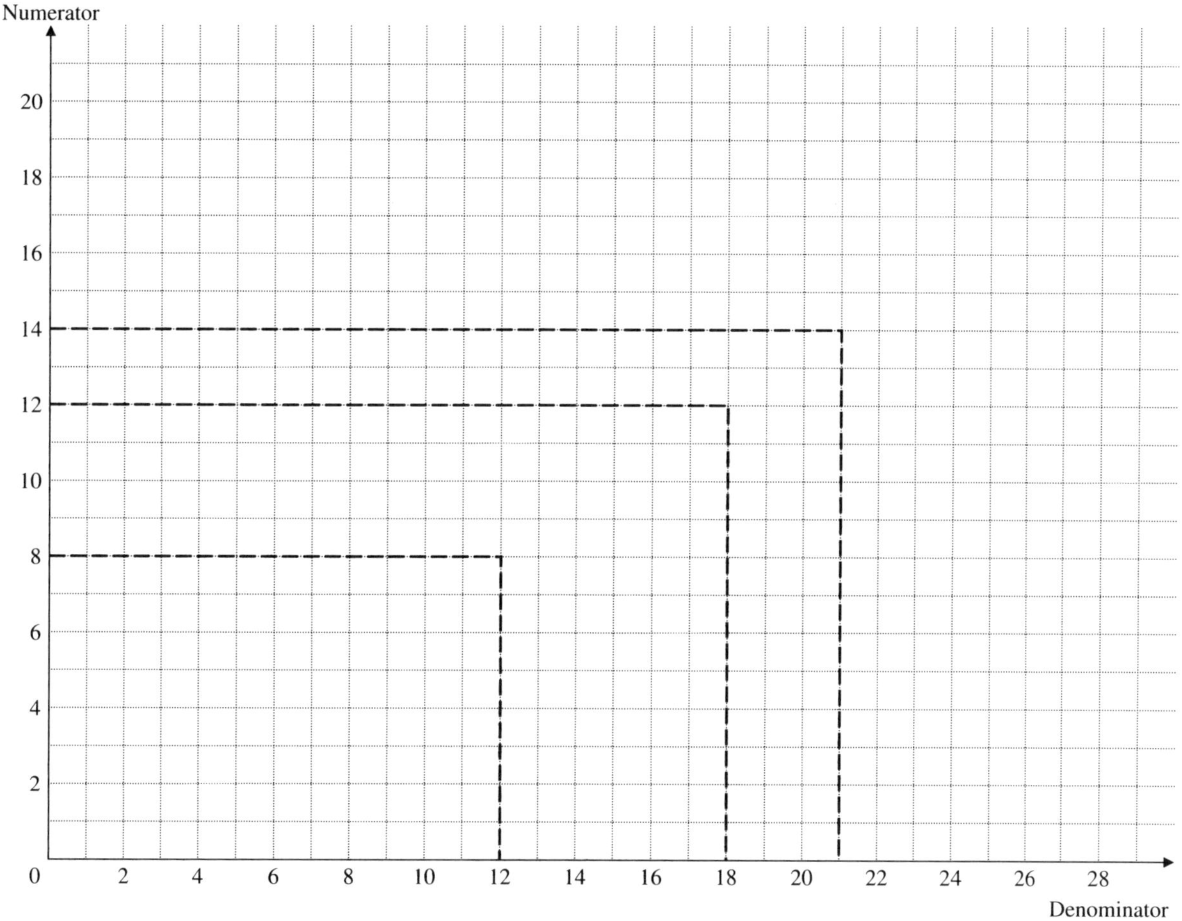

Draw a graph of numerator against denominator for each set of equivalent fractions.

1 $\frac{3}{4}$ $\frac{6}{8}$ $\frac{12}{16}$ $\frac{18}{24}$ $\frac{21}{28}$
From your graph find two more equivalent fractions.

2 $\frac{4}{5}$ $\frac{8}{10}$ $\frac{20}{25}$
From your graph find two more equivalent fractions.

3 $\frac{1}{3}$ $\frac{2}{6}$ $\frac{4}{12}$ $\frac{6}{18}$ $\frac{8}{24}$
From your graph find four more equivalent fractions.

4.6 Working out fractional or percentage changes

Fractions and percentages are often used to describe how a quantity increases or decreases.

Example 1

This year the number of pupils at Hay Bank School is 800. Next year the total is expected to rise by one fifth.
How many pupils are expected next year?

The expected increase is $\frac{1}{5}$ of $800 = \frac{1}{5} \times 800 = 160$

Therefore the number of pupils expected next year $= 800 + 160 = 960$

Example 2

A sales representative earns £240 per week. He expects his earnings to increase by 10%.
What will his new weekly wage be?

The increase is 10% of £240 $= \frac{10}{100} \times 240 =$ £24

Therefore his new wage = £240 + £24 = £264

Exercise 4.30

1 Frooties are sold in tubes which contain 20 sweets. The manufacturer decides to make a longer tube which will contain one fifth as many sweets again.
How many sweets will the new tube contain?

2 Twenty years ago a tree was 20 metres tall. It is now one quarter as tall again.
How tall is it now?

3 Last year Kirsty's weight was 60 kg, but she is now one twelfth as heavy again.
What is her weight now?

4 Mr Smith's car can travel a distance of 15 kilometres on one litre of petrol. Mrs Smith's smaller car can travel one third as far again.
How far can Mrs Smith's car travel on one litre of petrol?

5 A bus carries 40 passengers, but after calling at a bus stop it carries one twentieth as many again.
How many is it carrying now?

6 At Rani's school the morning lessons each last 40 minutes, but the afternoon lessons each last one eighth as long again.
How long does each afternoon lesson last?

7 Last year Chantelle put £240 into her savings account at the bank and now her savings are 10% more.
How much are her savings now?

8 Mr Jones is paid £180 per week. He is promoted and gets a 15% rise.
What is his new weekly wage?

9 Last year a house was valued at £40 000 but it is now worth 8% more.
How much is it worth now?

10 Last year David could clear 150 cm at high jump, but he can now jump 12% higher.
What height can he clear now?

Example 3

This year the number of pupils at Sunny Hill School is 900. Next year the total is expected to fall by one twelfth.
How many pupils are expected next year?

The expected decrease is: $\frac{1}{12}$ of $900 = \frac{1}{12} \times 900 = 75$

The number of pupils expected next year is: $900 - 75 = 825$.

Example 4

A sales representative earns £250 per week, but owing to a trade recession her earnings drop by 6%.
What will her new weekly wage be?

The decrease is: 6% of £250 $= \frac{6}{100} \times 250 =$ £15

Her new wage is: £250 – £15 = £235.

Exercise 4.31

1 Daniel has 30 marbles, but he plays a game and loses one fifth of them.
How many has he left?

2 Mrs Patel has a packet which contains 36 biscuits. She puts one ninth of them on the table at tea time.
How many are left in the packet?

3 Last year James ran a four hundred metre race in 63 seconds, but this year his time is better by one seventh.
What is his time this year?

4 Mr Brown has 120 screws in a box, but he uses one sixth of them while repairing a cabinet.
How many are left in the box?

5 Mrs Clarke has a curtain which is 240 cm long but she has to cut off one twelfth of it in order to make it fit.
How long is the finished curtain?

6 There are 80 houses in West Street but one sixteenth of them are pulled down because they are in poor condition.
How many houses remain?

7 In June they have 16 hours of daylight in Paris, but by September they have 25% less.
How many hours of daylight do they have in September?

8 If Saleem goes to school by bus the journey takes 50 minutes. If he goes by train the journey time is 16% less.
How long does it take by train?

9 The price of a certain table is £80, but the shop manager reduces the price by 15% because it is scratched.
What is the reduced price?

10 Mr Carter buys a car for £8500, but after two years it is worth 40% less.
What is it worth after two years?

4.7 Converting fractions to decimals and percentages and finding one number as a percentage of another

The simplest way of converting a fraction to a decimal is to use your calculator to divide the top number by the bottom number.

Example 1

Convert each fraction to its decimal equivalent.

a $\frac{3}{4}$ **b** $\frac{2}{5}$ **c** $\frac{3}{50}$

a $\frac{3}{4} = 3 \div 4 = 0.75$ (Note that $\frac{3}{4} = \frac{3\times25}{4\times25} = \frac{75}{100}$)
b $\frac{2}{5} = 2 \div 5 = 0.4$ (Note that $\frac{2}{5} = \frac{2\times2}{5\times2} = \frac{4}{10}$)
c $\frac{3}{50} = 3 \div 50 = 0.06$ (Note that $\frac{3}{50} = \frac{3\times2}{50\times2} = \frac{6}{100}$)

Exercise 4.32

Convert each fraction to its decimal equivalent.

1 $\frac{9}{20}$	**2** $\frac{3}{20}$	**3** $\frac{19}{20}$	**4** $\frac{13}{20}$	**5** $\frac{13}{25}$	**6** $\frac{16}{25}$
7 $\frac{12}{25}$	**8** $\frac{21}{25}$	**9** $\frac{3}{25}$	**10** $\frac{7}{25}$	**11** $\frac{7}{40}$	**12** $\frac{29}{40}$
13 $\frac{1}{4}$	**14** $\frac{9}{50}$	**15** $\frac{17}{50}$	**16** $\frac{39}{50}$	**17** $\frac{49}{50}$	**18** $\frac{37}{50}$
19 $\frac{1}{5}$	**20** $\frac{1}{20}$	**21** $\frac{1}{25}$	**22** $\frac{3}{40}$	**23** $\frac{6}{125}$	**24** $\frac{9}{125}$

In order to change a decimal to a fraction the place value of the figures must be considered.

Example 2

Convert each decimal to its fraction equivalent.

a 0.16 **b** 0.03 **c** 0.006

a $0.16 = \frac{1}{10} + \frac{6}{100} = \frac{10}{100} + \frac{6}{100} = \frac{16}{100} = \frac{4}{25}$
b $0.03 = \frac{0}{10} + \frac{3}{100} = \frac{3}{100}$
c $0.006 = \frac{0}{10} + \frac{0}{100} + \frac{6}{1000} = \frac{6}{1000} = \frac{3}{500}$

Exercise 4.33

Convert each decimal to its fraction equivalent and then write it in its simplest form.

1 0.26	**2** 0.46	**3** 0.22	**4** 0.86	**5** 0.42	**6** 0.66
7 0.44	**8** 0.32	**9** 0.85	**10** 0.55	**11** 0.325	**12** 0.675
13 0.6	**14** 0.7	**15** 0.1	**16** 0.01	**17** 0.07	**18** 0.09
19 0.02	**20** 0.08	**21** 0.025	**22** 0.035	**23** 0.065	**24** 0.055

Example 3

Convert each decimal to its percentage equivalent.

a 0.21 **b** 0.50 **c** 0.03 **d** 0.7 **e** 0.125

a $0.21 = \frac{21}{100} = 21\%$
b $0.50 = \frac{50}{100} = 50\%$
c $0.03 = \frac{3}{100} = 3\%$
d $0.7 = \frac{7}{10} = \frac{70}{100} = 70\%$
e $0.125 = \frac{125}{1000} = \frac{12.5}{100} = 12.5\%$

Exercise 4.34

Convert each decimal to its percentage equivalent.

1 0.25	**2** 0.29	**3** 0.23	**4** 0.36	**5** 0.32	**6** 0.42
7 0.85	**8** 0.18	**9** 0.15	**10** 0.12	**11** 0.04	**12** 0.05
13 0.08	**14** 0.07	**15** 0.4	**16** 0.9	**17** 0.2	**18** 0.1
19 0.765	**20** 0.165	**21** 0.245	**22** 0.945	**23** 0.645	**24** 0.015

The simplest way of converting a percentage to a decimal is to divide the percentage figure by 100.

Example 4

Convert each percentage to its decimal equivalent.

a 45% **b** 2% **c** 95.5%

a $45\% = 45 \div 100 = 0.45$
b $2\% = 2 \div 100 = 0.02$
c $95.5\% = 95.5 \div 100 = 0.955$

Exercise 4.35

Convert each percentage to its decimal equivalent.

1 67%	**2** 53%	**3** 49%	**4** 76%	**5** 84%	**6** 16%
7 30%	**8** 60%	**9** 80%	**10** 9%	**11** 6%	**12** 2%
13 73.5%	**14** 93.5%	**15** 7.5%	**16** 2.5%	**17** 3.5%	**18** 9.5%

Example 5

Find each of the following:

a 1 as a percentage of 5 **b** 5 as a percentage of 20
c 20 as a percentage of 50 **d** 7 as a percentage of 40

a '1 out of 5' $= \frac{1}{5} = \frac{20}{100} = 20\%$ (Note $\frac{20}{100} = 0.20$)
b '5 out of 20' $= \frac{5}{20} = \frac{25}{100} = 25\%$ (Note $\frac{25}{100} = 0.25$)
c '20 out of 50' $= \frac{20}{50} = \frac{40}{100} = 40\%$ (Note $\frac{40}{100} = 0.40$)
d '7 out of 40' $= \frac{7}{40} = \frac{175}{1000} = \frac{17.5}{100} = 17.5\%$ (Note $\frac{7}{40} = 0.175$)

Exercise 4.36

Find the 'odd answer out':

1 a 4 as a percentage of 20
b 5 as a percentage of 25
c 3 as a percentage of 12

2 a 15 as a percentage of 50
b 7 as a percentage of 20
c 9 as a percentage of 30

3 a 7 as a percentage of 70
b 6 as a percentage of 40
c 9 as a percentage of 90

4 a 20 as a percentage of 50
b 14 as a percentage of 40
c 21 as a percentage of 60

5 a 48 as a percentage of 60
b 30 as a percentage of 40
c 27 as a percentage of 36

6 a 56 as a percentage of 80
b 21 as a percentage of 30
c 15 as a percentage of 20

7 a 14 as a percentage of 80
b 7 as a percentage of 56
c 3 as a percentage of 24

8 a 18 as a percentage of 48
b 13 as a percentage of 40
c 12 as a percentage of 32

Example 6

a This year my weekly pay is £160. I am told it is going to increase by £40 per week.
Express the increase as a percentage of my current pay.
b If instead my pay actually decreases by £8 per week, express this decrease as a percentage of my current pay.

a £40 as a fraction of £160 $= \frac{40}{160} = \frac{1}{4}$ or 0.25, and $0.25 = 25\%$.
Therefore my expected increase is 25%.
b £8 as a fraction of £160 $= \frac{8}{160} = \frac{1}{20}$ or 0.05, and $0.05 = 5\%$
Therefore my pay decrease is 5%.

Exercise 4.37

For Questions 1–6 express the increase as a percentage of the original number or quantity.

1 Last month there were 60 boys in the local scout group, but 12 more have joined.

2 One morning there were 20 boats in a harbour, but by midday there were 6 more.

3 Last year Joanna's baby sister weighed 15 kg, but now she is 12 kg heavier.

4 Last month there were 50 girls in the local guide group, but 4 more have joined.

5 On Friday night 240 people went to a local concert, but on Saturday night 36 more attended.

6 A carpenter charges £80 for the cabinets which he makes if they are left unpainted, but if he has to paint them he charges £28 more.

For Questions 7–10 express the decrease as a percentage of the original.

7 Last year Mr White earned £220 per week, but he now earns £44 a week less because of a trade depression.

8 Tyrone arranges to go on a package holiday which costs £120, but the tour company charges him £18 less because he books early.

9 The train fare from London to Penzance is normally £48, but the off-peak fare is £12 less.

10 Last year Mrs Green paid £250 to insure her car, but this year she has paid £75 less because she now has a garage for the car.

Example 7

Which is the better discount and by how many percent, '$\frac{1}{20}$ off' or '6% off'?

$\frac{1}{20}$ of ALL marked prices

6% off EVERYTHING

$\frac{1}{20} = \frac{5}{100} = 0.05$, and $0.05 = 5\%$.

6% is a bigger price reduction, by 1%.

Exercise 4.38

Find which is the better discount and by how many percent.

1 $\frac{1}{4}$ off' or '20% off' **2** '$\frac{1}{5}$ off' or '15% off'
3 '$\frac{1}{25}$ off' or '3% off' **4** '$\frac{1}{5}$ off' or '25% off'
5 '$\frac{1}{4}$ off' or '30% off' **6** '$\frac{1}{25}$ off' or '6% off'

Example 8

The catalogue price of a television set is £300.

There are two discount offers: '$\frac{1}{4}$ off' or '20% off'.
Which is the better discount offer and by how much?

$\frac{1}{4}$ of £300 = $\frac{1}{4}$ × £300 = 1 × 300 ÷ 4 = £75
20% of £300 = $\frac{20}{100}$ × £300 = £300 × 20 ÷ 100 = £60

Therefore '$\frac{1}{4}$ off' is the better discount by £15.

Exercise 4.39

Find which is the better discount and by how much.

Article for sale	Catalogue price	First discount	Second discount
1 Video	£320	$\frac{1}{4}$ off	30% off
2 Tumble dryer	£200	$\frac{1}{5}$ off	15% off
3 Cassette player	£60	$\frac{1}{5}$ off	25% off
4 Moped	£480	$\frac{1}{8}$ off	10% off
5 Dishwasher	£400	$\frac{1}{10}$ off	$12\frac{1}{2}$% off

5 Negative numbers in context

5.1 Using negative numbers in context

When measuring temperatures we usually describe temperatures above the freezing point (0° Celsius) (0° C) as positive and temperatures below the freezing point as negative.

Five temperatures are shown.

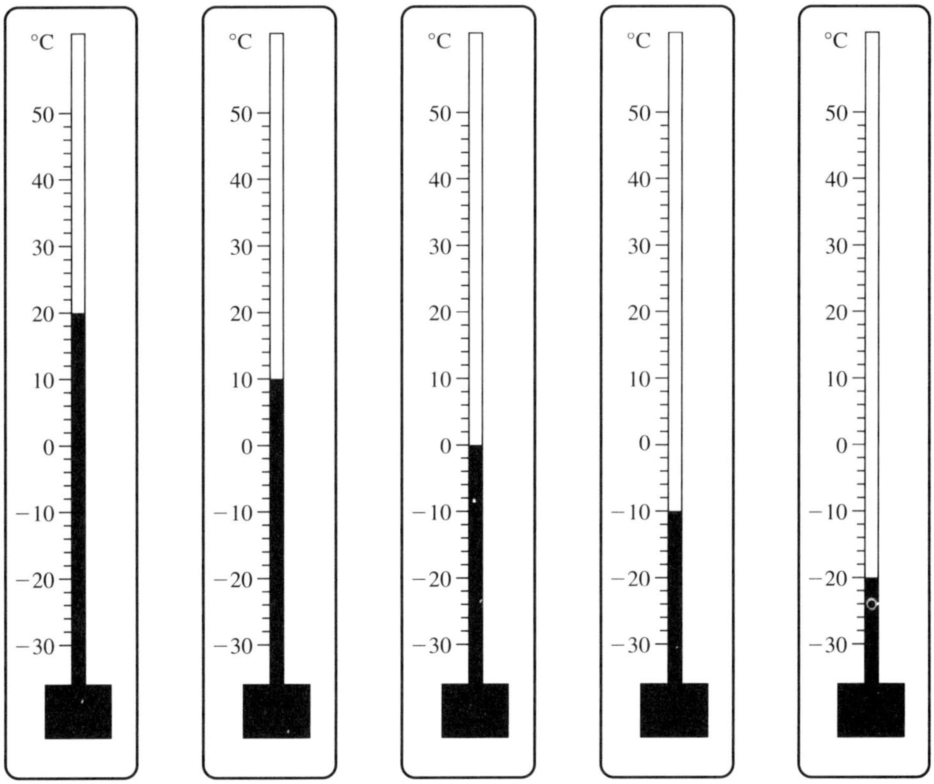

20° C Temperature on a warm day	10° C Temperature on a cool day	0° C Temperature on a cold day	−10° C Temperature on a very cold day	−20° C Temperature on an extremely cold day

Example 1

Put the following temperatures in order from warmest to coldest:

10° C	−40° C	30° C	−50° C	40° C	−20° C

The order from warmest to coldest is:

40° C	30° C	10° C	−20° C	−40° C	−50° C

Exercise 5.1

For Questions 1–5 put the temperatures in order from warmest to coldest.

1	20° C	−40° C	30° C	−50° C	10° C	−30° C
2	30° C	−20° C	50° C	−50° C	10° C	−40° C
3	40° C	−30° C	10° C	−50° C	50° C	−20° C
4	30° C	−10° C	20° C	−50° C	40° C	−30° C
5	30° C	−20° C	40° C	−50° C	50° C	−10° C

For Questions 6–10 put the temperatures in order from coldest to warmest.

6	−40° C	20° C	−30° C	10° C	−50° C	40° C
7	−40° C	50° C	−10° C	20° C	−50° C	30° C
8	−30° C	20° C	−50° C	10° C	−20° C	40° C
9	−30° C	50° C	−50° C	20° C	−10° C	40° C
10	−20° C	20° C	−30° C	10° C	−40° C	50° C

When we borrow money from a bank we usually have an overdraft or 'go into the red'.
An overdraft of £100 could be described as having a negative balance, whereas a credit of £100 would be a positive balance.

Example 2

Put the following in order of preference for the balance in your bank account.

£200, a credit of £300, an overdraft of £100, £50, an overdraft of £300, −£50, a credit of £150, −£250, an overdraft of £75

The order of preference is:

£300 (credit), £200, £150 (credit), £50, −£50, £75 (overdraft), £100 (overdraft), −£250, £300 (overdraft).

Exercise 5.2

Put in order of preference for the balance in your bank account:

1 £150, −£150, £200, −£250, £50 (credit), £100 (overdraft), £300 (credit), £200 (overdraft)

2 £200, −£250, £300, −£300, £50 (credit), £200 (overdraft), £100 (credit), £150 (overdraft)

3 £300, −£200, £150, −£100, £200 (credit), £300 (overdraft), £250 (credit), £50 (overdraft)

4 £100, −£100, £50, −£200, £250 (credit), £250 (overdraft), £300 (credit), £150 (overdraft)

5 £150 (credit), £100 (overdraft), £50 (credit), £250 (overdraft), £300, −£200, £100, −£150.

Example 3

On a certain day the temperature at 12 noon is 10° C, but by 6 p.m. it has dropped by 7° C.
By 9 p.m. it has dropped a further 6° C and by 12 midnight 5° C further still.

Find the temperature at:

a 6 p.m.
b 9 p.m.
c 12 midnight.

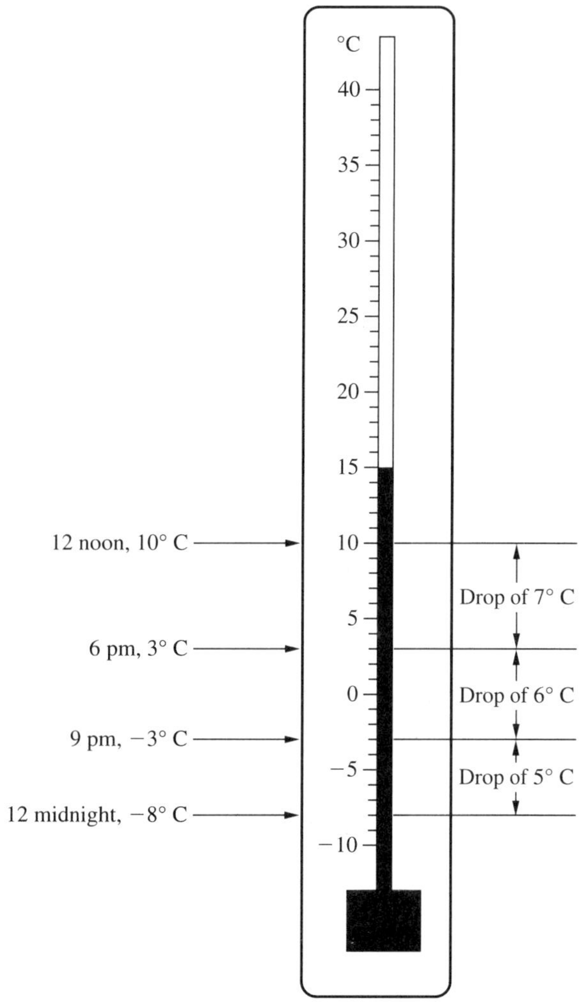

a 10° C − 7° C = 3° C
b 3° C − 6° C = −3° C (see diagram)
c −3° C − 5° C = −8° C (see diagram)

Exercise 5.3

1 The table below shows how the temperature dropped between 12 noon and 6 p.m. on each day during a certain week in April. Copy and complete the table.

Day	Temperature at 12 noon	Drop in temperature	Temperature at 6 p.m.
Monday	15° C	5° C	
Tuesday	17° C	6° C	
Wednesday	13° C	8° C	
Thursday	12° C	9° C	
Friday	11° C	7° C	
Saturday	14° C	8° C	
Sunday	16° C	9° C	

2 The table below shows how the temperature dropped between 12 noon and 6 p.m. on each day during a two-week period in January. Copy and complete the table.

Day	Temperature at 12 noon	Drop in temperature	Temperature at 6 p.m.
Monday	7° C	9° C	
Tuesday	6° C	9° C	
Wednesday	6° C	8° C	
Thursday	5° C	9° C	
Friday	5° C	8° C	
Saturday	5° C	7° C	
Sunday	4° C	8° C	
Monday	4° C	7° C	
Tuesday	7° C	8° C	
Wednesday	6° C	7° C	
Thursday	4° C	9° C	
Friday	3° C	9° C	
Saturday	3° C	8° C	
Sunday	3° C	7° C	

3 The table below shows the amount of money in each person's bank account and how much they each withdraw. Copy and complete the table.

Account holder	Initial balance	Money withdrawn	New balance
Mr Smith	£350	£125	£300 – £125 = £225
Miss Holmes	£500	£150	
Mrs Chaudray	£300	£50	
Mr Robinson	£100	£200	£100 – £200 = £100 A £100 overdraft
Ms Jones	£150	£300	
Mrs Jackson	£100	£175	
Mr Trueman	–£150 or a £150 overdraft	£25	–£150 – £25 = £175 A £175 overdraft
Mrs Bailey	–£100 or a £100 overdraft	£50	

Example 4

a Keeley Simpson has an overdraft of £75 on her bank account and she pays in £90.
What is her balance now?

b Andrew Saunders also has an overdraft of £75 on his bank account and he pays in £50.
What is his balance now?

a Her new balance is: $-£75 + £90 = £15$.
She is now £15 in credit.

b His new balance is: $-£75 + £50 = -£25$.
He now has an overdraft of £25.

Exercise 5.4

1 The table below shows how the temperature rose between 6 a.m. and 12 noon on each day over a two-week period in December.
Copy and complete the table.

Day	Temperature at 6 a.m.	Rise in temperature	Temperature at 12 noon
Monday	−1° C	4° C	
Tuesday	−1° C	5° C	
Wednesday	−1° C	9° C	
Thursday	−1° C	3° C	
Friday	−1° C	2° C	
Saturday	−2° C	5° C	
Sunday	−2° C	4° C	
Monday	−4° C	6° C	
Tuesday	−5° C	6° C	
Wednesday	−7° C	9° C	
Thursday	−6° C	8° C	
Friday	−6° C	7° C	
Saturday	−5° C	10° C	
Sunday	−4° C	10° C	

2 All of the people in the table below have a bank overdraft and they each pay some money in.
Copy and complete the table and state clearly whether each new balance is in credit or overdraft.

Account holder	Initial balance	Money paid in	New balance
Cerys Evans	−£125	£150	
Marcus Harper	−£135	£180	
Drashma Patel	−£135	£105	
Fiona Campbell	−£120	£90	
Dean Stubbs	−£110	£155	
Lucy Andrews	−£130	£145	
Andrew Megson	−£105	£90	
Harry Morgan	−£145	£130	

Example 5

The temperature at 12 noon is 5° C. By 6 p.m. it has dropped to −1° C, and by 9 p.m. has dropped further to −6° C.

Find the drop in temperature between
a 12 noon and 6 p.m.
b 6 p.m. and 9 p.m.

a Drop in temperature $= 5°\,C - (-1°\,C)$
$= 5°\,C + 1°\,C$
$= 6°\,C$ (see diagram)

b Drop in temperature $= -1°\,C - (-6°\,C)$
$= -1°\,C + 6°\,C$
$= 5°\,C$ (see diagram)

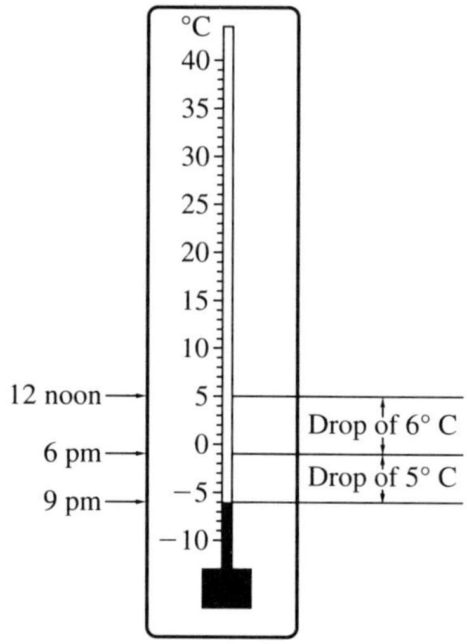

Exercise 5.5

1 The table below shows how the temperature dropped between 12 noon and 6 p.m. on each day during a two-week period in February.
Copy and complete the table.

Day	Temperature at 12 noon	Drop in temperature	Temperature at 6 p.m.
Monday	5° C		−3° C
Tuesday	6° C		−4° C
Wednesday	7° C		−2° C
Thursday	6° C		−5° C
Friday	8° C		−2° C
Saturday	4° C		−3° C
Sunday	5° C		−2° C
Monday	3° C		−1° C
Tuesday	3° C		−5° C
Wednesday	2° C		−4° C
Thursday	2° C		−6° C
Friday	4° C		−5° C
Saturday	3° C		−4° C
Sunday	1° C		−3° C

2 All of the people in the table below have a bank overdraft, but they each pay some money in.
Copy and complete the table.

Account holder	Initial balance	Money paid in	New balance
Mr Woods	£100 (overdraft)		£50 (credit)
Mr Walton	£100 (overdraft)		£100 (credit)
Mrs Smith	£150 (overdraft)		£200 (credit)
Ms Jones	£125 (overdraft)		£175 (credit)
Mr Brown	£100 (overdraft)		£50 (overdraft)
Ms Green	£150 (overdraft)		£125 (overdraft)
Mr Bennet	£125 (overdraft)		£50 (overdraft)
Mrs White	£75 (overdraft)		£50 (overdraft)

Example 6

When I woke up early this morning the temperature was $-4°$ C, but by lunch time it had risen to $11°$ C and by tea time it had dropped again to $-2°$ C.
Find:

a the rise in temperature between early morning and lunch time
b the fall in temperature between lunch time and tea time
c the net change in temperature between early morning and tea time.

a The rise in temperature is:
$11°\text{C} - (-4°\text{C}) = 11°\text{C} + 4°\text{C} = 15°\text{C}$
b The fall in temperature is:
$11°\text{C} - (-2°\text{C}) = 11°\text{C} + 2°\text{C} = 13°\text{C}$
c The net change in temperature is:
$-2°\text{C} - (-4°\text{C}) = -2°\text{C} + 4°\text{C} = 2°\text{C}$

Exercise 5.6

1 The table shows how the temperature varied between early morning, lunch time and tea time on each day during a certain week in February.
Copy and complete the table.

Day	Temperature at 6 a.m.	Rise during morning	Temperature at 12 noon	Fall during afternoon	Temperature at 6 p.m.	Net rise between 6 a.m. and 6 p.m.
Monday	$-5°$ C		$7°$ C		$-3°$ C	
Tuesday	$-4°$ C		$9°$ C		$-2°$ C	
Wednesday	$-3°$ C		$8°$ C		$-1°$ C	
Thursday	$-6°$ C		$4°$ C		$-2°$ C	
Friday	$-2°$ C		$10°$ C		$4°$ C	
Saturday	$-1°$ C		$11°$ C		$3°$ C	
Sunday	$-3°$ C		$9°$ C		$1°$ C	

2 All the people in the table have a bank overdraft. They each make two payments into the bank.
Copy and complete the table.

Account holder	Initial balance	First payment	Balance after 1st payment	Second payment	Balance after 2nd payment	Net improvement in balance
Liam Jones	£75 (o/draft)		£25 (o/draft)		£35 (credit)	
Annie Smith	£45 (o/draft)		£20 (o/draft)		£30 (credit)	
Jack White	£90 (o/draft)		£55 (o/draft)		£20 (credit)	
Nathan Green	£65 (o/draft)		£15 (o/draft)		£25 (credit)	
Jolene Brown	£60 (o/draft)		£45 (o/draft)		£10 (credit)	

Example 7

One day in March the temperature in London was $-2°$ C. In Chicago the temperature was 7 times further below the freezing point than this.
What was the temperature in Chicago?

The temperature in Chicago was $-2° \text{C} \times 7 = -14° \text{C}$

Example 8

Rajesh has a bank overdraft of £105. Natalie has a bank overdraft which is only $\frac{1}{3}$ as large.
What is Natalie's overdraft?

Natalie's overdraft is: $\frac{1}{3}$ of £105 = £105 ÷ 3 = £35

Exercise 5.7

1 One day in February the temperature in London was $-3°$ C.
Calculate the temperature in each of the following places:

a Vienna if the temperature is 3 times further below freezing point.

b Prague if the temperature is 5 times further below freezing point.

c Stockholm if the temperature is 6 times further below freezing point.

d Belgrade if the temperature is 4 times further below freezing point.

e Warsaw if the temperature is 7 times further below freezing point.

2 Sally has a bank overdraft of £15.
Find the following overdrafts:
- **a** David's if his is 3 times greater.
- **b** Jenny's if hers is 5 times greater.
- **c** Laura's if hers is 7 times greater.
- **d** Paul's if his is 4 times greater.
- **e** Robert's if his is 6 times greater.

3 The Zuider Zee in the Netherlands has an altitude of −4 metres.
That is to say, it is 4 metres below sea level.
Find the altitude of the following:
- **a** Lake Eyre (Australia), which is 3 times further below sea level.
- **b** The Caspian Sea (Russia), which is 7 times further below sea level.
- **c** Death Valley (USA), which is 21 times further below sea level.
- **d** The Danakin Depression (Ethiopia), which is 29 times further below sea level.
- **e** The Dead Sea (Middle East), which is 99 times further below sea level.

4 One day in January the temperature in Moscow was −36° C.
Find the temperature in each of the following places:
- **a** Budapest if the temperature is only $\frac{1}{4}$ as far below freezing point.
- **b** Vienna if the temperature is only $\frac{1}{9}$ as far below freezing point.
- **c** Helsinki if the temperature is only $\frac{1}{3}$ as far below freezing point.
- **d** Paris if the temperature is only $\frac{1}{12}$ as far below freezing point.
- **e** Berlin if the temperature is only $\frac{1}{6}$ as far below freezing point.

5 Peter has a bank overdraft of £120.
Find the following overdrafts:
- **a** Aisha's if hers is only $\frac{1}{4}$ of this.
- **b** Marie's if hers is only $\frac{1}{3}$ of this.
- **c** Arthur's if his is only $\frac{1}{8}$ of this.
- **d** Emlyn's if his is only $\frac{1}{6}$ of this.
- **e** Jodie's if hers is only $\frac{1}{5}$ of this.

6 Indices

6.1 Using index numbers to express powers of whole numbers

100	is the same as 10×10, therefore 100 is written 10^2
1000	is the same as $10 \times 10 \times 10$, therefore 100 is written as 10^3.
10 000	is the same as $10 \times 10 \times 10 \times 10$, therefore 10 000 is written as 10^4.

The raised numbers 2, 3 and 4 are known as **index numbers**.
10^2 is read '10 to the power 2' (or '10 squared').
10^3 is read '10 to the power 3' (or '10 cubed').
10^4 is read '10 to the power 4'.

We can write other whole numbers similarly by using index powers.
$8 = 2 \times 2 \times 2$, therefore $8 = 2^3$.
$81 = 3 \times 3 \times 3 \times 3$, therefore $81 = 3^4$.

Example 1

Evaluate: **a** 70^2 **b** 6^5

a $70^2 = 70 \times 70 = 4900$
b $6^5 = 6 \times 6 \times 6 \times 6 \times 6 = 7776$

Exercise 6.1

Evaluate:

1 4^2	**2** 8^2	**3** 15^2	**4** 25^2	**5** 40^2	**6** 60^2
7 15^3	**8** 30^3	**9** 2^4	**10** 4^4	**11** 5^4	**12** 6^4
13 5^6	**14** 10^6	**15** 4^7	**16** 5^7	**17** 2^8	**18** 3^8

Example 2

Write in index form:
a 32 **b** 125

a $32 = 2 \times 2 \times 2 \times 2 \times 2 = 2^5$ (2 to the power 5)
b $125 = 5 \times 5 \times 5 = 5^3$ (5 to the power 3)

Exercise 6.2

Write in index form:

1 4	**2** 25	**3** 36	**4** 49
5 121	**6** 144	**7** 400	**8** 900
9 2500	**10** 27	**11** 216	**12** 343
13 1728	**14** 8000	**15** 243	**16** 3125

Example 3

Copy and complete:
a $16 = 2^? = 4^?$
b $625 = 5^? = 25^?$

a $16 = 2 \times 2 \times 2 \times 2 = 2^4$ also $16 = 4 \times 4 = 4^2$
b $625 = 5 \times 5 \times 5 \times 5 = 5^4$ also $625 = 25 \times 25 = 25^2$

Exercise 6.3

Copy and complete:

1 $64 = 2^? = 4^? = 8^?$
2 $256 = 2^? = 4^? = 16^?$
3 $512 = 2^? = 8^?$
4 $1024 = 2^? = 4^? = 32^?$
5 $4096 = 2^? = 4^? = 8^? = 16^? = 64^?$
6 $729 = 3^? = 9^? = 27^?$
7 $6561 = 3^? = 9^? = 81^?$
8 $19\,683 = 3^? = 27^?$
9 $1296 = 6^? = 36^?$
10 $2401 = 7^? = 49^?$
11 $15\,625 = 5^? = 25^? = 125^?$
12 $10\,000 = 10^? = 100^?$
13 $1\,000\,000 = 10^? = 100^? = 1000^?$
14 $1\,000\,000\,000 = 10^? = 1000^?$

7 Ratio and scale

7.1 Using unitary ratios

One way of describing a mixture is to use a ratio.

In mixing mortar for bricklaying, one part of cement is used for every three parts of sand.
We say that the cement and sand are mixed in the ratio of 1 : 3.

For a weaker mix the ratio might be 1 : 4.
In this case, for every part of cement, four parts of sand are used.

Example 1

A bricklayer using a 1 : 3 mortar mix has 4 bags of cement.
How many equivalent sized bags of sand will he need?

The ratio 1 : 3 means that for every one bag of cement, three bags of sand are required.
So for 4 bags of cement, 4×3 or 12 bags of sand are required.

Exercise 7.1

1 Plaster powder has to be mixed with water in the ratio 1 : 5.
If 20 grams of plaster powder are to be used, how many grams of water must be added?

2 Flour has to be mixed with water in the ratio 1 : 8 to make a paste.
If 15 grams of flour are to be used, how many grams of water must be added?

3 In the coolant system of a lorry engine, antifreeze and water have to be mixed in the ratio 1 : 6.
If 2 kilograms of antifreeze are to be used, how many kilograms of water must be added?

4 At Elm Park school the ratio of teachers to pupils is 1 : 12.
If there are 15 teachers, how many pupils are there?

5 During a dry summer, from May to September, the ratio of wet to dry days was 1 : 8.
If there were 17 wet days, how many dry days were there?

Example 2

A bricklayer using a 1 : 4 mortar mix has 20 bags of sand.
How many equivalent sized bags of cement will he require?

The ratio 1 : 4 means that for every one bag of cement, 4 bags of sand are required.
Therefore the number of bags of cement required is $\frac{1}{4} \times 20 = 5$ bags

Exercise 7.2

1 On a supermarket shelf the ratio of packets of self-raising flour to packets of plain flour is 1 : 5.
If there are 20 packets of plain flour, how many packets of self-raising flour are there?

2 During a successful season a football team conceded and scored goals in the ratio 1 : 3.
If they scored 105 goals, how many did they concede?

3 Vans and cars are parked in a car park in the ratio of 1 : 6.
If there are 150 cars, how many vans are there?

4 On a baker's tray the ratio of brown loaves to white loaves is 1 : 5.
If there are 65 white loaves, how many brown loaves are there?

5 In a wood the ratio of holly bushes with berries to those without is 1 : 9.
If 144 bushes have no berries, how many have berries?

Example 3

A cake mixture requires 1 pound (lb) of currants for every 6 pounds (lb) of flour. So the weights of the currants and flour are in the ratio 1 : 6.
How many pounds of currants are required for:
a a large cake which uses 12 lb of flour?
b a small cake which uses 3 lb of flour?

a Weight of currants required is: $\frac{1}{6} \times 12\,\text{lb} = 2\,\text{lb}$
b Weight of currants required is: $\frac{1}{6} \times 3\,\text{lb} = 0.5\,\text{lb}$ $\left(\text{or } \frac{1}{2}\,\text{lb}\right)$

Exercise 7.3

1 A mortar is made by mixing cement and sand in the ratio 1 : 4.
How many kilograms of cement are required for each of the following masses of sand?
a 8 kg **b** 20 kg **c** 28 kg **d** 14 kg **e** 2 kg

2 A chemistry teacher has to mix caustic soda pellets with water in the ratio 1 : 12 in order to make a dilute solution.
How many grams of pellets must she dissolve in each of the following masses of water?
a 60 g **b** 48 g **c** 45 g **d** 27 g **e** 9 g

3 Pastry and jam are weighed out in the ratio 1 : 2 for making jam tarts.
How much pastry is required for each of the following masses of jam?
a 50 g **b** 30 g **c** 70 g **d** 45 g **e** 25 g

4 Coloured dye is mixed with alcohol in the ratio 1 : 50 for use in a thermometer.
Find the mass of dye which must be mixed with each of the following masses of alcohol:
a 15 g **b** 10 g **c** 12 g **d** 8 g **e** 9 g

The ratio 1 : 3 can be rewritten as the ratio 4 : 12 and vice versa.
Here each quantity is simply multiplied by 4.

In the same way the ratio 3 : 18 can be rewritten as the ratio 1 : 6.
Here each quantity is simply divided by 3.

Example 4

Fill in the missing number:
a 1 : 3 is the same ratio as 2 : ?
b 1 : 4 is the same ratio as 3 : ?
c 3 : 21 is the same ratio as 1 : ?

a 1 : 3 is the same ratio as 2 : 6 (each quantity is multiplied by 2).
b 1 : 4 is the same ratio as 3 : 12 (each quantity is multiplied by 3).
c 3 : 21 is the same ratio as 1 : 7 (each quantity is divided by 3).

Exercise 7.4

Fill in the missing number:

1 1 : 5 = 2 : ? = 3 : ? = 5 : ? = 7 : ?
2 1 : 8 = 2 : ? = 3 : ? = 5 : ? = 7 : ?
3 1 : 3 = 4 : ? = 6 : ? = 8 : ? = 9 : ?
4 1 : 9 = 4 : ? = 6 : ? = 8 : ? = 9 : ?
5 1 : 10 = 4 : ? = 6 : ? = 8 : ? = 9 : ?
6 1 : 50 = 2 : ? = 3 : ? = 5 : ? = 7 : ?
7 1 : 4 = ? : 8 = ? : 12 = ? : 20 = ? : 28
8 1 : 5 = ? :20 = ? :30 = ? : 40 = ? : 45
9 1 : 8 = ? : 80 = ? : 32 = ? : 48 = ? : 64
10 1 : 12 = ? : 48 = ? : 72 = ? : 120 = ? : 96
11 1 : 20 = ? : 80 = ? : 120 = ? : 160 = ? : 180
12 1 : 50 = ? : 200 = ? : 300 = ? : 400 = ? : 500

7.2 Understanding the notion of scale in maps and drawings

The scale of a map is often described by a ratio.
If a map has a scale of 1 : 1000 it means that every length of 1 unit on the map represents a length of 1000 units on the ground.

Example 1

The map shows four places in a village and the roads that connect them.
The scale of the map is 1 : 5000.

For each case below find:
(i) the distance between the two places on the map by measuring
(ii) the real distance between the two places in centimetres
(iii) The real distance between the two places in metres.

a The station and the Grand Hotel
b The station and the post office
c The station and the church
d The Grand Hotel and the post office.

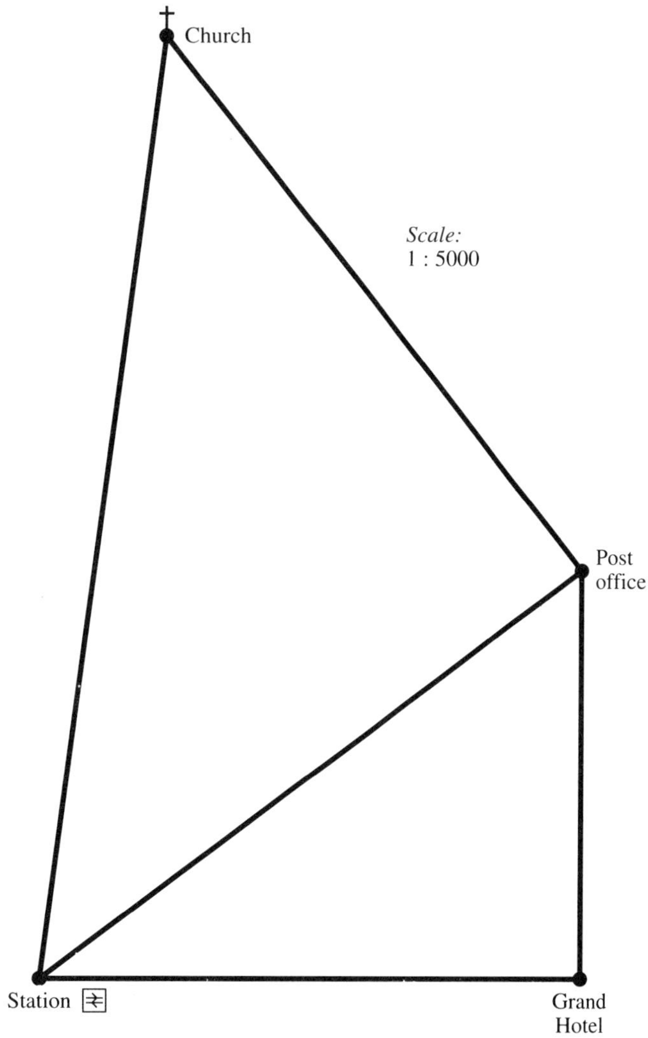

a (i) 8 cm (ii) (8×5000) cm = 40 000 cm (iii) $(40\,000 \div 100)$ m = 400 m
b (i) 10 cm (ii) (10×5000) cm = 50 000 cm (iii) $(50\,000 \div 100)$ m = 500 m
c (i) 14 cm (ii) (14×5000) cm = 70 000 cm (iii) $(70\,000 \div 100)$ m = 700 m
d (i) 6 cm (ii) (6×5000) cm = 30 000 cm (iii) $(30\,000 \div 100)$ m = 300 m

Exercise 7.5

1 This is a map of a village to a scale of 1 : 10 000. This is the village where Jaswant lives.

For all parts of this question find:

(i) the distance between the two places on the map in centimetres.

(ii) the real distance between the two places in centimetres.

(iii) the real distance between the two places in metres.

a Jaswant's house and his school
b Jaswant's house and the supermarket
c Jaswant's house and the swimming pool
d Jaswant's house and the church.
e Jaswant's school and the supermarket
f The supermarket and the swimming pool
g The swimming pool and the church

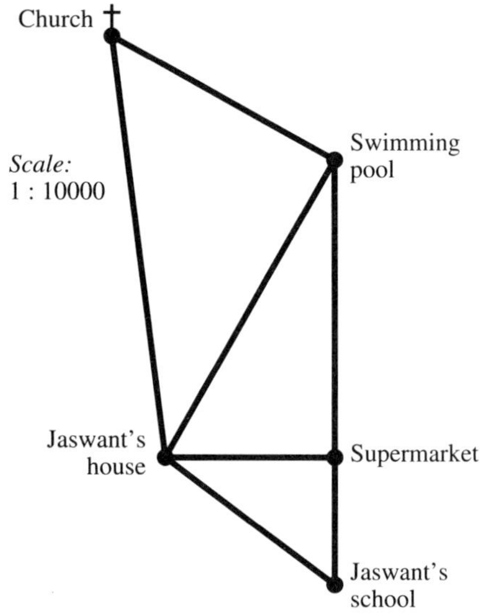

2 This is a map of where Melanie lives. The scale is 1 : 5000.

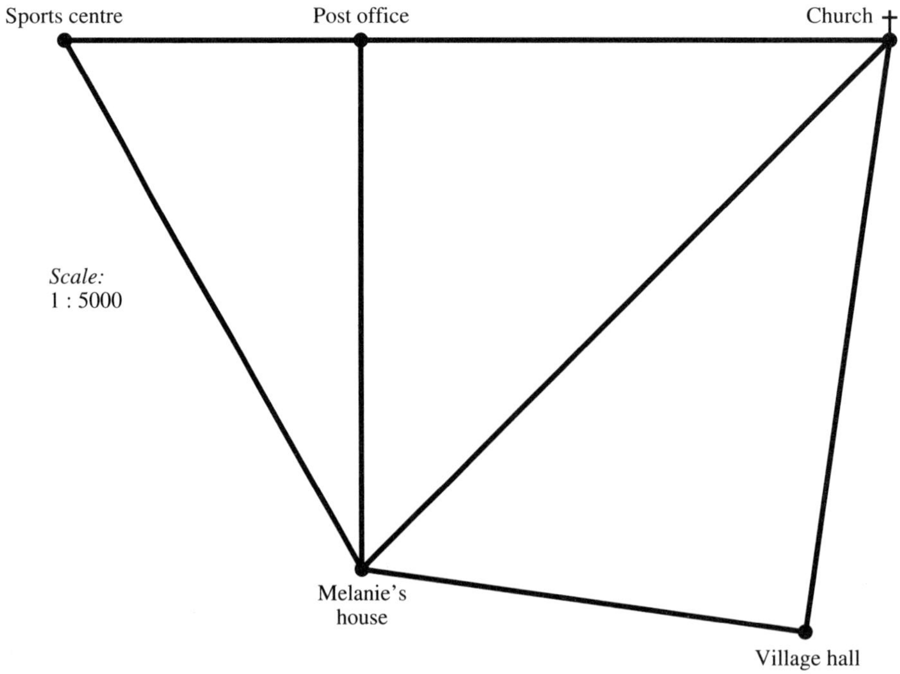

For all parts of this question find:

(i) the distance between the two places on the map in centimetres
(ii) the real distance between the two places in centimetres
(iii) the real distance between the two places in metres.

a Melanie's house and the village hall
b Melanie's house and the church
c Melanie's house and the post office
d Melanie's house and the sports centre
e The village hall and the church
f The church and the post office
g The post office and the sports centre

Example 2

A model of a car is constructed using a scale of 1 : 20.
If the length of the real car is 6 metres, what is the length of the model?

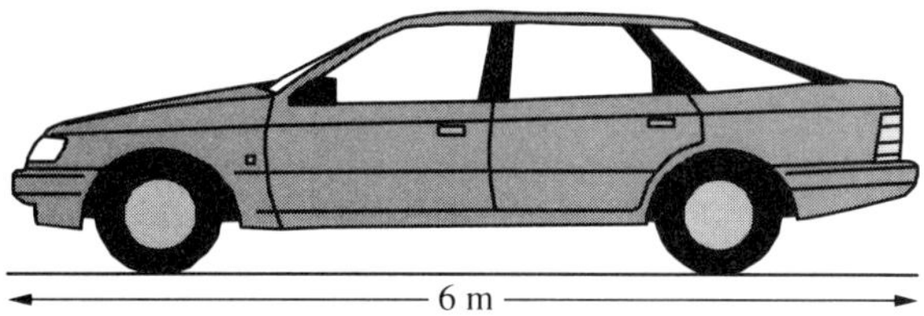

The length of the real car is: 6 metres
$= (6 \times 100)$ centimetres
$= 600$ centimetres

Therefore the length of the model is: $\left(\frac{1}{20} \times 600\right)$ cm
$= 30$ cm

Exercise 7.6

1 A model of a van is constructed using a scale of 1 : 40.
If the length of the real van is 8 metres, what is the length of the model?

2 A model of a church is to be built using a scale of 1 : 120.
If the height of the tower on the real church is 60 metres, what will it be on the model?

3 David is drawing a plan of his classroom using a scale of 1 : 50.
If the length and width of the classroom are 8 metres and 6 metres, how long will they each be on his drawing?

4 Jodie is drawing a plan of her garden using a scale of 1 : 60.
If the length and the width of her garden are 15 metres and 9 metres, how long will they each be on her drawing?

5 Kerry has a foot stool whose dimensions are 60 cm, 48 cm and 36 cm. She decides to make a smaller replica for her baby sister using a scale of 1 : 3.
What will the dimensions of the replica be?

Example 3

The model of the car in Example 2 has a bonnet of length 10 cm.
What is the length of the bonnet on the real car? (The scale is 1 : 20.)

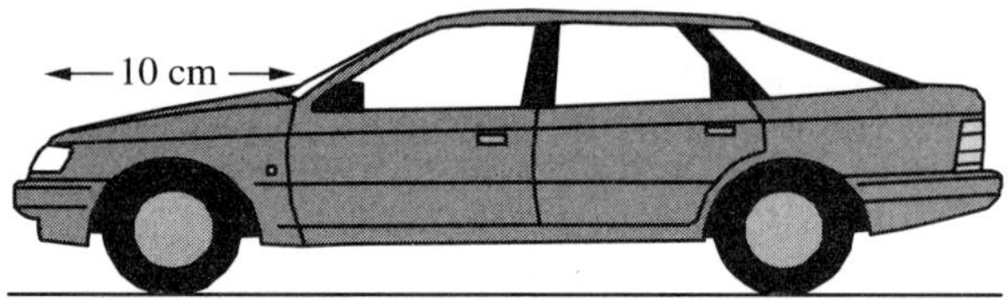

The length of the bonnet on the real car is:
$10 \times 20 = 200$ cm or 2 m

Exercise 7.7

1 A model aeroplane has length 50 cm and it is built to a scale of 1 : 40.
What is the length of the real aeroplane?

2 A coal wagon on a model railway has a length of 15 cm. It is built to a scale of 1 : 80.
What is the length of the real wagon?

3 Samantha is drawing a plan of the gymnasium at her school using a scale of 1 : 75.
If the dimensions of the gymnasium on her plan are 28 cm and 16 cm, what are the dimensions of the real gymnasium?

4 Roger is drawing a plan of a door that he has to make, using a scale of 1 : 5.
If the dimensions of the door on his plan are 20 cm and 40 cm, what will the dimensions of the real door be?

5 Carly has made a replica table for a doll's house by using a scale of 1 : 15.
If the dimensions of the toy table are 10 cm, 8 cm and 5 cm, what are the dimensions of the real table?

Example 4

On a model of a house the height of the door is 4 cm, whereas on the actual house it is 200 cm.
What is the scale of the model?

4 cm on the model represents 200 cm on the actual house, therefore the ratio is 4 : 200.
4 : 200 can be simplified to 1 : 50 by dividing both quantities by 4.

Therefore the scale of the model is 1 : 50.

Exercise 7.8

1 The track gauge of a miniature railway is 24 cm and the standard gauge of British Rail tracks is 144 cm.
What is the scale of the miniature railway?

2 Kamaljit has a toy sewing machine of length 28 cm which is a replica of her mother's real one, of length of 56 cm.
To what scale was the toy built?

3 A model boat has length 25 cm, whereas its real counterpart has length 4 m (400 cm).
To what scale was the model built?

4 At a Tourist Information Office there is a model of Salisbury Cathedral on which the spire is 60 cm high.
If the height of the spire on the real cathedral is 120 m (12 000 cm), what is the scale of the model?

5 Andrew has a model of the Eiffel Tower which is 16 cm high.
If the height of the real tower is 320 m, what is the scale of the model?

7.3 Calculating by using ratios in a variety of situations

Example 1

a A square photograph has a side length of 8 cm. For an enlargement the side length is fractionally increased by the ratio 1 : 4.
Find (i) the increase in side length, and (ii) the side length of the enlargement

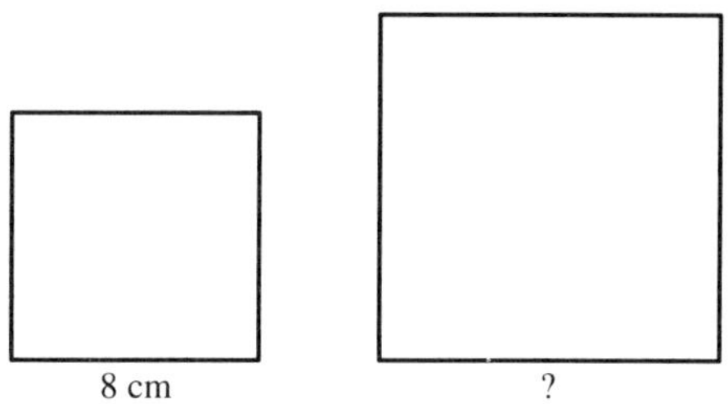

b A lorry has 135 litres of diesel in its tank. After a short journey the quantity is fractionally reduced by the ratio 1 : 5.
Find (i) the quantity of diesel consumed, and (ii) the quantity remaining in the tank.

a (i) The increase is: $\frac{1}{4}$ of 8 cm $= \left(\frac{1}{4} \times 8\right)$ cm $= 2$ cm
(ii) The side length of the enlargement is $(8 + 2)$ cm $= 10$ cm
b (i) The quantity consumed is $\left(\frac{1}{5} \times 135\right)$ litres $= 27$ litres
(ii) The quantity remaining is $(135 - 27)$ litres $= 108$ litres

Exercise 7.9

1 Last year Martin was 153 cm tall, but his height has now fractionally increased by the ratio 1 : 9.
Find **a** by how much he has grown, and **b** his height now.

2 A road is 24 metres wide. It is to be widened fractionally by the ratio 1 : 8.
Find **a** the increase in width, and **b** the new width of the road.

3 An elastic band is 14 cm long. It is stretched to fractionally increase its length by the ratio 1:2.
Find **a** the increase in length of the band, and **b** the stretched length of the band.

4 A station platform is 240 metres long. In order to accommodate longer trains its length is to be fractionally increased by the ratio 1:6.
Find **a** the length by which the platform is to be extended, and **b** the new length of the platform.

5 The pressure in the water mains where John lives is normally 12 atmospheres, but last summer there was a drought and the pressure was fractionally reduced by the ratio 1:4.
Find **a** the drop in pressure, and **b** the pressure during the drought.

6 In June last year there were 18 wet days, but the July figure was a fractional decrease by the ratio 1:3.
Find **a** the number of wet days fewer in July than June, and **b** the actual number of wet days in July.

7 The mains power is normally delivered at 240 volts, but one cold evening a great demand for power caused a fractional drop in voltage by the ratio 1:12.
Find **a** the voltage drop, and **b** the actual voltage that evening.

8 A football ground used to have a crowd capacity of 21 000, but safety regulations have enforced a fractional reduction by the ratio 1:15.
Find **a** the reduction in crowd capacity, and **b** the new crowd capacity.

Example 2

A carton containing 180 millilitres of orange juice costs 24 p.
How many millilitres will there be in a carton which costs:
a 16 p **b** 36 p?

a 180 millilitres must be proportionally changed by the ratio 16:24.
Therefore the contents $= \left(180 \times \frac{16}{24}\right)$ millilitres $= (180 \times 16 \div 24)$ millilitres
$= 120$ millilitres

b 180 millilitres must be proportionally changed by the ratio 32:24.
Therefore the contents $= \left(180 \times \frac{36}{24}\right)$ millilitres $= (180 \times 36 \div 24)$ millilitres
$= 270$ millilitres

Example 3

A can containing 250 millilitres of lemonade costs 30 p.
What is the price of a can which contains: **a** 150 millilitres, **b** 400 millilitres?

a 30 p must be proportionally changed by the ratio 150 : 250.
Therefore the cost = $\left(30 \times \frac{150}{250}\right)$ p = $(30 \times 150 \div 250)$ p = 18 p

b 30 p must be proportionally changed by the ratio 400 : 250.
Therefore the cost = $\left(30 \times \frac{400}{250}\right)$ p = $(30 \times 400 \div 250)$ p = 48 p

Exercise 7.10

1 Lucy lives 1320 metres from her school and it takes her 12 minutes to walk this distance.
How far could she walk in each of the following times?
a 5 minutes **b** 7 minutes **c** 9 minutes **d** 15 minutes
e 20 minutes **f** 21 minutes **g** 25 minutes

2 Pencils cost 90 p for 6.
Find the cost of:
a 2 pencils **b** 3 pencils **c** 5 pencils **d** 9 pencils
e 11 pencils **f** 13 pencils **g** 16 pencils

3 The taxi firm 'Cablifts' charges a fare of £6 (600 p) for an 8 km journey.
What would be the charge for a journey of:
a 3 km **b** 5 km **c** 7 km **d** 12 km
e 15 km **f** 18 km **g** 20 km?

4 Nancy has used 350 grams of flour in order to make a fruit pie for 7 people.
How many grams of flour would she have to use if she made a fruit pie for:
a 2 people **b** 5 people **c** 6 people **d** 8 people
e 9 people **f** 10 people **g** 12 people?

5 90 metres of electric cable costs £6.
What length could be bought for:
a £2 **b** £5 **c** £2.40 **d** £3.60
e £8 **f** £9 **g** £10.80 **h** £13.20?

6 Nathan lives 1800 metres from his school and it takes him 16 minutes to walk this distance.
How long would it take him to walk to each of the following places?
a the post office which is 450 metres from his house
b the supermarket which is 675 metres from his house
c the railway station which is 1350 metres from his house
d the swimming pool which is 2025 metres from his house
e the library which is 2250 metres from his house

f the football ground which is 2700 metres from his house
g the leisure centre which is 3150 metres from his house.

7 Pens cost £3.60 (360 p) for 8.
How many pens can be bought for:
a 90 p b £1.35 c £2.25 d £3.15
e £4.05 f £5.40 g £6.30 h £9?

8 The taxi firm 'U2 Limited' charge a fare of £8 for a 12 km journey.
How far would they take you for:
a £2 b £3 c £5 d £7
e £11 f £14 g £18 h £20?

9 A packet contains 960 grams of dry soup powder and is marked '8 servings'.
How many servings can you make from:
a 240 grams b 360 grams c 600 grams d 720 grams
e 1080 grams f 1200 grams g 1440 grams?

10 An 8 metre length of clothes line costs £3.20.
What would be the cost of each of the following lengths of clothes line?
a 3 metres b 5 metres c 7 metres d 12 metres
e 15 metres f 18 metres g 20 metres.

Example 4

On a bus there are 24 seats downstairs and 27 seats upstairs.
Find the ratio of seats in its simplest form.

The ratio is $24:27 = \frac{24}{27} = \frac{8}{9} = 8:9$.

Exercise 7.11

Find the ratio in its simplest form in each of the following.

1 In a factory car park there are 15 vans and 18 cars.

2 While David was collecting train numbers he noticed that 12 goods trains and 21 passenger trains went past.

3 In Class 1A there are 12 boys and 16 girls.

4 A farmer has 16 black sheep and 28 white sheep.

5 At Willow Lane School 20 third-year boys chose rugby and 25 chose football.

6 At North Grange School 25 first-year girls chose netball and 30 chose hockey.

7 At a motorway service area there are 5 diesel pumps and 15 petrol pumps.

8 At Greenmead School 12 fifth-year boys said they supported United and 18 said they supported Rovers.

9 In a certain town there are 24 pay phone kiosks and 30 phonecard kiosks.

10 In North Avenue 16 houses have wooden gates and 24 have metal ones.

11 In South Avenue 32 homes have oil-fired central heating and 40 have gas-fired central heating.

12 At Whitemoss School 36 sixth-formers study arts subjects and 45 study science subjects.

Example 5

Share 39 marbles between Khalid and Robert in the ratio 5 : 8.

As 5 and 8 added together give 13, one share is $39 \div 13 = 3$.
Therefore Khalid's share: $5 \times 3 = 15$
and Robert's share is: $8 \times 3 = 24$ (Check: $15 + 24 = 19$)

Exercise 7.12

1 Share £50 between Joshua and Liam in the ratio 2 : 3.

2 Share 450 kg of sand between two builders in the ratio 2 : 7.

3 Share 350 grams of sweets between Alice and Claire in the ratio 3 : 4.

4 A piece of wood of length 275 cm is cut into two parts in the ratio 3 : 8.

5 180 millilitres of milk is shared between a kitten and a puppy in the ratio 4 : 5.

6 10.5 metres of electric wire is cut into two parts in the ratio 4 : 11.

7 A stud farm accommodates 55 horses and the ratio of mares to stallions is 5 : 6.

8 A home for the elderly accommodates 70 people and the ratio of women to men is 5 : 9.

9 420 grams of jam is distributed between a small jar and a large jar in the ratio 1 : 3.

10 180 millilitres of soup are shared between Richard and his older brother James in the ratio 1 : 2.

11 900 kg of grit for a road are made by mixing salt and shingle in the ratio 1 : 5.

12 120 millilitres of ink are distributed between a small bottle and a large bottle in the ratio 1 : 4.

SECTION B: ALGEBRA

8 Understanding patterns and relationships between numbers

8.1 Generalising, mainly in words, patterns which arise in various situations

Numbers can have several properties. Look at the examples below:

Odd numbers 1 3 5 7 etc.
Even numbers 2 4 6 8 etc. (even numbers are multiples of 2)

Multiples of 3

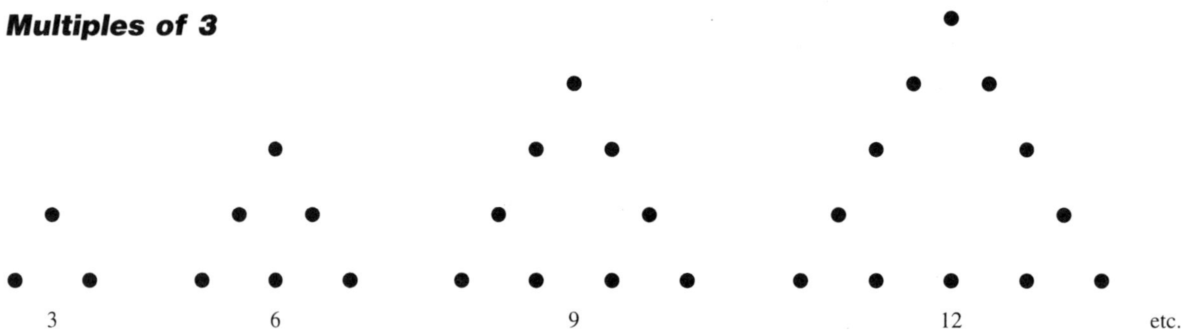

Multiples of 4

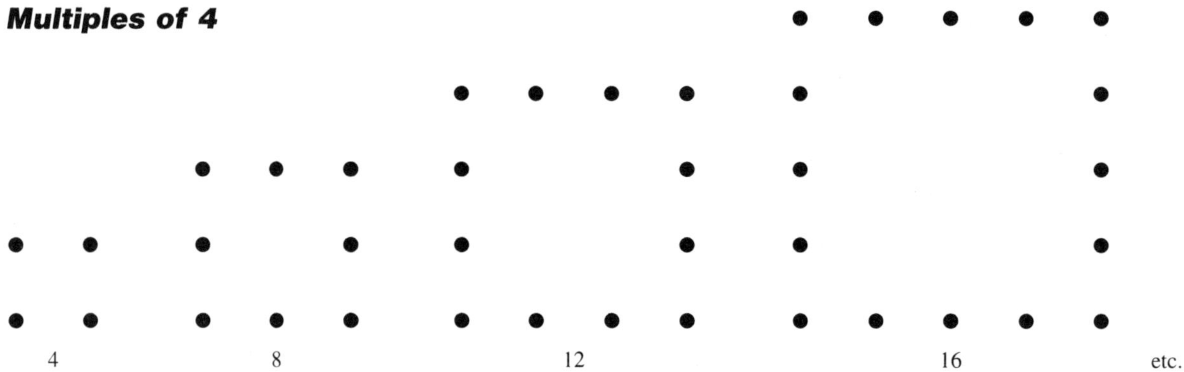

Square numbers

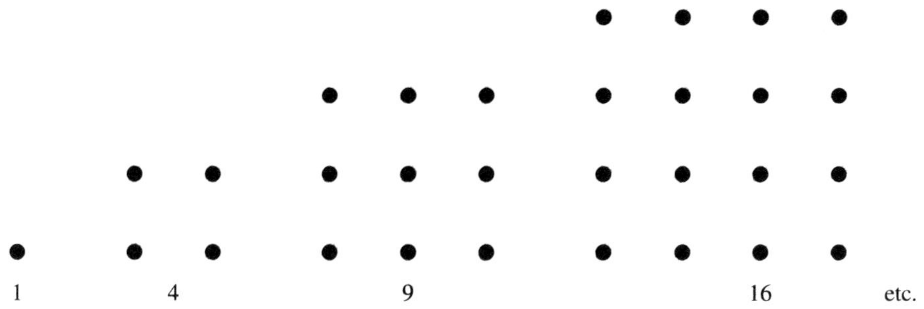

Triangular numbers

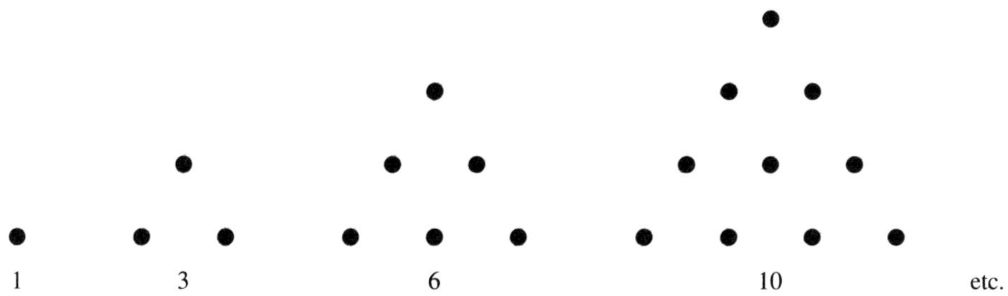

Exercise 8.1

1 Write down the first ten odd numbers.

2 Write down the first ten even numbers.

3 Write down the first ten multiples of 3.

4 Write down the first ten multiples of 4.

5 Write down the first ten multiples of 5.

6 Write down the first ten multiples of 10.

7 Write down the first ten multiples of 8.

8 Write down the first ten multiples of 6.

9 Write down the first ten multiples of 9.

10 Write down the first ten multiples of 7.

11 Write down the first eight square numbers.

12 Write down the first eight triangular numbers.

Example 1

a Copy and complete the addition table.
b Write down what you notice about the numbers along the marked diagonal.
c Is the diagonal an axis of symmetry?
Explain your answer.
d Write a statement or statements about any other patterns that you find.

+	1	2	3	4	5	6	7	8
1								
2								
3								
4								
5								
6								
7								
8								

a

+	1	2	3	4	5	6	7	8
1	2	3	4	5	6	7	8	9
2	3	4	5	6	7	8	9	10
3	4	5	6	7	8	9	10	11
4	5	6	7	8	9	10	11	12
5	6	7	8	9	10	11	12	13
6	7	8	9	10	11	12	13	14
7	8	9	10	11	12	13	14	15
8	9	10	11	12	13	14	15	16

b The numbers along the marked diagonal are all even numbers and they appear in ascending order downwards.
c The diagonal illustrated is an axis of symmetry axis because the arrays of numbers on either side of it are the same.
d (i) In each row and column successive numbers differ by 1.
(ii) The numbers along each left to right diagonal are the same.

Exercise 8.2

For each of the following:
a Copy and complete the addition table.
b Write down what you notice about the numbers along the marked diagonal.
c Is the diagonal an axis of symmetry?
Explain your answer.
d Write a statement or statements about any other patterns that you find.

1

+	2	3	4	5	6	7	8
1							
2							
3							
4							
5							
6							
7							

2

+	1	2	3	4	5	6	7
2							
3							
4							
5							
6							
7							
8							

3

+	1	2	3	4	5	6	7
3							
4							
5							
6							
7							
8							
9							

4

+	2	3	4	5	6	7	8
2							
3							
4							
5							
6							
7							
8							

Example 2

a Copy and complete the multiplication table.
b Write down what you notice about the numbers along the marked diagonal.
c Is the diagonal an axis of symmetry? Explain your answer.
d Write a statement or statements about any other patterns that you find.

×	1	2	3	4	5	6	7
1							
2							
3							
4							
5							
6							
7							

a

×	1	2	3	4	5	6	7
1	1	2	3	4	5	6	7
2	2	4	6	8	10	12	14
3	3	6	9	12	15	18	21
4	4	8	12	16	20	24	28
5	5	10	15	20	25	30	35
6	6	12	18	24	30	36	42
7	7	14	21	28	35	42	49

b The numbers along the marked diagonal are the square numbers. They appear in ascending order downwards.
c Yes, the diagonal illustrated is an axis of symmetry because the arrays of numbers on either side of it are the same.
d The numbers in successive rows or columns are multiples of the successive numbers from 1 to 7.

Exercise 8.3

For each question:

a Copy and complete the multiplication table.
b Write down what you notice about the numbers along the marked diagonal.
c Is the diagonal an axis of symmetry?
Explain your answer.
d Write a statement or statements about any other patterns that you find.

1

×	1	2	4	8	16	32
1						
2						
4						
8						
16						
32						

2

×	1	3	9	27	81
1					
3					
9					
27					
81					

3

×	1	4	16	64
1				
4				
16				
64				

4

×	1	2	4	8	16	32
2						
4						
8						
16						
32						
64						

5

×	1	3	9	27
3				
9				
27				
81				

6

×	3	5	7	9
1				
2				
3				
4				

The numbers 1, 2, 3, 6, 9 and 18 can be described as the **factors** of 18 because they are all the numbers which divide exactly into 18.

The factors of 16 are: 1, 2, 4, 8 and 16
The factors of 21 are: 1, 3, 7 and 21

Exercise 8.4

List the factors of:

1 6 **2** 10 **3** 8 **4** 14 **5** 22 **6** 26
7 15 **8** 27 **9** 33 **10** 35 **11** 55 **12** 9
13 4 **14** 25 **15** 49

Example 3

a Copy and complete the multiplication table.

×	28	14	7
1			
2			
4			

b Write down the number which repeatedly appears along the marked diagonal.

c From the table, list the factors of the number in **b** in ascending order.

a

×	28	14	7
1	28	14	7
2	56	28	14
4	112	56	28

b 28 **c** 1, 2, 4, 7, 14 and 28

Exercise 8.5

For each question:

a Copy and complete the multiplication table.

b Write down the number which repeatedly appears along the marked diagonal.

c From the table, list the factors of the number in **b** in ascending order.

1

×	20	10	5
1			
2			
4			

2

×	32	16	8
1			
2			
4			

3

×	44	22	11
1			
2			
4			

4

×	66	33	22	11
1				
2				
3				
6				

5

×	40	20	10	8
1				
2				
4				
5				

6

×	56	28	14	8
1				
2				
4				
7				

7

×	100	50	25	20	10
1					
2					
4					
5					
10					

8

×	81	27	9
1			
3			
9			

9

×	64	32	16	8
1				
2				
4				
8				

Example 4

Find without drawing the pattern:
a the number of squares, **b** the number of dots, and **c** the number of lines in each of the next four cases of the sequence.

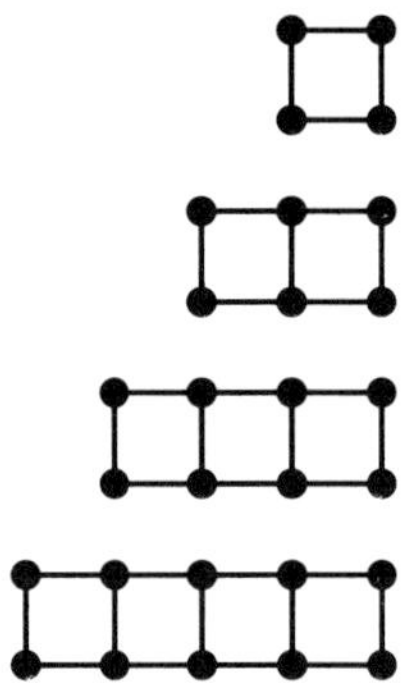

a Number of squares	**b** Number of dots	**c** Number of lines
1	4	4
2	6	7
3	8	10
4	10	13

In the next four cases the numbers are as follows:

a Number of squares	**b** Number of dots	**c** Number of lines
5	12	16
6	14	19
7	16	22
8	18	25

The number of squares increases successively by 1
the number of dots increases successively by 2
the number of lines increases successively by 3.

Any of the above numbers can be checked by drawing lines and dots on squared paper.

Exercise 8.6

Find without drawing the pattern:

a the number of squares, **b** the number of dots, and **c** the number of lines in each of the next four cases of the sequence.

1

	a Number of squares	**b** Number of dots	**c** Number of lines
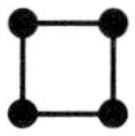	1	4	4
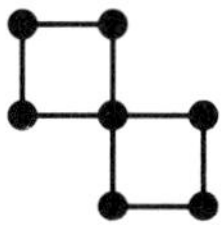	2	7	8
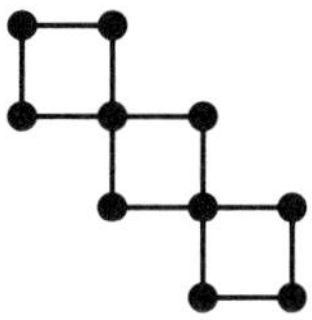	3	10	12

2

	a Number of squares	**b** Number of dots	**c** Number of lines
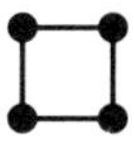	1	4	4
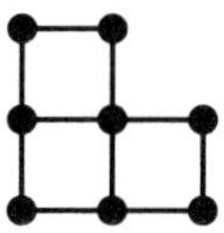	3	8	10
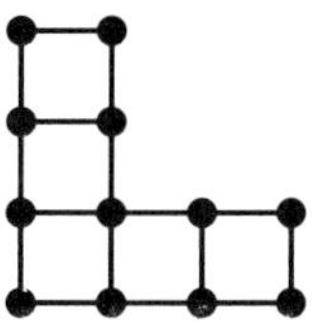	5	12	16

3

	a Number of small squares	**b** Number of dots	**c** Number of lines
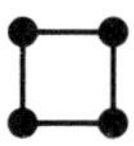	1	4	4
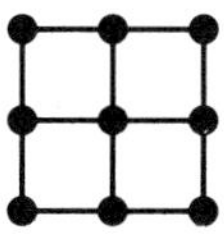	4	9	12
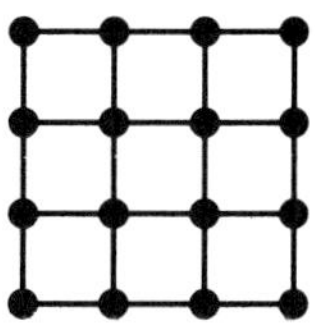	9	16	24

Example 5

Find the next four sums of the sequence below:

$$
\begin{aligned}
1 &= 1 \\
1+2+1 &= 4 \\
1+2+3+2+1 &= 9 \\
1+2+3+4+3+2+1 &= 16
\end{aligned}
$$

The next four sums are 25, 36, 49 and 64.
This can easily be seen because the four sums given are the first four square numbers.

Exercise 8.7

For each sequence find the next four sums:

1 $1 = 1$
$1+3+1 = 5$
$1+3+5+3+1 = 13$
$1+3+5+7+5+3+1 = 25$

2 $2 = 2$
$2+4+2 = 8$
$2+4+6+4+2 = 18$
$2+4+6+8+6+4+2 = 32$

3 $1 = 1$
$1+1+1 = 3$
$1+1+2+1+1 = 6$
$1+1+2+2+2+1+1 = 10$
$1+1+2+2+3+2+2+1+1 = 15$

8.2 Using doubling and halving to explore the properties of numbers

A quick way of multiplying a number by 2 is simply to double the number.
Two simple examples are:

$12 \times 2 = 24$ (24 is 12 doubled) $\qquad$ $101 \times 2 = 202$ (202 is 101 doubled)

We can use the same principle in order to multiply by 4.
If we want to multiply by 4, we simply double and then double again.

Example 1

Find 101×4 using the 'doubling' method.

$$101 \times 4 = 101 \times 2 \times 2 = 202 \times 2 = 404$$

In order to multiply by 8 we have to double three times.

Example 2

Find 71×8 using the 'doubling' method:

$71 \times 8 = 71 \times 2 \times 2 \times 2 = 142 \times 2 \times 2 = 284 \times 2 = 568$

Exercise 8.8

Copy the table and use the 'doubling' method to complete it.

Number	Number ×2	Number ×4	Number ×8
1 11			
2 15			
3 25			
4 32			
5 75			
6 125			

Dividing a number by 2 is the same as halving the number.
Two simple examples are:

$12 \div 2 = 6$ (6 is 12 halved) $104 \div 2 = 52$ (52 is 104 halved)

We can use the same principle in order to divide by 4.
If we want to divide by 4 we simply halve and then halve again.

Example 3

Find $76 \div 4$ using the 'halving' method.

$76 \div 4 = 76 \div 2 \div 2 = 38 \div 2 = 19$

In order to divide by 8 we have to halve three times.

Example 4

Find $104 \div 8$ using the 'halving' method.

$104 \div 8 = 104 \div 2 \div 2 \div 2 = 52 \div 2 \div 2 = 26 \div 2 = 13$

Exercise 8.9

Copy the table and use the 'halving' method to complete it.

Number	Number ÷2	Number ÷4	Number ÷8
1 80			
2 240			
3 400			
4 424			
5 680			
6 840			

Exercise 8.10

1 When Nathan was born he weighed 4 kilograms. When he was two months old he was twice as heavy as his birth weight; when he was one year old he was four times his birth weight, and when he was five years old he was eight times his birth weight.
Find his weight at each of these three ages.

2 Amy walks at 6 kilometres per hour. She jogs at twice that speed and cycles at four times that speed.
Find her jogging and cycling speeds.

3 The train fare from London to Berkhamsted is £7. To Northampton it is twice as much, to Stafford it is four times as much, and to Glasgow it is eight times as much.
Find the fare to each of the other three places.

4 The distance from London to Cardiff is 240 kilometres. The distance from London to Swindon is only half as much. The distance from London to Reading is only a quarter as much, and the distance from London to Slough is only one eighth as much.
Find the distance from London to each of the other three places.

5 James has 72 marbles. Daniel has only half as many, Saleem has only one quarter as many, and Liam has only one eighth as many.
Find how many each of the other three boys have.

Example 5

What are the first ten numbers that are obtained by doubling 2 and then doubling the result?

$2 \times 2 = 4$ $4 \times 2 = 8$ $8 \times 2 = 16$ $16 \times 2 = 32$
$32 \times 2 = 64$ $64 \times 2 = 128$ $128 \times 2 = 256$ $256 \times 2 = 512$
$512 \times 2 = 1024$ $1024 \times 2 = 2048$

Exercise 8.11

1 What are the first eight numbers that are obtained by doubling 3 and then doubling the result?

2 What are the first seven numbers that are obtained by doubling 5 and then doubling the result?

3 What are the first eight numbers that are obtained by doubling 7 and then doubling the result?

4 What are the first six numbers that are obtained by doubling 11 and then doubling the result?

5 What are the first eight numbers that are obtained by doubling 13 and then doubling the result?

Example 6

Use repeated division by 2 to express as a product of its prime factors:

a 24 **b** 32 **c** 40

a $24 = 12 \times 2 = 6 \times 2 \times 2 = 3 \times 2 \times 2 \times 2$
b $32 = 16 \times 2 = 8 \times 2 \times 2 = 4 \times 2 \times 2 \times 2 = 2 \times 2 \times 2 \times 2 \times 2$
c $40 = 20 \times 2 = 10 \times 2 \times 2 = 5 \times 2 \times 2 \times 2$

Exercise 8.12

Use repeated division by 2 to express as a product of its prime factors:

1 12 **2** 28 **3** 20 **4** 44 **5** 52 **6** 68
7 152 **8** 16 **9** 48 **10** 112 **11** 176 **12** 80

A quick way of multiplying by 5 is to multiply by 10 and then halve the result.

Example 7

Use the above method to multiply **a** 14 and **b** 72 by 5.

a $14 \times 5 = 14 \times 10 \div 2 = 140 \div 2 = 70$
b $72 \times 5 = 72 \times 10 \div 2 = 720 \div 2 = 360$

A quick way of multiplying by 25 is to multiply by 100 and then halve the result twice.

Example 8

Use the above method to multiply **a** 14 and **b** 72 by 25.

a $14 \times 25 = 14 \times 100 \div 2 \div 2 = 1400 \div 2 \div 2 = 700 \div 2 = 350$
b $72 \times 25 = 72 \times 100 \div 2 \div 2 = 7200 \div 2 \div 2 = 3600 \div 2 = 1800$

Exercise 8.13

Copy the table and use the methods of Examples 7 and 8 to complete it.

Number	Number ×5	Number ×25
1 12		
2 40		
3 16		
4 160		

5 In Elm Avenue there are only 8 houses. In Birch Avenue there are 5 times as many, and in Oak Avenue there are 25 times as many. Use the methods of Examples 7 and 8 to find the number of houses in each of the other two avenues.

6 Jodie's pet mouse is 11 cm long. Her dog is 5 times longer and her horse is 25 times longer.
Use the methods of Examples 7 and 8 to find the length of each of her other two animals.

A quick method of dividing by 5 is to double the number and then divide by 10.

Example 9

Use the above method to divide **a** 750 and **b** 1800 by 5.

a $750 \div 5 = 750 \times 2 \div 10 = 1500 \div 10 = 150$
b $1800 \div 5 = 1800 \times 2 \div 10 = 3600 \div 10 = 360$

A quick method of dividing by 25 is to double the number twice and then divide by 100.

Example 10

Use the above method to divide **a** 750 and **b** 1800 by 25.

a $750 \div 25 = 750 \times 2 \times 2 \div 100 = 1500 \times 2 \div 100 = 3000 \div 100 = 30$
b $1800 \div 25 = 1800 \times 2 \times 2 \div 100 = 3600 \times 2 \div 100 = 7200 \div 100 = 72$

Exercise 8.14

Copy the table and use the methods of Examples 9 and 10 to complete it.

Number	Number ÷5	Number ÷25
1 150		
2 250		
3 800		
4 1600		

5 Inverness is 900 kilometres from London. Birmingham is only $\frac{1}{5}$ as far away, and Hemel Hempstead is only $\frac{1}{25}$ as far away.
Find the distance from London to each of the other two places by using the methods of Examples 9 and 10.

6 In Dean's village there are 1500 inhabitants. In Kirsty's village there are only $\frac{1}{5}$ as many, and Vicky lives in a very small village which has only $\frac{1}{25}$ as many.
Use the methods of Examples 9 and 10 to find the number of inhabitants in Kirsty and Vicky's villages.

Example 11

Starting with 2, follow the sequence described below:
a Double the number.
b Subtract 1.
c Repeat the process until the result exceeds 100.

$2 \times 2 - 1 = 3$ $2 \times 3 - 1 = 5$ $2 \times 5 - 1 = 9$ $2 \times 9 - 1 = 17$
$2 \times 17 - 1 = 33$ $2 \times 33 - 1 = 65$ $2 \times 65 - 1 = 129$

Exercise 8.15

1 Repeat the process explained in Example 11 starting with each of the following numbers:
a 3 **b** 4 **c** 5 **d** 6 **e** 7 **f** 8 **g** 9

2 a There are three numbers in Question 1 that generate the same sequence. Which numbers are they?
b There are two other numbers in Question 1 which also generate the same sequence. Which numbers are they?

8.3 Recognising that multiplication and division are inverse operations and using this to check calculations

If I multiply a number by 2 I must divide the result by 2 in order to get back to the number I started with.
For example: $15 \times 2 = 30$ and $30 \div 2 = 15$

We therefore say that dividing by 2 is the inverse of multiplying by 2 and vice versa.

In the same way if I multiply a number by 7, I must divide the result by 7 in order to get back to the original number.
For example: $111 \times 7 = 777$ and $777 \div 7 = 111$

Example 1

Write down the inverse operation of each of the following.
Check each operation and its inverse by starting with the number 42.

a Multiply by 2 **b** Multiply by 3 **c** Multiply by 5
d Divide by 2 **e** Divide by 3 **f** Divide by 7

a Divide by 2: $42 \times 2 = 84$, $84 \div 2 = 42$
b Divide by 3: $42 \times 3 = 126$, $126 \div 3 = 42$
c Divide by 5: $42 \times 5 = 210$, $210 \div 5 = 42$
d Multiply by 2: $42 \div 2 = 21$, $21 \times 2 = 42$
e Multiply by 3: $42 \div 3 = 14$, $14 \times 3 = 42$
f Multiply by 7: $42 \div 7 = 6$, $6 \times 7 = 42$

Exercise 8.16

Write down the inverse operation of each of the following.
Check each operation and its inverse by starting with the number shown.

1 Multiply by 6 (check by starting with the number 35)

2 Multiply by 8 (check by starting with the number 45)

3 Multiply by 12 (check by starting with the number 28)

4 A school has 25 classes with 16 pupils in each class.
How many pupils has the school altogether?
Check your answer with an inverse operation.

5 The coaches on an excursion train have 56 seats each and the train consists of 12 coaches.
If every seat is occupied, how many passengers are on the train?
Check your answer with an inverse operation.

6 Paul's younger sister has a counting frame and the bars each hold 15 beads. If there are 8 bars on the frame, how many beads can it hold altogether? Check your answer with an inverse operation.

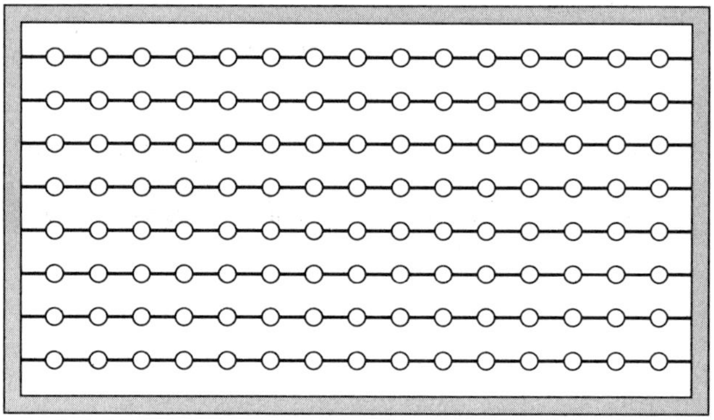

Write down the inverse operation of each of the following and check each operation and its inverse by starting with the number shown.

7 Divide by 4 (check by starting with the number 56).

8 Divide by 5 (check by starting with the number 85).

9 Divide by 20 (check by starting with the number 360).

10 Divide by 75 (check by starting with the number 900).

11 A new estate has 1080 houses and there are the same number of houses along each of the 18 avenues.
How many houses are there in each avenue?
Check your answer with an inverse operation.

12 An apple-picker has picked 1296 apples and he has to pack them into boxes which each hold 144 apples.
How many boxes will he require?
Check your answer with an inverse operation.

Example 2

a Evaluate 763×42
b Divide the result of **a** by 42. Does this give 763?
c Divide the result of **a** by 763. Does this give 42?

a $763 \times 42 = 32\,046$
b $32\,046 \div 42 = 763$, yes
c $32\,046 \div 763 = 42$, yes

Exercise 8.17

1 **a** Evaluate 720×45
b Divide the result of **a** by 45.
Does this give 720?
c Divide the result of **a** by 720.
Does this give 45?

2 **a** Evaluate 750×16
b Divide the result of **a** by 16.
Does this give 750?
c Divide the result of **a** by 750.
Does this give 16?

3 **a** Evaluate 840×25
b Divide the result of **a** by 25.
Does this give 840?
c Divide the result of **a** by 840.
Does this give 25?

4 One Saturday at 3 p.m. 46 football league matches kicked off with 22 players playing at each match.
How many players were there altogether?

Divide the total number of players by 22.
Does this give the number of matches?

Divide the total number of players by 46.
Does this give the number of players at each match?

5 48 egg cartons which each contain 12 eggs packed into a large box.
How many eggs are there in the box?

Divide the total number of eggs by 48.
Does this give the number of eggs per carton?

Divide the total number of eggs by 12.
Does this give the number of cartons?

Example 3

a Evaluate $13\,113 \div 31$
b Multiply the result of **a** by 31. Does this give 13 113?

a $13\,113 \div 31 = 423$ **b** $423 \times 31 = 13\,113$, yes.

Exercise 8.18

1 a Evaluate $1440 \div 36$
b Multiply the result of **a** by 36. Does this give 1440?

2 a Evaluate $3240 \div 24$
b Multiply the result of **a** by 24. Does this give 3240?

3 a Evaluate $7000 \div 125$
b Multiply the result of **a** by 125 Does this give 7000?

4 504 cars have to be moved from a production factory on transporters which can each carry 9 cars.
How many transporter journeys are required?

Multiply the number of journeys by 9.
Does this give the total number of cars?

8.4 Understanding and using terms such as 'prime', 'cube', 'square root' and 'cube root'

We are already familiar with 'square' numbers. These numbers occur very often when dealing with the areas of square shapes.

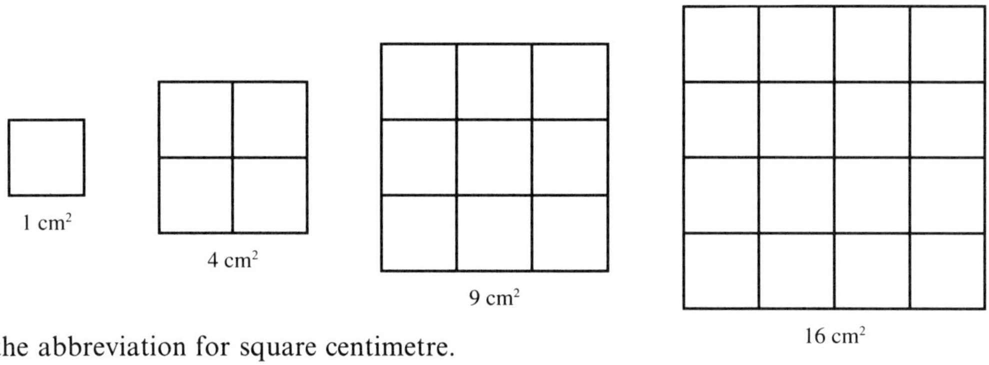

cm^2 is the abbreviation for square centimetre.

In a similar way 'cube' numbers can be found from the volumes of different cubes.

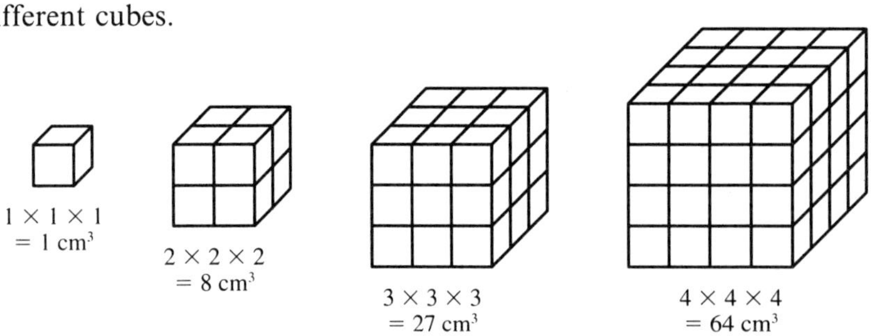

cm^3 is the abbreviation for cubic centimetre.

The numbers 1, 8, 27 and 64 are the first four 'cubic' numbers.
They can also be written as 1^3, 2^3, 3^3, and 4^3.
The raised 3 shows that the number has been raised to the power of 3.

Exercise 8.19

1 Find the volume of a cube if the edge length is:
a 5 cm **b** 6 cm **c** 8 cm **d** 10 cm **e** 9 cm **f** 7 cm

2 Use your calculator to find each of the following:
a 11^3 (or $11 \times 11 \times 11$) **b** 13^3 **c** 21^3 **d** 31^3
e 75^3

3 Use your calculator to find each of the following:
a 1.5^3 **b** 2.5^3 **c** 2.4^3 **d** 3.2^3 **e** 0.3^3 **f** 0.1^3

The square of a number is found by multiplying the number by itself.
We say that the square of 3 is $3 \times 3 = 9$, or $9 = 3^2$.
We also say that 3 is the square root of 9 or we use the square root sign, $\sqrt{9} = 3$.

Example 1

Find the square root of: **a** 16 **b** 36 **c** 3600

a $\sqrt{16} = 4$ because $4 \times 4 = 16$
b $\sqrt{36} = 6$ because $6 \times 6 = 36$
c $\sqrt{3600} = 60$ because $60 \times 60 = 3600$

Exercise 8.20

Find the square root of each of the following:

1 25 **2** 64 **3** 49 **4** 81 **5** 4
6 1 **7** 1600 **8** 4900 **9** 2500 **10** 6400
11 8100 **12** 10 000

Example 2

Use your calculator to evaluate $\sqrt{441}$

Press your calculator keys in this sequence:

[4] [4] [1] [√] 21

Therefore $\sqrt{441} = 21$

Exercise 8.21

Use your calculator to evaluate each of the following:

1 $\sqrt{576}$ **2** $\sqrt{625}$ **3** $\sqrt{484}$ **4** $\sqrt{784}$ **5** $\sqrt{676}$
6 $\sqrt{729}$ **7** $\sqrt{1024}$ **8** $\sqrt{1225}$ **9** $\sqrt{1521}$ **10** $\sqrt{2025}$
11 $\sqrt{1369}$ **12** $\sqrt{5.29}$ **13** $\sqrt{8.41}$ **14** $\sqrt{2.89}$ **15** $\sqrt{3.61}$
16 $\sqrt{2.25}$ **17** $\sqrt{12.96}$ **18** $\sqrt{16.81}$ **19** $\sqrt{14.44}$ **20** $\sqrt{24.01}$
21 $\sqrt{0.81}$ **22** $\sqrt{0.36}$ **23** $\sqrt{0.49}$ **24** $\sqrt{0.64}$ **25** $\sqrt{0.16}$
26 $\sqrt{0.09}$ **27** $\sqrt{0.04}$ **28** $\sqrt{0.01}$

Example 3

Find the side length of a square wooden board of area $1.21\,\text{m}^2$.

Side length is: $\sqrt{1.21}\,\text{m} = 1.1\,\text{m}$

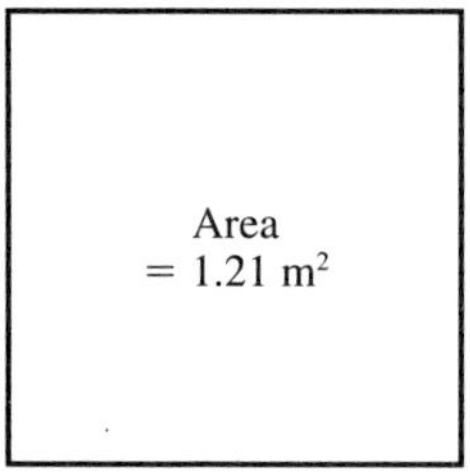

Exercise 8.22

Find the side length of each of the following:

1 A square window pane of area $1296\,\text{cm}^2$.

2 A square tray of area $1156\,\text{cm}^2$.

3 A square mat of area $1.44\,\text{m}^2$.

4 A square paving stone of area $1.69\,\text{m}^2$.

5 A square lawn of area $4225\,\text{m}^2$.

6 A square field of area $40\,000\,\text{m}^2$.

7 A square courtyard of area $3025\,\text{m}^2$.

8 A square pond of area $529\,\text{m}^2$.

We already know that the sequence 1, 4, 9, 16, . . . is the sequence of square numbers.
The sequence 2, 3, 5, 7, is the sequence of prime numbers (see p.122).
Any other number (i.e. a number which is neither prime nor square) is said to be **rectangular**.

Look at this list of the numbers from 1 to 10.

1 is a square number

2 is a prime number

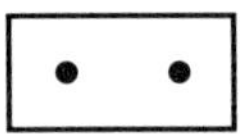

3 is a prime number

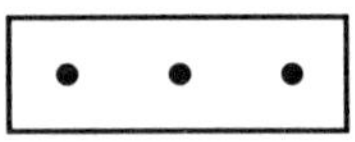

4 is a square number

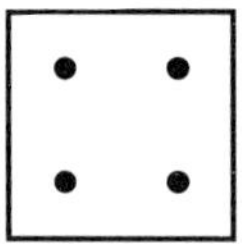

5 is a prime number

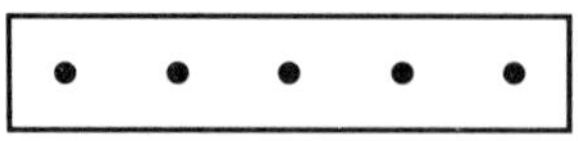

6 is a rectangular number

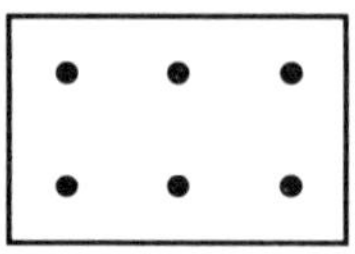

7 is a prime number

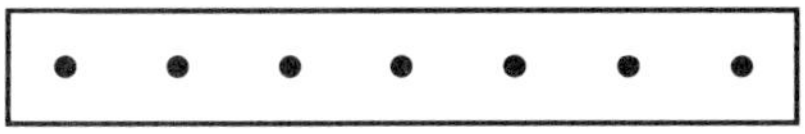

9 is a square number

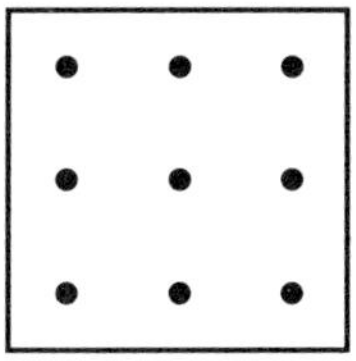

8 is a rectangular number

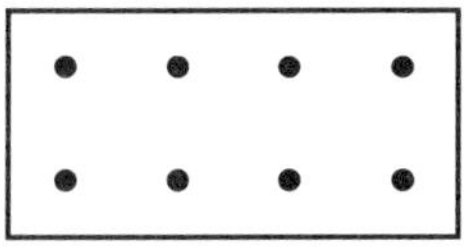

10 is a rectangular number

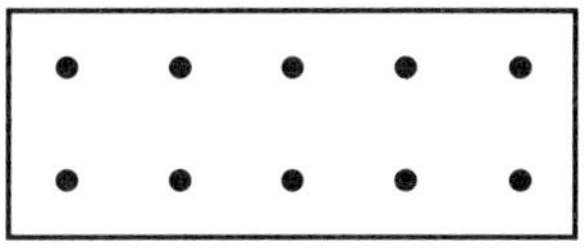

Exercise 8.23

List the numbers in each of the following ranges.
State whether each number is prime, square or rectangular.

1 11 to 20 inclusive.
2 21 to 30 inclusive.
3 41 to 50 inclusive.
4 61 to 70 inclusive.

The cube of a number is found by multiplying a number by itself and then by itself again.
We say that the cube of 7 is $7 \times 7 \times 7 = 343$ or $7^3 = 343$
We also say that 7 is the cube root of 343, or we use the cube root sign:
$\sqrt[3]{343} = 7$

Example 4

a Find the cube root of (i) 8 (ii) 64 (iii) 216
b Find (i) $\sqrt[3]{27}$ (ii) $\sqrt[3]{125}$ (iii) $\sqrt[3]{125\,000}$

a (i) $8 = 2 \times 2 \times 2$, therefore the cube root of 8 is 2.
(ii) $64 = 4 \times 4 \times 4$, therefore the cube root of 64 is 4.
(iii) $216 = 6 \times 6 \times 6$, therefore the cube root of 216 is 6.
b (i) $\sqrt[3]{27} = 3$ because $3 \times 3 \times 3 = 27$
(ii) $\sqrt[3]{125} = 5$ because $5 \times 5 \times 5 = 125$
(iii) $\sqrt[3]{125\,000} = 50$ because $50 \times 50 \times 50 = 125\,000$

Example 5

Use your calculator to evaluate $\sqrt[3]{1.331}$

Press your calculator keys in this sequence:

[1] [.] [3] [3] [1] [$x^{1/y}$] [3] [=] 1.1

Therefore $\sqrt[3]{1.331} = 1.1$

Exercise 8.24

1 Find the cube root of each of the following.
Check your answer by multiplication.
a 8000 **b** 64 000 **c** 216 000 **d** 512 000

2 Use your calculator to evaluate each of the following:
a $\sqrt[3]{4913}$ **b** $\sqrt[3]{6859}$ **c** $\sqrt[3]{5832}$ **d** $\sqrt[3]{10\,648}$

3 Use your calculator to evaluate each of the following:
a $\sqrt[3]{4.096}$ **b** $\sqrt[3]{6.859}$ **c** $\sqrt[3]{4.913}$ **d** $\sqrt[3]{9.261}$

Example 6

Find the side length of a cubic box which has a volume of $1000\,cm^3$.

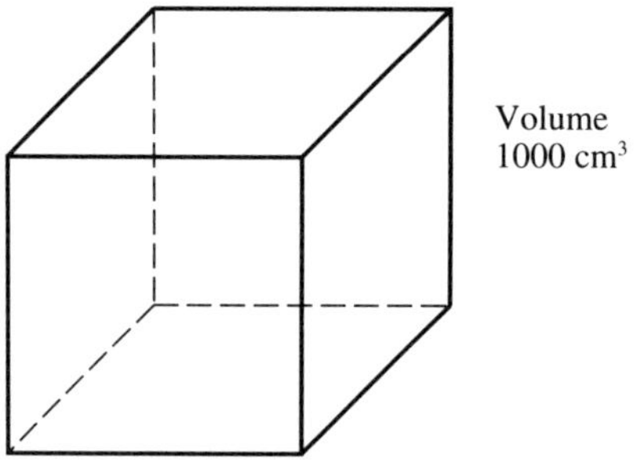

Side length is:
$\sqrt[3]{1000}\,cm = 10\,cm$.

Exercise 8.25

Find the side length for each of the following:

1 A cubic gift box of volume $1728\,cm^3$.

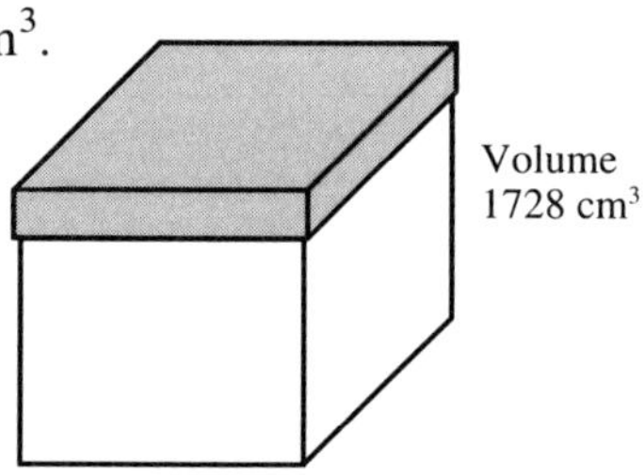

2 A cubic toy brick which has a volume of $42.875\,cm^3$.

3 An eraser which is cubic in shape and has a volume of $4.913\,cm^3$.

4 A domestic water tank which is cubic in shape and has a volume of $2.197\,m^3$.

9 Patterns, sequences and functional relationships

9.1 Generating sequences

Example 1

A sequence of numbers is formed by starting with 1, adding 2 to obtain the next term and then repeatedly adding 2 to obtain all the other terms. The first three terms of the sequence are 1, 3, 5. Write down the next five terms of the sequence and name the sequence of terms that you have found.

The next five terms are 7, 9, 11, 13, 15.
The sequence 1, 3, 5, 7, 9, 11, 13, 15, ... is the sequence of odd numbers.

Exercise 9.1

For Questions 1 and 2 write down the next five terms of the sequence.

1 Start with 1 and repeatedly add 3.
The first three terms are 1, 4, 7.

2 Start with 1 and repeatedly add 4.
The first three terms are 1, 5, 9.

For Questions 3 and 4 write down the next five terms of the sequence and name the sequence of terms that you have found.

3 Start with 2 and repeatedly add 2.
The first three terms are 2, 4, 6.

4 Start with 3 and repeatedly add 3.
The first three terms are 3, 6, 9.

Example 2

A sequence of numbers is formed by starting with 1, multiplying by 2 to obtain the next term and then repeatedly multiplying by 2 to obtain all the other terms.
The first three terms are 1, 2, 4.
Write down the next five terms of the sequence and name the sequence of terms that you have found.

The next five terms are 8, 16, 32, 64, 128.
The sequence 1, 2, 4, 8, 16, 32, 64, 128, ... is known as the sequence of 'powers of 2'

$2 = 2 = 2^1$ (2 to the power 1)
$4 = 2 \times 2 = 2^2$ (2 to the power 2)
$8 = 2 \times 2 \times 2 = 2^3$ (2 to the power 3)

and so on. (1 can be regarded as 2 to the power 0).

Exercise 9.2

For Questions 1 and 2 write down the next five terms of the sequence and name the sequence of terms that you have found.

1 Start with 1 and repeatedly multiply by 3.
The first three terms are 1, 3, 9.

2 Start with 1 and repeatedly multiply by 4.
The first three terms are 1, 4, 16.

3 Start with 1 and repeatedly multiply by 10.
The first three terms are 1, 10, 100.
Write down the next four terms of the sequence and name the sequence of terms that you have found.

Example 3

A sequence of numbers is formed by starting with 0, first adding 1, then adding 2, then 3 and so on.
The first four terms of this sequence are 0, 1, 3, 6.
Write down the next five terms of this sequence and name the sequence of the terms that you have found.

The next five terms are 10, 15, 21, 28, 36.
The sequence 1, 3, 6, 10, 15, 21, 28, 36, ... is known as the sequence of 'triangular' numbers,

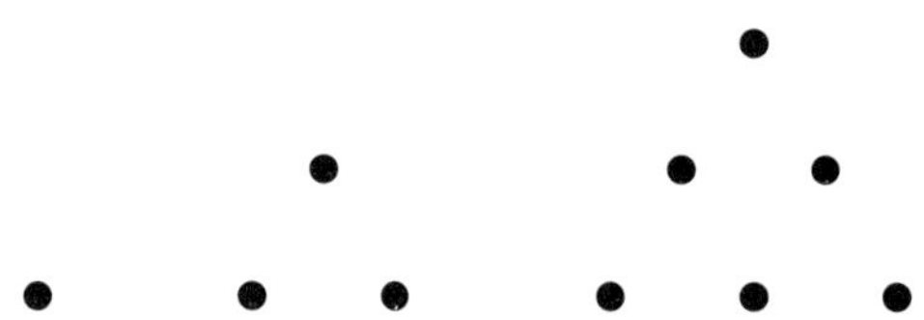

1 3 6 and so on.

Exercise 9.3

For questions 1 and 2 write down the next five terms of the sequence.

1 Start with 0, first add 2, then add 4, then 6 and so on.
The first four terms are 0, 2, 6, 12.

2 Start with 0, first add 3, then add 6, then 9 and so on.
The first four terms are 0, 3, 9, 18.

3 Start with 0, first add 1, then add 3, then 5 and so on.
The first four terms are 0, 1, 4, 9. Write down the next five terms of the sequence and name the sequence of terms that you have found.

Example 4

A sequence of numbers is formed by starting with the numbers 0 and 1 and then adding these together to get the third term. The second and third terms are then added together to get the fourth term and so on.
The first five terms of this sequence are 0, 1, 1, 2, 3, 5.
Write down the next five terms.

The next five terms are 8, 13, 21, 34, 55.
The sequence 0, 1, 1, 2, 3, 5, 8, 13, 21, 34, 55,... is known as the Fibonacci sequence.

Exercise 9.4

For all questions find the next five terms of the sequence.
In each sequence any term is the sum of the two previous terms.

1 Start with 1 and 3.
The first five terms are 1, 3, 4, 7, 11.

2 Start with 2 and 3.
The first five terms are 2, 3, 5, 8, 13.

3 Start with 0 and 3.
The first five terms are 0, 3, 3, 6, 9.

Prime numbers

A prime number is a number with two and only two factors.
Look at the examples below:

2 is a prime number	$2 = 2 \times 1$ only
3 is a prime number	$3 = 3 \times 1$ only
11 is a prime number	$11 = 11 \times 1$ only
6 is not a prime number	$6 = 6 \times 1$ or 3×2
9 is not a prime number	$9 = 9 \times 1$ or 3×3
15 is not a prime number	$15 = 15 \times 1$ or 5×3

Note. The two factors of a prime number are always the number itself and one.

1 is not counted as a prime number as it has only one factor.

Exercise 9.5

1 You can find the prime numbers between 1 and 100 using a 'one hundred' square.

~~1~~	②	③	~~4~~	⑤	~~6~~	⑦	~~8~~	~~9~~	~~10~~
11	12	13	14	15	16	17	18	19	20
21	22	23	24	25	26	27	28	29	30
31	32	33	34	35	36	37	38	39	40
41	42	43	44	45	46	47	48	49	50
51	52	53	54	55	56	57	58	59	60
61	62	63	64	65	66	67	68	69	70
71	72	73	74	75	76	77	78	79	80
81	82	83	84	85	86	87	88	89	90
91	92	93	94	95	96	97	98	99	100

(i) Cross out 1

(ii) Ring 2 and then cross out all multiples of 2.

(iii) Ring 3 and then cross out all multiples of 3.

(iv) Ring 5 and then cross out all multiples of 5.

(v) Ring the next number which is not crossed out and then cross out all its multiples.

(vi) Repeat (v) until all the numbers are ringed or crossed out.

The ringed numbers are the prime numbers.
List all the prime numbers between 1 and 100.
How many are there?

2 Draw a 'one hundred' square and ring all the prime numbers. Shade all the multiples of 6 and state what you notice about all the prime numbers except 2 and 3.

3 Draw a 'one hundred' square and ring all the prime numbers. Shade all the multiples of 4 and state what you notice about all the prime numbers except 2.

4 Draw a 'one hundred' square and ring all the prime numbers. What do you notice about the number in the line directly below every prime number except 67 if it is not another prime number?

5 Draw a 'one hundred' square and ring all the prime numbers. What do you notice about the number in the line directly above every prime number except 11 and 59 if it is not another prime number?

6 Draw a 'one hundred' square but instead of the figures from 1 to 100, fill in the number from 101 to 200.
Can you find the prime numbers between 101 and 200? How many are there?.

9.2 Recognising patterns in numbers through spatial arrangements

Example 1

Look at the patterns below.
Draw the next two patterns and write down the number of dots in each.
Write down the next three numbers in the sequence.

3 8 15

The next two patterns are:

24

35

The next three numbers in the sequence are:

$7 \times 7 - 1 = 48$ $8 \times 8 - 1 = 63$ $9 \times 9 - 1 = 80$

The numbers in the sequence: 3 8 15 24 35 48 63 80 are all one less than a square number.

Example 2

Look at the patterns below.

a Draw the next two patterns and state number of lines and dots.
b State the number of lines and dots for the next three patterns.
c What is the sequence of
(i) the number of lines and
(ii) the number of dots?

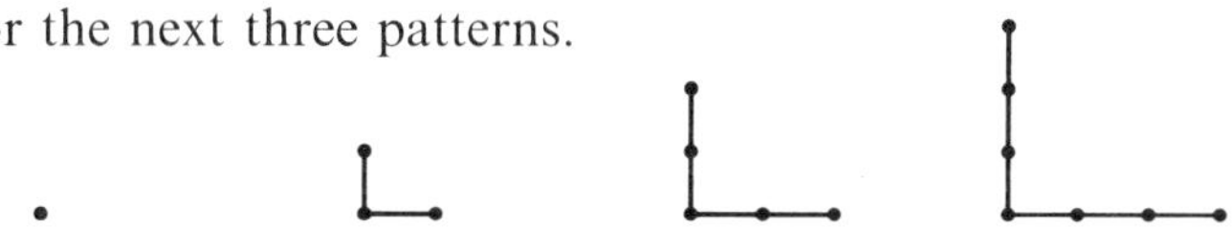

0 lines, 1 dot; 2 lines, 3 dots; 4 lines, 5 dots; 6 lines, 7 dots

a The next two patterns are:

8 lines
9 dots

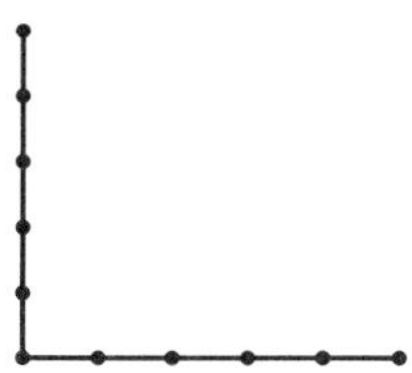

10 lines
11 dots

b The next three patterns have:
Lines: 12, 14, 16
Dots: 13, 15, 17

c (i) The numbers in the sequence 0, 2, 4, 6, 8, 10, 12, 14, 16, ... are the even numbers.
(ii) The numbers in the sequence 1, 3, 5, 7, 9, 11, 13, 15, 17, ... are the odd numbers.

Exercise 9.6

For Questions 1 and 2 write down the number of **a** lines and **b** dots in the next four patterns.
Do not draw the patterns.

1

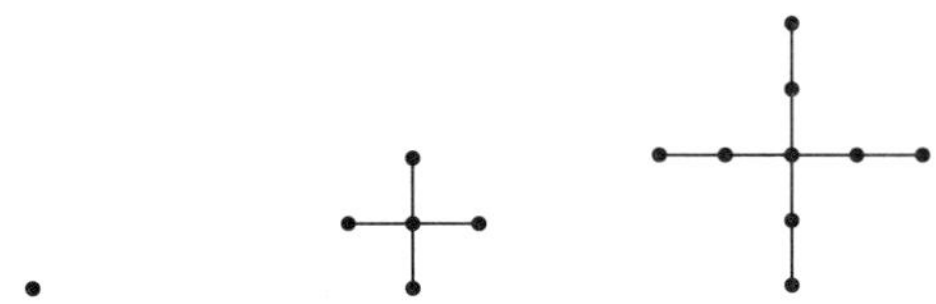

Lines	0	4	8
Dots	1	5	9

2

Lines	0	3	6
Dots	1	4	7

3 Write down the number of **a** triangles **b** dots **c** lines in the next four patterns.
Do not draw the patterns.

Triangles	1	2	3
Dots	3	4	5
Lines	3	5	7

4 Write down the number of **a** small triangles **b** squares **c** dots **d** lines in the next four patterns.
Do not draw the patterns.

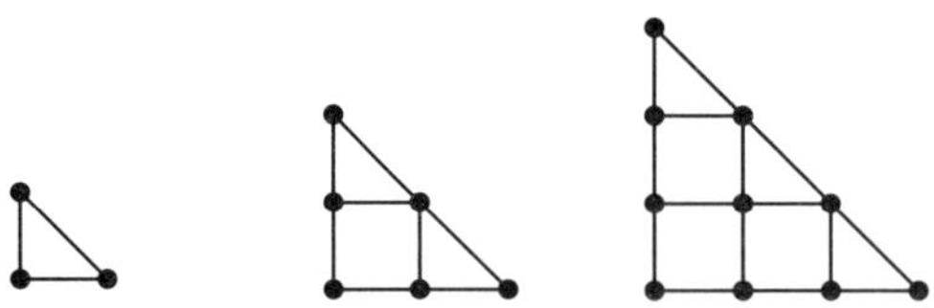

Triangles	1	2	3
Squares	0	1	3
Dots	3	6	10
Lines	3	8	15

All the patterns in Questions 1–4 can be checked by drawing lines and dots on squared paper.

For Questions 5 and 6 copy and complete the table.

5

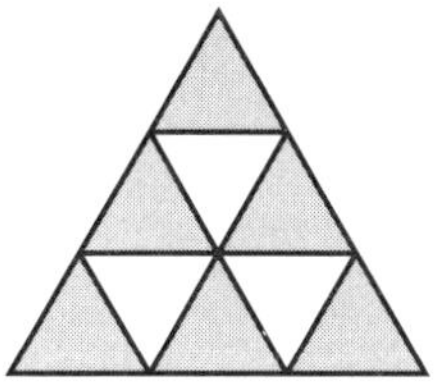

Row	**a** Number of shaded triangles	**b** Number of white triangles	**c** Total number of triangles
1	1	0	1
2	2	1	3
3	3	2	5
4			
5			
6			
7			
8			
9			
10			
20			
50			

6

	a Side length of square	**b** Number of small squares	**c** Number of dots	**d** Number of squares of any size
	1 unit	1	4	1
	2 units	4	9	5 (= 4 + 1)
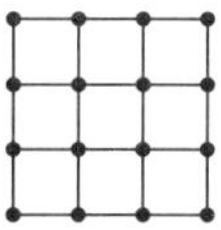	3 units	9	16	14 (= 9 + 4 + 1)
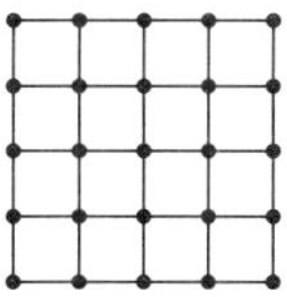	4 units	16	25	30 (= 16 + 9 + 4 + 1)
	5 units			
	6 units			
	7 units			
	8 units			

9.3 Dealing with inputs to and outputs from simple function machines

This function machine adds 7 to any number that is fed into its input. For example, if the INPUT is 2, then the OUTPUT is $2 + 7 = 9$.

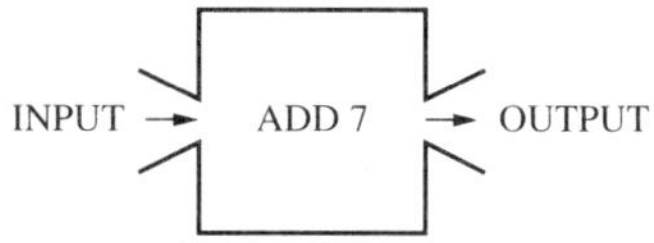

Example 1

Use the ADD 7 function machine and write down:

a the OUTPUT number when the INPUT number is (i) 3, (ii) 8, (iii) 15 and (iv) 20.

b the INPUT number when the OUTPUT number is (i) 10, (ii) 17, (iii) 21 and (iv) 30.

a (i) $3 + 7 = 10$ (ii) $8 + 7 = 15$ (iii) $15 + 7 = 22$ (iv) $20 + 7 = 27$

b (i) $10 - 7 = 3$ (ii) $17 - 7 = 10$ (iii) $21 - 7 = 14$ (iv) $30 - 7 = 23$

Note that to find the INPUT number we use the inverse of ADD 7 which is SUBTRACT 7.

Exercise 9.7

For each question copy the table and use the function machine to complete it.

1

INPUT	OUTPUT
1	
3	
4	
6	
	7
	10
	13
	14

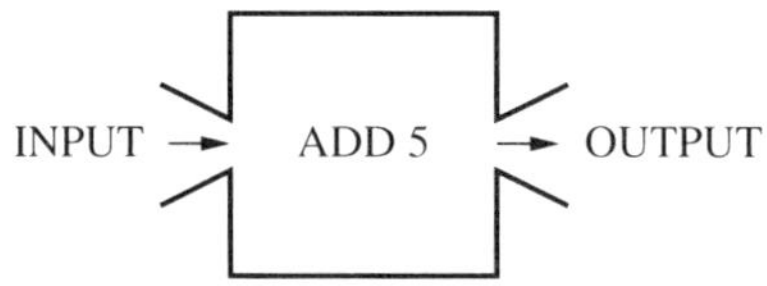

2

INPUT	OUTPUT
5	
25	
35	
55	
	45
	65
	75
	80

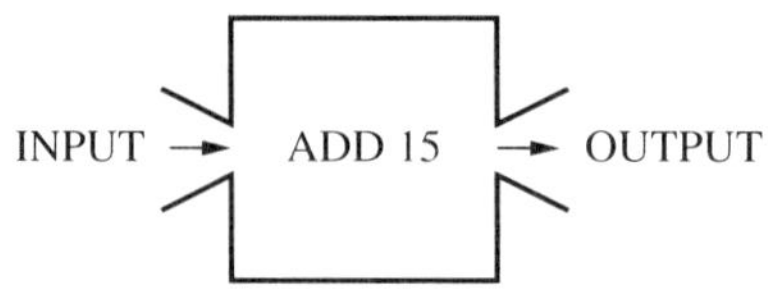

3

INPUT	OUTPUT
5	
7	
8	
10	
	2
	5
	7
	9

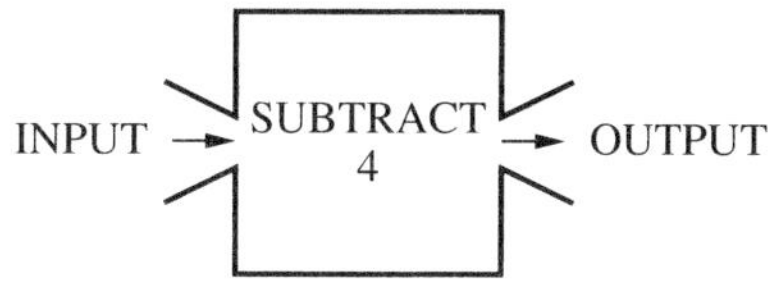

4

INPUT	OUTPUT
15	
20	
25	
40	
	23
	33
	38
	58

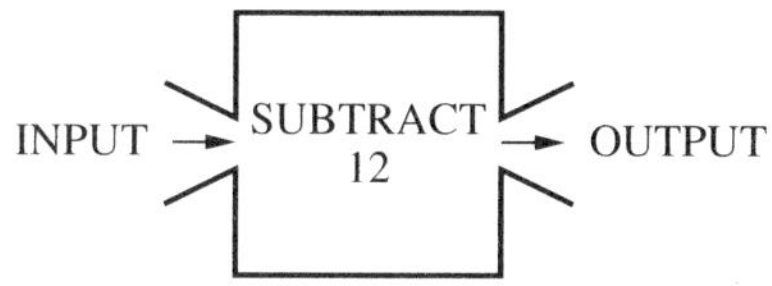

This function machine multiplies any number which is fed into its input by 3.

For example, if the INPUT is 5, then the OUTPUT is 3 × 5 or 15.

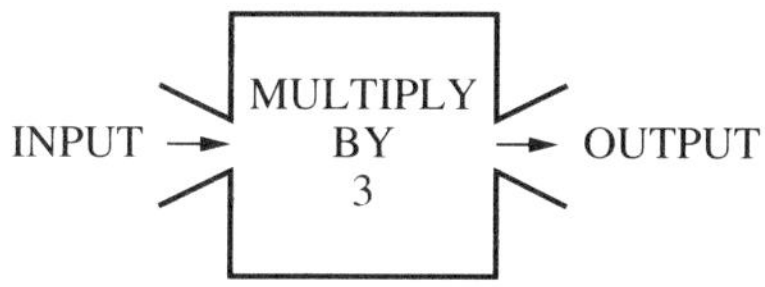

Example 2

Use the MULTIPLY-by-3 function machine and write down:

a the OUTPUT number when the INPUT number is (i) 4, (ii) 7, (iii) 15, and (iv) 20.

b the INPUT number when the OUTPUT number is (i) 15, (ii) 18, (iii) 27, and (iv) 90.

a (i) $4 \times 3 = 12$ (ii) $7 \times 3 = 21$ (iii) $15 \times 3 = 45$
(iv) $20 \times 30 = 60$

b (i) $15 \div 3 = 5$ (ii) $18 \div 3 = 6$ (iii) $27 \div 3 = 9$
(iv) $90 \div 3 = 30$

Note that to find the INPUT number we use the inverse of MULTIPLY by 3 which is DIVIDE by 3.

Exercise 9.8

For each question copy the table and use the function machine to complete it.

1

INPUT	OUTPUT
1	
2	
4	
7	
8	
12	
	18
	30
	36
	54
	60
	90

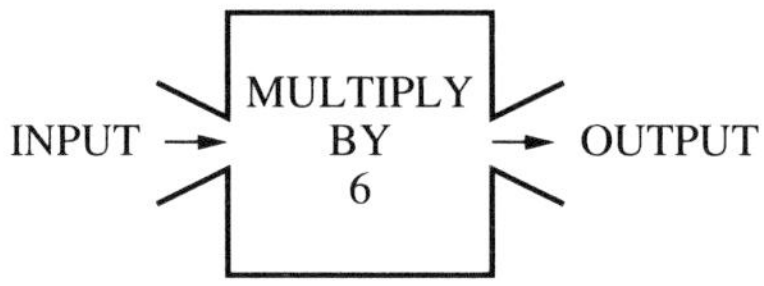

2

INPUT	OUTPUT
4	
7	
8	
10	
12	
20	
	30
	45
	75
	90
	135
	225

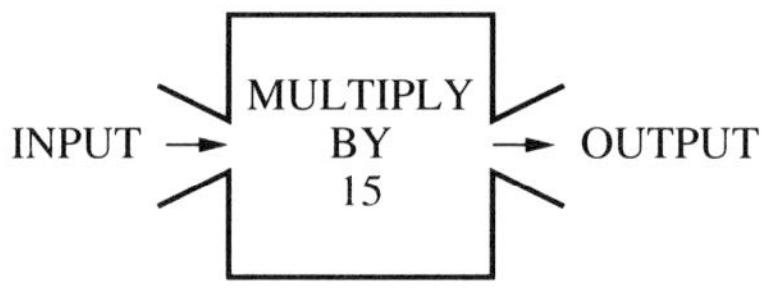

3

INPUT	OUTPUT
3	
9	
12	
18	
24	
45	
	2
	5
	7
	9
	10
	36

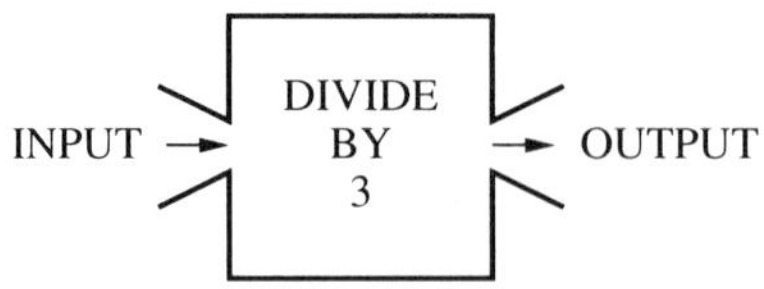

This function machine multiplies any number which is fed into its input by 2 and then adds 3 to the result.

For example, if the INPUT is 5, the OUTPUT is $2 \times 5 + 3$, or 13.

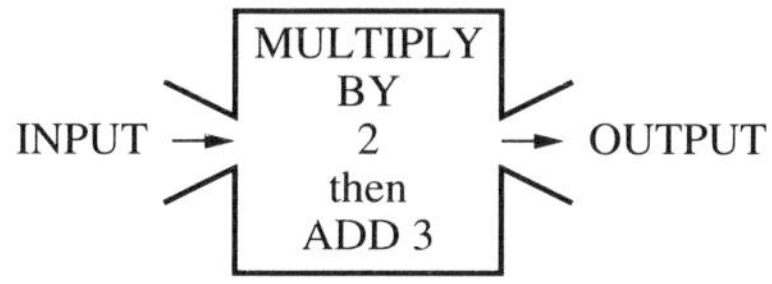

Example 3

Use the above function machine and write down:

a the OUTPUT number when the INPUT number is (i) 4, (ii) 7, (iii) 15 and (iv) 20

b the INPUT number when the OUTPUT number is (i) 13, (ii) 23, (iii) 35 and (iv) 47.

a (i) $4 \times 2 + 3 = 11$ (ii) $7 \times 2 + 3 = 17$ (iii) $15 \times 2 + 3 = 33$ (iv) $20 \times 2 + 3 = 43$

b (i) $(13 - 3) \div 2 = 5$ (ii) $(23 - 3) \div 2 = 10$ (iii) $(35 - 3) \div 2 = 16$ (iv) $(47 - 3) \div 2 = 22$

Note that to find the INPUT number we first SUBTRACT 3 (the inverse of ADD 3) and then DIVIDE by 2 (the inverse of MULTIPLY by 2).

This function machine subtracts 6 from any number which is fed into its input and then divides the result by 2.

For example, if the INPUT is 12, the OUTPUT is $(12 - 6) \div 2$, or 3.

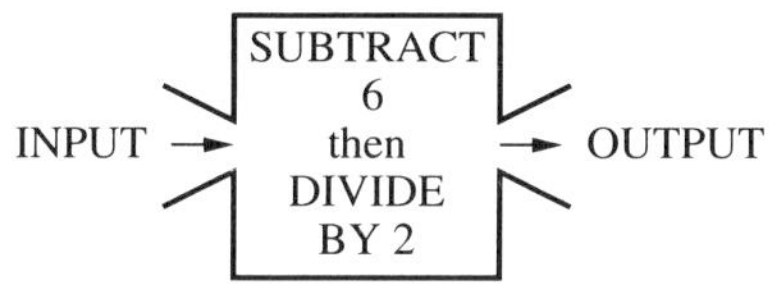

Example 4

Use the above function machine and write down:

a the OUTPUT number when the INPUT number is (i) 8, (ii) 14, (iii) 24 and (iv) 30

b the INPUT number when the OUTPUT number is (i) 2, (ii) 8, (iii) 11 and (iv) 22.

a (i) $(8 - 6) \div 2 = 1$ (ii) $(14 - 6) \div 2 = 4$ (iii) $(24 - 6) \div 2 = 9$ (iv) $(30 - 6) \div 2 = 12$

b (i) $2 \times 2 + 6 = 10$ (ii) $8 \times 2 + 6 = 22$ (iii) $11 \times 2 + 6 = 28$ (iv) $22 \times 2 + 6 = 50$

Note that to find the INPUT number we first MULTIPLY by 2 (the inverse of DIVIDE by 2) and then ADD 6 (the inverse of SUBTRACT 6).

Exercise 9.9

For each question copy the table and then use the function machine to complete it.

1

INPUT	OUTPUT
1	
3	
4	
6	
	8
	14
	18
	20

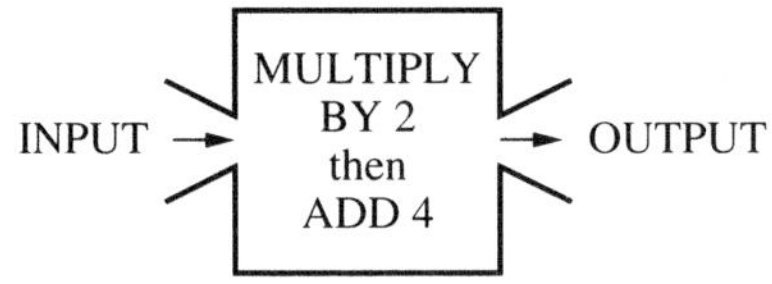

2

INPUT	OUTPUT
2	
4	
5	
8	
	3
	9
	11
	17

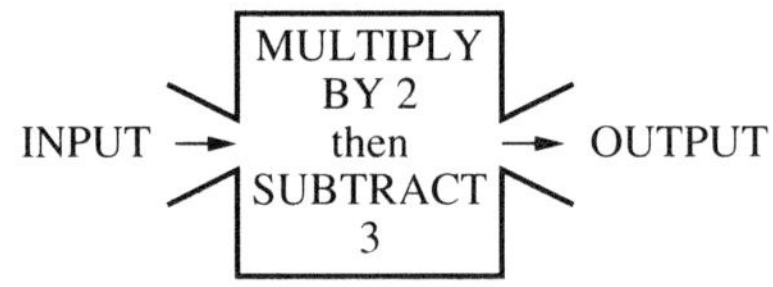

3

INPUT	OUTPUT
1	
4	
5	
8	
	8
	11
	20
	23

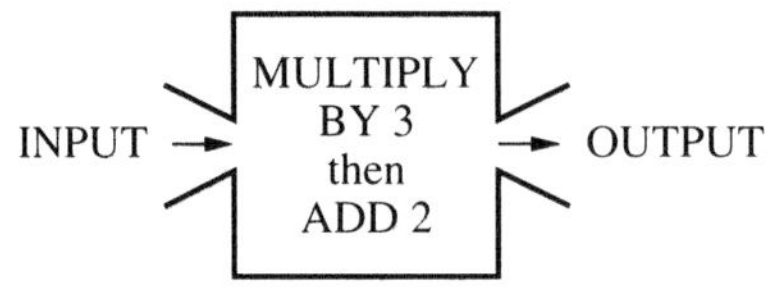

Example 5

The tables show the INPUT and OUTPUT numbers for two function machines.
For each table state what function the machine executes.
Draw the machine and give the OUTPUT number when 6 is the INPUT number.

a

INPUT	OUTPUT
1	5
2	10
3	15
4	20
5	25
6	

b

INPUT	OUTPUT
1	7
2	12
3	17
4	22
5	27
6	

a The machine MULTIPLIES by 5.
Check:
$1 \times 5 = 5$
$2 \times 5 = 10$
$3 \times 5 = 15$
$4 \times 5 = 20$
$5 \times 5 = 25$

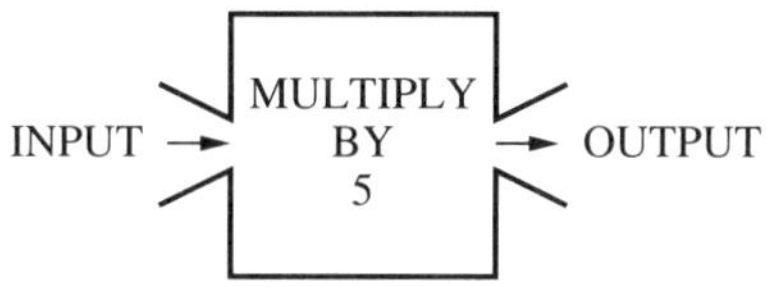

If the INPUT number is 6, the OUTPUT number is 6×5, or 30.

b The machine MULTIPLIES by 5 and then ADDS 2.
Check:
$1 \times 5 + 2 = 7$
$2 \times 5 + 2 = 12$
$3 \times 5 + 2 = 17$
$4 \times 5 + 2 = 22$
$5 \times 5 + 2 = 27$

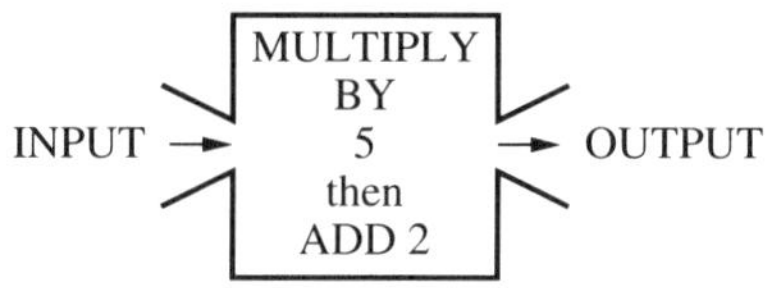

If the INPUT number is 6, the OUTPUT number is $6 \times 5 + 2$, or 32.

Exercise 9.10

For each table write down the machine's function.
Draw the machine and fill in the missing values.

1 a

INPUT	OUTPUT
1	2
2	4
3	6
4	8
5	10
6	
7	
8	

b

INPUT	OUTPUT
1	3
2	5
3	7
4	9
5	11
6	
7	
8	

c

INPUT	OUTPUT
1	4
2	6
3	8
4	10
5	12
6	
8	
10	

d

INPUT	OUTPUT
1	1
2	3
3	5
4	7
5	9
6	
8	
10	

2

INPUT (number of weights added)	OUTPUT (length of spring)
0	5 cm
1	7 cm
2	9 cm
3	11 cm
4	
5	
6	

3

INPUT (number of hours after low tide)	OUTPUT (depth of water in harbour)
0	10 m
1	12 m
2	14 m
3	16 m
4	
5	
6	

4

INPUT (number of minutes after heater is turned on)	OUTPUT (temperature of water)
0	10°C
1	15°C
2	20°C
3	25°C
4	
5	
6	

9.4 Expressing simple functions symbolically

Example 1

This function machine multiplies each INPUT number by 3.
Copy and complete the table.
Describe the function of the machine symbolically.

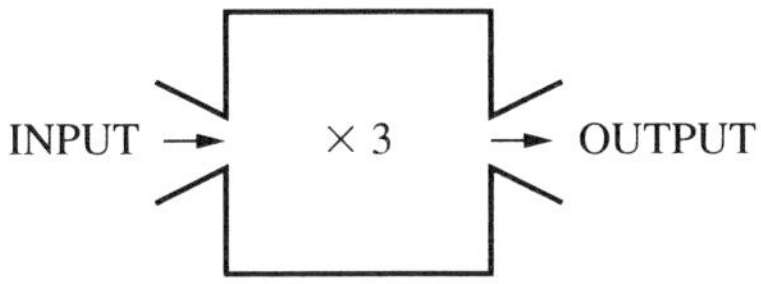

INPUT	OUTPUT
5	
8	
20	
n	

If the INPUT is 5, the OUTPUT is 15 (i.e. 5×3)
If the INPUT is 8, the OUTPUT is 24 (i.e. 8×3)
If the INPUT is 20, the OUTPUT is 60 (i.e. 20×3)
If the INPUT is n, the OUTPUT is $3n$ (i.e. $n \times 3$)

INPUT	OUTPUT
5	15
8	24
20	60
n	$3n$

The function of the machine can be expressed symbolically as $n \rightarrow 3n$.

Exercise 9.11

For each question copy and complete the table.
Describe the function of the machine symbolically as $n \rightarrow \ldots$.

1

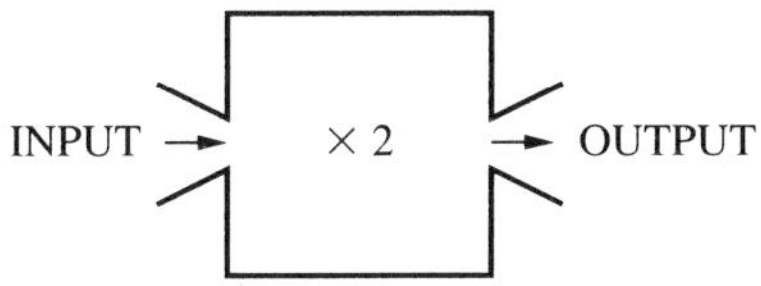

INPUT	OUTPUT
5	
6	
7	
10	
n	

2

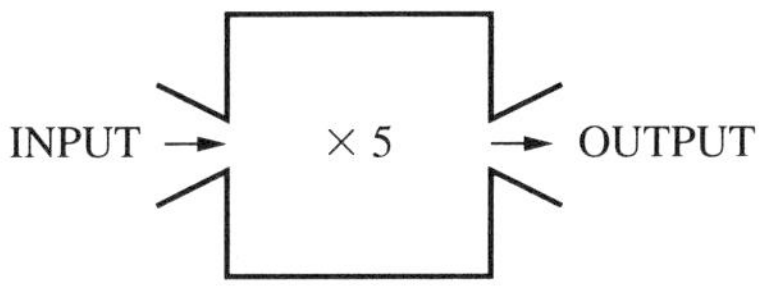

INPUT	OUTPUT
6	
7	
10	
20	
n	

3

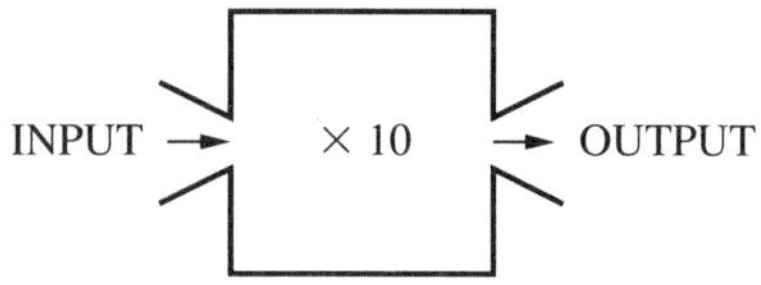

INPUT	OUTPUT
2	
3	
4	
6	
15	
25	
n	

Example 2

If one cake costs 15 p, use the function machine $n \rightarrow 15n$ to find the cost of:

3 cakes	4 cakes	5 cakes	8 cakes
10 cakes	12 cakes	a cakes	b cakes
c cakes	d cakes.		

Copy the table and write your answers in the output column.

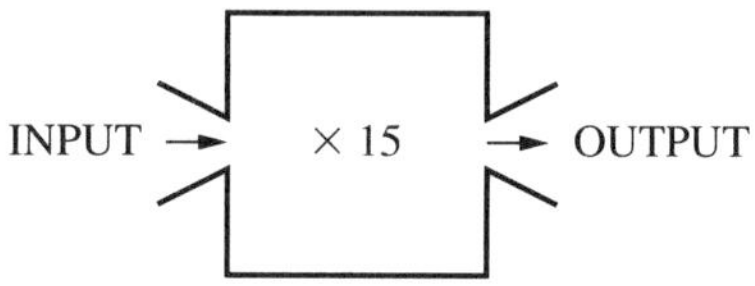

INPUT	OUTPUT
(i)	
(ii)	
(iii)	
(iv)	
(v)	
(vi)	
(vii)	
(viii)	
(ix)	
(x)	

(i) The cost of 3 cakes is 45 p (3 × 15 p)
(ii) The cost of 4 cakes is 60 p (4 × 15 p)
(iii) The cost of 5 cakes is 75 p (5 × 15 p)
(iv) The cost of 8 cakes is 120 p (or £1.20) (8 × 15 p)
(v) The cost of 10 cakes is 150 p (or £1.50) (10 × 15 p)
(vi) The cost of 12 cakes is 180 p (or £1.80) (12 × 15 p)
(vii) The cost of a cakes is $15a$ p ($a \times 15$ p)
(viii) The cost of b cakes is $15b$ p ($b \times 15$ p)
(ix) The cost of c cakes is $15c$ p ($c \times 15$ p)
(x) The cost of d cakes is $15d$ p ($d \times 15$ p)

	INPUT	OUTPUT
(i)	3	45 p
(ii)	4	60 p
(iii)	5	75 p
(iv)	8	£1.20
(v)	10	£1.50
(vi)	12	£1.80
(vii)	a	$15a$ p
(viii)	b	$15b$ p
(ix)	c	$15c$ p
(x)	d	$15d$ p

Exercise 9.12

1 If one pencil costs 12 p, use the function machine $n \rightarrow 12n$ to find the cost of:

(i) 2 pencils (ii) 3 pencils (iii) 4 pencils (iv) 5 pencils
(v) 8 pencils (vi) 10 pencils (vii) *a* pencils (viii) *b* pencils
(ix) *c* pencils (x) *d* pencils

Copy the table and write your answers in the output column.

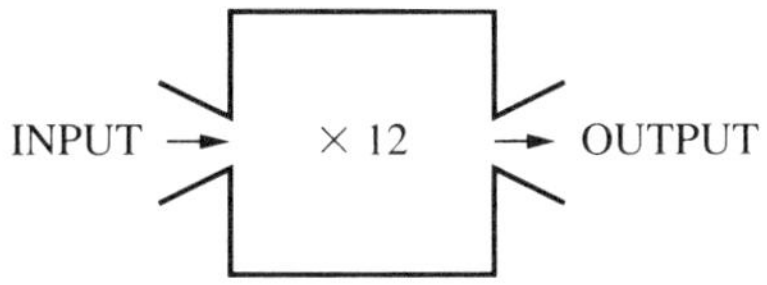

	INPUT	OUTPUT
(i)	2	
(ii)	3	
(iii)	4	
(iv)	5	
(v)	8	
(vi)	10	
(vii)	*a*	
(viii)	*b*	
(ix)	*c*	
(x)	*d*	

2 If one pen costs 20 p, use the function machine $n \to 20n$ to find the cost of:
(i) 2 pens (ii) 3 pens (iii) 4 pens (iv) 6 pens
(v) 10 pens (vi) 12 pens (vii) *a* pens (viii) *b* pens
(ix) *c* pens (x) *d* pens

Copy the table and write your answers in the output column.

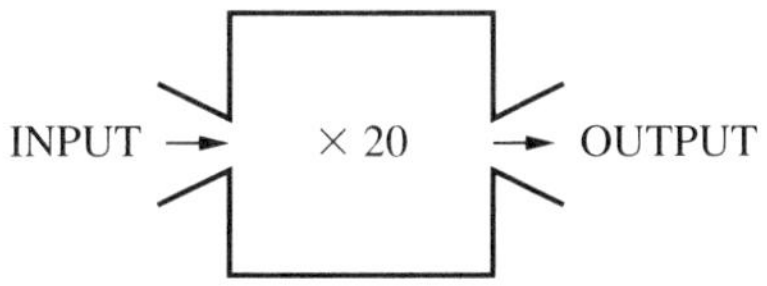

	INPUT	OUTPUT
(i)	2	
(ii)	3	
(iii)	4	
(iv)	6	
(v)	10	
(vi)	12	
(vii)	*a*	
(viii)	*b*	
(ix)	*c*	
(x)	*d*	

3 If one loaf of bread costs 50 p, use the function machine $n \to 50n$ to find the cost of:
(i) 3 loaves (ii) 4 loaves (iii) 6 loaves (iv) 8 loaves
(v) 9 loaves (vi) 10 loaves (vii) *p* loaves (viii) *q* loaves
(ix) *r* loaves (x) *s* loaves

Example 3

This function machine multiplies each INPUT number by 3 and then adds 2.
Copy and complete the table and describe the function of the machine symbolically.

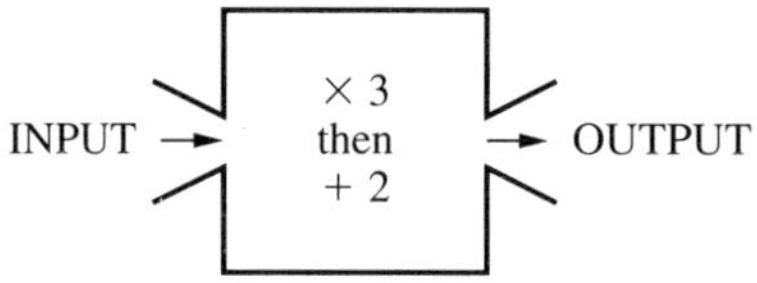

INPUT	OUTPUT
5	
8	
20	
n	

If the INPUT is 5 the OUTPUT is 17 $(5 \times 3 + 2)$
If the INPUT is 8 the OUTPUT is 26 $(8 \times 3 + 2)$
If the INPUT is 20 the OUTPUT is 62 $(20 \times 3 + 2)$
If the INPUT is n the OUTPUT is $3n + 2$ $(n \times 3 + 2)$

INPUT	OUTPUT
5	17
8	26
20	62
n	$3n + 2$

The function of the machine can be expressed symbolically as $n \to 3n + 2$

Exercise 9.13

For each question copy and complete the table and describe the machine's function symbolically.

1

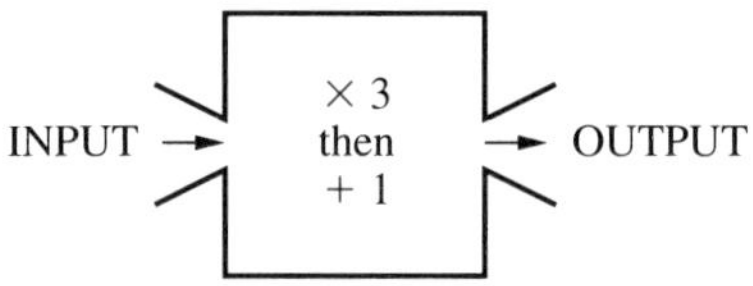

INPUT	OUTPUT
5	
8	
10	
15	
20	
30	
n	

2

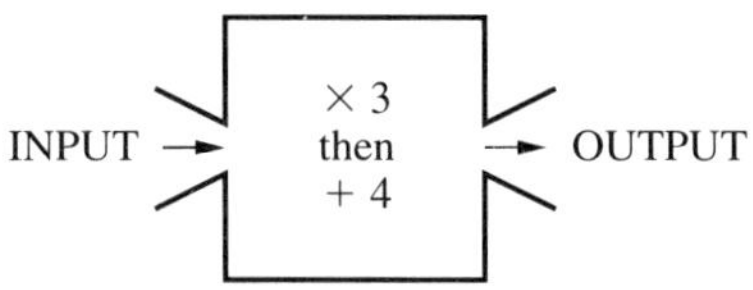

INPUT	OUTPUT
5	
8	
9	
12	
25	
40	
n	

Example 4

A large cylindrical drum has water inside it to a depth of 25 cm. One day it rains and the depth increases by 4 cm every hour. Use the function machine $n \rightarrow 4n + 25$ to find the depth of water after:

(i) 2 hours (ii) 3 hours (iii) 4 hours (iv) 5 hours
(v) 8 hours (vi) a hours (vii) b hours

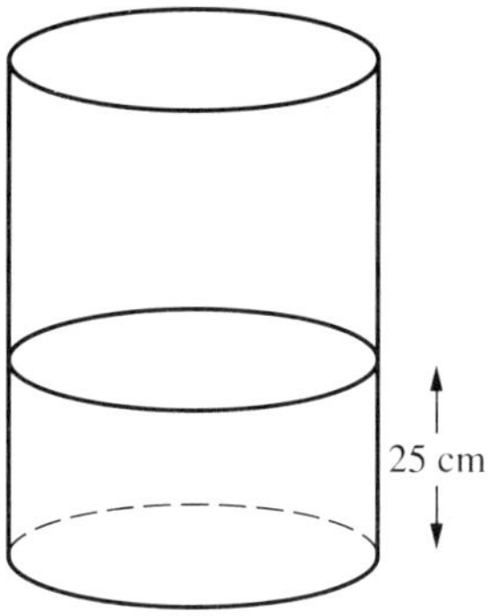

Copy the table and write your answers in the output column.

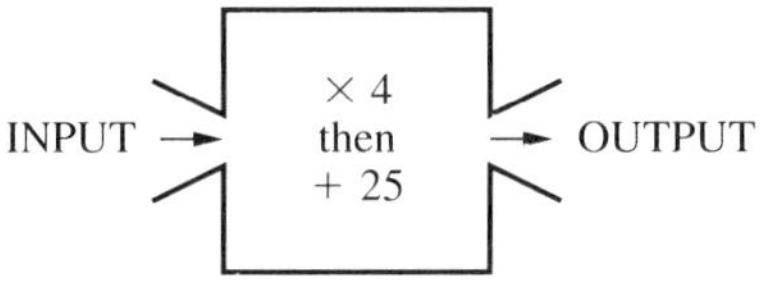

	INPUT	OUTPUT
(i)	2	
(ii)	3	
(iii)	4	
(iv)	5	
(v)	8	
(vi)	a	
(vii)	b	

(i) The depth after 2 hours is 33 cm $(2 \times 4 + 25)$
(ii) The depth after 3 hours is 37 cm $(3 \times 4 + 25)$
(iii) The depth after 4 hours is 41 cm $(4 \times 4 + 25)$
(iv) The depth after 5 hours is 45 cm $(5 \times 4 + 25)$
(v) The depth after 8 hours is 57 cm $(8 \times 4 + 25)$
(vi) The depth after a hours is $(4a + 25)$ cm $(a \times 4 + 25)$
(vii) The depth after b hours is $(4b + 25)$ cm $(b \times 4 + 25)$

	INPUT	OUTPUT
(i)	2	33 cm
(ii)	3	37 cm
(iii)	4	41 cm
(iv)	5	45 cm
(v)	8	57 cm
(vi)	a	$(4a + 25)$ cm
(vii)	b	$(4b + 25)$ cm

Exercise 9.14

1 A club starts with 20 members but 3 new members join every month.
Use the function machine $n \rightarrow 3n + 20$ to find the number of members after:
(i) 2 months (ii) 3 months (iii) 4 months (iv) 5 months
(v) 8 months (vi) a months (vii) b months

Copy the table and write your answers in the output column.

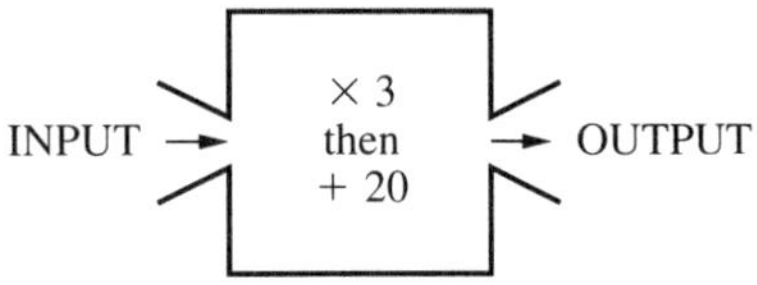

	INPUT	OUTPUT
(i)	2	
(ii)	3	
(iii)	4	
(iv)	5	
(v)	8	
(vi)	a	
(vii)	b	

2 Peter buys a bicycle by paying a deposit of £25 and then paying £6 every month until the purchase price is paid.
Use the function machine $n \rightarrow 6n + 25$ to find how much he has paid after:
(i) 2 months (ii) 3 months (iii) 5 months (iv) 8 months
(v) 10 months (vi) a months (vii) b months

Copy the table and write your answers in the output column.

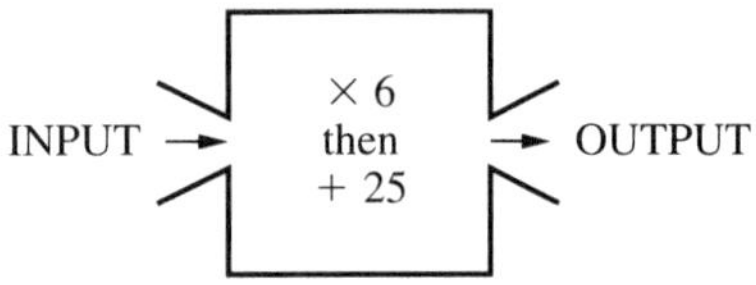

	INPUT	OUTPUT
(i)	2	
(ii)	3	
(iii)	5	
(iv)	8	
(v)	10	
(vi)	a	
(vii)	b	

3 The fire escape at the side of a building starts from a concrete base of height 100 cm. Each rung is separated from the next one by a gap of 25 cm.
Use the function machine $n \rightarrow 25n + 100$ to find the height above the ground of:
(i) the 2nd rung (ii) the 4th rung (iii) the 5th rung
(iv) the 10th rung (v) the pth rung (vi) the qth rung

Copy the table and write your answers in the output column.

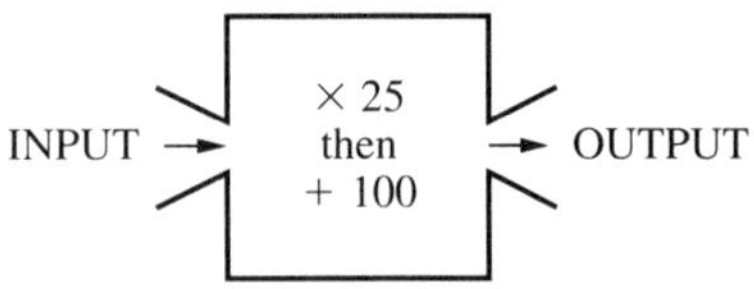

	INPUT	OUTPUT
(i)	2	
(ii)	4	
(iii)	5	
(iv)	10	
(v)	p	
(vi)	q	

Example 5

This function machine adds 4 to each INPUT number and then multiplies the result by 3. Copy and complete the table and describe the function of the machine symbolically as $n \to \dots$.

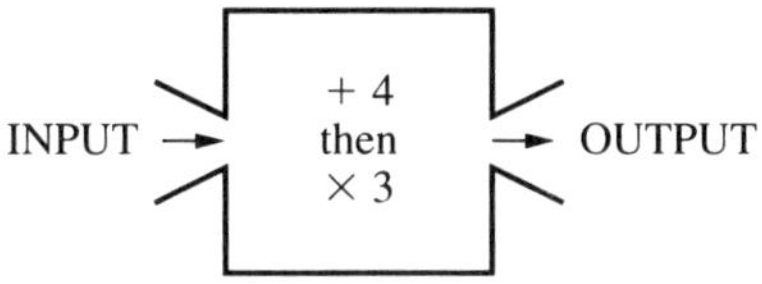

INPUT	OUTPUT
5	
8	
20	
n	

If the INPUT is 5, the OUTPUT is 27 $(3\times[5+4])$
If the INPUT is 8, the OUTPUT is 36 $(3\times[8+4])$
If the INPUT is 20, the OUTPUT is 72 $(3\times[20+4])$
If the INPUT is n, the OUTPUT is $3(n+4)$

INPUT	OUTPUT
5	27
8	36
20	72
n	$3(n+4)$

The function of the machine can be expressed symbolically as $n \to 3(n+4)$

Exercise 9.15

For Questions 1–4 copy and complete the table and describe the function of the machine symbolically as $n \to \dots$

1

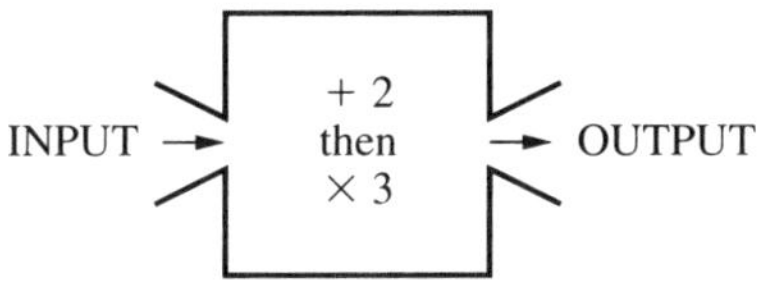

INPUT	OUTPUT
5	
6	
7	
8	
10	
12	
n	

2

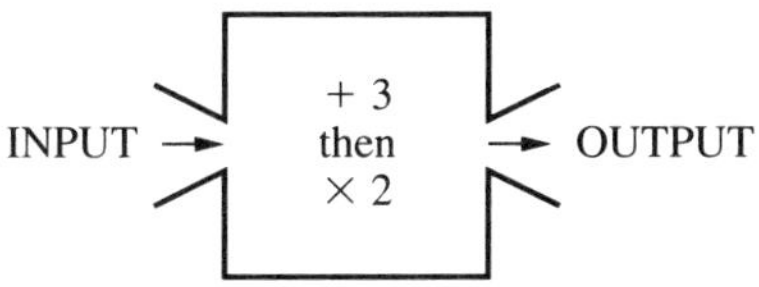

INPUT	OUTPUT
5	
6	
8	
9	
10	
12	
n	

3 The temperature in degrees Celsius (°C) can be approximately converted to degrees Fahrenheit (°F) by adding 15 and then doubling the result.
Use the function machine $n \rightarrow 2(n + 15)$ to find, an approximation in °F for:
(i) 10°C (ii) 15°C (iii) 20°C (iv) 30°C (v) 5°C
(vi) p°C (vii) q°C (viii) r°C (ix) s°C

Copy the table and write your answers in the output column.

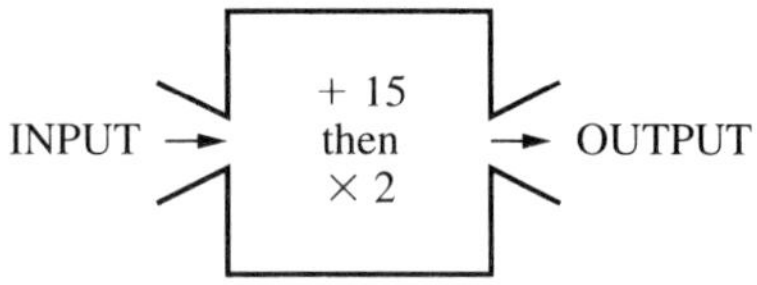

	INPUT	OUTPUT
(i)	10°C	
(ii)	15°C	
(iii)	20°C	
(iv)	30°C	
(v)	5°C	
(vi)	p°C	
(vii)	q°C	
(viii)	r°C	
(ix)	s°C	

4 The perimeter of a rectangle of length 5 cm can be found by adding 5 cm to the width and then doubling the result.
Use the function machine $n \rightarrow 2(n + 5)$ to find the perimeter of each of the following:
(i) width = 2 cm (ii) width = 3 cm (iii) width = 4 cm
(iv) width = 2.5 cm (v) width = 1.5 cm (vi) width = 3.5 cm
(vii) width = p cm (viii) width = q cm

Copy the table and write your answers in the output column.

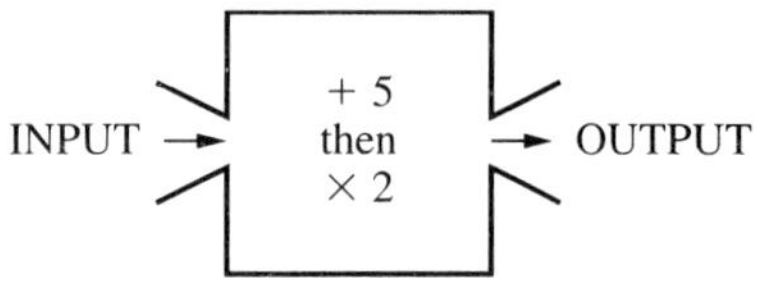

	INPUT	OUTPUT
(i)	2 cm	
(ii)	3 cm	
(iii)	4 cm	
(iv)	2.5 cm	
(v)	1.5 cm	
(vi)	3.5 cm	
(vii)	p cm	
(viii)	q cm	
(ix)	r cm	
(x)	s cm	

10 Formulae, equations and graphs

10.1 Use of simple formulae and equations expressed in words

Example 1

If the price of a packet of crisps is 22 p, find the cost of:
a 3 packets **b** 7 packets

a 3 packets cost 22 p × 3 = 66 p
b 7 packets cost 22 p × 7 = 154 p = £1.54

Exercise 10.1

1 If the price of a can of orangeade is 24 p, find the cost of:
a 3 cans **b** 5 cans **c** 8 cans

2 If the price of one ice cream cone is 40 p, find the cost of:
a 3 cones **b** 4 cones **c** 7 cones

3 If the price of one pen is 35 p, find the cost of:
a 5 pens **b** 6 pens **c** 8 pens

4 If the price of one ruler is 60 p, find the cost of:
a 5 rulers **b** 7 rulers **c** 8 rulers

5 If lemons cost 25 p each, find the cost of:
a 4 lemons **b** 6 lemons **c** 10 lemons

Example 2

Camera films cost £1.50 to develop into negatives (regardless of length). The cost of producing one print is 10 p.
Find the cost of developing and printing a film with **a** 12 prints, **b** 24 prints, **c** 36 prints.

a Cost = £1.50 + (10 p × 12) = £1.50 + £1.20 = £2.70
b Cost = £1.50 + (10 p × 24) = £1.50 + £2.40 = £3.90
c Cost = £1.50 + (10 p × 36) = £1.50 + £3.60 = £5.10

Exercise 10.2

1 A kitten was 15 cm long when she was born. She grew by 2 cm every month until she was 1 year old.
Find her length after **a** 1 month, **b** 2 months, **c** 3 months, **d** 5 months, **e** 10 months.

2 A spring is 20 cm long and it extends by 6 cm for each weight hung on its end.
Find the length when **a** 1 weight, **b** 2 weights, **c** 4 weights, **d** 5 weights and **e** 7 weights, are hung on its end.

3 A puppy weighed 1 lb when he was born. His weight increased by 2 lb every month until he was a year old.
Find his weight after **a** 1 month, **b** 2 months, **c** 6 months, **d** 8 months, **e** 9 months, **f** 1 year.

4 A new housing estate was built and 100 people moved in as soon as possible. After that 50 more people moved in each month for two years.
Find the number of people who had moved in after: **a** 1 month, **b** 2 months, **c** 4 months, **d** 10 months, **e** 1 year, **f** 1 year 6 months.

5 One Saturday 540 people arrived at a football ground at 1.30 p.m. and found the ground empty. 400 more people arrived during each 15-minute interval from 1.30 p.m until the game started at 3.00 p.m.
Find the number of people who were at the ground by **a** 1.45 p.m. **b** 2.00 p.m. **c** 2.15 p.m. **d** 2.30 p.m. **e** 3.00 p.m.

Example 3

This machine multiplies any input number by 5 and then adds 2.
Copy and complete the table.

	INPUT	OUTPUT
	2	12
a	3	?
b	7	?
c	?	22
d	?	42

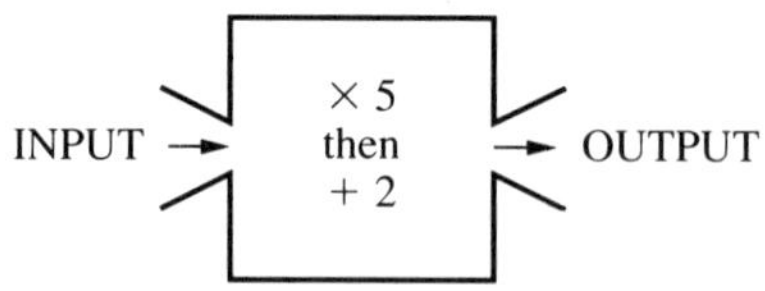

a $3 \times 5 + 2 = 17$
b $7 \times 5 + 2 = 37$
c $4 \times 5 + 2 = 22$
(or $(22 - 2) \div 5 = 4$)
d $8 \times 5 + 2 = 42$
(or $(42 - 2) \div 5 = 8$)

	INPUT	OUTPUT
	2	23
a	3	17
b	7	37
c	4	22
d	8	42

Exercise 10.3

Copy and complete the table for each machine.

1

	INPUT	OUTPUT
	1	8
a	2	
b	3	
c	5	
d	7	
e		23
f		33
g		43

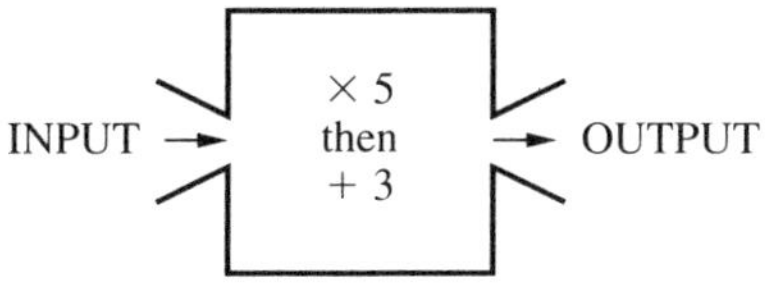

This machine multiplies any number by 5 and then adds 3.

2

	INPUT	OUTPUT
	1	2
a	2	
b	4	
c	5	
d	8	
e		12
f		27
g		32

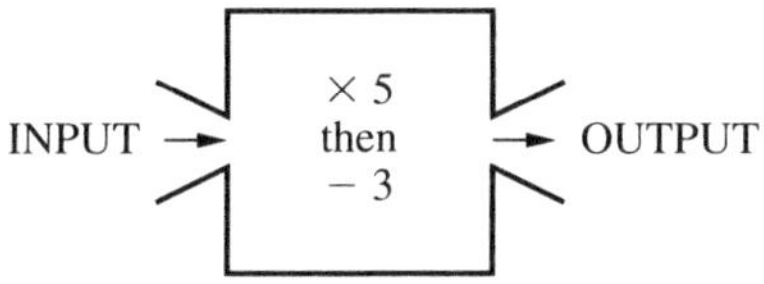

This machine multiplies any number by 5 and then subtracts 3.

3

	INPUT	OUTPUT
	4	13
a	2	
b	3	
c	5	
d	6	
e		7
f		19
g		21

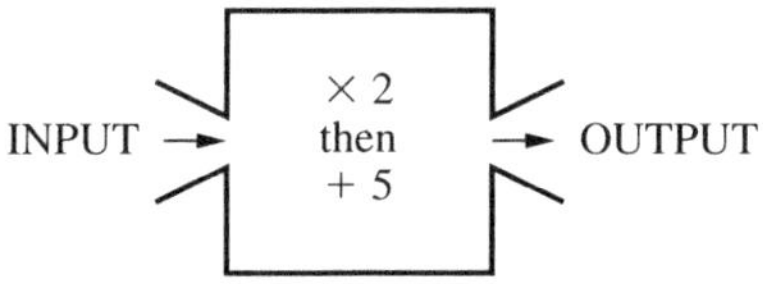

This machine multiplies any number by 2 and then adds 5.

4

	INPUT	OUTPUT
	4	4
a	7	
b	8	
c	2	
d	10	
e		2
f		6
g		8

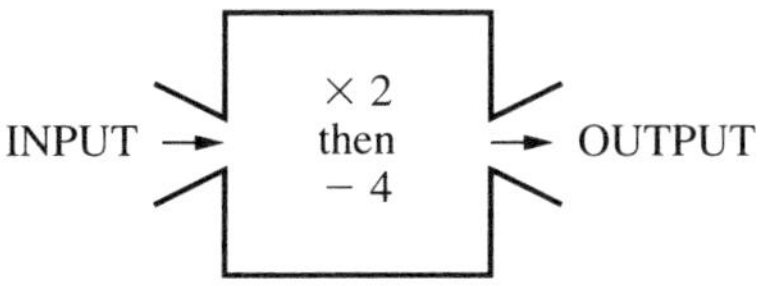

This machine multiplies any number by 2 and then subtracts 4.

Example 4

If I double a number and then add 1, the result is 49.
What is the number?

If double the number +1 is 49, then double the number is 48 (that is, 49 − 1), so the number itself is 24 (that is, 48 ÷ 2).

Exercise 10.4

For each of the following find the number.

1 Double the number and then add 1 gives 31.

2 Double the number and then add 3 gives 35.

3 Double the number and then subtract 1 gives 17.

4 Double the number and then subtract 2 gives 28.

5 Triple the number and then add 1 gives 19.

6 Triple the number and then add 2 gives 26.

7 Triple the number and then subtract 1 gives 11.

8 Triple the number and then subtract 3 gives 21.

10.2 Understanding and using simple formulae and equations

Example 1

The perimeter of a triangle with side lengths a, b and c is given by the formula $p = a + b + c$.

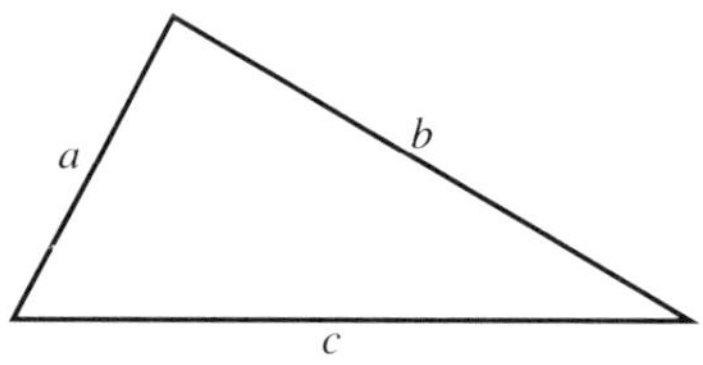

Find the perimeter of a triangle with $a = 3\,\text{cm}$, $b = 5\,\text{cm}$ and $c = 7\,\text{cm}$.

The perimeter,
$p = (3 + 5 + 7)\,\text{cm} = 15\,\text{cm}$

Exercise 10.5

Find the perimeter of the triangles with these sides.

	a	b	c
1	5 cm	9 cm	12 cm
2	8 cm	11 cm	16 cm
3	6 cm	12 cm	14 cm
4	15 mm	16 mm	25 mm
5	1.5 m	2.5 m	3 m

Example 2

The perimeter of an equilateral triangle is given by the formula $p = a + a + a$ or $p = 3a$

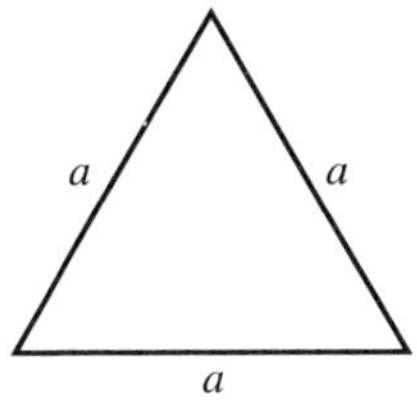

Equilateral triangle (all three sides are the same length)

Find the perimeter of an equilateral triangle whose side length a is 5 cm.

The perimeter, $p = (3 \times 5)$ cm = 15 cm

Exercise 10.6

Find the perimeter of these equilateral triangles.

1 Side length $a = 4$ cm

2 Side length $a = 8$ cm

3 Side length $a = 9$ cm

4 Side length $a = 21$ mm

5 Side length $a = 2.5$ m

6 Side length $a = 1.5$ m

Example 3

The perimeter of an isosceles triangle is given by the formula $p = a + a + b$ or $p = 2a + b$

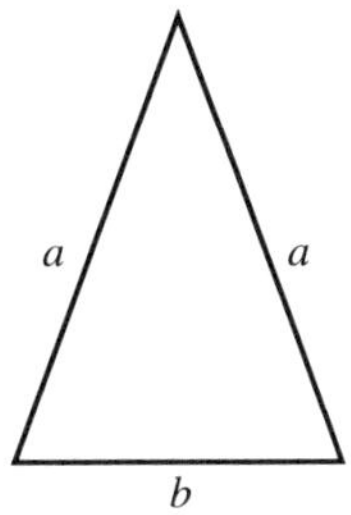

Isosceles triangle (two sides are equal in length)

Find the perimeter of an isosceles triangle with $a = 3$ cm and $b = 5$ cm

The perimeter,
$p = (2 \times 3 + 5)$ cm = $(6 + 5)$ cm = 11 cm

Exercise 10.7

Find the perimeter of these isosceles triangles.

	a	b
1	6 cm	8 cm
2	8 cm	9 cm
3	11 mm	15 mm
4	16 mm	25 mm
5	2.5 mm	3 m

Example 4

The perimeter of a square is given by the formula $p = l + l + l + l$ or $p = 4l$

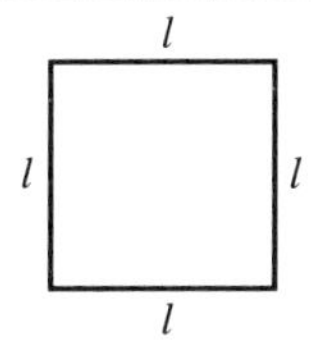

Find the perimeter of a square whose side length is 7 cm.

The perimeter, $p = (4 \times 7)$ cm = 28 cm

(l is the side length of the square)

Exercise 10.8

Find the perimeter of these squares.

1 Side length $l = 5$ cm **2** Side length $l = 6$ cm

3 Side length $l = 8$ cm **4** Side length $l = 20$ mm

5 Side length $l = 3.5$ m **6** Side length $l = 2.5$ m

Example 5

The perimeter of a rectangle is given by the formula $p = l + w + l + w$ or $p = 2l + 2w$ or $p = 2(l + w)$

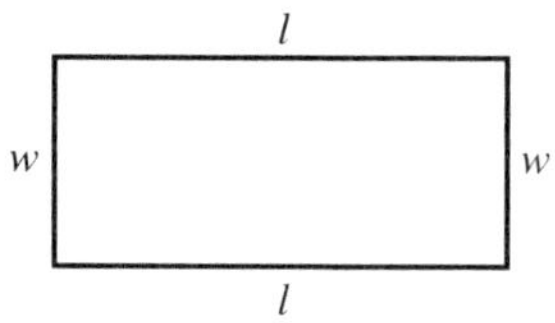

(l is the length of the rectangle and w is the width)

Find the perimeter of a rectangle with $l = 5$ cm and $w = 3$ cm

The perimeter, $p = 2(5 + 3)$ cm $= (2 \times 8)$ cm $= 16$ cm

Exercise 10.9

Find the perimeter of these rectangles.

1 Length $l = 9$ cm, width $w = 4$ cm

2 Length $l = 8$ cm, width $w = 5$ cm

3 Length $l = 12$ cm, width $w = 7$ cm

4 Length $l = 15$ cm, width $w = 8$ cm

5 Length $l = 35$ mm, width $w = 15$ mm

6 Length $l = 30$ mm, width $w = 16$ mm

Example 6

The area of a rectangle is given by the formula $A = lw$

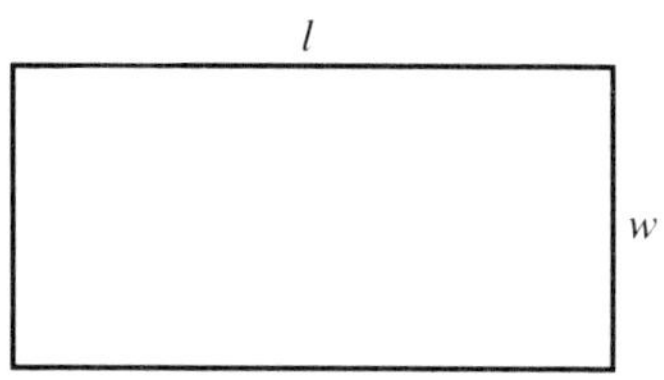

Find the area of a rectangle with $l = 5$ cm and $w = 4$ cm.

The area, $A = (5 \times 4)\,\text{cm}^2 = 20\,\text{cm}^2$

Exercise 10.10

Find the area of these rectangles:

1 Length $l = 9\,\text{cm}$, width $w = 3\,\text{cm}$

2 Length $l = 8\,\text{cm}$, width $w = 6\,\text{cm}$

3 Length $l = 12\,\text{cm}$, width $w = 8\,\text{cm}$

4 Length $l = 15\,\text{cm}$, width $w = 7\,\text{cm}$

5 Length $l = 3.6\,\text{m}$, width $w = 2.5\,\text{m}$

6 Length $l = 2.4\,\text{m}$, width $w = 1.25\,\text{m}$

Example 7

The area of a parallelogram is given by the formula $A = bh$

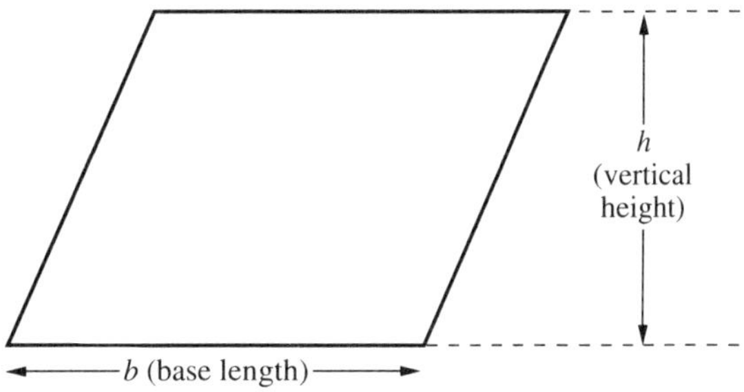

Find the area of a parallelogram for which $b = 7\,\text{cm}$ and $h = 4\,\text{cm}$

The area, $A = (7 \times 4)\,\text{cm}^2 = 28\,\text{cm}^2$

Exercise 10.11

Find the area of these parallelograms.

1 Base $b = 8\,\text{cm}$, height $h = 3\,\text{cm}$

2 Base $b = 9\,\text{cm}$, height $h = 7\,\text{cm}$

3 Base $b = 3.5\,\text{m}$, height $h = 2\,\text{m}$

4 Base $b = 8.5\,\text{m}$, height $h = 6\,\text{m}$

5 Base $b = 35\,\text{mm}$, height $h = 24\,\text{mm}$

6 Base $b = 45\,\text{mm}$, height $h = 16\,\text{mm}$

Example 8

The area of a square of side length l is given by the formula $A = l \times l$ or $A = l^2$

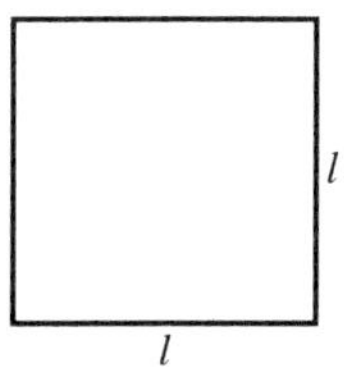

Find the area of a square of side length 9 cm.

The area, $A = (9 \times 9)\,\text{cm}^2 = 81\,\text{cm}^2$

Exercise 10.12

Find the area of these squares.

1 Side length $l = 7\,\text{cm}$ **2** Side length $l = 6\,\text{cm}$

3 Side length $l = 4\,\text{cm}$ **4** Side length $l = 1.5\,\text{m}$

5 Side length $l = 2.5\,\text{m}$ **6** Side length $l = 1.2\,\text{m}$

Example 9

The area of a triangle is given by the formula $A = \frac{1}{2}\,bh$

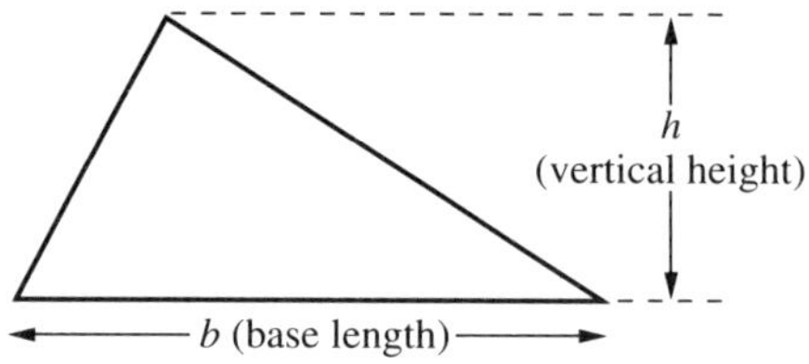

Find the area of a triangle with $b = 6\,\text{cm}$ and $h = 4\,\text{cm}$

The area, $A = \left(\frac{1}{2} \times 6 \times 4\right)\text{cm}^2 = (0.5 \times 24)\,\text{cm}^2 = 12\,\text{cm}^2$

Exercise 10.13

Find the area of these triangles.

1 Base length $b = 8\,\text{cm}$, vertical height $h = 5\,\text{cm}$

2 Base length $b = 10\,\text{cm}$, vertical height $h = 4\,\text{cm}$

3 Base length $b = 16\,\text{cm}$, vertical height $h = 8\,\text{cm}$

4 Base length $b = 20\,\text{mm}$, vertical height $h = 25\,\text{mm}$

5 Base length $b = 12\,\text{mm}$, vertical height $h = 16\,\text{mm}$

6 Base length $b = 1.6\,\text{m}$, vertical height $h = 2.5\,\text{m}$

Example 10

The area of a trapezium is given by the formula $A = \frac{1}{2}h(a + b)$

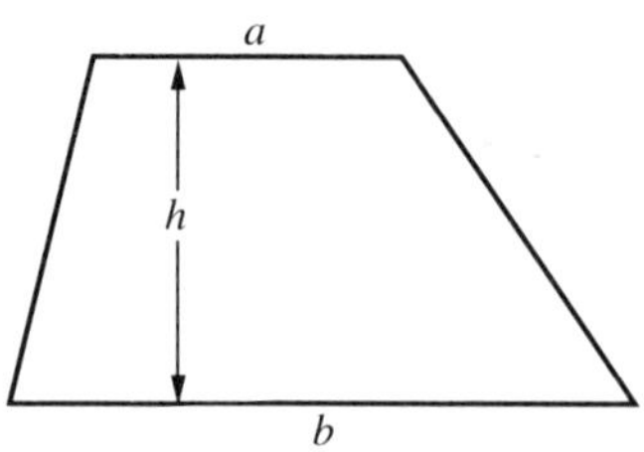

(a and b are the lengths of the parallel sides of the trapezium.
h is the vertical height.)

Find the area of a trapezium with $h = 8\,\text{cm}$, $a = 9\,\text{cm}$ and $b = 3\,\text{cm}$.

The area, $A = \left(\frac{1}{2} \times 8 \times (9 + 3)\right)\text{cm}^2 = (0.5 \times 8 \times 12)\,\text{cm}^2 = 48\,\text{cm}^2$

Exercise 10.14

Find the area of each of these trapeziums.

	h	a	b
1	6 cm	7 cm	2 cm
2	10 cm	6 cm	4 cm
3	16 cm	20 cm	15 cm
4	20 mm	25 mm	15 mm
5	24 mm	50 mm	30 mm
6	4 m	4.5 m	2.5 m
7	6 m	3.5 m	1.5 m
8	5 m	7.5 m	2.5 m

Example 11

The volume of a cuboid is given by the formula $V = lwh$

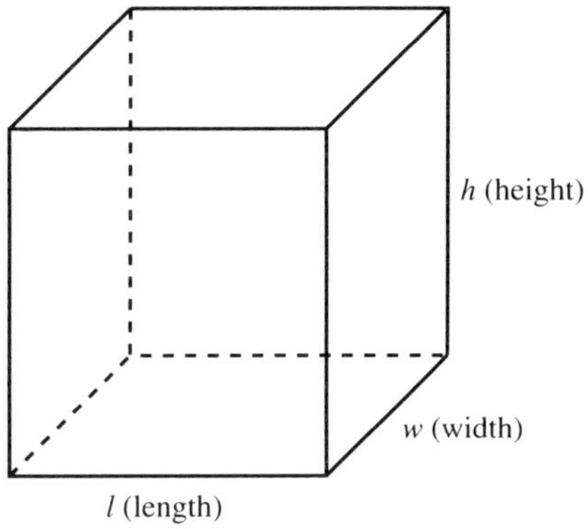

Find the volume of a cuboid whose dimensions are 12 cm, 10 cm and 5 cm.

The volume $V = (12 \times 10 \times 5)\,\text{cm}^3 = 600\,\text{cm}^3$

Exercise 10.15

Find the volume of each of these cuboids.

	l	w	h
1	15 cm	8 cm	6 cm
2	24 cm	7 cm	5 cm
3	30 cm	16 cm	15 cm
4	36 mm	20 mm	15 mm
5	50 mm	36 mm	32 mm
6	7.5 m	4.5 m	4 m
7	3.5 m	3 m	2 m
8	5.5 m	7.5 m	8 m

10.3 Solving simple equations

Example 1

Solve the equation $x + 10 = 15$

$$x + 10 = 15$$

Subtract 10 from both sides $$x + 10 - 10 = 15 - 10$$
$$x = 5$$

The solution is $x = 5$

Example 2

Solve the equation $7 - y = 3$

$$7 - y = 3$$

Add y to both sides $$7 - y + y = 3 + y$$
$$7 = 3 + y$$

Subtract 3 from both sides $$7 - 3 = 3 + y - 3$$
$$4 = y$$

The solution is $y = 4$

Exercise 10.16

Solve these equations:

1 $x + 4 = 9$	**2** $x + 8 = 20$	**3** $x + 12 = 30$	**4** $y + 15 = 50$
5 $y + 25 = 40$	**6** $y + 45 = 100$	**7** $t + 35 = 75$	**8** $t - 5 = 8$
9 $t - 4 = 7$	**10** $m - 12 = 13$	**11** $m - 15 = 25$	**12** $n - 8 = 15$
13 $n - 11 = 29$	**14** $5 - x = 2$	**15** $8 - x = 6$	**16** $9 - y = 3$

Example 3

Solve the equation $7x = 63$

$$7x = 63$$

Divide both sides by 7 $$x = 63 \div 7$$
$$x = 9$$

The solution is $x = 9$

Example 4

Solve the equation $\frac{x}{5} = 13$

$$\frac{x}{5} = 13$$

Multiply both sides by 5 $$x = 13 \times 5$$
$$x = 65$$

The solution is $x = 65$

Exercise 10.17

Solve these equations

1 $4x = 60$	**2** $5x = 80$	**3** $3x = 51$	**4** $6y = 84$
5 $2y = 54$	**6** $9t = 117$	**7** $8t = 136$	**8** $7m = 105$
9 $12m = 192$	**10** $15n = 360$	**11** $25p = 275$	**12** $45p = 540$
13 $\frac{x}{3} = 8$	**14** $\frac{x}{4} = 9$	**15** $\frac{y}{6} = 7$	**16** $\frac{y}{5} = 3$
17 $\frac{t}{12} = 6$	**18** $\frac{t}{15} = 4$	**19** $\frac{m}{25} = 8$	**20** $\frac{m}{18} = 5$
21 $\frac{n}{2} = 25$	**22** $\frac{n}{10} = 19$	**23** $\frac{f}{25} = 40$	**24** $\frac{f}{91} = 7$

Example 5

Solve the equation $3y + 4 = 16$

$3y + 4 = 16$

Subtract 4 from both sides $3y + 4 - 4 = 16 - 4$

$3y = 12$

Divide both sides by 3 $y = 12 \div 3$

$y = 4$

The solution is $y = 4$

Example 6

Solve the equation $15 - 2t = 7$

$15 - 2t = 7$

Add $2t$ to both sides $15 = 7 + 2t$

Subtract 7 from both sides $8 = 2t$

Divide both sides by 2 $4 = t$

The solution is $t = 4$

Example 7

Solve the equation $\frac{m}{2} - 3 = 1$

$\frac{m}{2} - 3 = 1$

Add 3 to both sides $\frac{m}{2} = 4$

Multiply both sides by 2 $m = 8$

The solution is $m = 8$

Exercise 10.18

Solve these equations:

1 $2x+3=9$	**2** $2x+4=16$	**3** $2x+7=25$	**4** $3x+2=14$
5 $6y-5=31$	**6** $3m-2=22$	**7** $2m-9=13$	**8** $4m-7=33$
9 $6m-45=15$	**10** $9-2y=5$	**11** $16-3y=7$	**12** $7-x=3$
13 $\frac{x}{4}+6=13$	**14** $\frac{y}{5}+2=14$	**15** $\frac{y}{6}+4=7$	**16** $\frac{t}{10}+8=10$
17 $\frac{m}{4}-3=6$	**18** $\frac{n}{3}-2=10$	**19** $\frac{n}{6}-4=3$	**20** $\frac{p}{2}-9=1$

Example 8

A set square has a perimeter of 51 cm and the length of the unequal side is 21 cm.

Find x cm, the length of the equal sides.

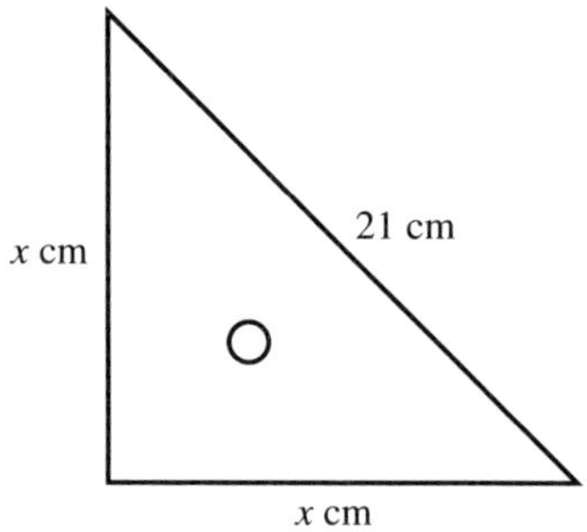

The perimeter is $x+x+21=51$

therefore $2x+21=51$

Subtracting 21 from both sides $2x=30$

Dividing both sides by 2 $x=15$

The answer is 15 cm

Exercise 10.19

1 If I think of a number, double it and add 5 to the result, the answer is 21.
Find the number.

2 The isosceles triangle ABC has AB and AC equal to x cm and BC equal to 3 cm.
If the perimeter is 13 cm, find the value of x.

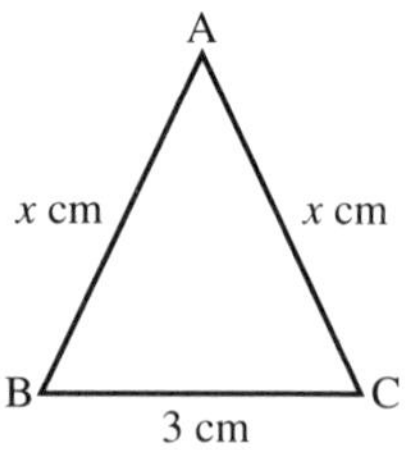

3 The isosceles triangle PQR has Q and R equal to x° and $P = 108^\circ$.
If the sum of the angles of any triangle is 180°, find the value of x.

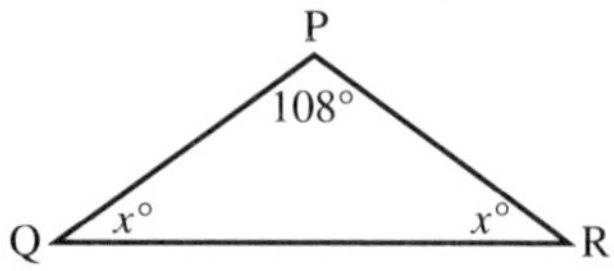

4 If the distance from London to Chippenham is 150 km, find the distance x km.

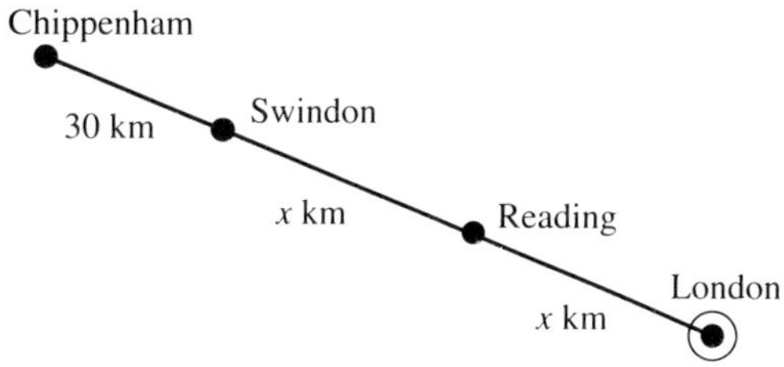

5 If the distance from London to Sheffield is 260 km, find the distance x km.

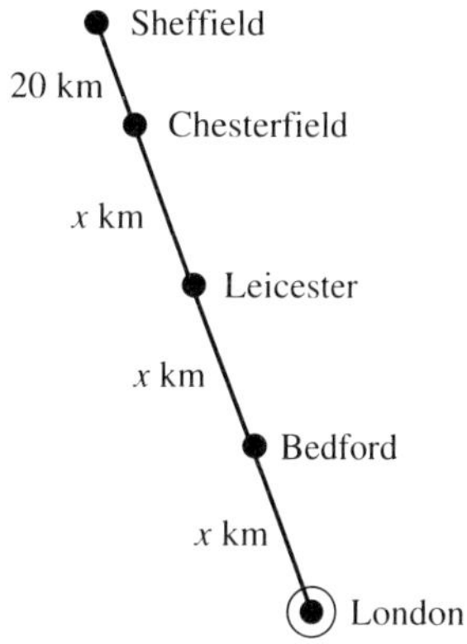

6 Joel has x marbles. He wins a game and finds that he as 3 times as many. He then plays another game and wins 4 more.
If he ends up with 22 marbles, find x.

7 The four-sided figure ABCD has AB, BC and CD all equal to x cm and DA equal to 16 cm.
If the perimeter of the figure is 40 cm, find the value of x.

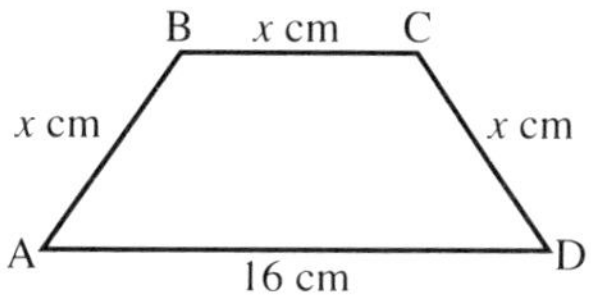

8 If I think of a number, multiply it by 3 and then subtract 4 from the result the answer is 14.
Find the number.

9 If Marissa's school is 250 m from her house and the Post Office is 50 m from her school, find the distance x metres.

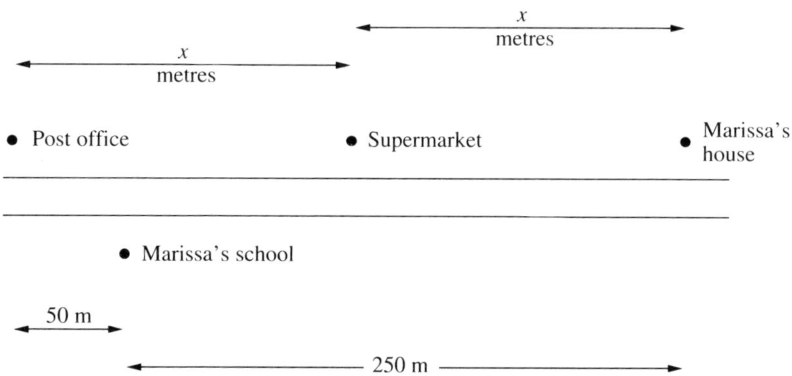

Example 9

Solve the equation $5x + 1 = 2x + 10$

	$5x + 1 = 2x + 10$
Subtract $2x$ from both sides	$3x + 1 = 10$
Subtract 1 from both sides	$3x = 9$
Divide both sides by 3	$x = 3$

The solution is $x = 3$

Example 10

Solve the equation $3y + 4 = 12 - y$

	$3y + 4 = 12 - y$
Add y to each side	$4y + 4 = 12$
Subtract 4 from each side	$4y = 8$
Divide both sides by 4	$y = 2$

The solution is $y = 2$

Exercise 10.20

Solve these equations:

1 $5x + 3 = 2x + 15$
2 $6x + 5 = 3x + 14$
3 $9x + 1 = 5x + 9$
4 $12t + 5 = 4t + 37$
5 $5x + 7 = 4x + 12$
6 $3x + 8 = 2x + 21$
7 $6b + 12 = b + 27$
8 $5b + 3 = b + 11$
9 $7m + 5 = m + 23$
10 $7x - 2 = 5x + 14$
11 $6x - 5 = 4x + 7$
12 $7y - 14 = 2y + 11$
13 $8n - 29 = 7n + 13$
14 $13n - 15 = 12n + 19$
15 $8t - 6 = t + 15$
16 $9q - 19 = 2q - 5$
17 $23q - 27 = 16q - 13$
18 $16a - 35 = 7a - 8$
19 $20c - 51 = 9c - 7$
20 $6x - 5 = 13 - 3x$
21 $4x - 3 = 9 - 2x$
22 $4t - 9 = 6 - t$
23 $6t - 5 = 16 - t$
24 $5m - 2 = 10 - m$

Example 11

Solve the equation $x^2 = 5$ by using your calculator and a trial and improvement method.
Give your answer correct to one decimal place.

We need x such that $x^2 = 5$
Try $x = 2$ $\quad x^2 = 4$ (too small)
Try $x = 3$ $\quad x^2 = 9$ (too big)
Try $x = 2.5$ $\quad x^2 = 6.25$ (too big)
Try $x = 2.3$ $\quad x^2 = 5.29$ (too big)
Try $x = 2.2$ $\quad x^2 = 4.84$ (too small)

4.84 is closer to 5 than 5.29 so the solution (correct to one decimal place) is $x = 2.2$

Exercise 10.21

Solve these equations by using your calculator and a trial and improvement method.
Give each answer correct to one decimal place.

1 $x^2 = 24$	**2** $x^2 = 22$	**3** $y^2 = 28$	**4** $y^2 = 15$
5 $q^2 = 2$	**6** $q^2 = 3$	**7** $m^2 = 17$	**8** $m^2 = 8$

Example 12

Solve the equation $x^3 = 20$ by using your calculator and a trial and improvement method.
Give your answer correct to one decimal place.

We need x such that $x^3 = 20$
Try $x = 2$, $x^3 = 8$ (too small)
Try $x = 3$, $x^3 = 27$ (too big, but closer)
Try $x = 2.7$, $x^3 = 19.683$ (too small, but very close)
Try $x = 2.8$, $x^3 = 21.952$ (too big, and further away)

19.683 is closer to 20 than 21.952, so the solution (correct to one decimal place) is $x = 2.7$

Exercise 10.22

Solve these equations by using your calculator and a trial and improvement method.
Give each answer correct to one decimal place.

1 $x^3 = 50$ **2** $x^3 = 68$ **3** $y^3 = 15$ **4** $y^3 = 2$

10.4 Exploring number patterns using difference methods

Looking at the difference between each pair of numbers in a sequence often helps us to find other numbers in the sequence.

Example 1

For each sequence find the difference between each pair of numbers.
Use this to find the next three numbers in the sequence.
4, 7, 10, 13, 16, 19

The sequence is:	4		7		10		13		16		19
The differences are:		3		3		3		3		3	

For each case the difference is 3.
Therefore the next three terms are 22, 25, 28.

Example 2

For each sequence find the difference between each pair of numbers.
Use this to find the next three numbers in the sequence.
1, 3, 6, 10, 15, 21

The sequence is: 1 3 6 10 15 21
The differences are: 2 3 4 5 6
The differences are successive numbers starting from 2.
Therefore the next three terms are 28 (21 + 7), 36 (28 + 8), 45 (36 + 9).

Example 3

For each sequence find the difference between each pair of numbers.
Use this to find the next three numbers in the sequence.
2, 5, 10, 17, 26, 37

The sequence is: 2 5 10 17 26 37
The differences are: 3 5 7 9 11
The differences are successive odd numbers starting from 3.
Therefore the next three terms are 50 (37 + 13), 65 (50 + 15), 82 (65 + 17).

Exercise 10.23

For each sequence find the difference between each pair of numbers.
Use this to find the next five numbers in the sequence.

1 **a** 5, 9, 13, 17, 21 **b** 2, 7, 12, 17, 22 **c** 1, 7, 13, 19, 25 **d** 3, 7, 11, 15, 19
e 2, 2.5, 3, 3.5, 4 **f** 1, 2.5, 4, 5.5, 7 **g** 0, 4.5, 9, 13.5, 18 **h** 2, 4.5, 7, 9.5, 12

2 **a** 1, 2, 4, 7, 11 **b** 3, 4, 6, 9, 13 **c** 3, 5, 8, 12, 17
d 1, 4, 8, 13, 19 **e** 2, 6, 11, 17, 24 **f** 0.5, 1.5, 2.5, 3.5, 4.5

3 **a** 1, 2, 6, 15, 31 **b** 4, 4, 5, 9, 18 **c** 6, 10, 19, 35, 60
d 1, 2, 5, 11, 21 **e** 0, 1, 4, 10, 20 **f** 11, 14, 20, 30, 45

Example 4

a Write down the complete letter sequence that starts: A, D, G, ...
b Write down the first twenty terms of the letter sequence that starts: A, Z, B, Y, ...

a The sequence includes every fourth letter of the alphabet.
The complete sequence is: A, D, G, J, M, P, S, V, Y
b The odd terms of the sequence are the letters of the alphabet in order, whereas the even terms of the sequence are the letter in reverse order.
The first twenty terms are: A, Z, B, Y, C, X, D, W, E, V, F, U, G, T, H, S, I, R, J, Q

Exercise 10.24

For Questions 1–6 write out the complete letter sequence.

1 ACE **2** AEI **3** AFK **4** ZXV **5** ZWT **6** ZVR

10.5 Suggesting possible rules for generating sequences

Using the method of differences can help us to find a rule for generating a sequence.

Example 1

For each sequence find the difference between each pair of numbers.
Write down the next two terms in the sequence and suggest a rule for generating the sequence.

a 3, 6, 9, 12, 15

b 4, 7, 10, 13, 16

a The sequence is: 3 6 9 12 15
The differences are: 3 3 3 3
The next two terms are 18 $(15 + 3)$ and 21 $(18 + 3)$.
As the differences are all 3, and the lowest term is three, the rule for this sequence is $n \to 3n + 0$ or $n \to 3n$.

b The sequence is: 4 7 10 13 16
The differences are: 3 3 3 3
The next two terms are 19 $(16 + 3)$ and 22 $(19 + 3)$.
As the differences are all 3, and the lowest term is four, the rule for this sequence is $n \to 3n + 1$.

Exercise 10.25

For each sequence, find the difference between each pair of numbers. Write down the next five terms in the sequence and suggest a rule for generating the sequence.

1 4, 8, 12, 16, 20
2 5, 9, 13, 17, 21
3 5, 10, 15, 20, 25
4 6, 11, 16, 21, 26
5 7, 13, 19, 25, 31
6 3, 9, 15, 21, 27

Example 2

Find the first and second row of differences for the sequence. Write down the next two terms and suggest a rule for generating the sequence.
1 4 9 16 25

The sequence is:	1		4		9		16		25
The first differences are:		3		5		7		9	
The second differences are:			2		2		2		

The next two terms are 36 (25 + 11) and 49 (36 + 13).

As the second differences are all two, this sequence is related to the square number sequence. The rule for this sequence is $n \rightarrow n^2$.

Example 3

Find the first and second row of differences for the sequence. Write down the next two terms and suggest a rule for generating the sequence.
2 5 10 17 26

The sequence is:	2		5		10		17		26
The first differences are:		3		5		7		9	
The second differences are:			2		2		2		

The next two terms are 37 (26 + 11) and 50 (37 + 13).

As the second differences are all two, this sequence is related to the square number sequence. Each term is one more than a square number so the rule for this sequence is $n \rightarrow n^2 + 1$.

Example 4

Find the first and second row of differences for the sequence. Write down the next two terms and suggest a rule for generating the sequence.
0 3 8 15 24

The sequence is:	0		3		8		15		24
The first differences are:		3		5		7		9	
The second differences are:			2		2		2		

The next two terms are 35 (24 + 11) and 48 (35 + 13).

As the second differences are all two and each term is one less than a square number, the rule for this sequence is $n \rightarrow n^2 - 1$.

Exercise 10.26

For each sequence find the difference between each pair of numbers. Write down the next five terms in the sequence and suggest a rule for generating the sequence.

1 3, 6, 11, 18, 27

2 4, 7, 12, 19, 28

3 10, 13, 18, 25, 34

4 −1, 2, 7, 14, 23

10.6 Drawing and interpreting simple mappings

Example 1

Complete a table for the mapping $x \rightarrow x + 4$ using values of x from 0 to 5.
Draw this mapping on a graph.

$x \rightarrow x + 4$
$0 \rightarrow 4$
$1 \rightarrow 5$
$2 \rightarrow 6$
$3 \rightarrow 7$
$4 \rightarrow 8$
$5 \rightarrow 9$

On an (x, y) graph plot the coordinates (0, 4), (1, 5), (2, 6), (3, 7), (4, 8) and (5, 9).

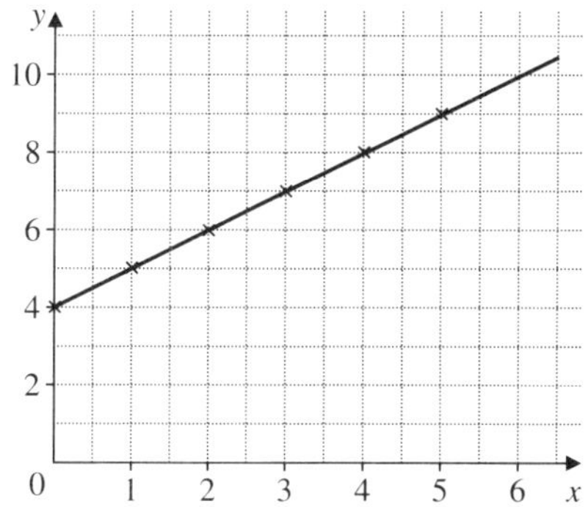

Note that the six points all lie on the same straight line.
The line is described by $y = x + 4$

Exercise 10.27

For this exercise you will need eight copies of this grid.

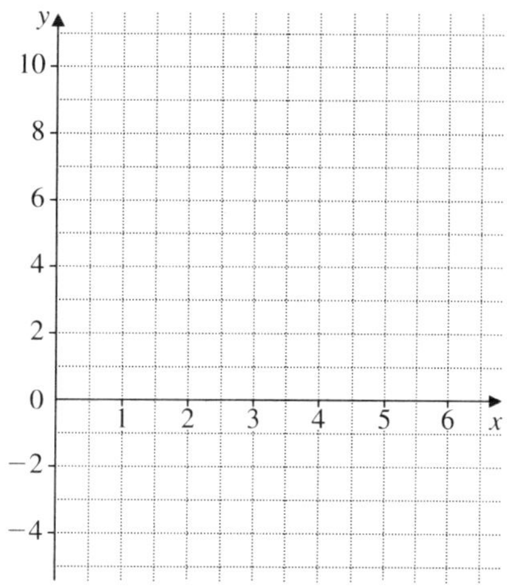

For each question complete a table for the mapping using values of x from 0 to 5 and draw the mapping on a graph.

1 $x \rightarrow x + 3$
2 $x \rightarrow x + 5$
3 $x \rightarrow x + 2$
4 $x \rightarrow x + 1$
5 $x \rightarrow x - 1$
6 $x \rightarrow x - 2$
7 $x \rightarrow x - 4$
8 $x \rightarrow x - 3$

Example 2

Complete a table for the mapping $x \to 2 - x$ using the values of x from 0 to 5.
Draw this mapping on a graph.

$x \to 2 - x$
$0 \to 2$
$1 \to 1$
$2 \to 0$
$3 \to -1$
$4 \to -2$
$5 \to -3$

On an (x, y) graph plot the coordinates (0, 2), (1, 1), (2, 0), (3, −1), (4, −2) and (5, −3).

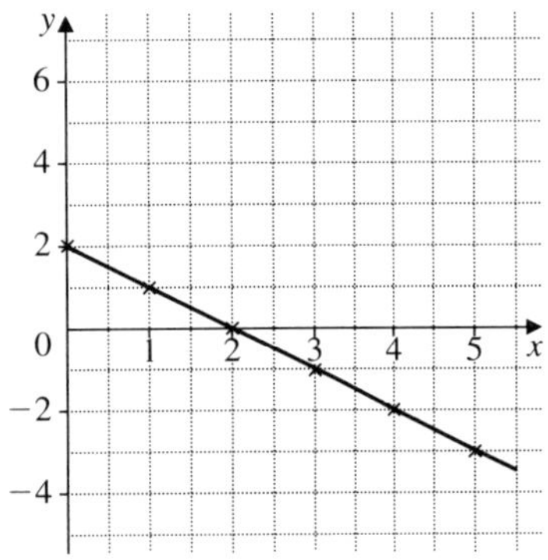

Note that the six points all lie on the same straight line.
The line is described by
$y = 2 - x$

Exercise 10.28

For this exercise you will need five copies of this grid.

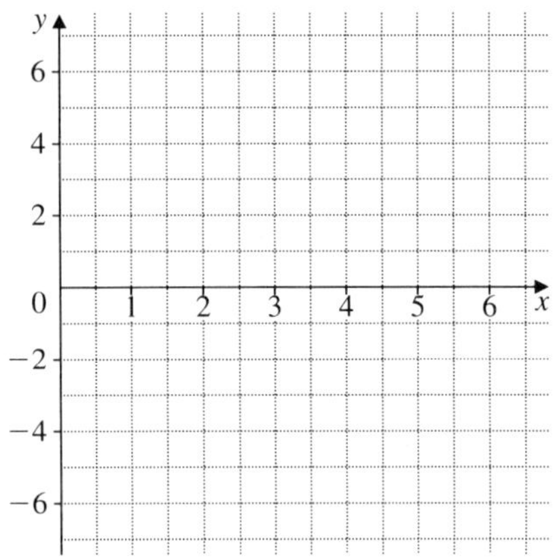

For each question complete a table for the mapping using values of x from 0 to 5 and draw the mapping on a graph.

1 $x \to 6 - x$
2 $x \to 5 - x$
3 $x \to 3 - x$
4 $x \to 4 - x$
5 $x \to 1 - x$

Example 3

Complete a table for the mapping $x \to x^2$ using values of x from -4 to $+4$.
Draw this mapping on a graph.

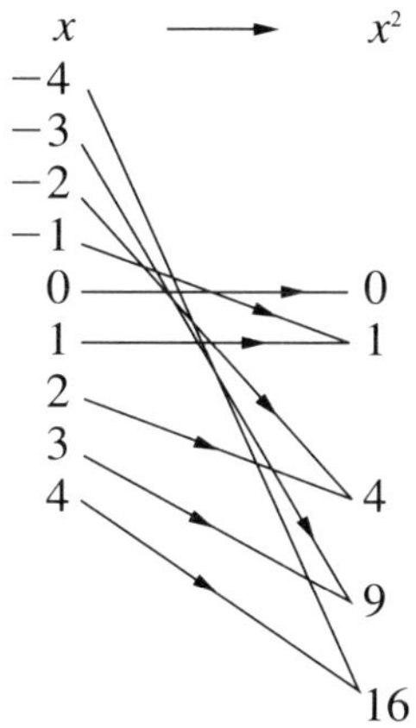

On an (x, y) graph plot the coordinates $(-4, 16)$, $(-3, 9)$, $(-2, 4)$, $(-1, 1)$, $(0, 0)$, $(1, 1)$, $(2, 4)$, $(3, 9)$ and $(4, 16)$.

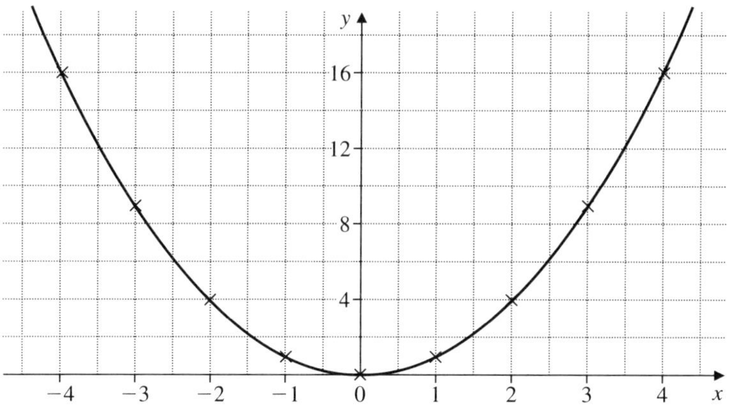

Note that the points all lie on a smooth curve.
The curve is described by $y = x^2$

Example 4

Complete the table for $y = x^2 - 2$ and plot the points on an (x, y) graph.

$x =$	-4	-3	-2	-1	0	1	2	3	4
$x^2 =$ $y = x^2 - 2 =$	16 14								

The completed table is:

$x =$	-4	-3	-2	-1	0	1	2	3	4
$x^2 =$ $y = x^2 - 2 =$	16 14	9 7	4 2	1 -1	0 -2	1 -1	4 2	9 7	16 14

On an (x, y) graph plot the coordinates $(-4, 14)$, $(-3, 7)$, $(-2, 2)$, $(-1, -1)$, $(0, -2)$, $(1, -1)$, $(2, 2)$, $(3, 7)$ and $(4, 14)$.

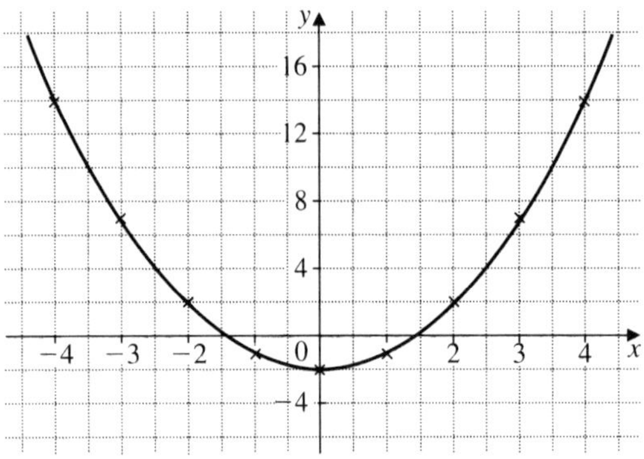

Note that the points all lie on a smooth curve.
The curve is described by $y = x^2 - 2$

Exercise 10.29

For this exercise you will need three copies of this grid.

For each question copy and complete the table and plot the points on an (x, y) graph.

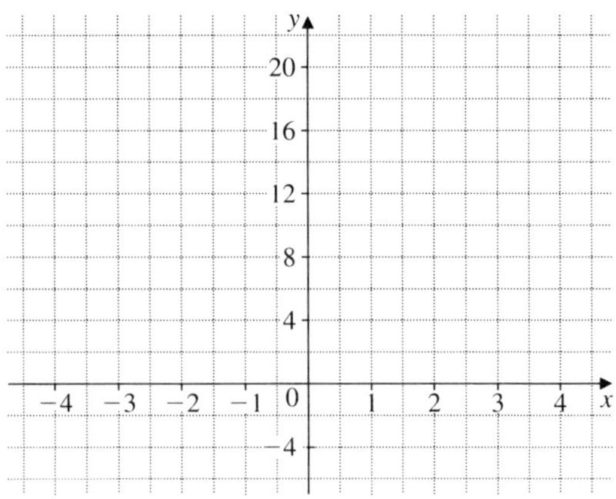

1 $y = x^2 + 2$

$x =$	−4	−3	−2	−1	0	1	2	3	4
$x^2 =$ $y = x^2 + 2 =$	16 18								

2 $y = x^2 + 3$

$x =$	−4	−3	−2	−1	0	1	2	3	4
$x^2 =$ $y = x^2 + 3 =$	16 19								

3 $y = x^2 - 1$

$x =$	−4	−3	−2	−1	0	1	2	3	4
$x^2 =$ $y = x^2 - 1 =$	16 15								

Example 5

The table below shows how far a cyclist has travelled after various times. Show this information on a graph.

Time (minutes)	0	10	20	30	40	50	60
Distance travelled (miles)	0	4	8	12	16	20	24

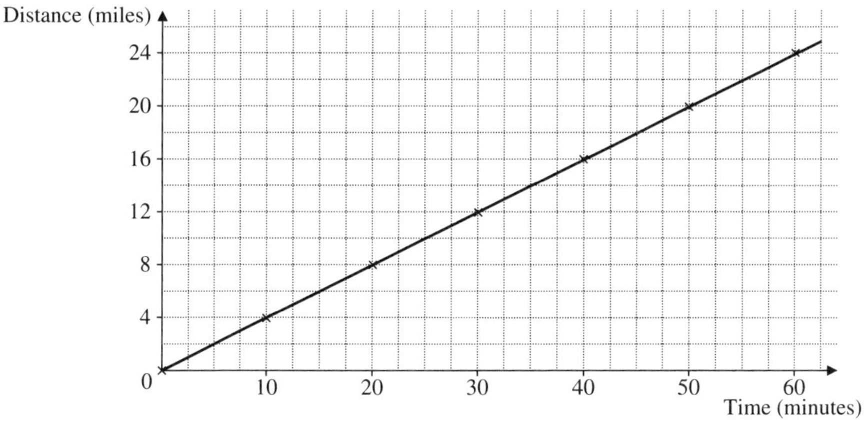

Note that all the points lie on a straight line.

Exercise 10.30

For each question show the given information on a graph.
Draw your graph on a copy of the grid illustrated.

1 The table shows the distance covered by a boy running in a 400-metre race after various times.

Time (seconds)	0	20	40	60	80
Distance covered (m)	0	100	200	300	400

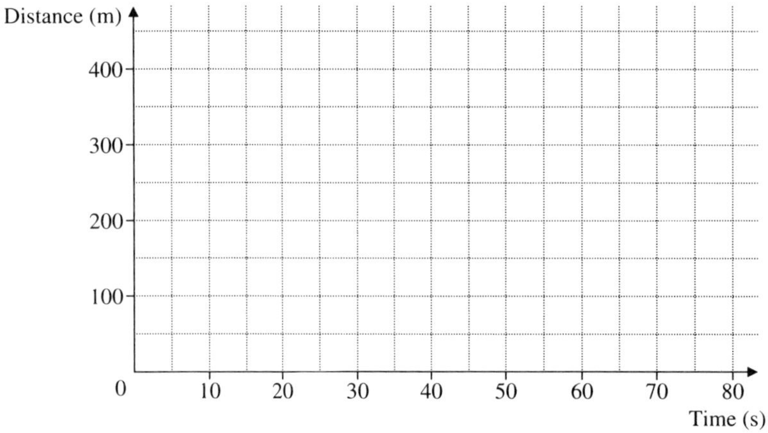

What distance has he covered after 10, 30, 50 and 70 seconds?

2 The table shows the distance covered by a girl swimming lengths of a pool after various times.

Time (seconds)	0	5	10	15	20	25	30
Distance covered (m)	0	10	20	30	40	50	60

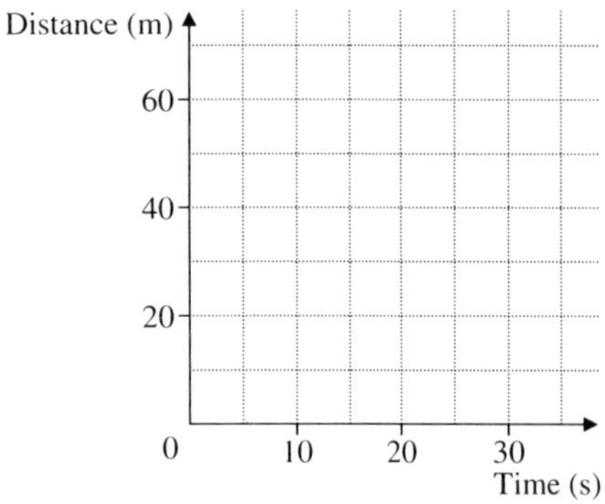

What distance has she covered after 2.5, 7.5, 12.5 and 22.5 seconds?

3 The table shows the altitude reached by a balloon after various times.

Time (minutes)	0	2	4	6	8	10
Altitude (m)	0	10	20	30	40	50

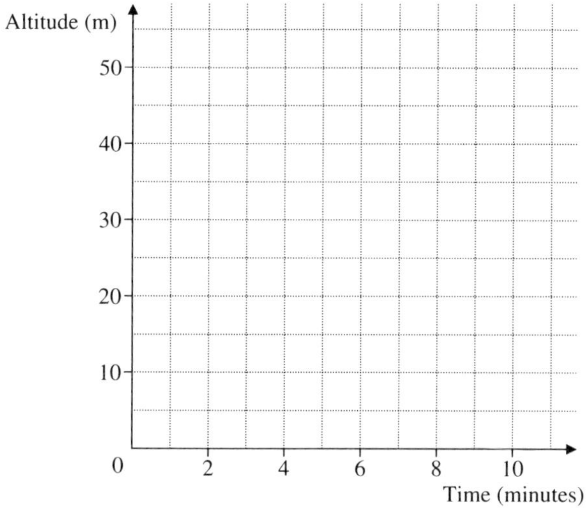

What altitude has it reached after 1, 3, 5, 7 and 9 minutes?

4 The table shows the distance covered by a jogger after various times.

Time (minutes)	0	10	20	30	40	50	60
Distance (km)	0	2	4	6	8	10	12

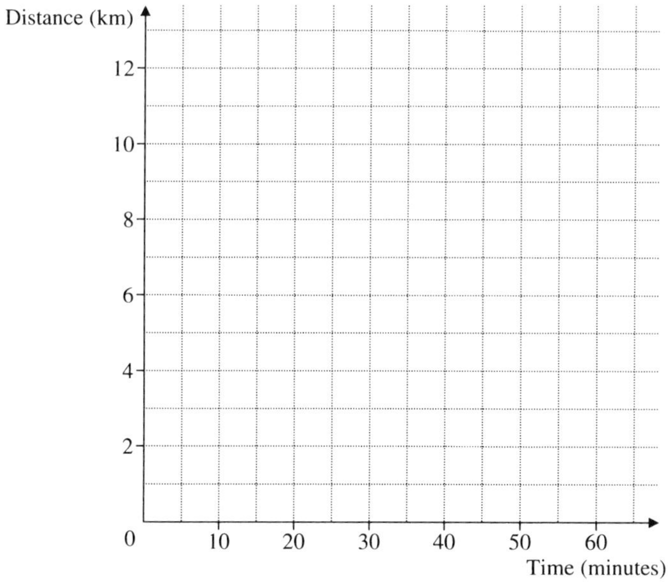

What distance has she covered after 5, 15, 45 and 55 minutes?

SECTION C: SHAPE, SPACE AND MEASURES

11 Constructing and recognising 2-D or 3-D shapes and knowing associated language

11.1 Construction of simple shapes

Example 1

A triangle ABC has its vertices at (1, 1), (1, 11) and (11, 1) respectively.
Draw this triangle on 1 cm squared paper and measure its perimeter.

Mark points A, B, C.
Then join them with straight lines.

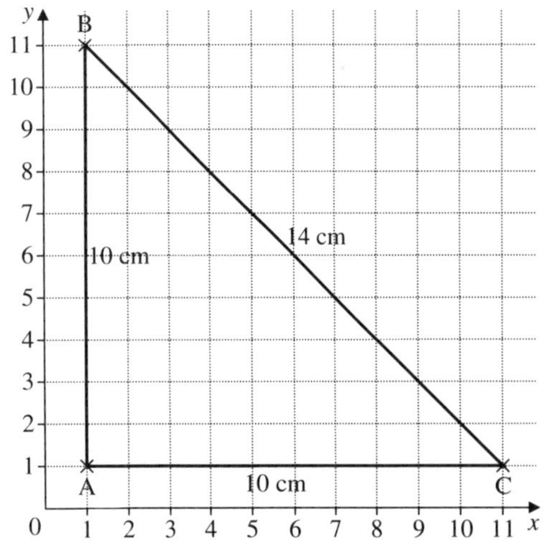

AB = 10 cm, BC = 14 cm and CA = 10 cm
The perimeter is: 10 + 14 + 10 = 34 cm.

Exercise 11.1

In each question give your answers correct to the nearest whole cm.

1 Draw these triangles on the same sheet of 1 cm squared paper. Measure the perimeter of each.

a Triangle ABC, with A at (1, 1), B at (1, 6) and C at (6, 1).
b Triangle LMN, with L at (7, 1), M at (7, 5) and N at (14, 1).
c Triangle PQR, with P at (16, 1), Q at (16, 5) and R at (19, 1).
d Triangle XYZ, with X at (20, 1), Y at (20, 8) and Z at (27, 1).
e Triangle DEF, with D at (3, 6), E at (3, 11) and F at (15, 6).
f Triangle GHK, with G at (1, 12), H at (1, 18) and K at (9, 12).
g Triangle UVW, with U at (10, 9), V at (10, 17) and W at (25, 9).

2 Draw the following triangles on the same sheet of 1 cm squared paper.
Measure the perimeter of each one.

a Triangle ABC, with A at (1, 1), B at (6, 13) and C at (11, 1).
b Triangle LMN, with L at (12, 1), M at (19, 8) and N at (26, 1).
c Triangle PQR, with P at (12, 9), Q at (19, 13) and R at (26, 9).
d Triangle XYZ, with X at (15, 2), Y at (19, 5) and Z at (23, 2).
e Triangle DEF, with D at (2, 2), E at (6, 9) and F at (10, 2).
f Triangle GHK, with G at (3, 3), H at (6, 7) and K at (9, 3).
g Triangle UVW, with U at (2, 18), V at (14, 13) and W at (26, 18).

3 Draw these four-sided figures on the same sheet of 1 cm squared paper and measure the perimeter of each one.

a The figure ABCD, with A at (4, 4), B at (4, 8), C at (8, 8) and D at (8, 4).
b The figure LMNP, with L at (3, 14), M at (3, 16), N at (7, 16) and P at (7, 14).
c The figure EFGH, with E at (14, 12), F at (14, 16), G at (20, 16) and H at (20, 12).
d The figure WXYZ, with W at (17, 3), X at (17, 7), Y at (19, 7) and Z at (19, 3).

Three-dimensional shapes can be made from nets.

The diagram shows the net of a cube.

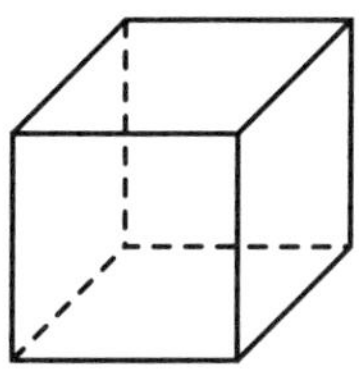

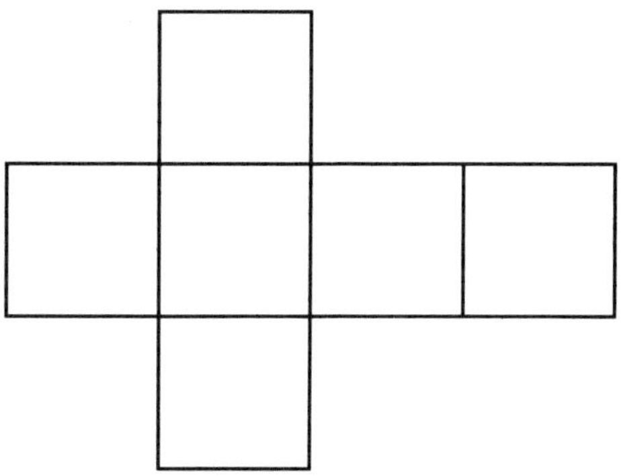

It can easily be seen that both the cube and the net have six square faces.

Other examples of simple solids are the cuboid, the triangular prism and the pyramid.
These are illustrated below.

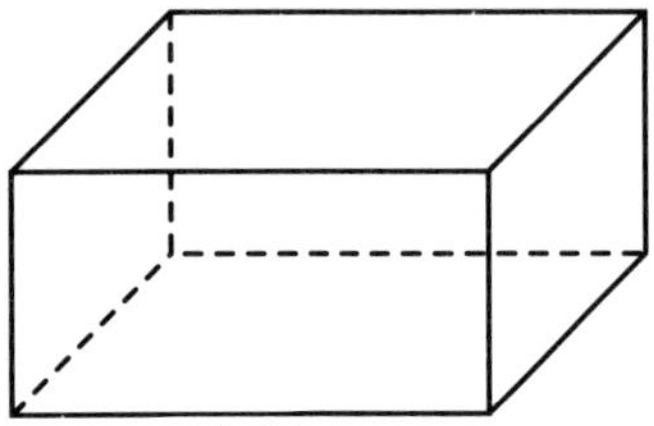

Cuboid

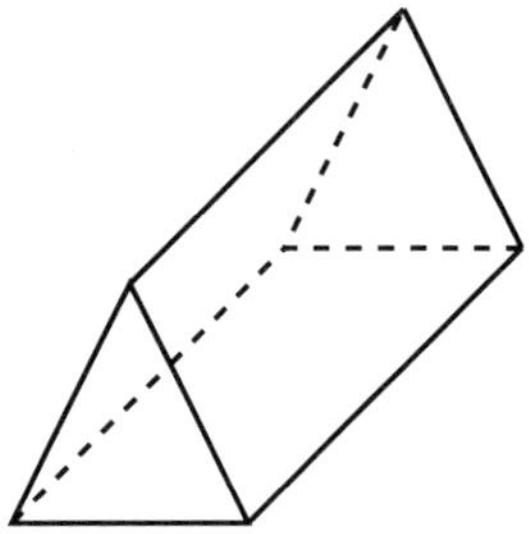

Triangular prism

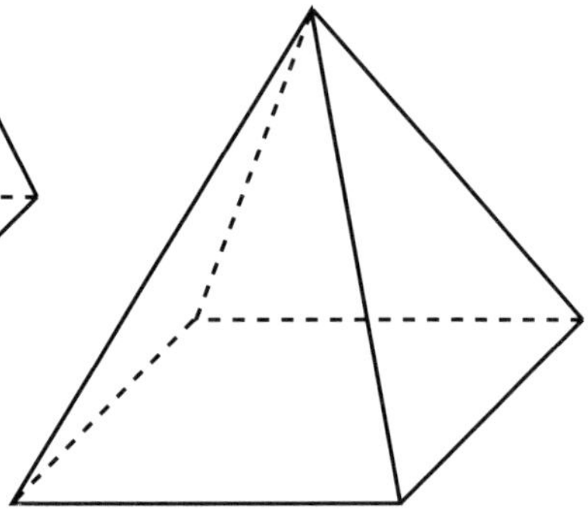

Pyramid

Exercise 11.2

Draw to scale each of these nets for cubes and cuboids.
Cut out the net and construct the solid.
(The scale is 1 square = 1 centimetre, and glue tabs are included on the nets.)

1

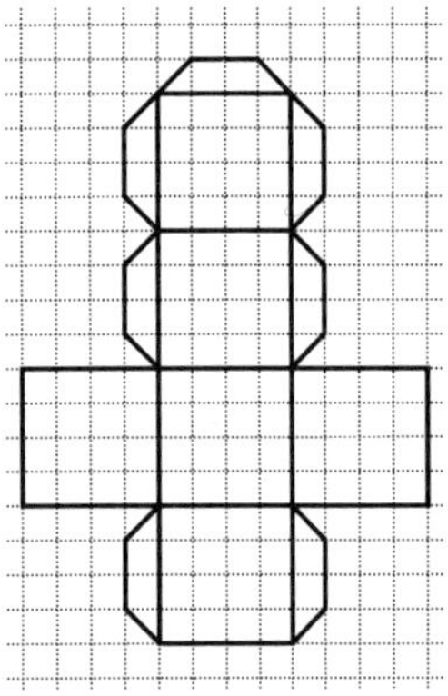

2

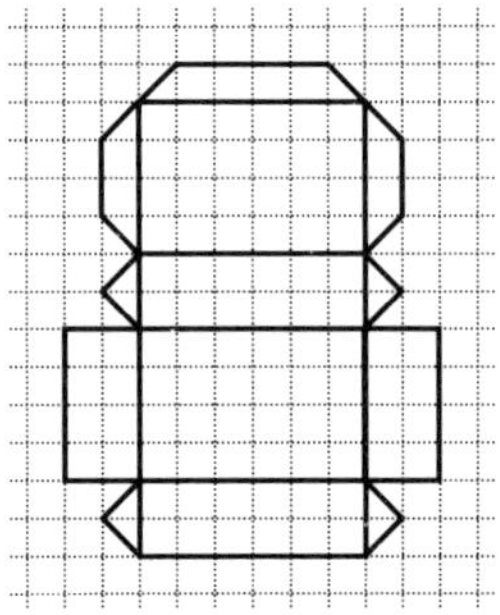

3

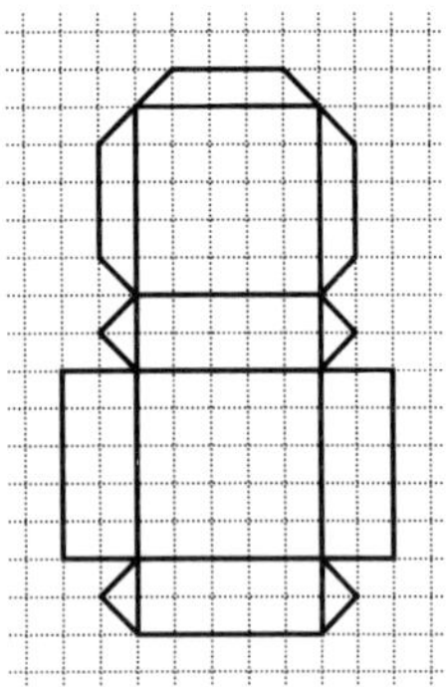

4

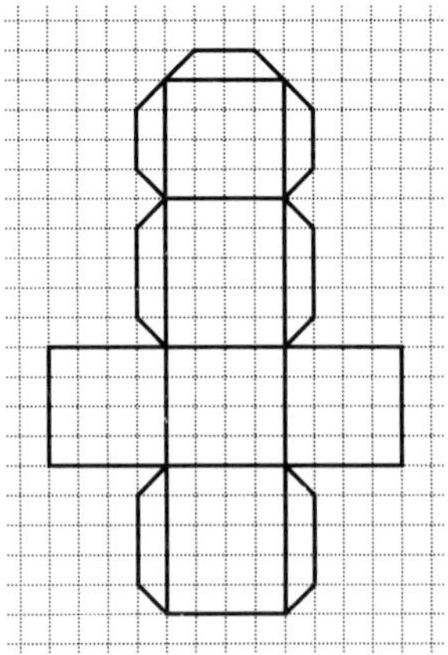

Exercise 11.3

Draw to scale these nets for triangular prisms and pyramids.
Cut out the net and construct the solid.
(The scale is 1 square = 1 centimetre, and glue tabs are included on the nets.)
Measure the vertical height of each solid.

1

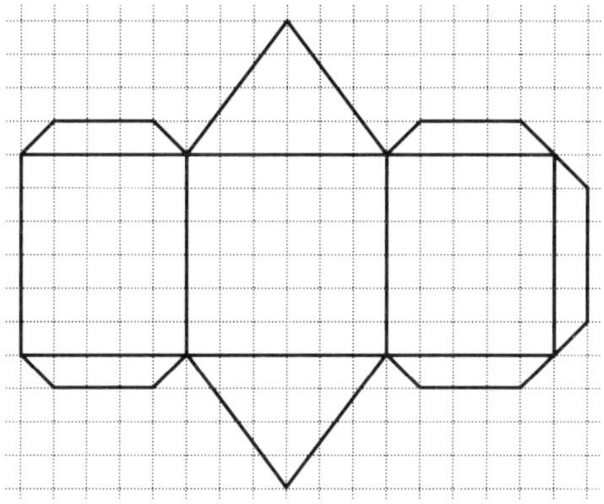

2

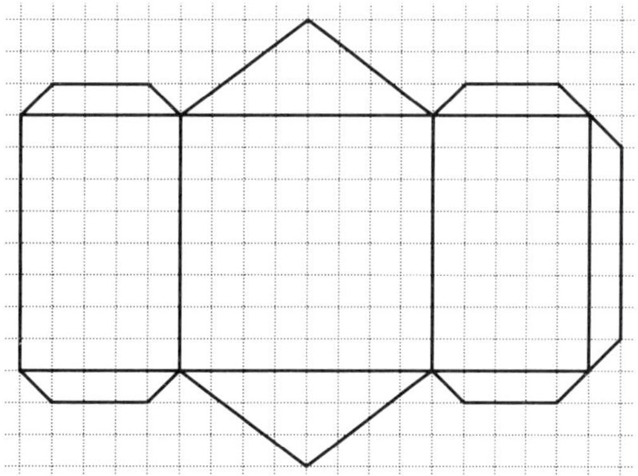

3

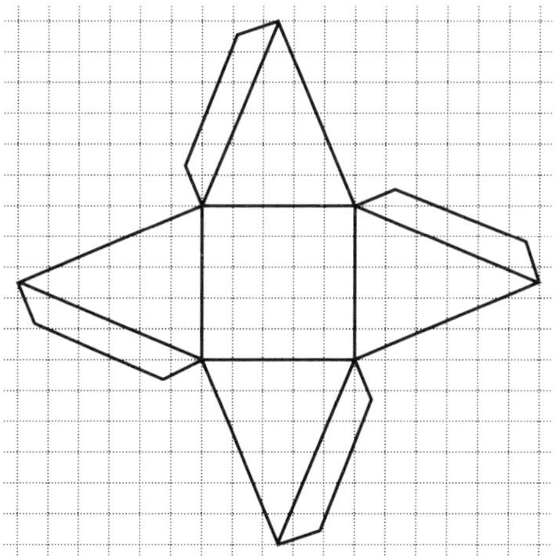

4

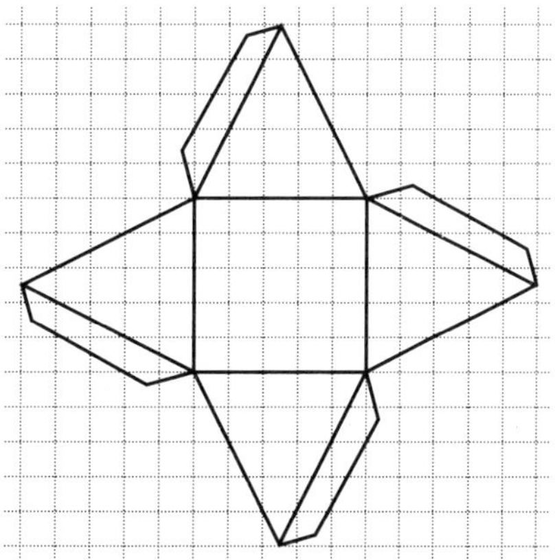

11.2 The congruence of simple shapes

These two triangles are **congruent** because when they are placed one on top of the other they coincide.

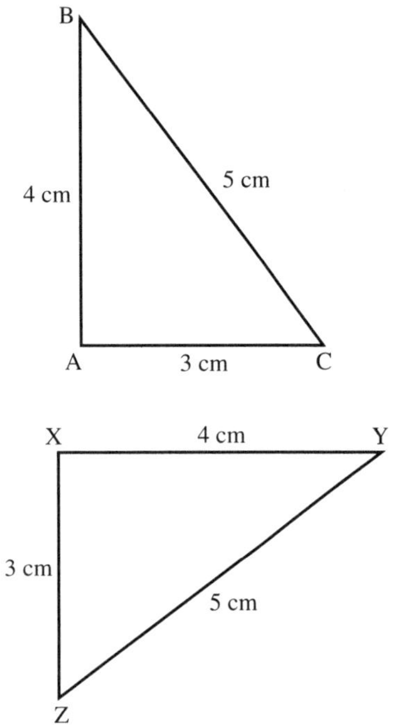

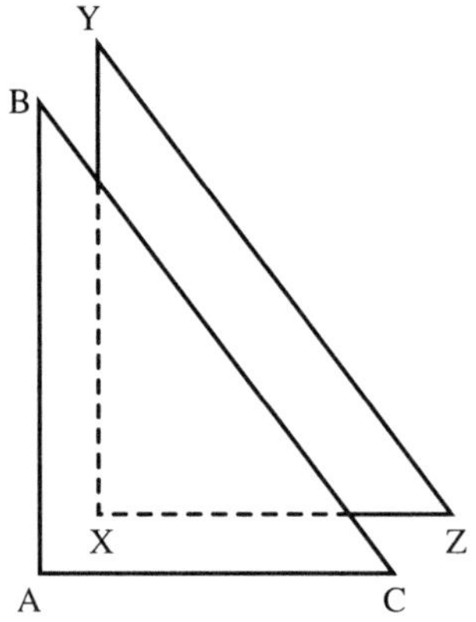

Exercise 11.4

For each question draw the given pair of triangles on 1 cm squared paper. Cut the triangles out and show that they are congruent by placing one on top of the other. Also measure the three side lengths for each triangle.

Questions 1–3 can be done on the same sheet of 1 cm squared paper.

1 a Triangle ABC, with A at (1, 1), B at (1, 5) and C at (8, 1).
b Triangle DEF, with D at (9, 1), E at (9, 8) and F at (13, 1).

2 a Triangle LMN, with L at (1, 8), M at (1, 12) and N at (4, 8).
b Triangle PQR, with P at (9, 9), Q at (13, 9) and R at (13, 12).

3 a Triangle UVW, with U at (1, 13), V at (7, 13) and W at (4, 17).
b Triangle XYZ, with X at (9, 12), Y at (9, 18) and Z at (13, 15).

4 a Triangle ABC, with A at (1, 1), B at (6, 6) and C at (11, 1).
b Triangle DEF, with D at (13, 1), E at (13, 8) and F at (20, 1).

5 a Triangle ABC, with A at (1, 1), B at (1, 13) and C at (6, 1).
b Triangle DEF, with D at (7, 1), E at (7, 6) and F at (19, 1).

Exercise 11.5

For each question draw the given pair of four-sided figures on 1 cm squared paper.
Cut the figures out and show that they are congruent by placing one on top of the other.
Also measure the four side lengths for each figure.

Questions 1–3 can be done on the same sheet of 1 cm squared paper.

1 a Figure ABCD, with A at (1, 1), B at (1, 6), C at (5, 3) and D at (5, 1).
b Figure LMNP, with L at (6, 1), M at (6, 5), N at (8, 5) and P at (11, 1).

2 a Figure PQRS, with P at (1, 7), Q at (1, 14), R at (4, 10), and S at (4, 7).
b Figure WXYZ, with W at (5, 7), X at (9, 10), Y at (12, 10) and Z at (12, 7).

3 a Figure EFGH, with E at (13, 1), F at (13, 3), G at (20, 7) and H at (20, 1).
b Figure KLMN, with K at (21, 1), L at (25, 8) M at (27, 8) and N at (27, 1).

4 a Figure ABCD, with A at (1, 4), B at (5, 7), C at (9, 4) and D at (5, 1).
b Figure KLMN, with K at (11, 5), L at (14, 9), M at (17, 5) and N at (14, 1).

11.3 Recognising and using 2-dimensional representations of 3-dimensional objects

Example 1

Make a 2-dimensional drawing of a cuboid of length 4 cm, width 2 cm and height 3 cm.

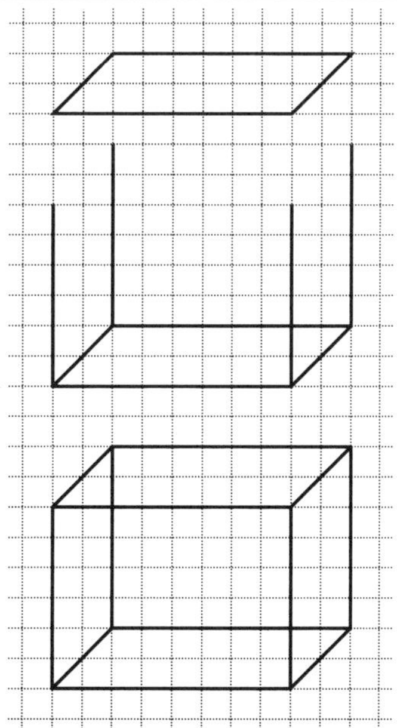

1 Draw a parallelogram to represent the base. The side representing the length is drawn to scale but the side representing the width is conveniently drawn to a length of 2 square diagonals rather than 2 real centimetres.

2 Draw the vertical edges to scale.

3 Draw another parallelogram for the top.

Exercise 11.6

Make a 2-dimensional drawing of:

1. A cuboid of length 4 cm, width 3 cm and height 2 cm.
2. A cuboid of length 5 cm, width 3 cm and height 2 cm.
3. A cube of side length 3 cm.
4. A cube of side length 2 cm.

Example 2

Make a 2-dimensional drawing of a triangular prism of length 4 cm and with a triangular face as shown.

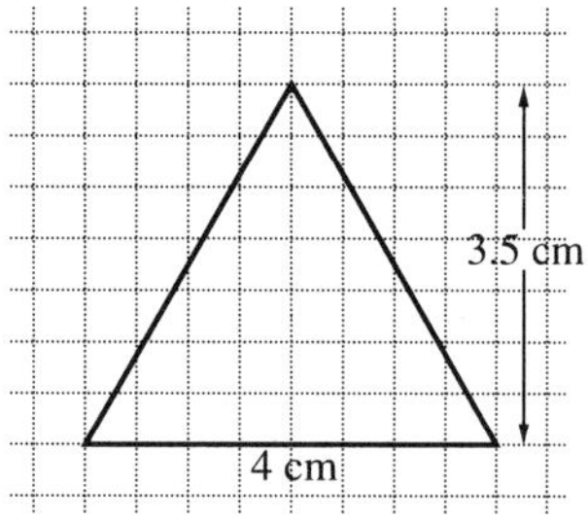

1 Make an accurate copy of the triangular face.

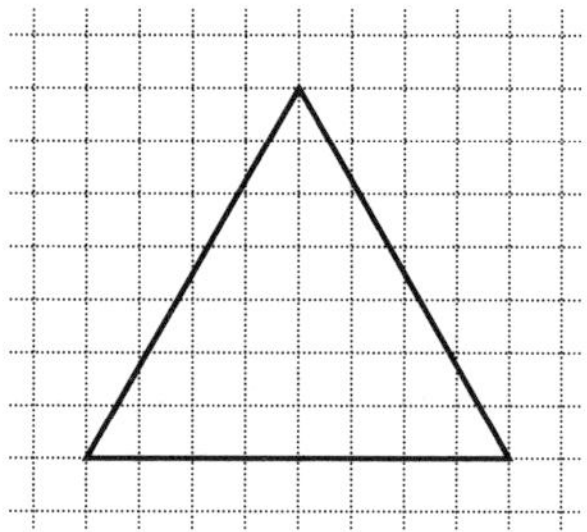

2 Draw the three 'length' edges as 4 square diagonals rather than 4 real centimetres.

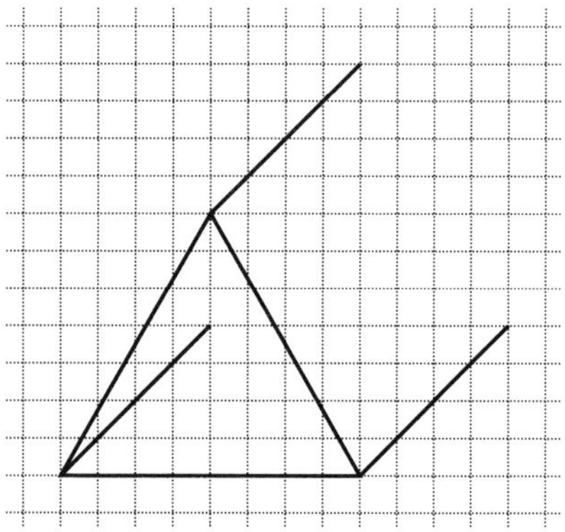

3 Draw another triangular end at the back.

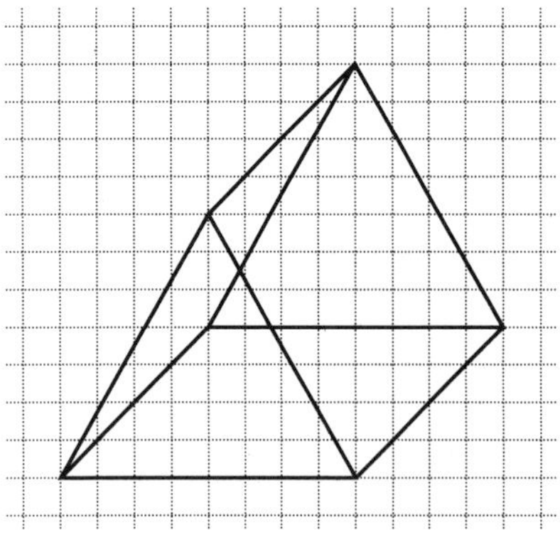

Exercise 11.7

Make a 2-dimensional drawing of each of these triangular prisms on squared paper.

1 Length 3 cm, face as shown.

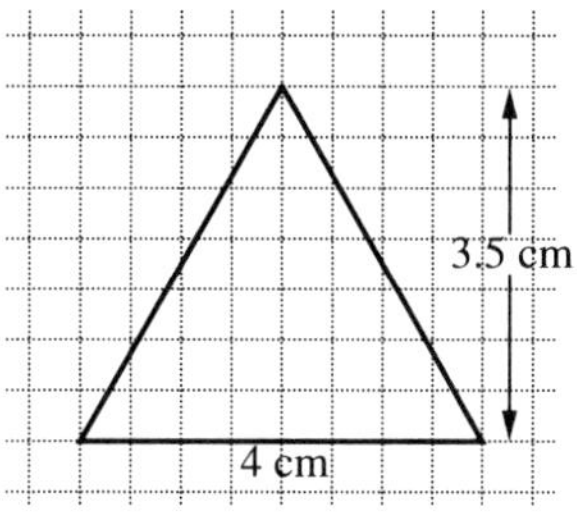

2 Length 2 cm, face as shown.

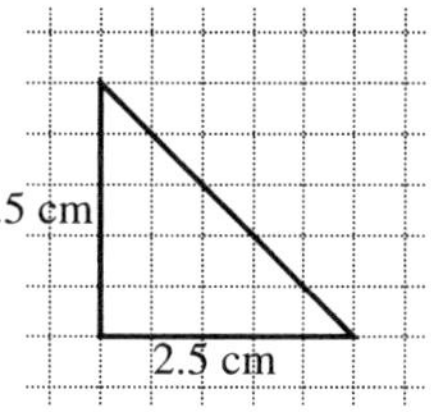

3 Length 4 cm, face as shown.

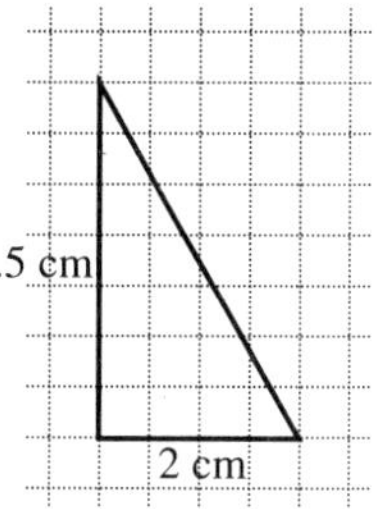

Example 3

Make a 2-dimensional drawing of a square-based pyramid which has base length 4 cm and vertical height 3 cm.

1 Draw a rhombus for the base by using exactly the same method as for the base in Example 1.

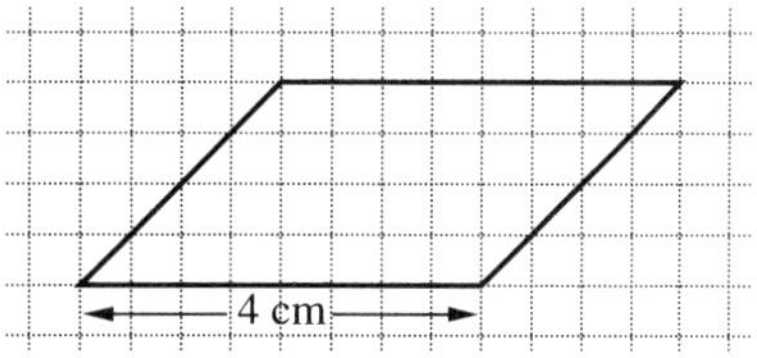

2 Draw in the diagonals to locate the centre of the base.
Use broken lines.

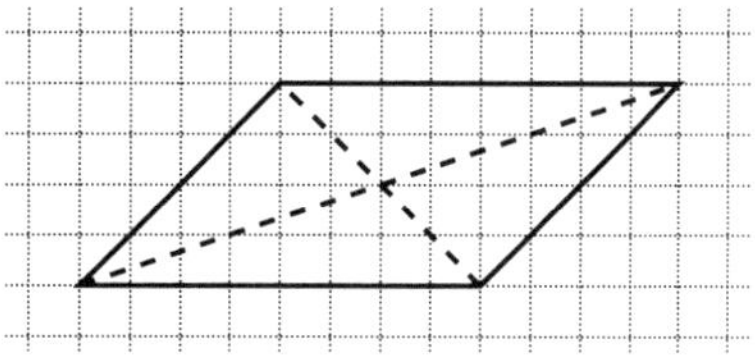

3 Draw in the vertical height to scale.
Use broken lines.

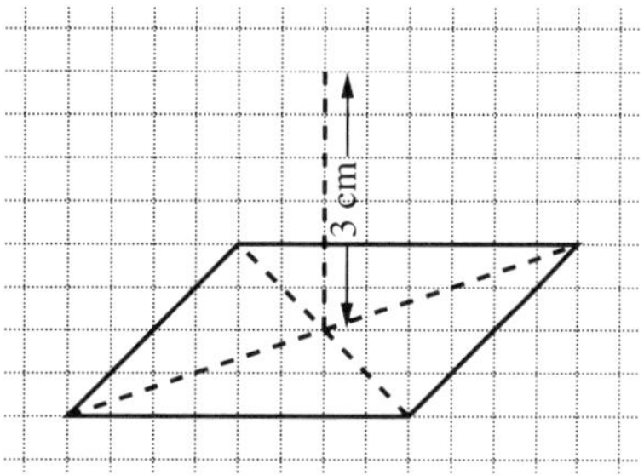

4 Draw in the four slant edges.

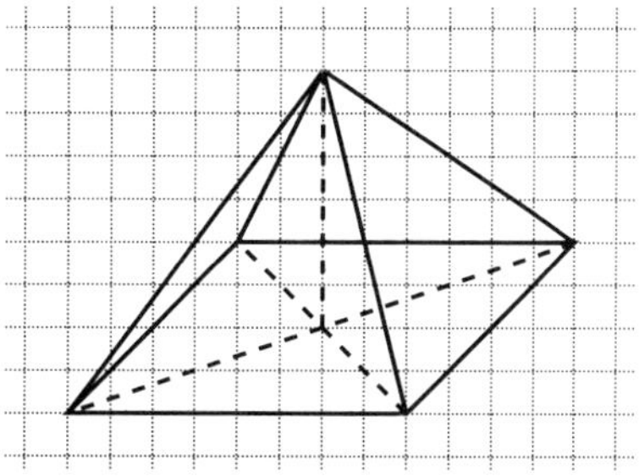

Exercise 11.8

Make a 2-dimensional drawing of each of these square pyramids on squared paper.

1 Base length 4 cm, vertical height 5 cm.

2 Base length 2 cm, vertical height 4 cm.

Example 4

Make a 2-dimensional drawing of a cylinder of diameter 2 cm and height 4 cm.

1 Draw an ellipse (or oval) shape for the circular base. Make its 'larger' diameter equal to the given diameter, but make its smaller diameter only half as much.

2 Draw in two vertical lines of length equal to the height.

3 Draw in another ellipse for the top.

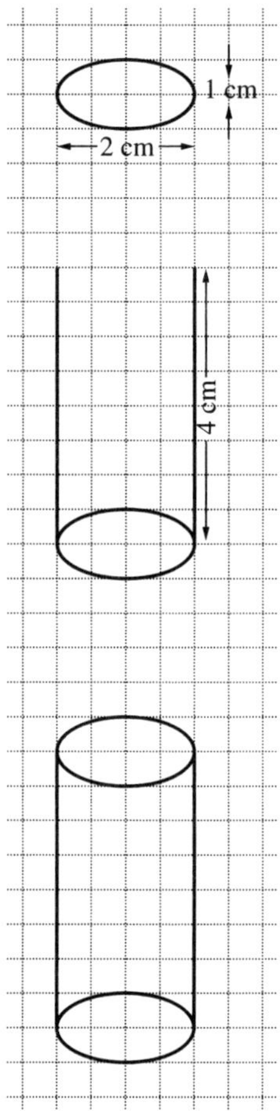

Exercise 11.9

Make a 2-dimensional drawing of each of these cylinders on squared paper.

1 Diameter 2 cm, height 3 cm. **2** Diameter 4 cm, height 5 cm.

Many 3-dimensional objects can be drawn 2-dimensionally on isometric paper.

Example 5

'Perspective' drawing of a 3-dimensional shape.

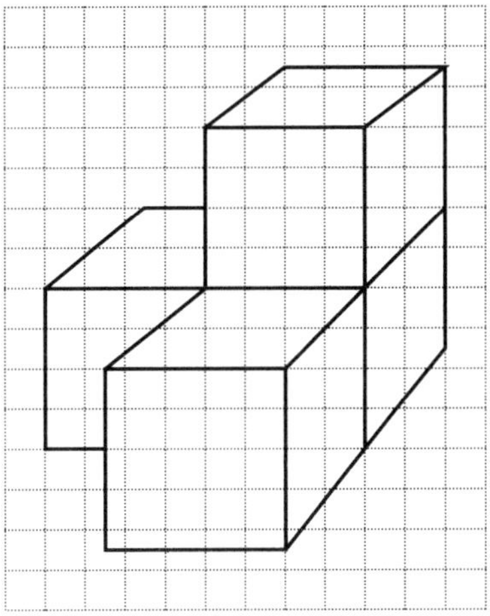

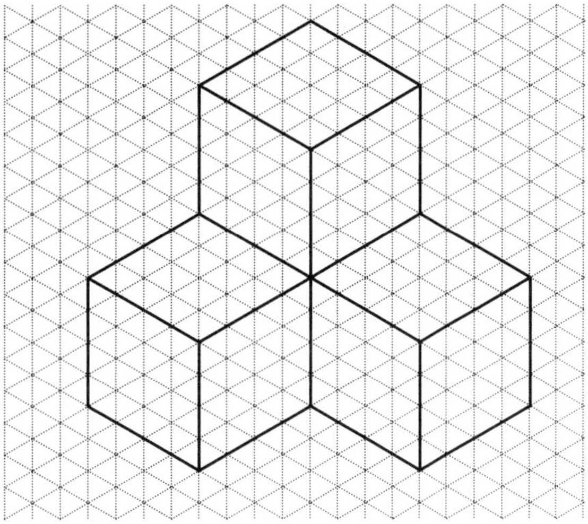

Drawing of the 3-dimensional shape on isometric paper.

Exercise 11.10

Make a drawing of each of these 3-dimensional shapes on isometric paper.

1 **2**

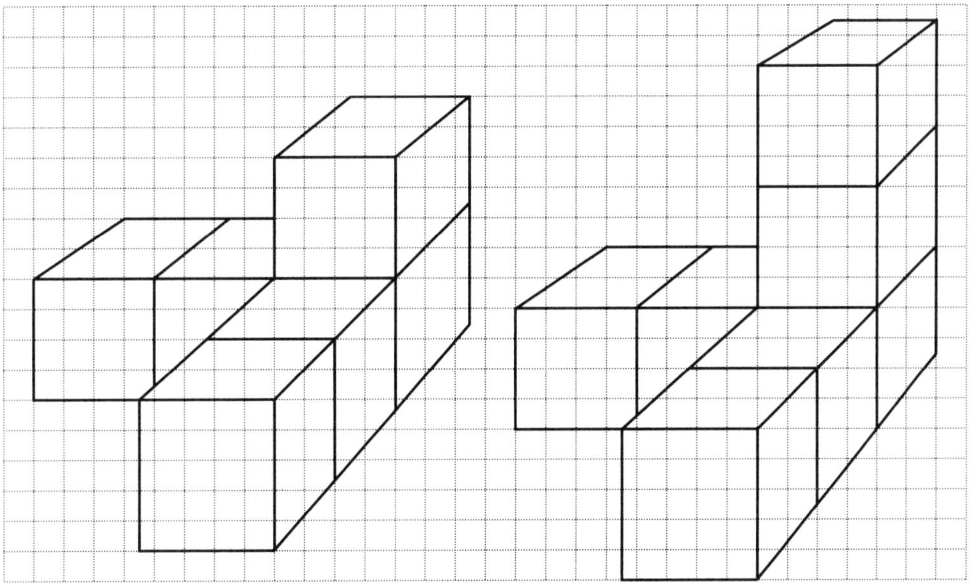

12 Symmetry and quadrilaterals

12.1 Reflecting simple shapes in a mirror line

Example 1

A triangle ABC has its vertices at (10, 2), (10, 16) and (2, 16) respectively.
Draw this triangle on squared paper.
Then draw its image formed by a reflection in the mirror line $x = 12$.

The points A, B and C are marked.
The lines AB, BC, CA are then drawn in.
The mirror line ($x = 12$) is also drawn in.

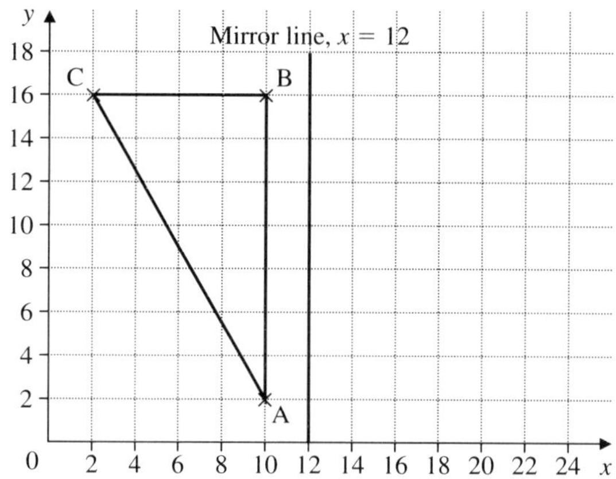

The image points are the same distance behind the mirror as the object points are in front.
The image triangle is $A'B'C'$.

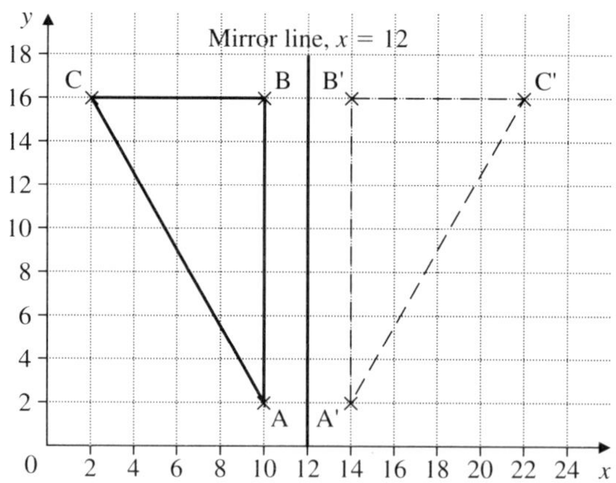

Exercise 12.1

1 Draw these three triangles on the same sheet of squared paper. Then draw the image of each triangle which is formed by a reflection in the line $x = 12$.

a Triangle ABC, with A at (11, 1), B at (11, 5) and C at (8, 1).

b Triangle LMN, with L at (11, 6), M at (11, 13) and N at (7, 6).

c Triangle PQR, with P at (11, 14), Q at (11, 17) and R at (7, 14).

2 Draw these three triangles on the same sheet of squared paper. Then draw the image of each triangle which formed by a reflection in the line $x = 12$.

a Triangle ABC, with A at (11, 5), B at (8, 5) and C at (8, 1).

b Triangle LMN, with L at (11, 9), M at (7, 9) and N at (7, 6).

c Triangle PQR, with P at (11, 14), Q at (4, 14) and R at (4, 10).

Exercise 12.2

1 Draw these four-sided figures on the same sheet of squared paper. Then draw the images which are formed by a reflection in the line $x = 12$.

a Figure ABCD, with A at (11, 1), B at (7, 1), C at (7, 3) and D at (11, 6).

b Figure PQRS, with P at (11, 7), Q at (8, 7), R at (8, 10) and S at (11, 14).

2 Draw these four-sided figures on the same sheet of squared paper. Then draw the images which are formed by a reflection in the line $x = 12$.

a Figure ABCD, with A at (11, 1), B at (7, 1), C at (7, 6) and D at (11, 3).

b Figure WXYZ, with W at (11, 7), X at (8, 7), Y at (8, 14) and Z at (11, 10).

3 Draw these four-sided figures on the same sheet of squared paper. Then draw the images which are formed by a reflection in the line $x = 12$.

a Figure ABCD, with A at (11, 1), B at (11, 8), C at (8, 8) and D at (4, 1).

b Figure EFGH, with E at (11, 10), F at (4, 10), G at (4, 15) and H at (8, 15).

In Exercises 12.1 and 12.2 the mirror line was always $x = 12$.
However, the mirror line can be anywhere.
Example 2 shows a reflection in the line $y = 9$.

Example 2

A triangle ABC has its vertices at (2, 2), (18, 2) and (10, 8) respectively. Draw this triangle on squared paper and then draw its image formed by a reflection in the mirror line $y = 9$.

The points A, B and C are marked.
The lines AB, BC and CA are then drawn in.
The mirror line ($y = 9$) is also drawn in.

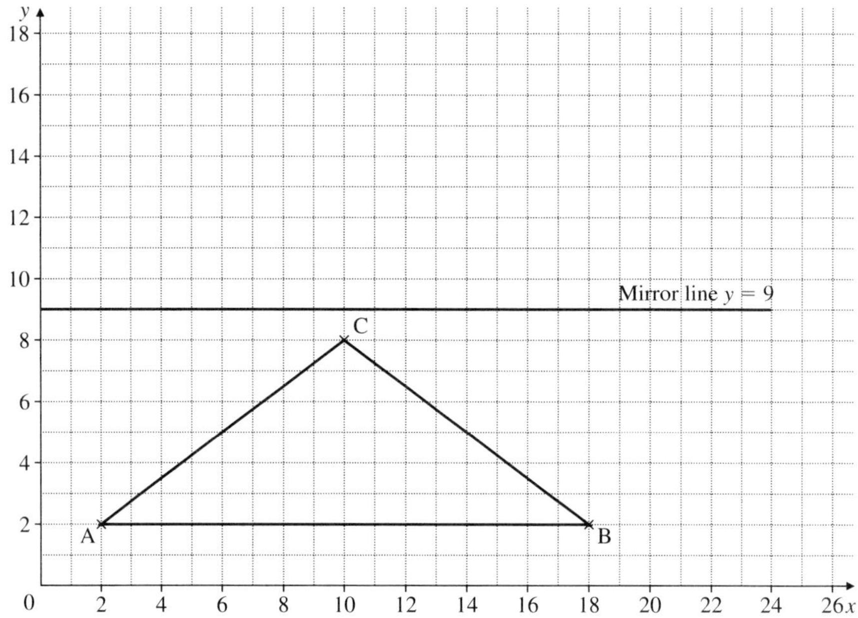

The image points are the same distance behind the mirror as the object points are in front.
The image triangle is A′B′C′.

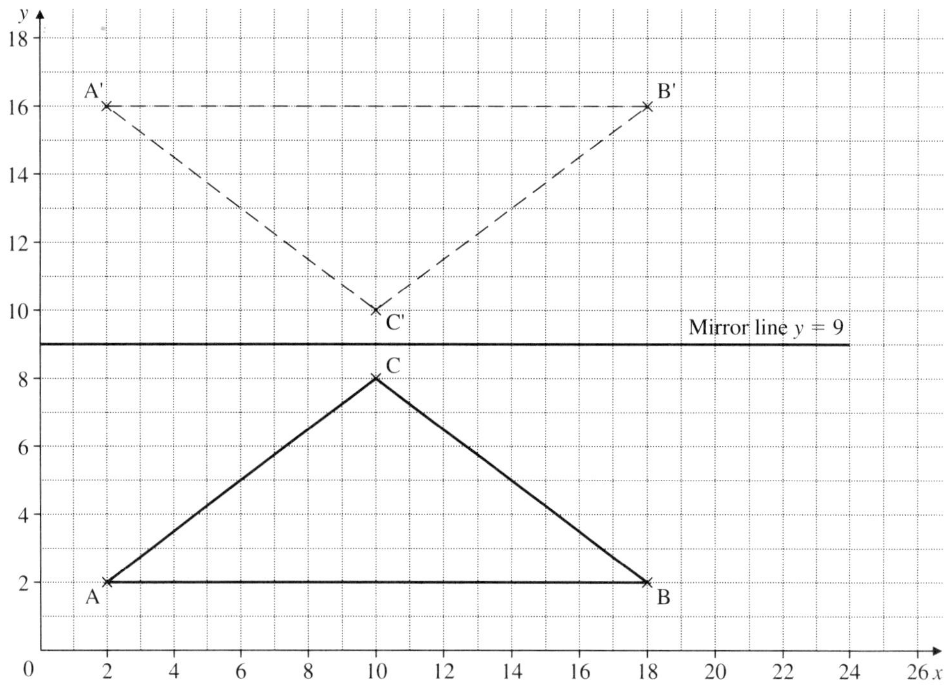

Exercise 12.3

1 Draw these triangles on the same sheet of squared paper and then draw the images which are formed by a reflection in the line $y = 9$.
  a Triangle ABC, with A at (1, 1), B at (9, 1) and C at (5, 8).
  b Triangle LMN, with L at (9, 4), M at (15, 4), and N at (12, 8).
  c Triangle PQR, with P at (16, 5), Q at (24, 5) and R at (20, 8).

2 Draw these triangles on the same sheet of squared paper and then draw the images which are formed by a reflection in the line $y = 9$.
  a Triangle ABC, with A at (1, 8), B at (9, 8) and C at (5, 1).
  b Triangle PQR, with P at (10, 8), Q at (16, 8) and R at (13, 4).
  c Triangle XYZ, with X at (17, 8), Y at (25, 8) and Z at (21, 5).

12.2 Recognising rotational symmetry

The shape illustrated has rotational symmetry order 3.

If a tracing is made of the shape it can be turned to fit exactly on to the shape in 3 different positions.

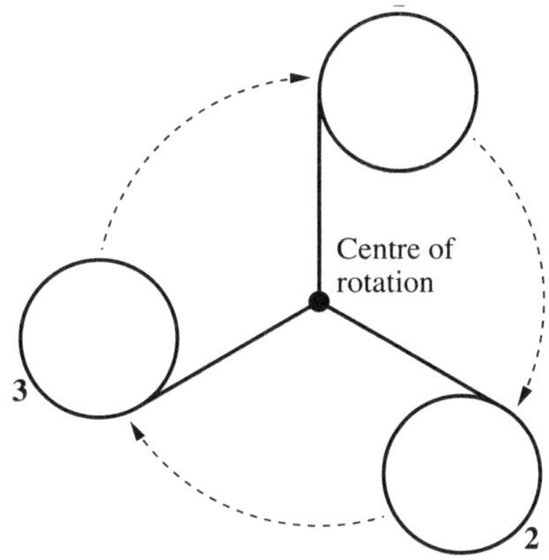

Example 1

State which of these shapes has rotational symmetry and the order of rotational symmetry.

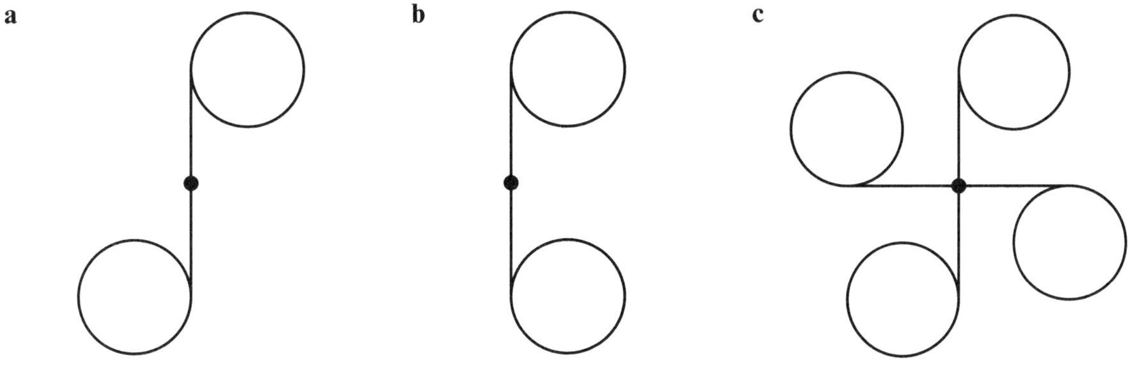

a Rotational symmetry, order 2. **b** No rotational symmetry. **c** Rotational symmetry, order 4.

Exercise 12.4

State which of these shapes has rotational symmetry and the order of rotational symmetry.

1

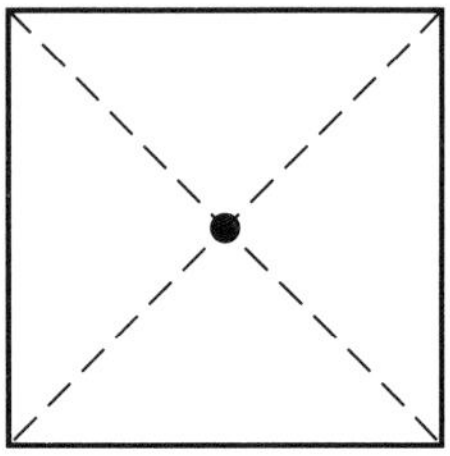

2

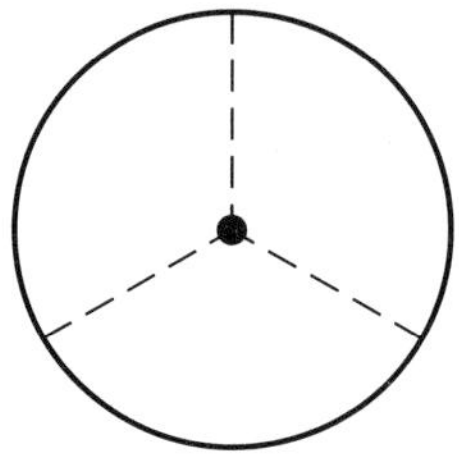

3

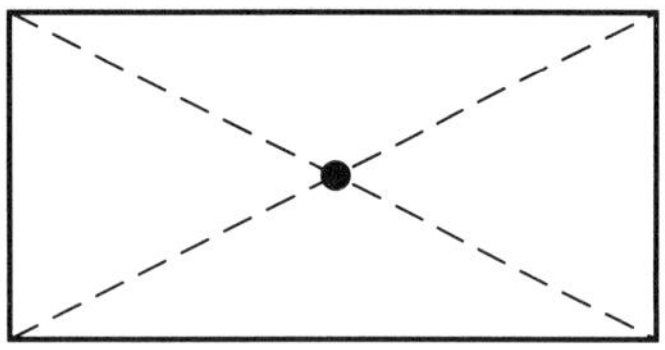

4

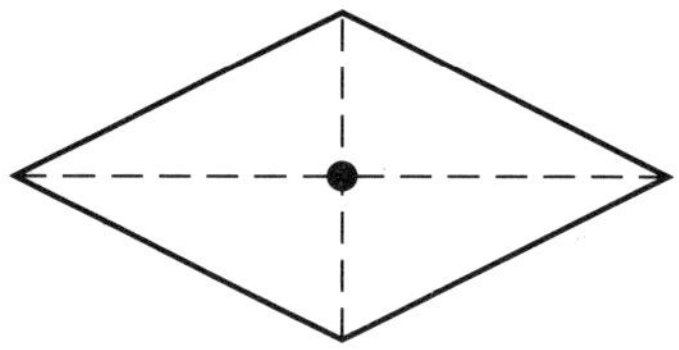

5

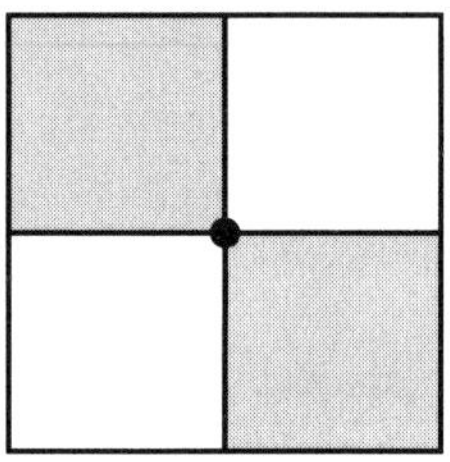

6

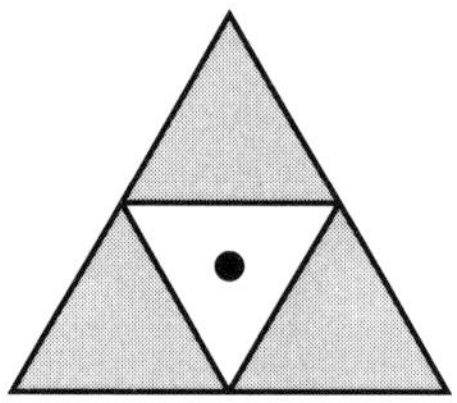

7

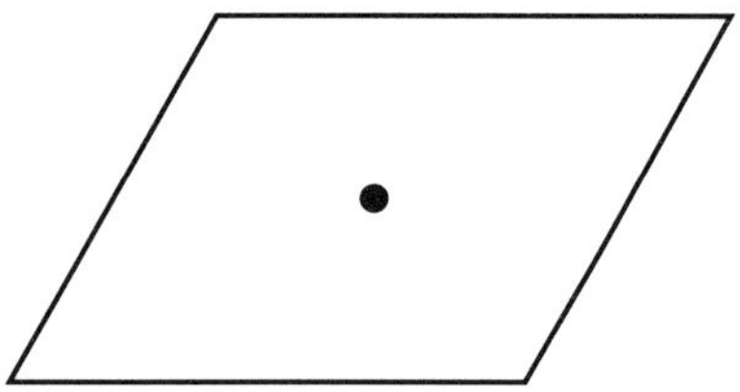

8

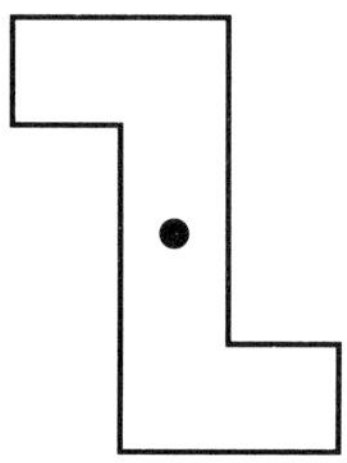

9

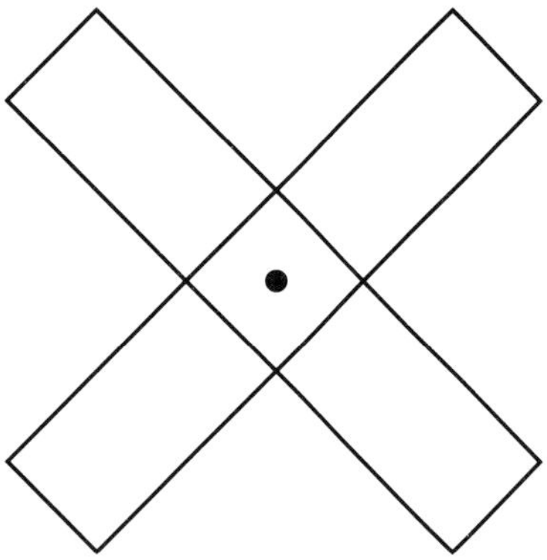

10

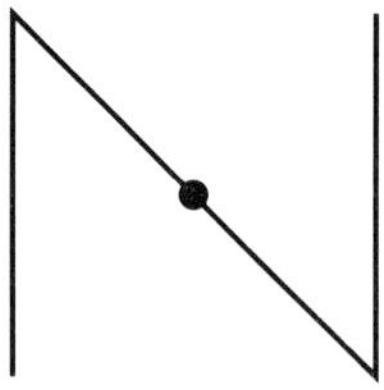

Example 2

Each diagram shows one half of a shape with rotational symmetry of order 2.
Draw the whole shape for each one.

a

b

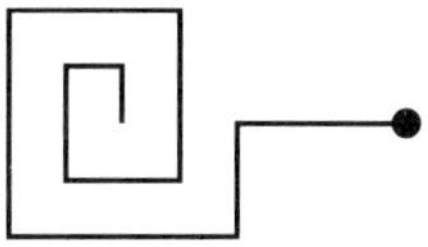

c

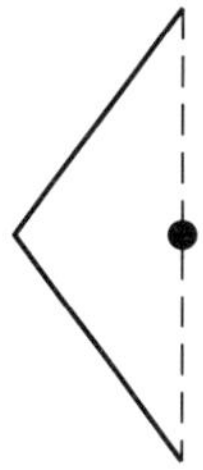

a

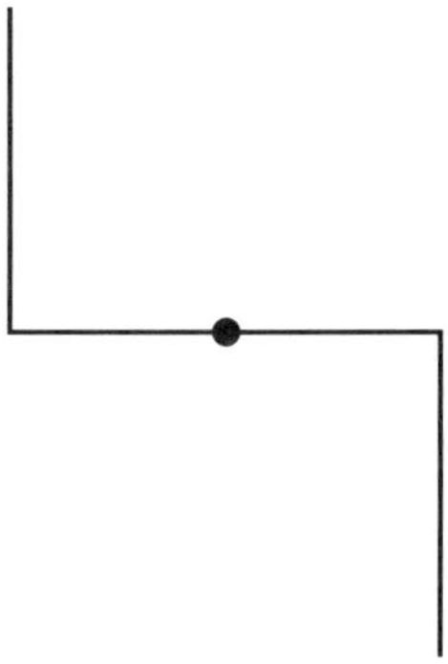

b

c

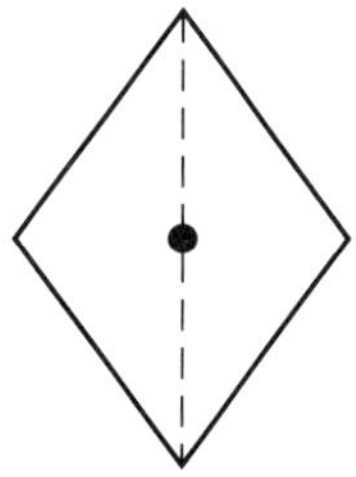

Exercise 12.5

Each diagram shows one half of a shape with rotational symmetry of order 2.
Draw the whole shape for each one.

1

2

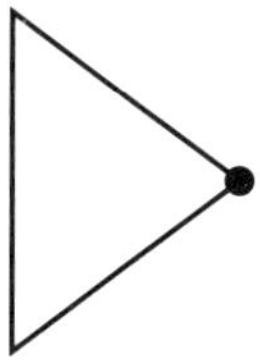

3

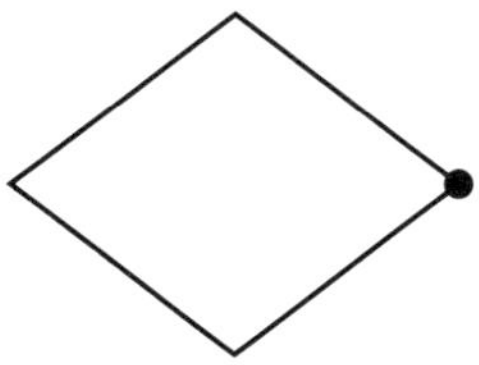

4

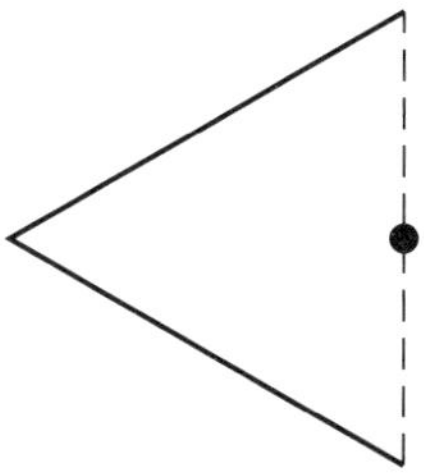

5

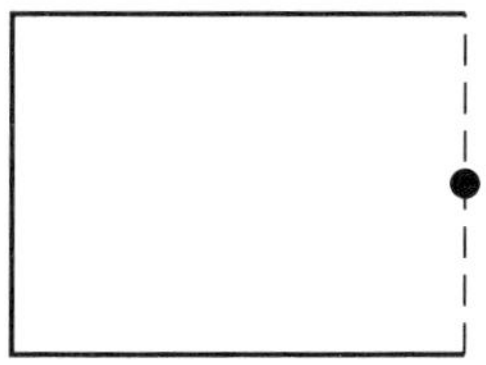

6

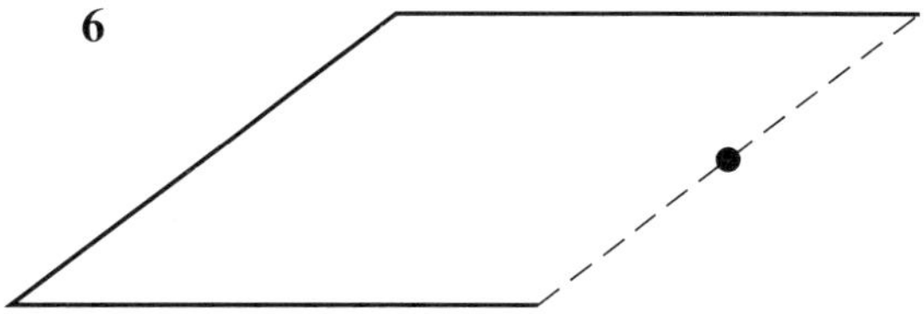

Example 3

Each diagram shows one quarter of a shape with rotational symmetry of order 4.
Draw the whole shape for each one.

a

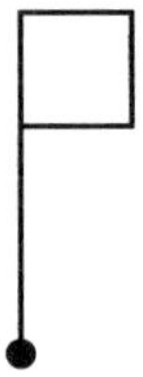

b

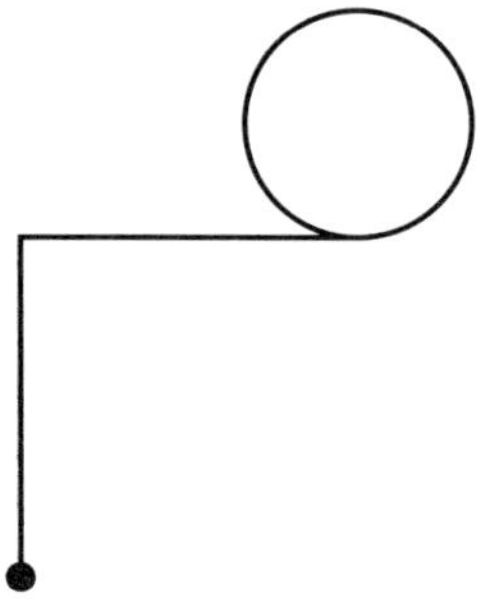

c

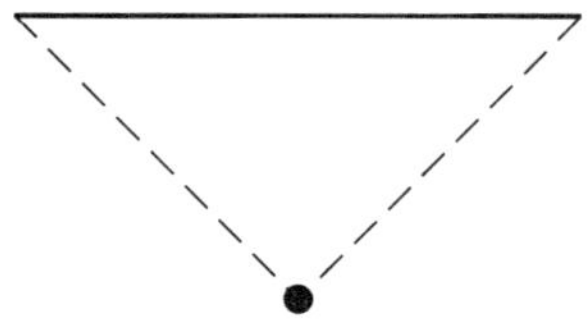

a

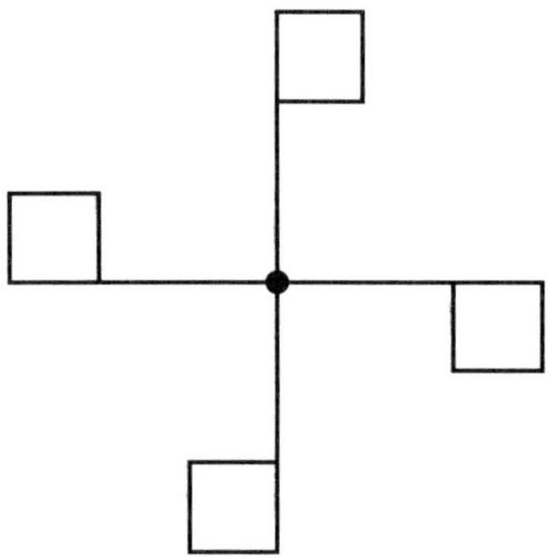

b

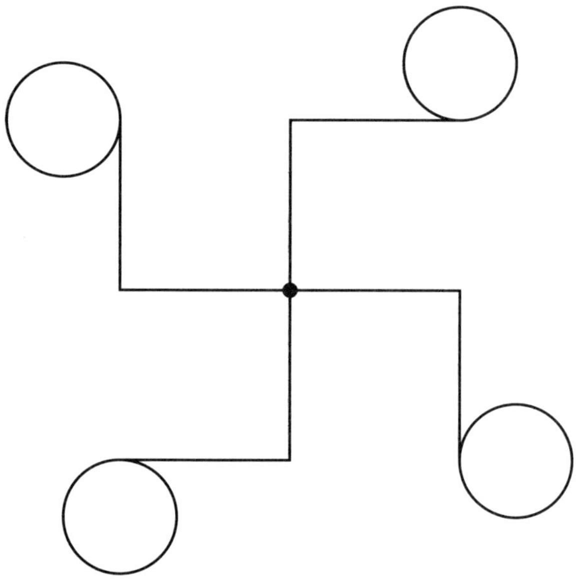

c

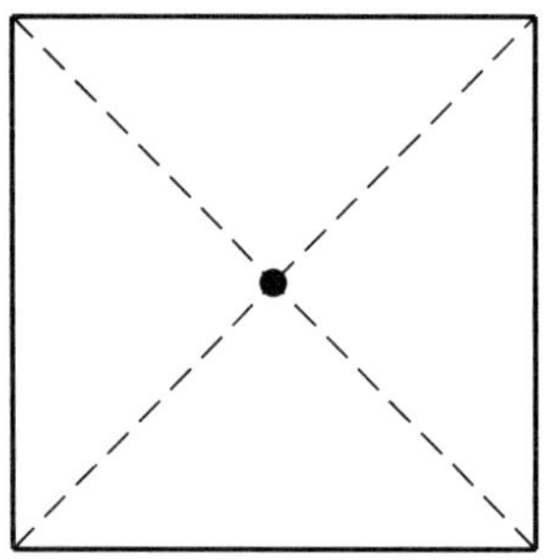

Exercise 12.6

Each diagram shows one quarter of a shape with rotational symmetry of order 4.
Draw the whole shape for each one.

1

2

3
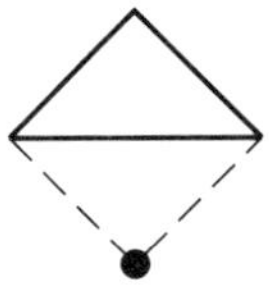
4
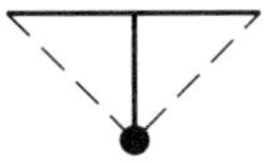

12.3 Identifying the symmetry of various shapes

Reminder: Two-dimensional symmetry

Reflective symmetry:
The dashed lines are lines of symmetry.

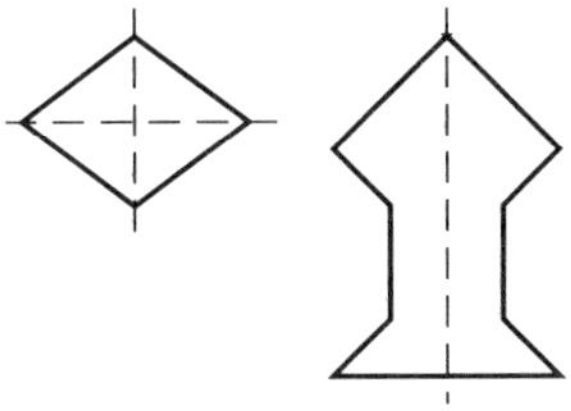

Rotational symmetry:
These two shapes have rotational symmetry of order 3.

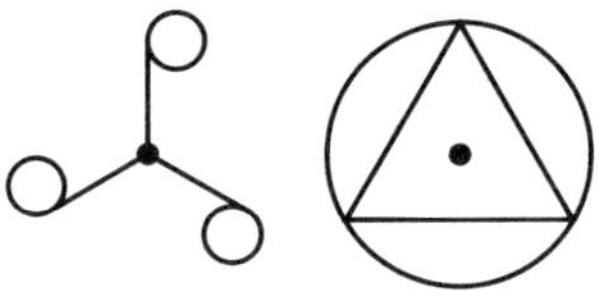

Example 1

On each diagram mark in any lines of symmetry and state the order of rotational symmetry.

a

b

a Rotational symmetry of order 4.

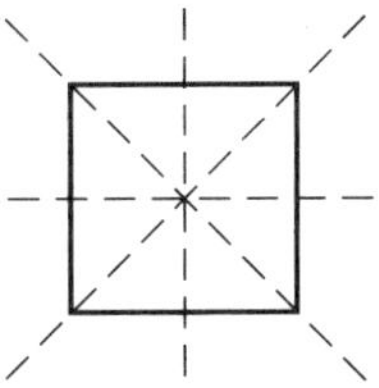

b Rotational symmetry of order 5.

Exercise 12.7

Copy these diagrams.
Mark in any lines of symmetry and state the order of rotational symmetry.

1

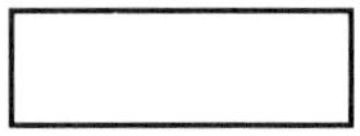

2

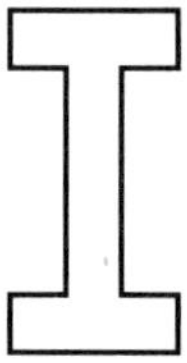

3

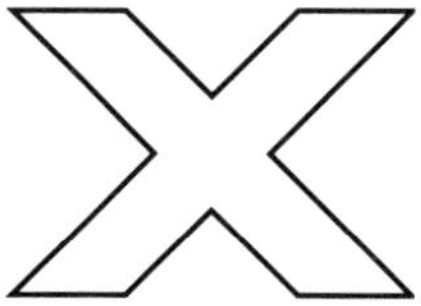

4

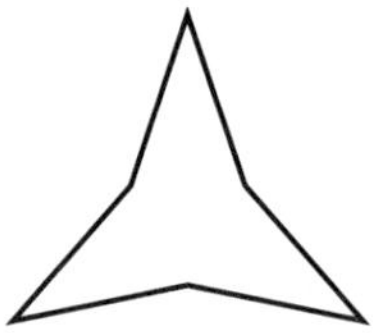

5

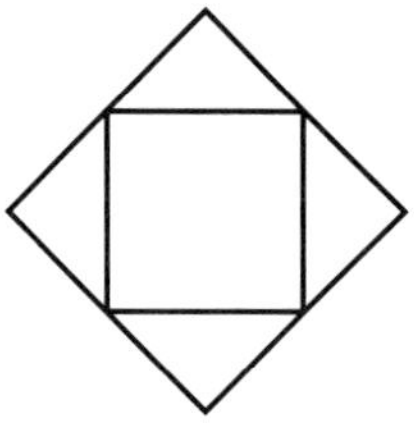

6

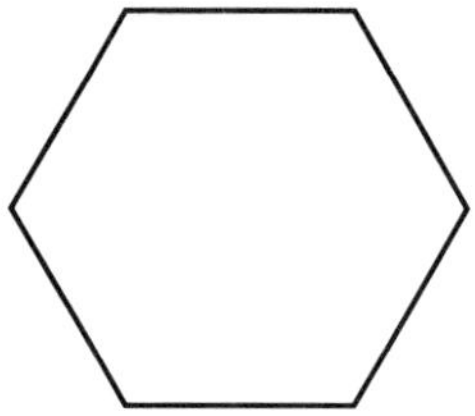

Solids may have planes of reflected symmetry and/or axes of rotational symmetry.

Example 2

State the number of planes of symmetry that the cuboid has. State the order of rotational symmetry about each of the axes shown.

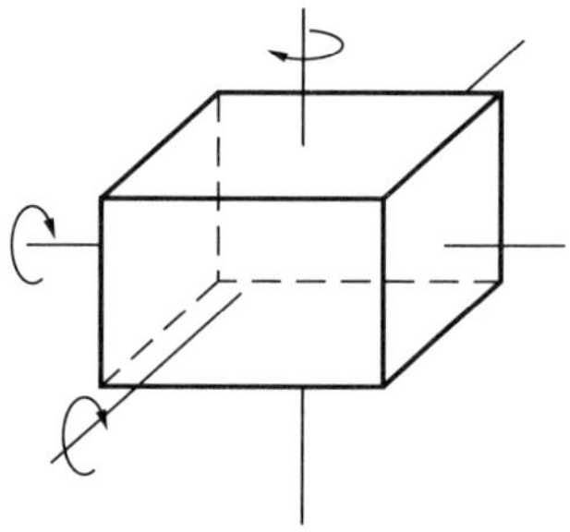

The cuboid has three planes of symmetry: ABCD, LMNP and WXYZ.

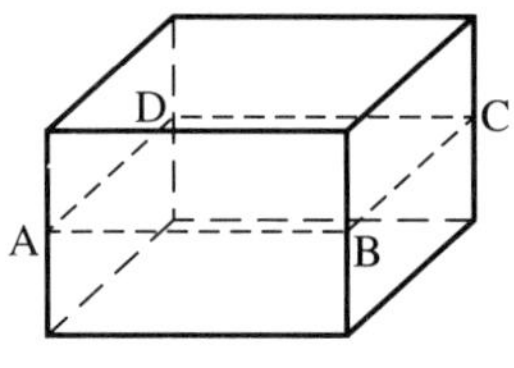

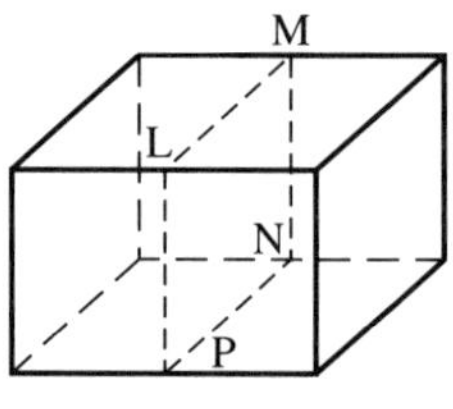

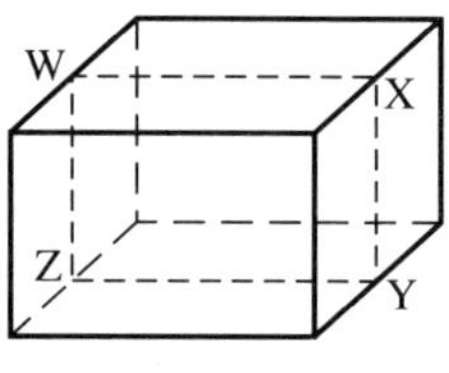

The order of rotational symmetry about each of the axes shown is 2, because each rectangular face has rotational symmetry of order 2.

Example 3

State the number of planes of symmetry that the pyramid has.
State the order of rotational symmetry about the axis shown.

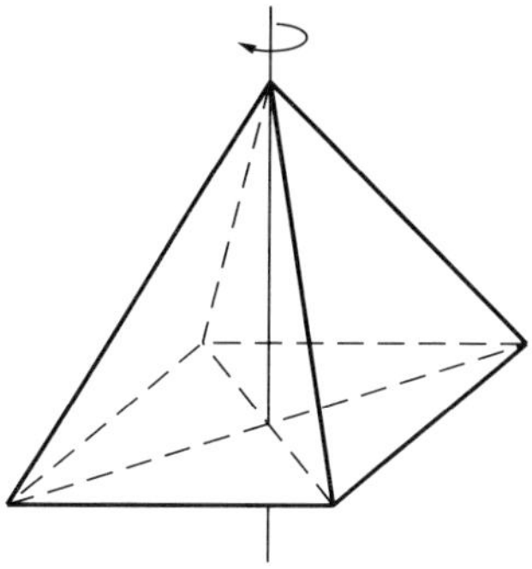

The pyramid has four planes of symmetry: ABC, DBE, LBM and PBQ.

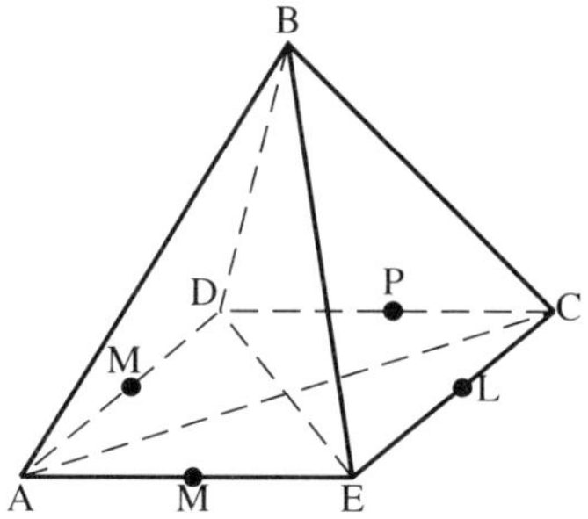

The order of rotational symmetry about the axis shown is 4 (the pyramid has a square base).

Exercise 12.8

For each of the following shapes state
a the number of planes of symmetry, and
b the order of rotational symmetry about the axis or axes shown.

1

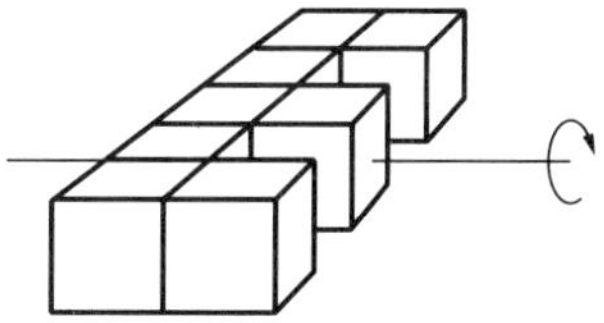

2

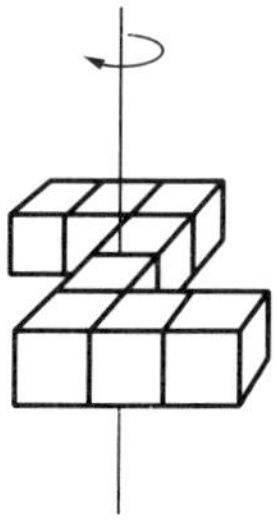

3

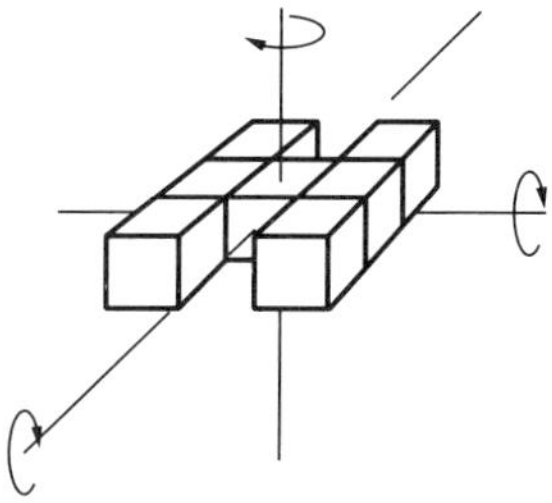

4

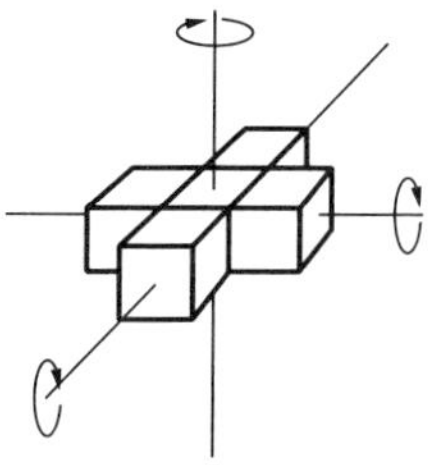

12.4 Classifying and defining types of quadrilaterals

Reminder

A quadrilateral is an enclosed shape with four straight edges.
A quadrilateral can be irregular in shape.
There are also seven kinds of special quadrilaterals.

They are all listed below together with their definitions and properties.

Irregular quadrilateral

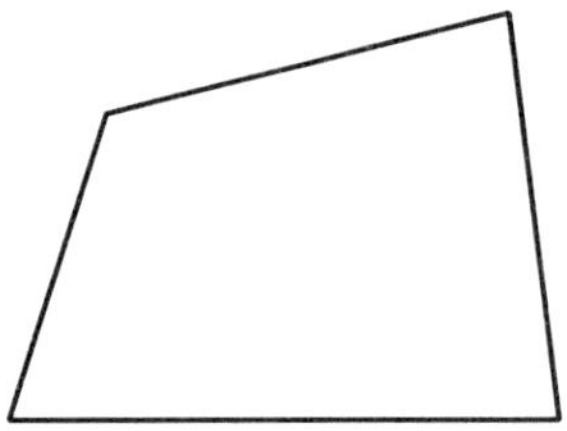

No symmetry

Trapezium

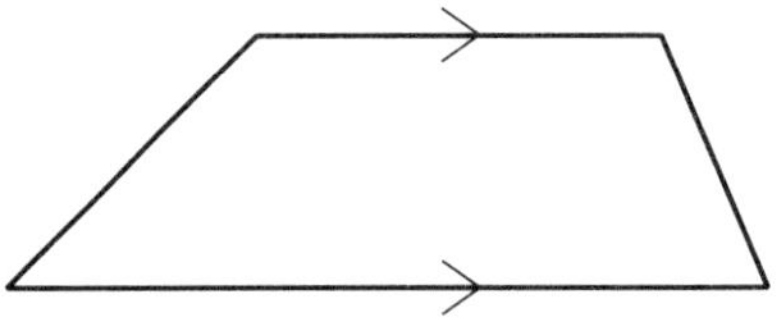

No symmetry, but one pair of parallel sides

Isosceles trapezium

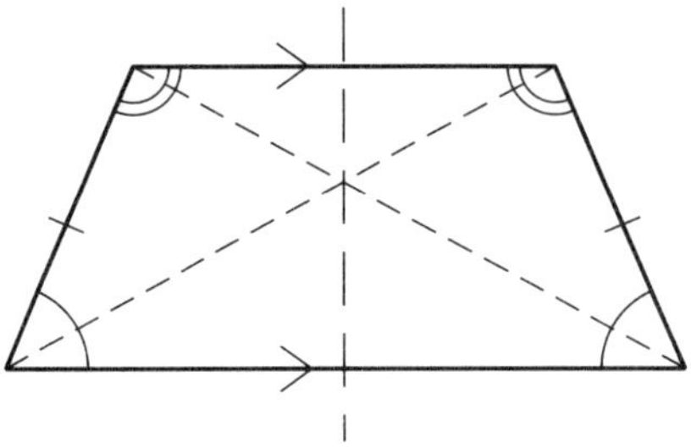

One pair of parallel sides
One pair of equal sides
Two pairs of equal angles
One axis of symmetry
Diagonals of equal length

Kite

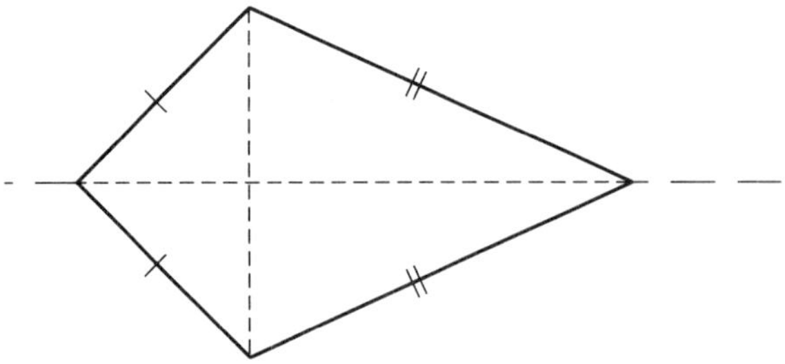

Two pairs of equal sides
One pair of equal angles
One symmetry axis
Diagonals cross at right angles

Parallelogram

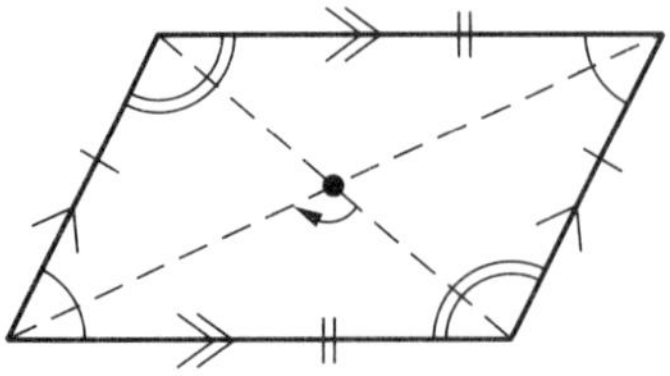

Two pairs of parallel sides
Two pairs of equal sides
Two pairs of equal angles
Rotational symmetry, order 2
Diagonals bisect each other

Rhombus

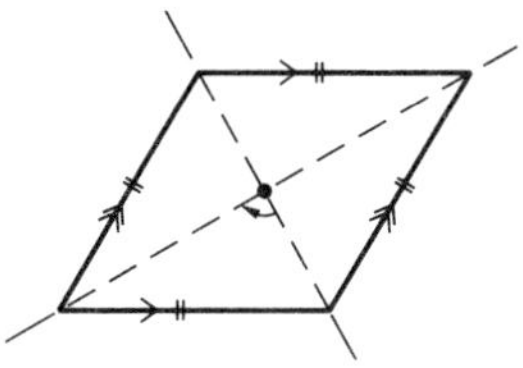

Two pairs of parallel sides
All four sides equal in length
Two pairs of equal angles
Two axes of symmetry
Rotational symmetry, order 2
Diagonals cross at right angles
Diagonals bisect each other

Rectangle

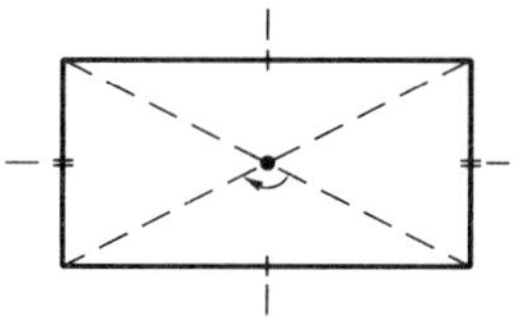

Two pairs of parallel sides
Two pairs of equal sides
All four angles equal to 90°
Two axes of symmetry
Rotational symmetry, order 2
Diagonals of equal length
Diagonals bisect each other

Square

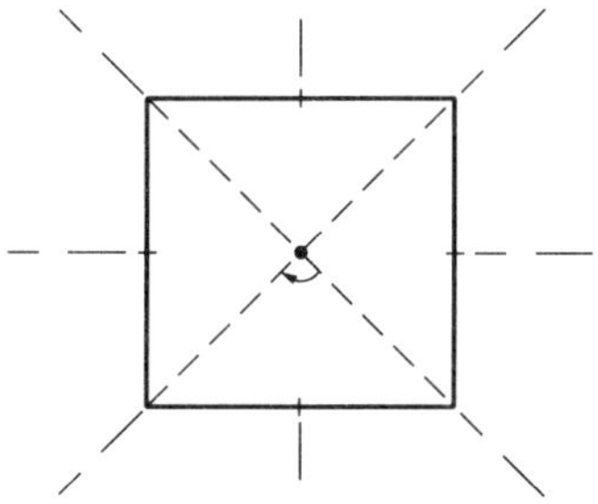

Two pairs of parallel sides
All four sides equal in length
All four angles equal to 90°
Four axes of symmetry
Rotational symmetry, order 4
Diagonals of equal length
Diagonals cross at right angles
Diagonals bisect each other

Example 1

How many quadrilateral shapes can you make from three isosceles, right-angled triangles? Illustrate your answers.

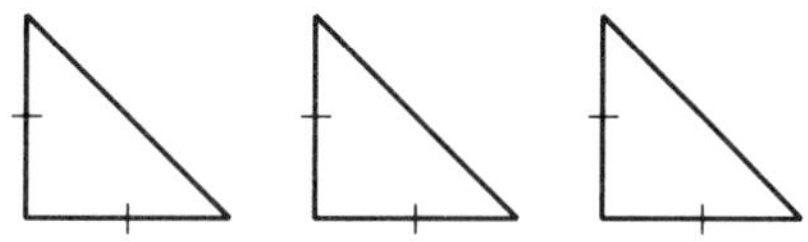

Two quadrilateral shapes can be made from three isosceles, right-angled triangles.

a Trapezium

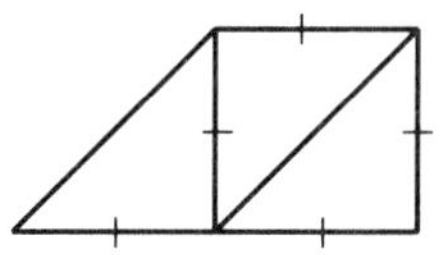

b Isosceles trapezium

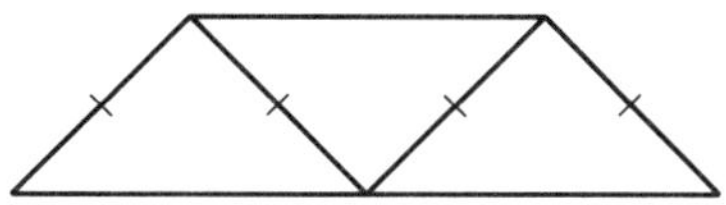

Exercise 12.9

What different quadrilateral shapes can you make using each of the following?
Illustrate your answers.

1 Two equilateral triangles.

2 Two isosceles, right-angled triangles.

3 Two non-isosceles, right-angled triangles.

4 Three non-isosceles, right-angled triangles.

5 Four equilateral triangles.

6 Four isosceles, right-angled triangles.

Example 2

Join the dots on the pattern together in order to make as many different quadrilateral shapes as possible.
Illustrate your answers.

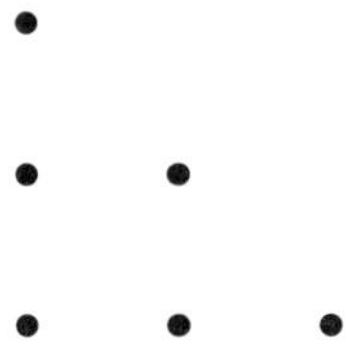

Four quadrilateral shapes are possible.

a A square

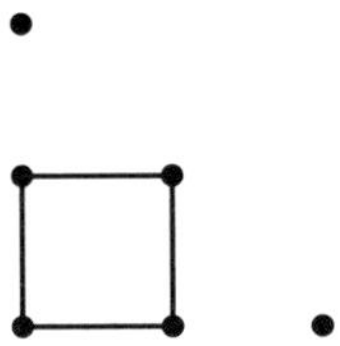

b A parallelogram

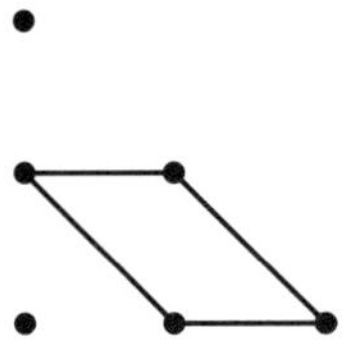

c An isosceles trapezium

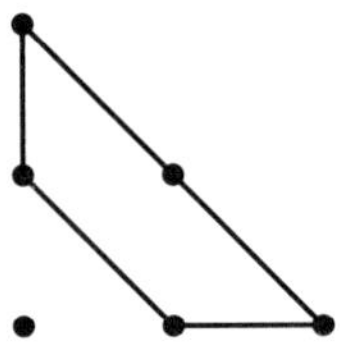

d A trapezium

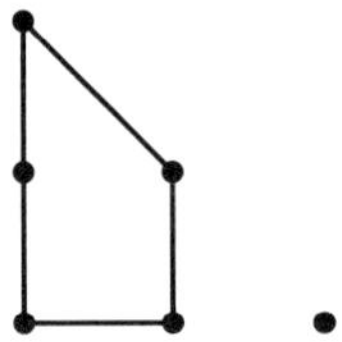

Exercise 12.10

For each question below join the dots on the pattern together in order to make as many different quadrilateral shapes as possible.
Illustrate your answers.

1

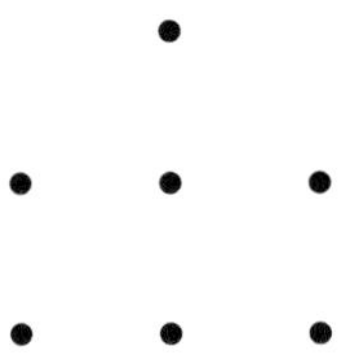

2

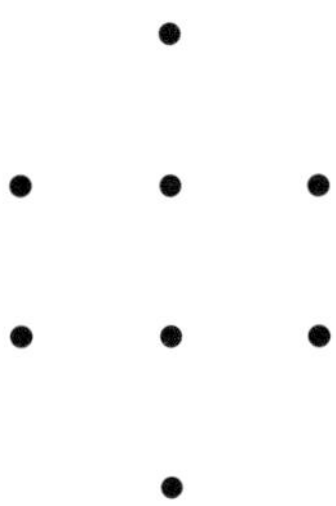

3

Example 3

Copy this square and join BE and LE.

What kind of quadrilateral shape has been made between the lines?

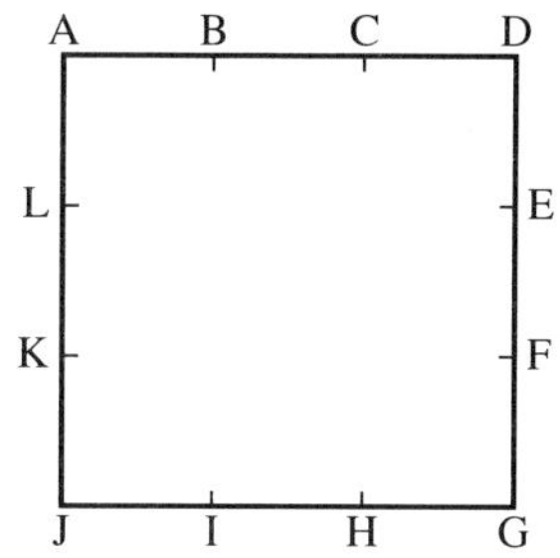

A trapezium has been made between the lines BE and LE.

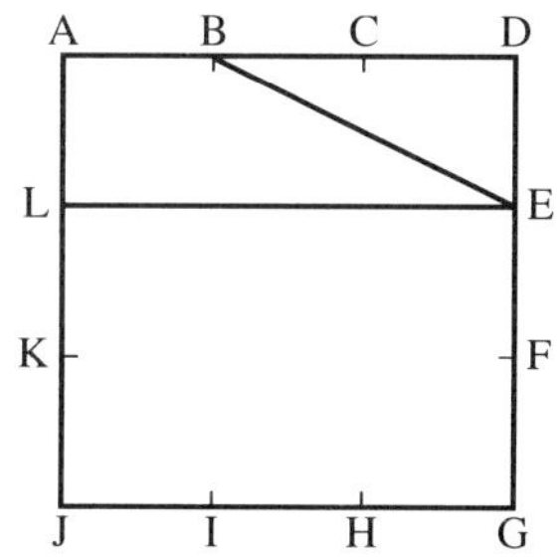

Exercise 12.11

For each question copy the square illustrated and join the points stated.

State what kind of quadrilateral shape is made between the lines.

1 LD and KF **2** LB and HF **3** BJ and CG

4 BF and KF **5** LG and BG **6** BI and CH

13 Angles, parallel lines and polygons

13.1 Understanding angles

An angle is how we measure turn or change of direction.
We can use a compass to find the North direction, and hence the other directions shown by the points on the compass.

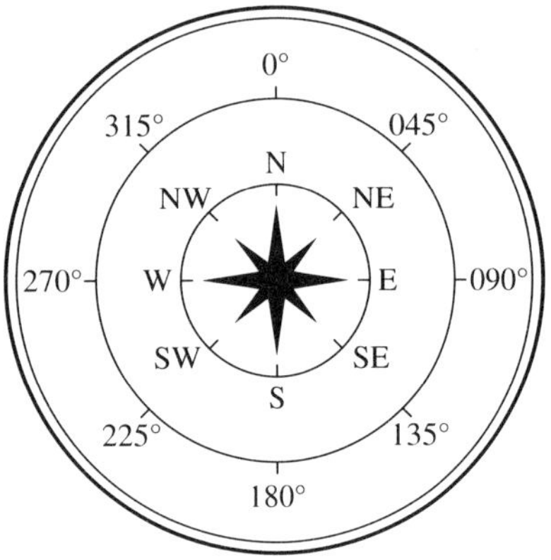

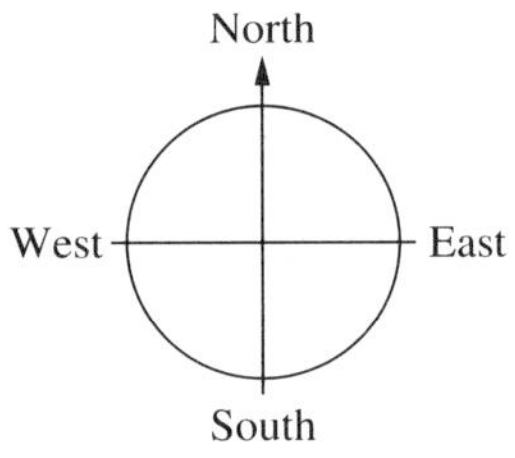

A complete turn is divided into 360 parts called degrees.
1 complete turn = 360°.

Example 1

Find the number of degrees of turn when we turn in a clockwise direction:

a From N to E

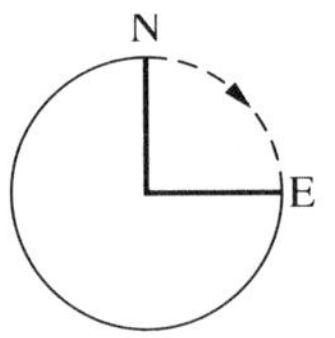

b From NE to W

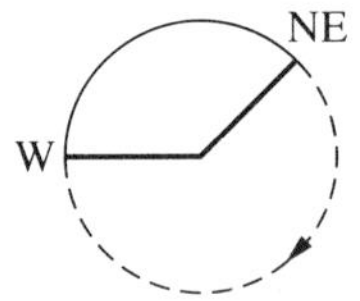

c From NW to SE

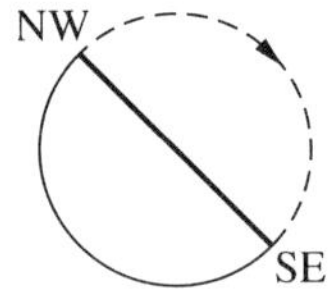

a From N to E = 90°

b NE to E = 45°
E to W = 180°
So NE to W = 45° + 180° = 225°

c NW to N = 45°
N to E = 90°
E to SE = 45°
So NW to SE =
45° + 90° = 45° = 180°

Exercise 13.1

Find the number of degrees of turn when we turn in a clockwise direction:

1 From N to E **2** From N to S **3** From N to NE **4** From N to W
5 From SE to W **6** From SE to NW **7** From W to NE **8** From NW to SW.

Example 2

Find the size of the smaller angle between the hands of the clock when the clock is showing the following times:

a 3 o'clock **b** 1 o'clock **c** 8 o'clock

a The angle is 90° **b** The angle is one-third of $90^\circ = 30^\circ$ **c** The angle is $90^\circ + 30^\circ = 120^\circ$

Exercise 13.2

Find the size of the smaller angle between the hands of a clock when the clock is showing the following times:

1 9 o'clock **2** 6 o'clock **3** 2 o'clock **4** 10 o'clock **5** 4 o'clock

A quarter turn (90°) is called a **right** angle.
An angle that is less than 90° is called an **acute** angle.
An angle that is greater than 90° but less than 180° is called an **obtuse** angle.
An angle of 180° is called a **straight** angle because it is a straight line.
An angle that is greater than 180° but less than 360° is called a **reflex** angle.

Example 3

This weighing machine in the school office can weigh up to 300 grams.

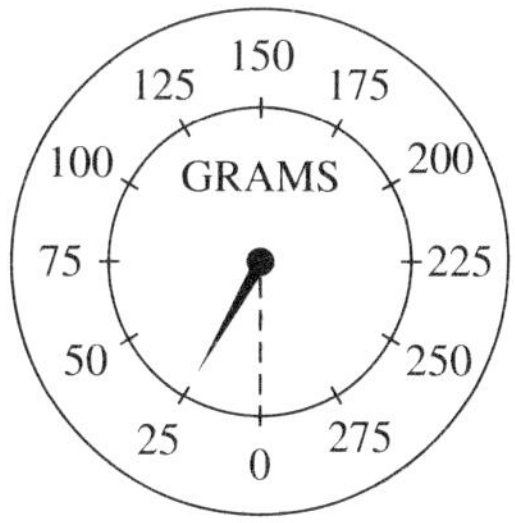

Find the angle through which the pointer turns when weighing a letter of 25 g. State the type of angle.

The pointer moves from 0 to 25.
This is $\frac{25}{300}$ of a complete turn.
angle $= \frac{25}{300} \times 360^\circ = 30^\circ$
This is an acute angle.

Example 4

This weighing machine in the school office can weigh up to 300 grams.

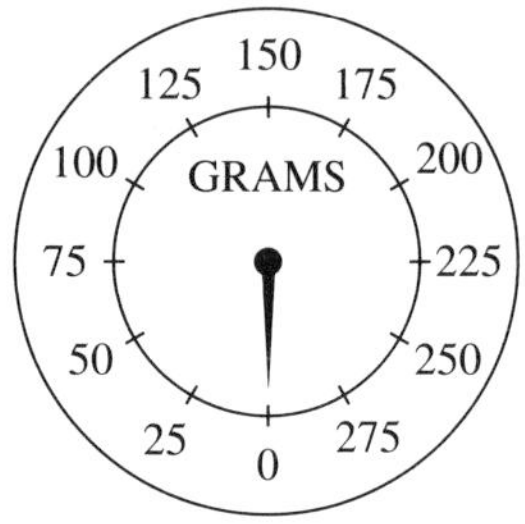

Find the angle through which the pointer turns when weighing a parcel of 150 g.
State the type of angle.
The pointer moves from 0 to 150
This is $\frac{150}{300}$ of a complete turn.
angle $= \frac{150}{300} = 360^\circ = 180^\circ$
This is a straight angle.

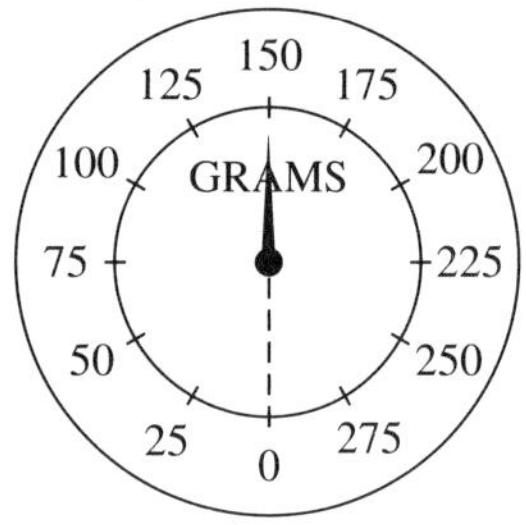

Exercise 13.3

This weighing machine can weigh up to 120 kg.

David weighs 60 kg.
Jamilla weighs 40 kg.
Jo weighs 30 kg.
Leroy weighs 20 kg.
Jamilla's dog Spot weighs 10 kg.

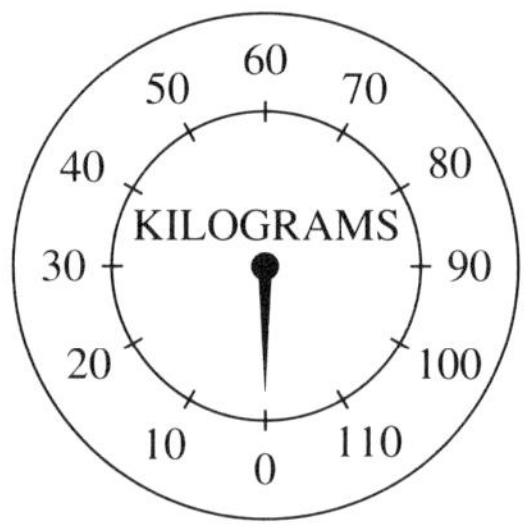

Find the angle through which the pointer turns when the following stand on the machine.
State the type of angle.

1 Leroy **2** Spot **3** Jo **4** Jamilla **5** David

6 Leroy and Spot together **7** Jo and Spot together

8 Jamilla and Spot together **9** Jamilla and Leroy together

Two angles whose sum is 90° are called **complementary** angles.
20° is the complement of 70°.

Two angles whose sum is 180° are called **supplementary** angles.
20° is the supplement of 160°.

Exercise 13.4

Find the complement of each angle:

1 10° **2** 20° **3** 59° **4** 45° **5** 38.5° **6** 62.5°

Find the supplement of each angle:

7 20° **8** 40° **9** 90° **10** 110° **11** 80.5° **12** 135.5°

The two straight lines AB and AC meet at A.
They form an acute angle BAC, which is written $B\hat{A}C$.

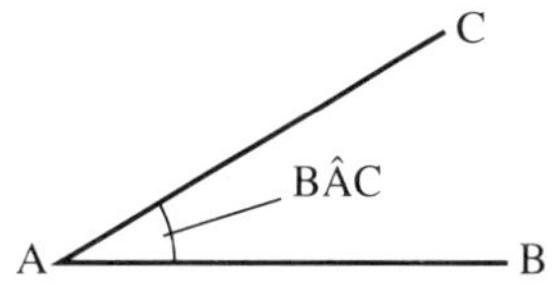

The two straight lines XY and XZ meet at X.
They form an obtuse angle $Y\hat{X}Z$.

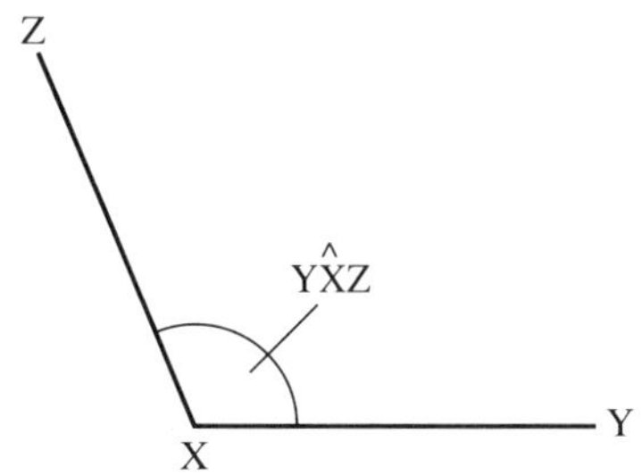

Example 4

Using three letters, name the angles marked (i) a (ii) b (iii) c

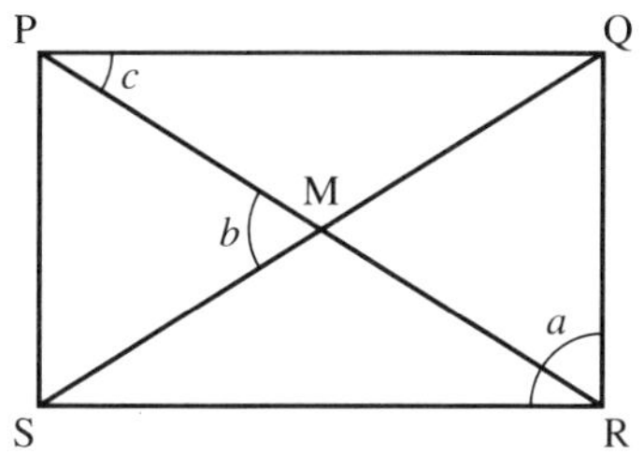

(i) $a = Q\hat{R}S$ (ii) $b = P\hat{M}S$ (iii) $c = Q\hat{P}R$

Exercise 13.5

Using three letters, name the marked angle or angles in each diagram.

1

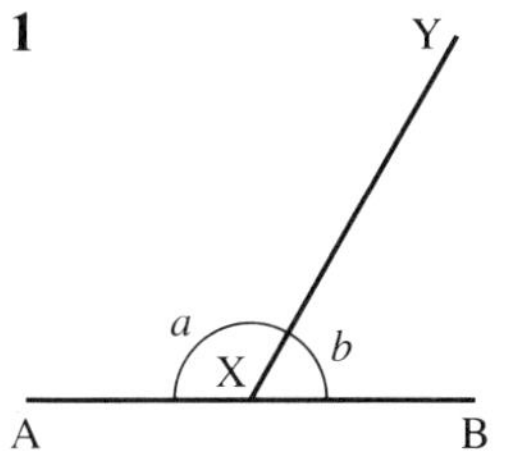

2

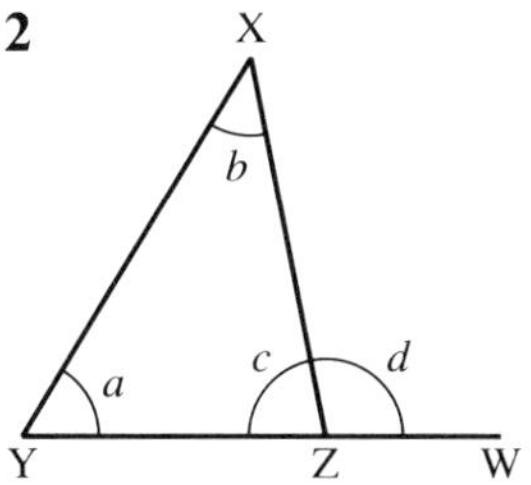

3

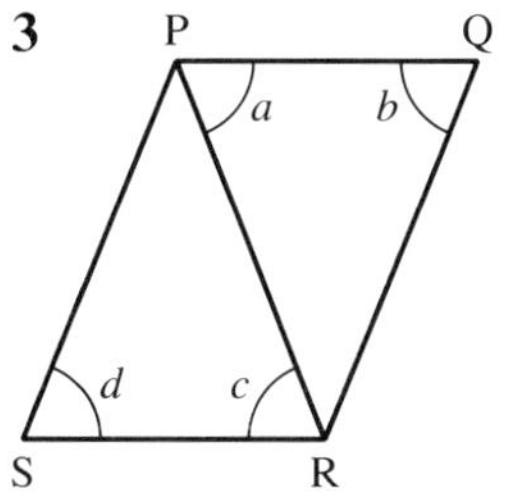

4

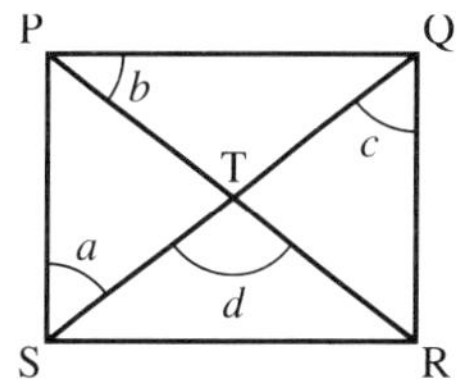

13.2 Measuring and drawing angles to the nearest degree

To measure an angle you need a protractor. The examples below show how to use one. AB shows the base line and O is the centre.

Example 1

Measure $K\hat{L}M$.

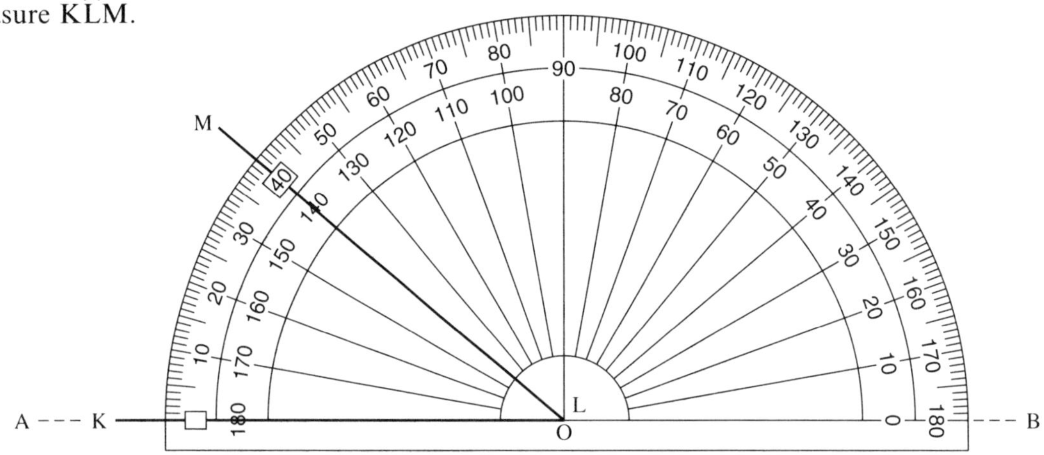

1 Place the protractor on the angle so that O is on L as shown, and OA lies on KL.
2 Where the line LM cuts the outer scale, read off the angle starting from 0° at K.
3 $K\hat{L}M = 40°$

Example 2

Measure $P\hat{Q}R$.

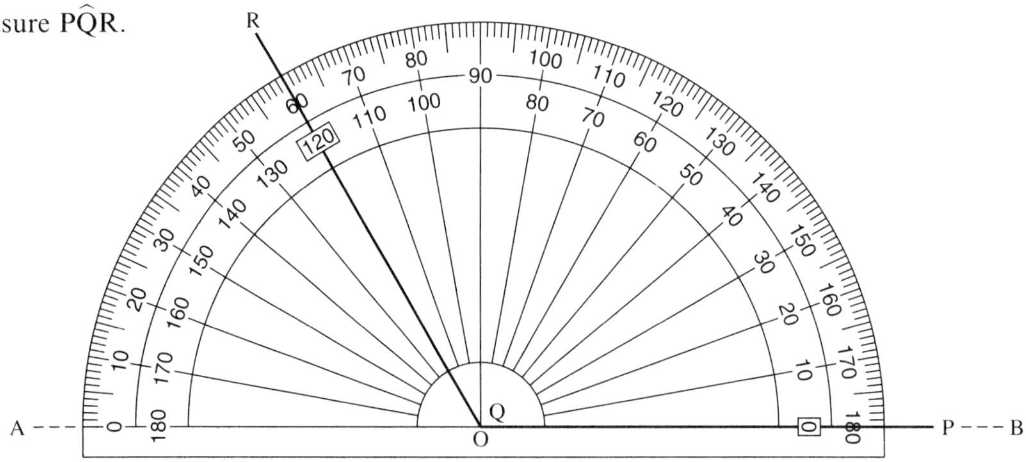

1 Place the protractor on the angle so that O is on Q as shown and OB lies on QP.
2 Where the line QR cuts the inner scale, read off the angle starting from 0° at P.
3 $P\hat{Q}R = 120°$

Exercise 13.6

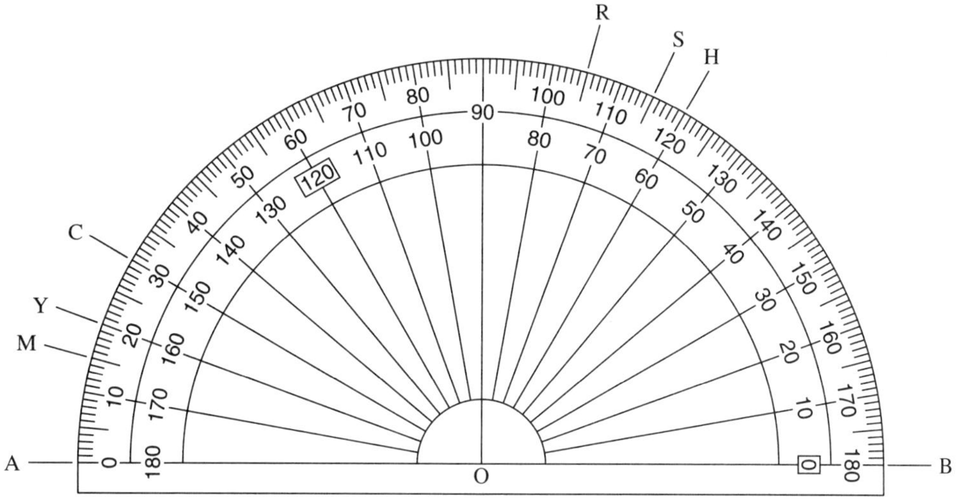

Find the angles on the diagram.

1 a $A\hat{O}H =$ **b** $A\hat{O}C =$ **c** $A\hat{O}Y =$
d $A\hat{O}M =$ **e** $A\hat{O}R =$ **f** $A\hat{O}S =$
g $B\hat{O}C =$ **h** $B\hat{O}Y =$ **i** $B\hat{O}H =$
j $B\hat{O}R =$ **k** $B\hat{O}S =$ **l** $B\hat{O}M =$

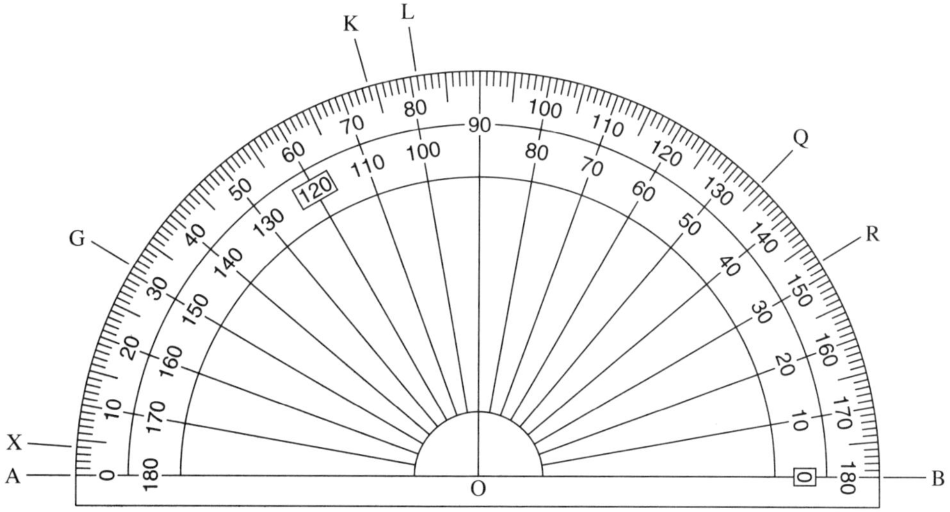

2 a $A\hat{O}G =$ **b** $A\hat{O}K =$ **c** $A\hat{O}X =$
d $A\hat{O}L =$ **e** $A\hat{O}Q =$ **f** $A\hat{O}R =$
g $B\hat{O}K =$ **h** $B\hat{O}G =$ **i** $B\hat{O}X =$
j $B\hat{O}Q =$ **k** $B\hat{O}R =$ **l** $B\hat{O}L =$

Exercise 13.7

Use a protractor to measure each of these angles.

1

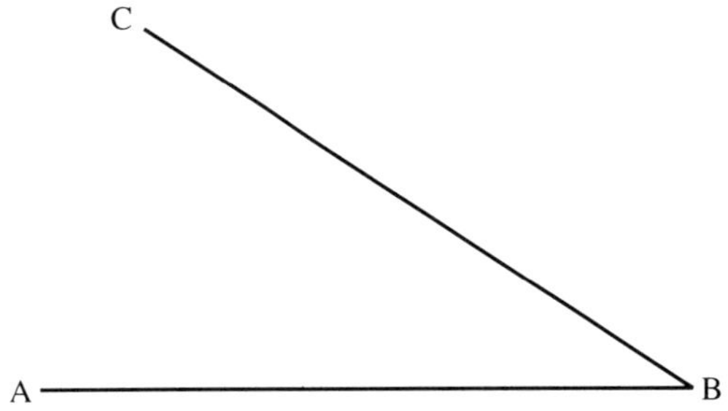

2

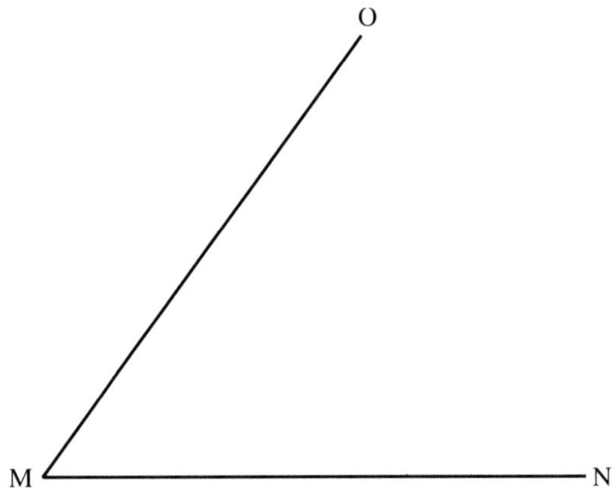

3

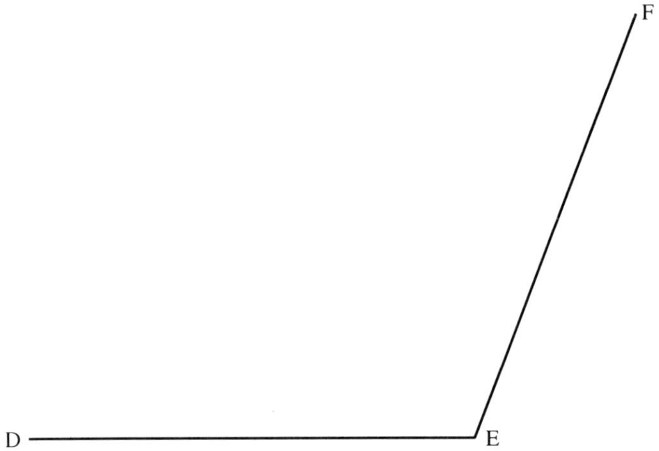

4

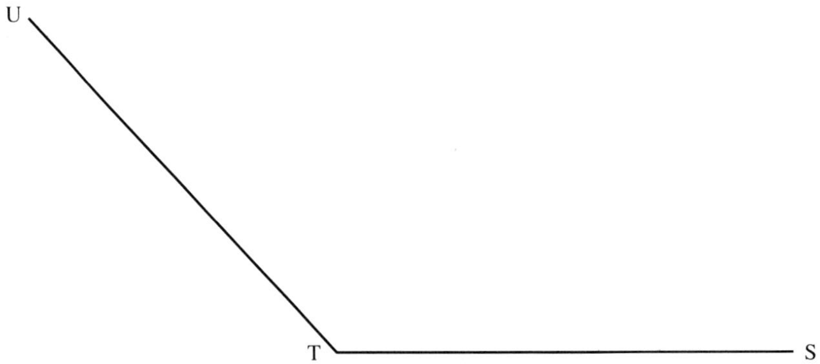

Example 3

Using a protractor draw **a** $A\hat{B}C = 50°$ **b** $X\hat{Y}Z = 170°$

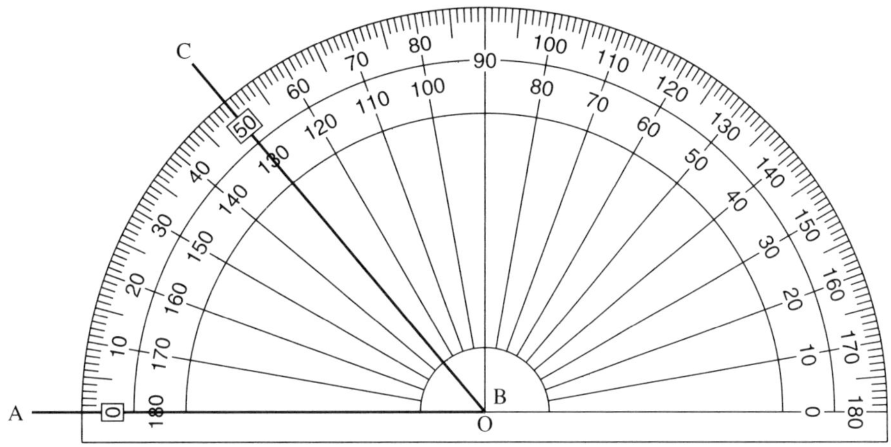

a **1** Draw the straight line AB. Make it 6 cm long.
2 Place the protractor with its base line on AB and its centre point O on B as shown.
3 Mark the point C at 50° on the scale starting from 0° at A.
4 Join BC. The angle 50° is now complete.

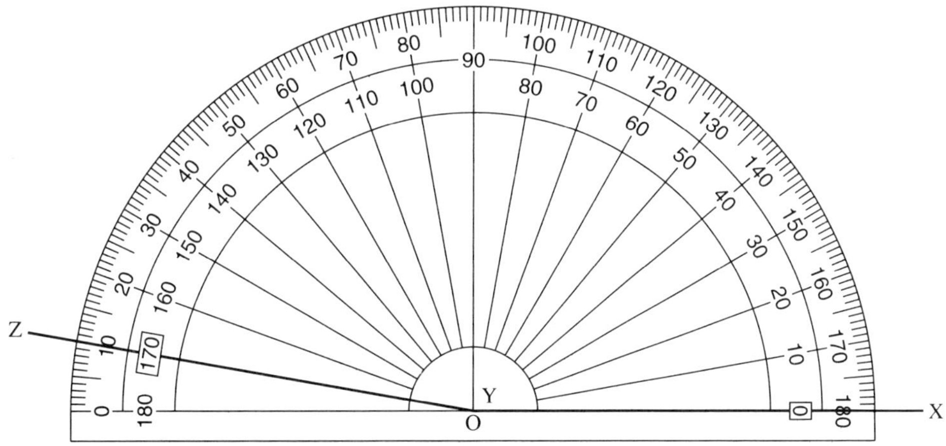

b **1** Draw the straight line XY. Make it 6 cm long.
2 Place the protractor with its base line on XY and its centre point O on Y as shown.
3 Mark the point Z at 170° on the scale starting from 0° at X.
4 Join YZ. The angle of 170° is now complete.

Exercise 13.8

For each question draw a line AB 5 cm long.
Then draw the angle.

1 Draw $A\hat{B}C = 90°$ **2** Draw $A\hat{B}C = 60°$ **3** Draw $A\hat{B}C = 30°$

4 Draw $B\hat{A}C = 40°$ **5** Draw $A\hat{B}C = 120°$ **6** Draw $A\hat{B}C = 170°$

Example 4

Draw a triangle with AB = 6 cm,
$\hat{A} = 70°$ and $\hat{B} = 50°$.
Measure the third angle.

By measurement, $\hat{C} = 60°$

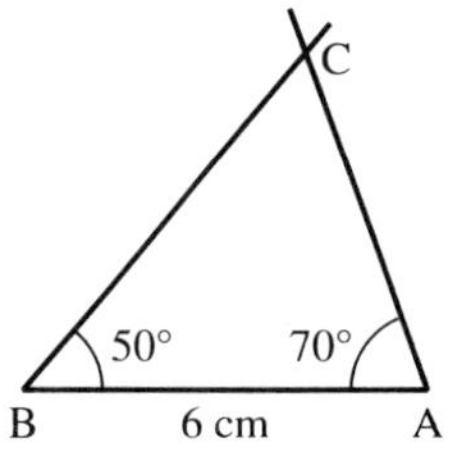

Exercise 13.9

For each question draw the triangle ABC from the details given.
Then measure angle C with a protractor.

1 AB = 6 cm, $\hat{A} = 30°$, $\hat{B} = 60°$ **2** AB = 6 cm, $\hat{A} = 40°$, $\hat{B} = 50°$

3 AB = 6 cm, $\hat{A} = 40°$, $\hat{B} = 40°$ **4** AB = 6 cm, $\hat{A} = 30°$, $\hat{B} = 30°$

5 AB = 5 cm, $\hat{A} = 30°$, $\hat{B} = 120°$ **6** AB = 5 cm, $\hat{A} = 40°$, $\hat{B} = 100°$

13.3 Angle properties of straight lines and triangles

Reminder 1

The angles on a straight line add up to 180°.
If there are only two angles on a straight line they are called **adjacent** angles.

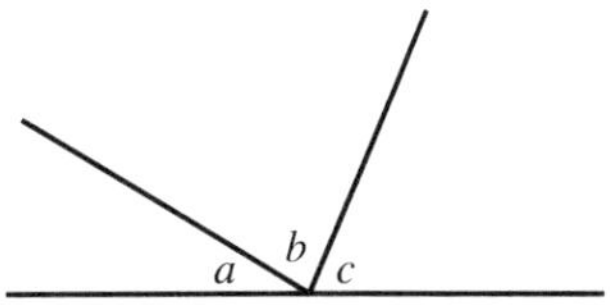

$a + b + c = 180°$

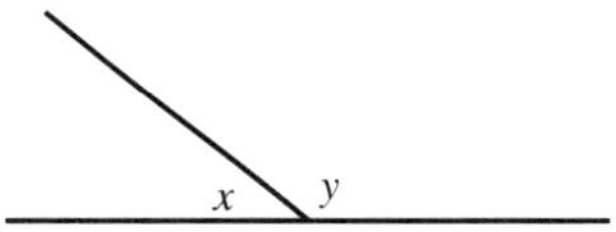

$x + y = 180°$

(x and y are adjacent angles)

Example 1

1 Find the marked angle.

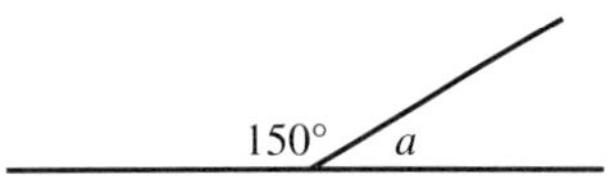

$a = 180° - 150° = 30°$.

2 Find the marked angle.

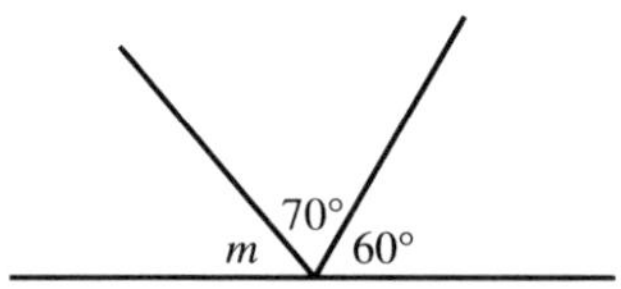

$m = 180° - 70° - 60° = 50°$.

Exercise 13.10

Find the marked angle.

1

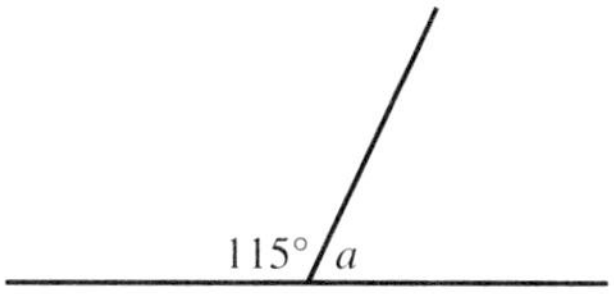

2

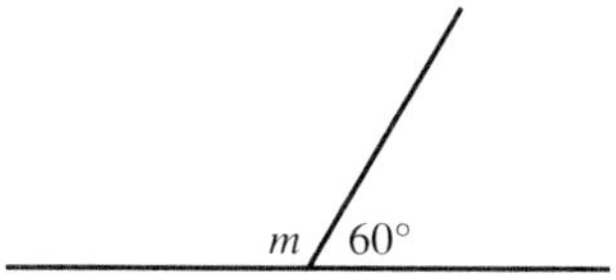

3

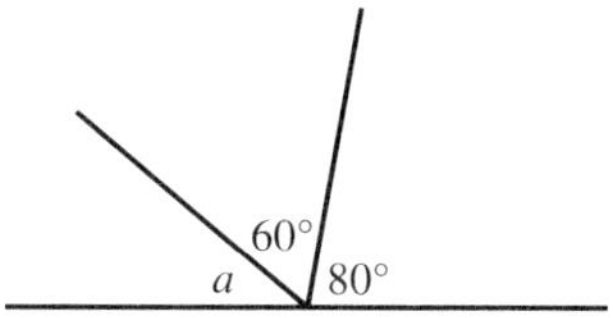

4

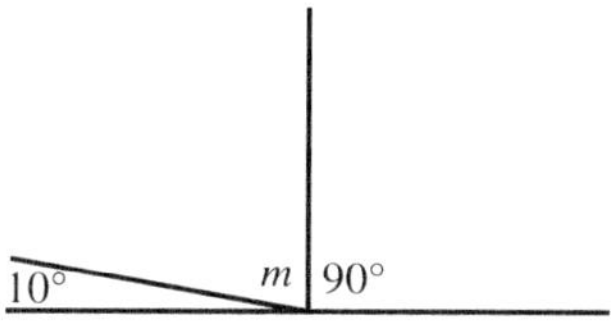

5

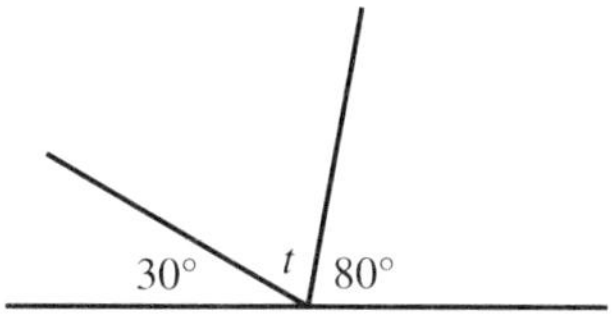

6

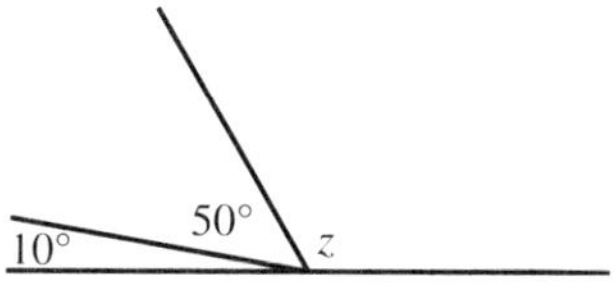

Example 2

Find the marked angles.

$$n + 8n = 180^\circ$$

Therefore $9n = 180^\circ$

Therefore $n = 20^\circ$

Hence $8n = 160^\circ$

The marked angles are 20° and 160°

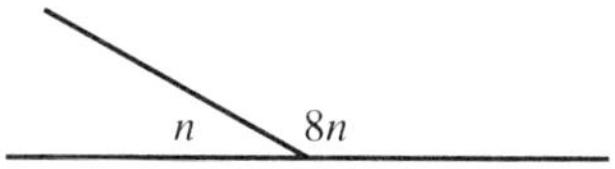

Exercise 13.11

Find the marked angles.

1

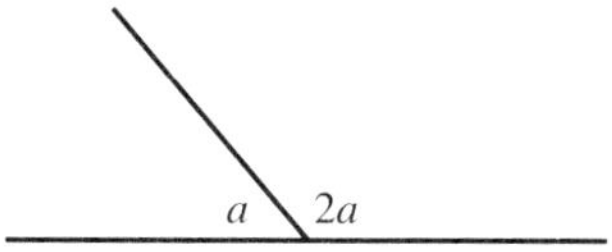

2

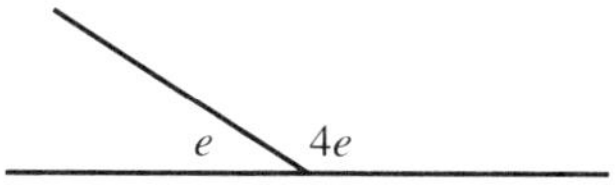

3

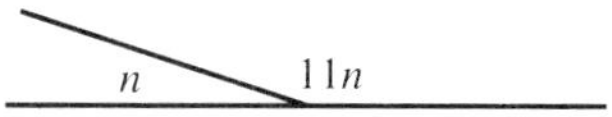

4

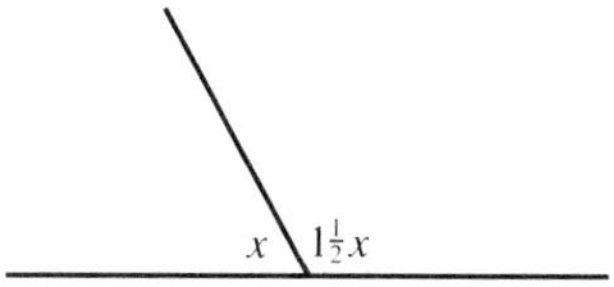

Two angles which total 90° are called **complementary** angles.

Example 3

Find the marked angle.

$a = 90° - 65° = 25°$

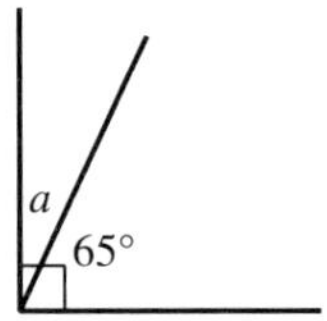

Exercise 13.12

Find the marked angles.

1

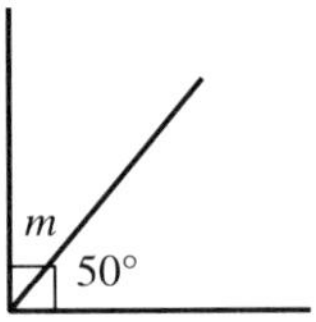

2

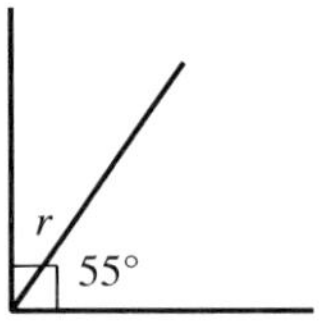

3

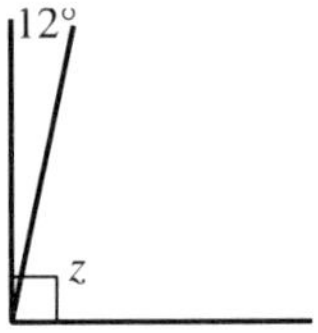

4

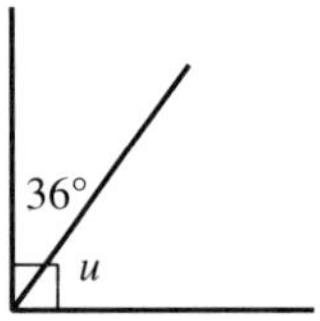

5

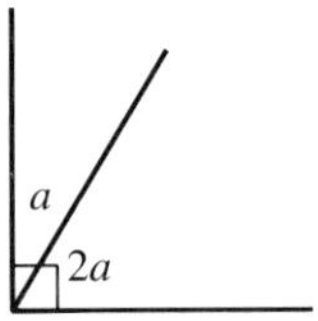

6

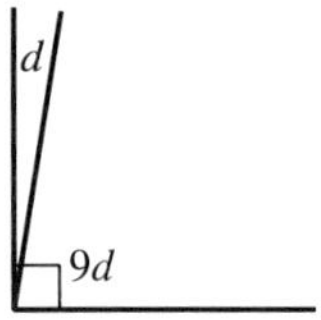

Reminder 2

The angles at a point total 360°.

$a + b + c = 360°$

Example 4

Find the marked angle.

$x = 360° - 120° - 80° - 60°$
$x = 100°$

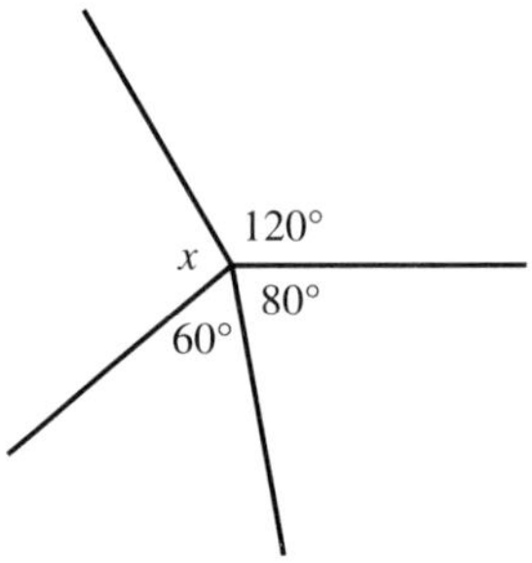

Exercise 13.13

Find the marked angles:

1

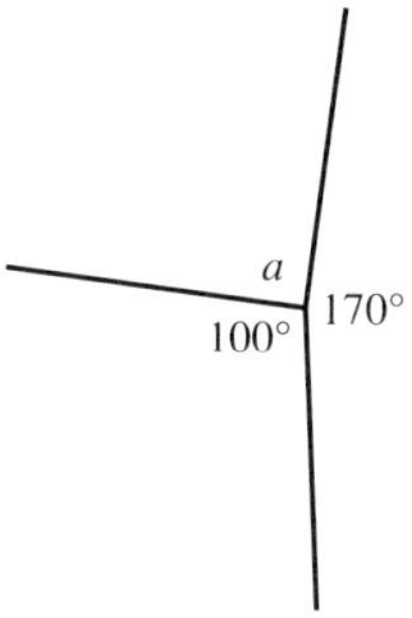

2

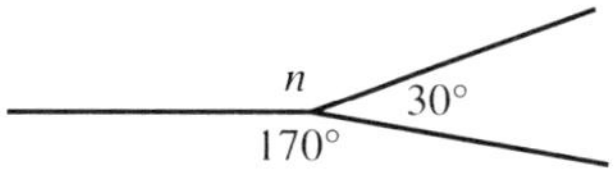

3

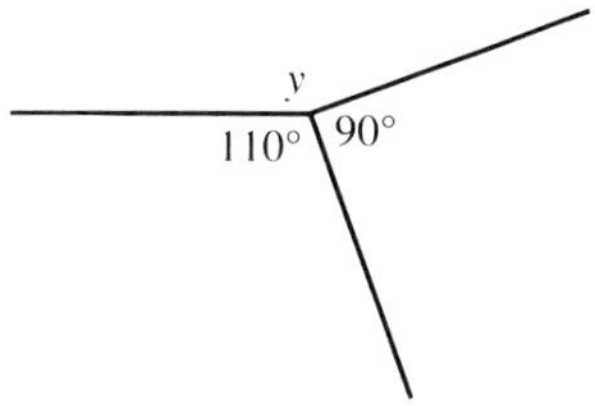

4

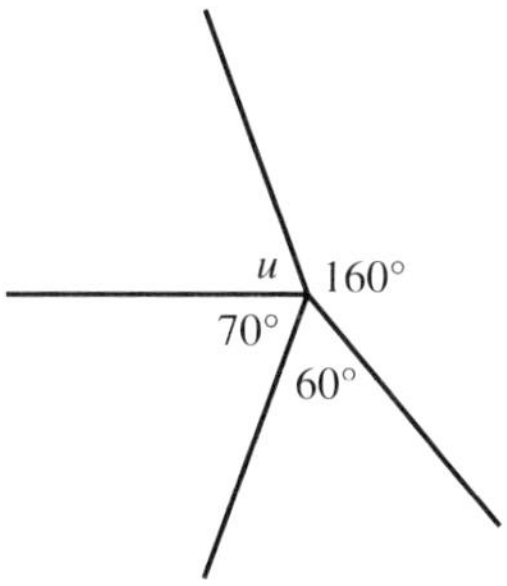

5

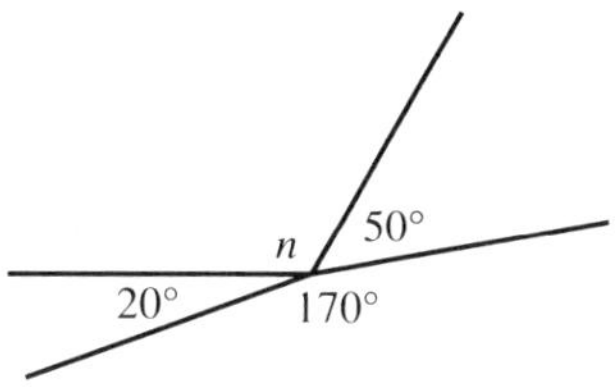

6

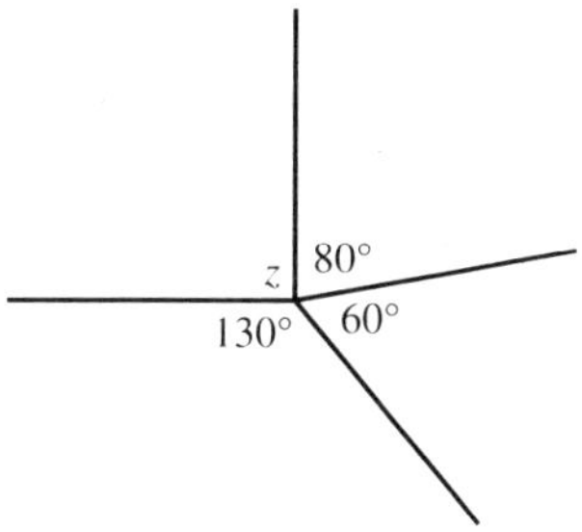

Reminder 3

Vertically opposite angles are equal in size.

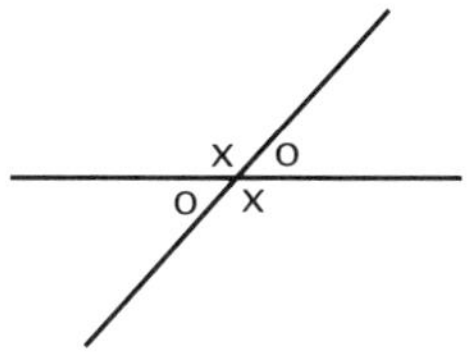

Reminder 4

Corresponding angles on parallel lines are equal in size.

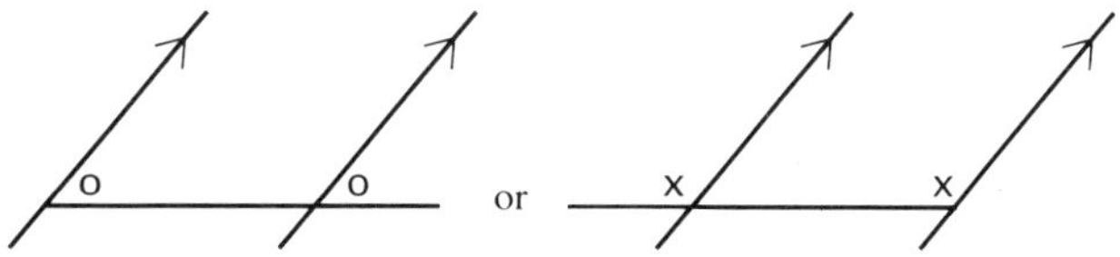

Reminder 5

Alternate angles on parallel lines are equal in size.

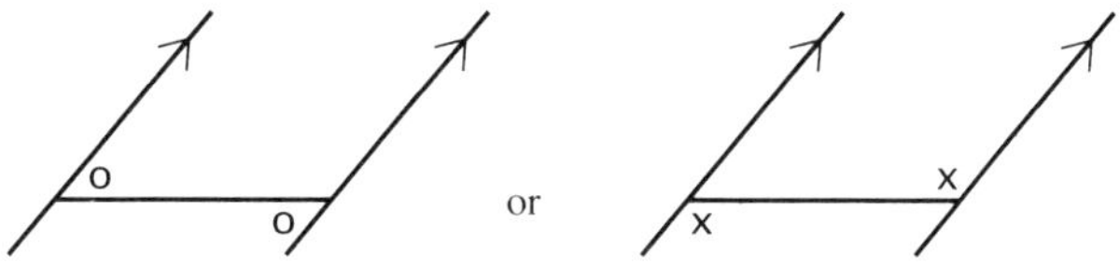

Reminder 6

Allied angles on parallel lines total 180° (i.e. they are supplementary).

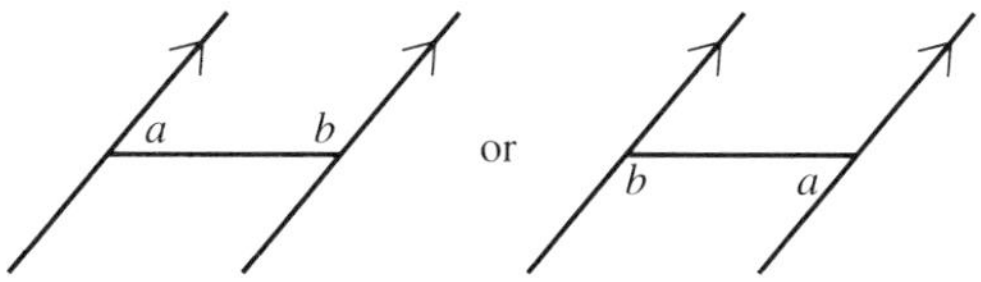

Example 5

Find the marked angles, giving your reasons.

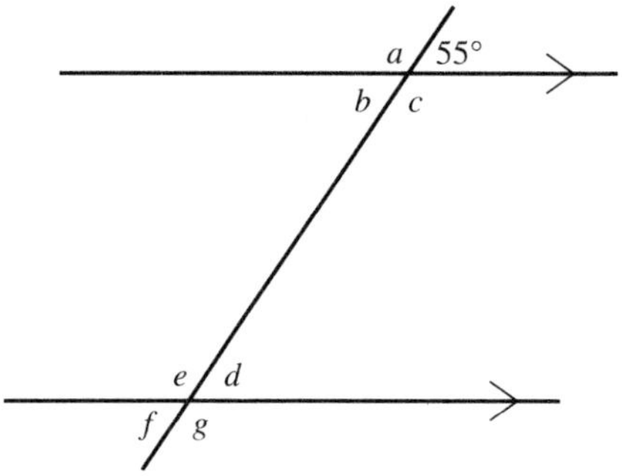

$a = 125°$ (adjacent to 55°)
$b = 55°$ (vertically opposite 55°, or adjacent to a)
$c = 125°$ adjacent to 55°, or vertically opposite a)
$d = 55°$ (alternate to b, or allied to c, or corresponding to 55°)
$e = 125°$ (adjacent to d, or allied to b, or corresponding to a, or alternate to c)
$f = 55°$ (adjacent to e, or vertically opposite d, or corresponding to b)
$g = 125°$ (vertically opposite e, or adjacent to both f and d, or corresponding to c)

Exercise 13.14

Find the marked angles.
Give reasons for your answers.

1

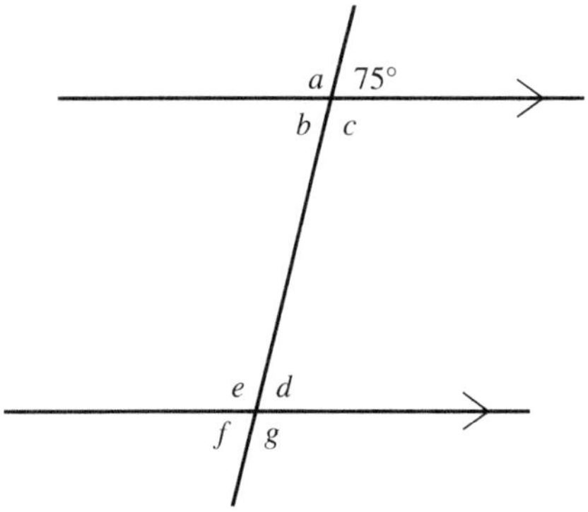

2

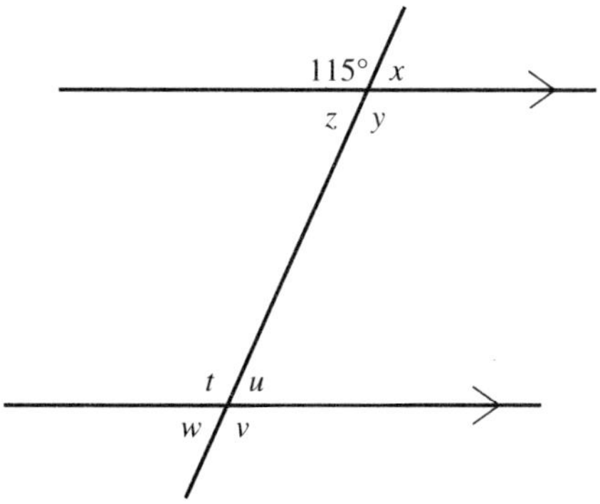

3

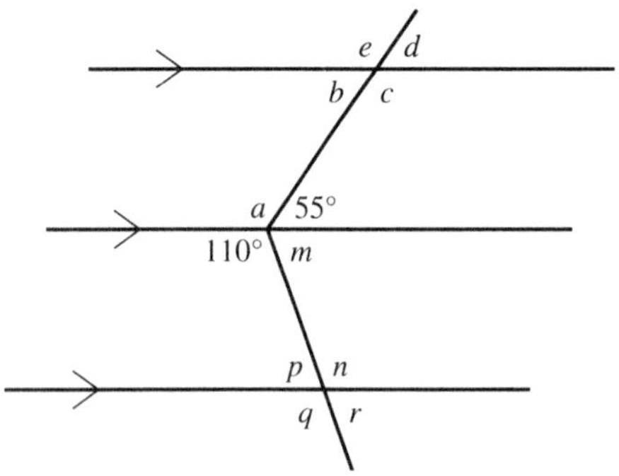

4

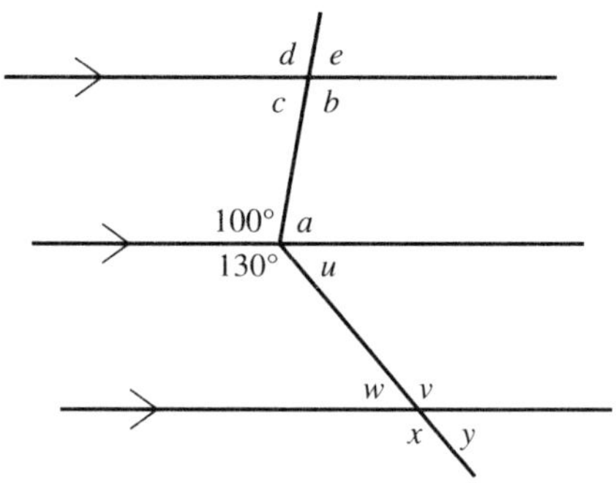

5

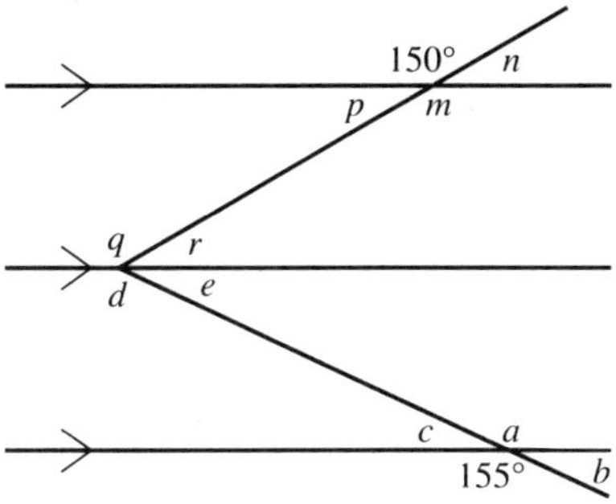

6

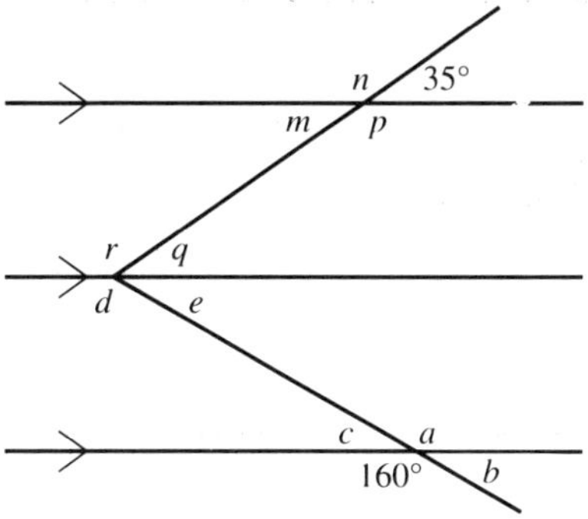

Reminder 7

The three angles of a triangle add up to 180°

$a + b + c = 180°$

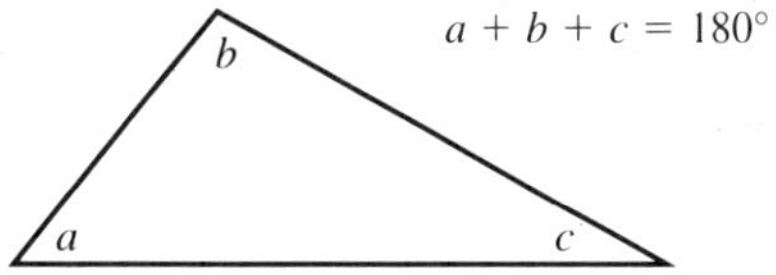

Reminder 8

An isosceles triangle has two equal sides and two equal angles.

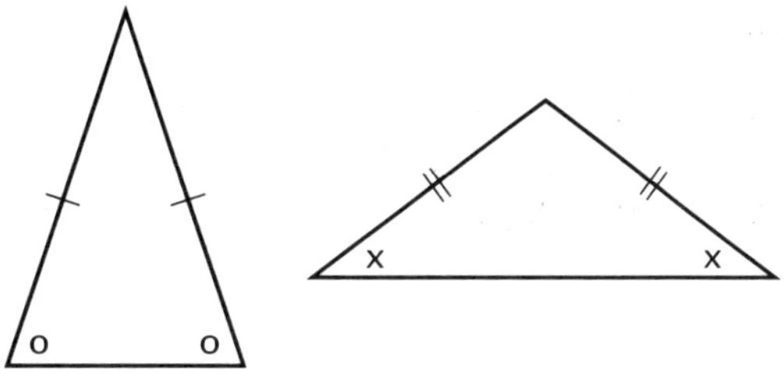

Reminder 9

An equilateral triangle has three equal sides and all three angles equal to 60°.

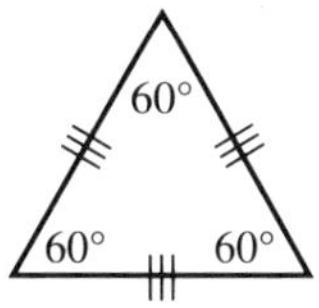

Reminder 10

A right-angled triangle has one angle equal to 90° and the other two angles add up to 90°

$x + y = 90°$

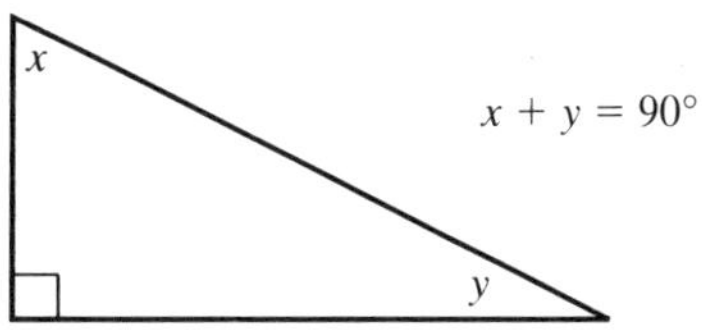

Example 6

Find the marked angles.
(i) $a = 180° - 70° - 50° = 60°$
(ii) $t = 90° - 36° = 54°$

(i)

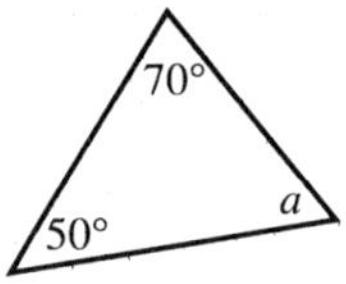

(ii)

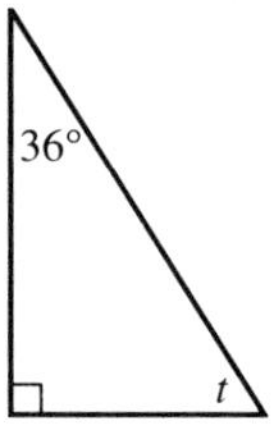

Exercise 13.15

1 Find the angle $\widehat{C}$ for each of the following cases:

a $\widehat{A} = 110°$, $\widehat{B} = 40°$ **b** $\widehat{A} = 75°$, $\widehat{B} = 25°$
c $\widehat{A} = 95°$, $\widehat{B} = 35°$ **d** $\widehat{A} = 85°$, $\widehat{B} = 40°$
e $\widehat{A} = 65°$, $\widehat{B} = 60°$ **f** $\widehat{A} = 125°$, $\widehat{B} = 30°$

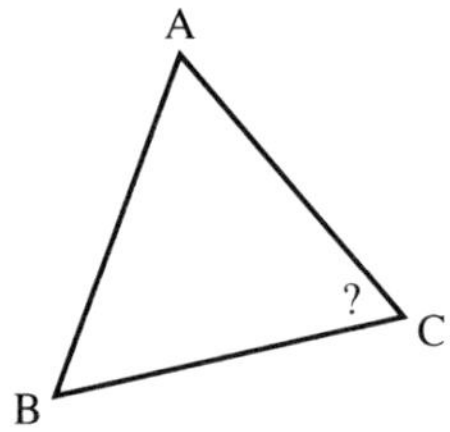

2 Find the angles $\widehat{A}$, $\widehat{B}$ and $\widehat{C}$ for each of the following cases:

a $\widehat{A} = x$, $\widehat{B} = 2x$, $\widehat{C} = 12x$ **b** $\widehat{A} = y$, $\widehat{B} = 3y$, $\widehat{C} = 11y$
c $\widehat{A} = x$, $\widehat{B} = 3z$, $\widehat{C} = 5z$ **d** $\widehat{A} = m$, $\widehat{B} = 2m$, $\widehat{C} = 6m$
e $\widehat{A} = n$, $\widehat{B} = 2n$, $\widehat{C} = 3n$ **f** $\widehat{A} = p$, $\widehat{B} = 4p$, $\widehat{C} = 5p$

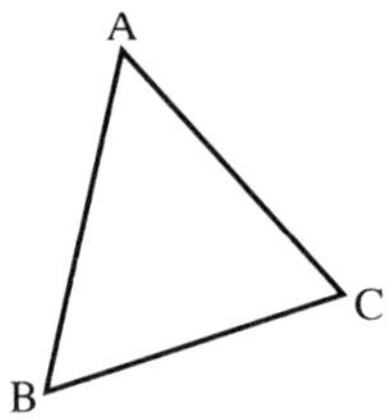

3 Find the three angles of the isosceles triangle PQR for each of the following cases:

a $\widehat{P} = 2x$, $\widehat{Q} = \widehat{R} = x$ **b** $\widehat{P} = 3y$, $\widehat{Q} = \widehat{R} = y$
c $\widehat{P} = 4z$, $\widehat{Q} = \widehat{R} = z$ **d** $\widehat{P} = 7t$, $\widehat{Q} = \widehat{R} = t$
e $\widehat{P} = a$, $\widehat{Q} = \widehat{R} = 2a$ **f** $\widehat{P} = b$, $\widehat{Q} = \widehat{R} = 4b$

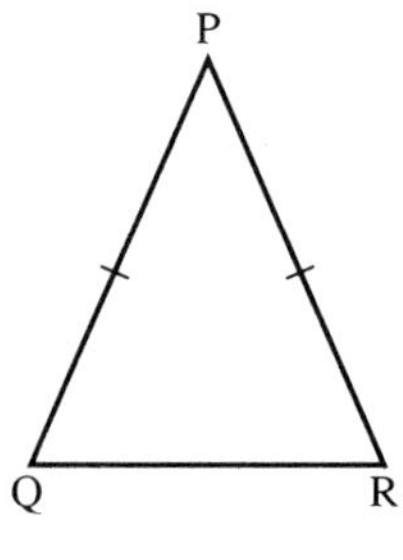

4 Find the $\widehat{B}$ in the right-angled triangle ABC for each of the following cases:

a $\widehat{A} = 26°$ **b** $\widehat{A} = 32°$ **c** $\widehat{A} = 41°$ **d** $\widehat{A} = 13°$
e $\widehat{A} = 39°$ **f** $\widehat{A} = 81°$ **g** $\widehat{A} = 62°$ **h** $\widehat{A} = 69°$

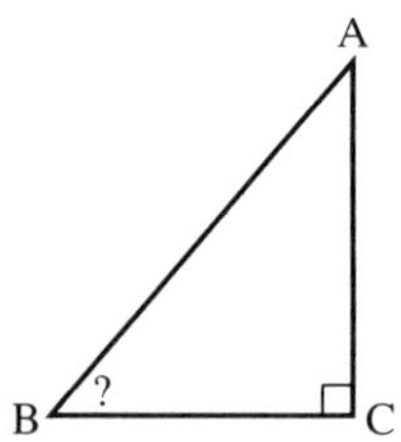

Example 7

Find the marked angles.

$a = 180° - 98° = 82°$
$b = 180° - 53° - 82° = 45°$

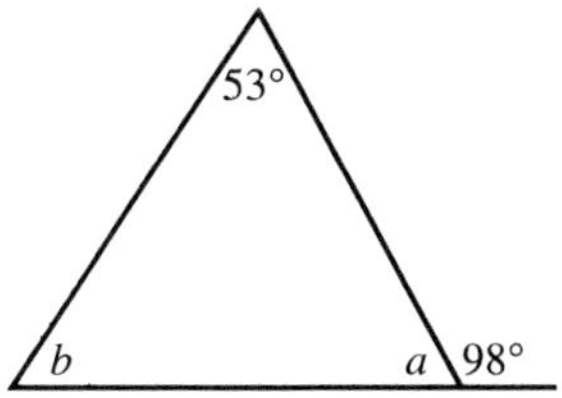

Exercise 13.16

Find the marked angles.

1

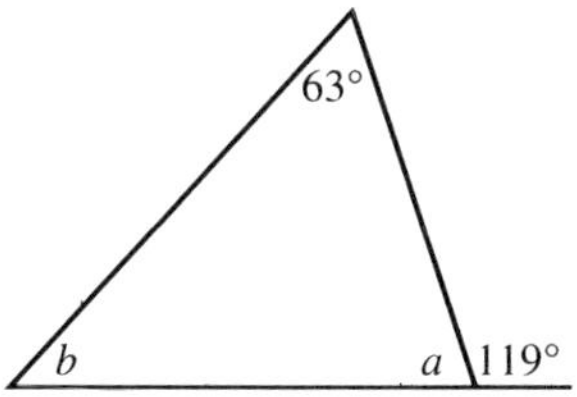

2

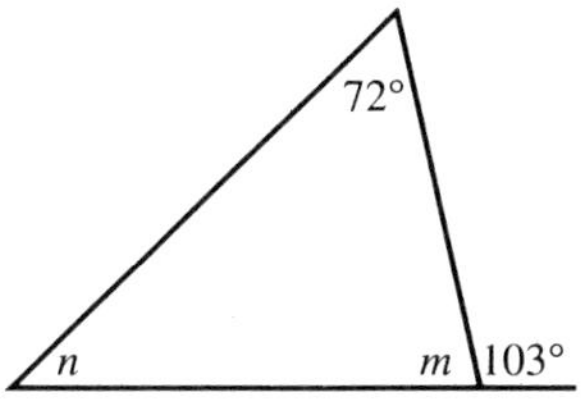

3

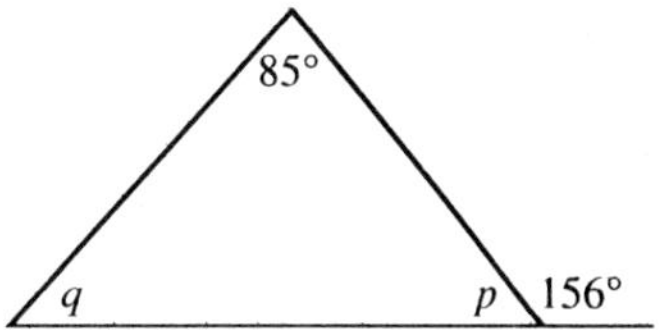

4

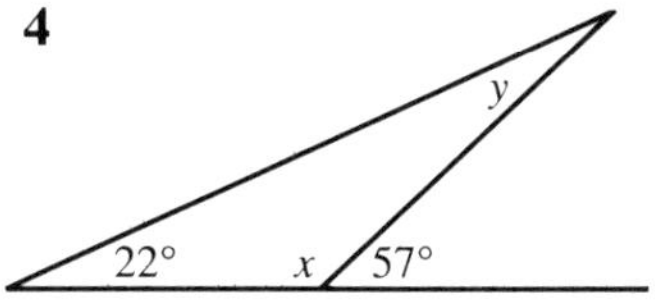

5

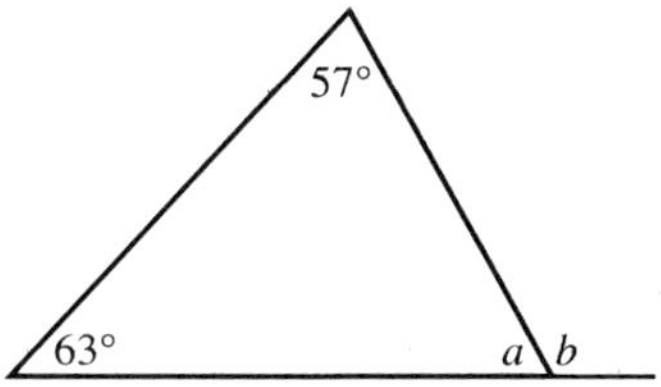

6

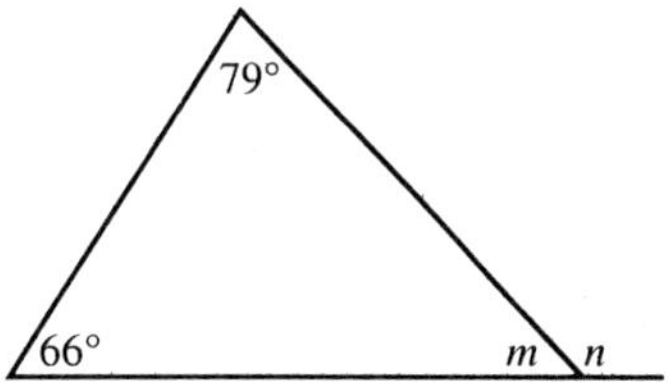

Example 8

(i) Find the marked angles.

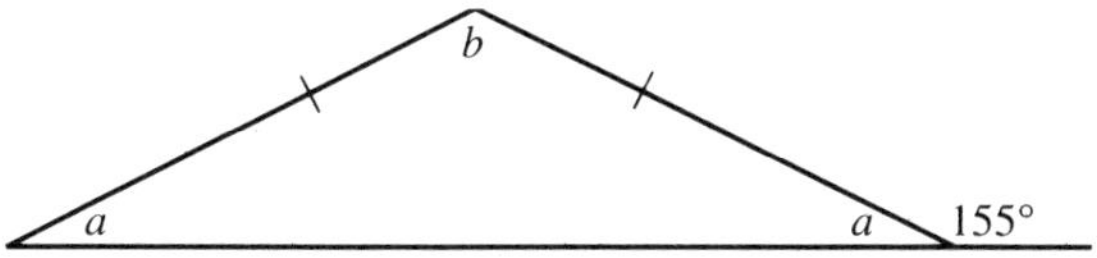

$a = 180° - 155° = 25°$
$b = 180° - 25° - 25° = 130°$

(ii) Find the marked angles.

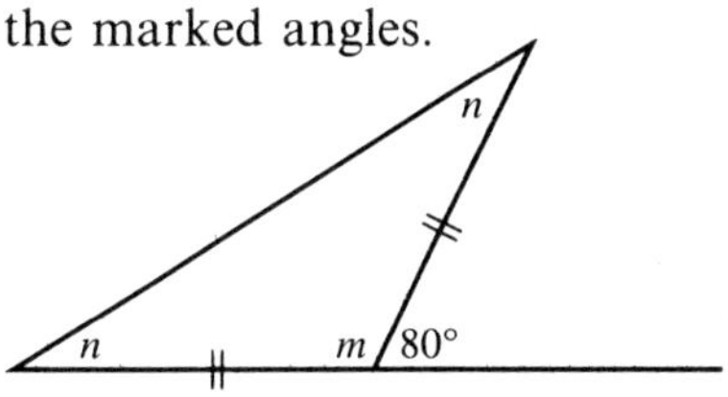

$m = 180° - 80° = 100°$
$n = (180° - 100°) \div 2 = 40°$

Exercise 13.17

Find the marked angles.

1

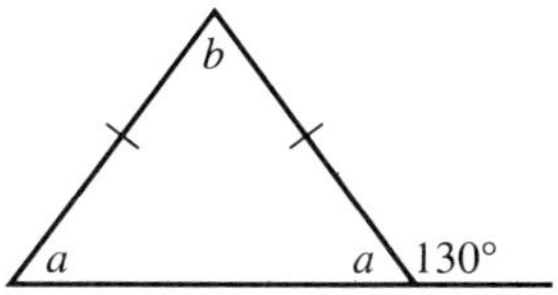

2

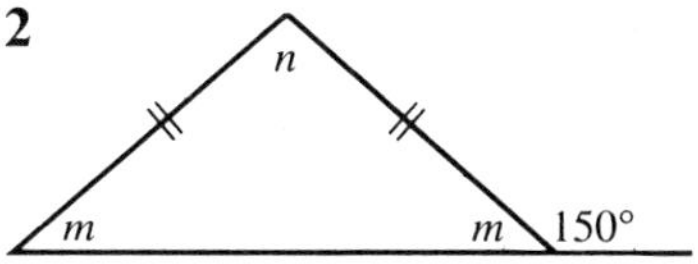

3

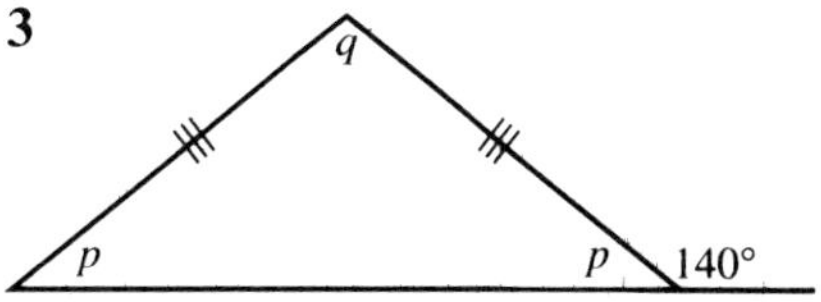

Example 9

Find the marked angles.

$a = 180° - 116° - 44° = 20°$
$b = 44°$ (alternate to given angle)
$c = 20°$ (alternate to a)
$d = 180° - 44° - 20° = 116°$

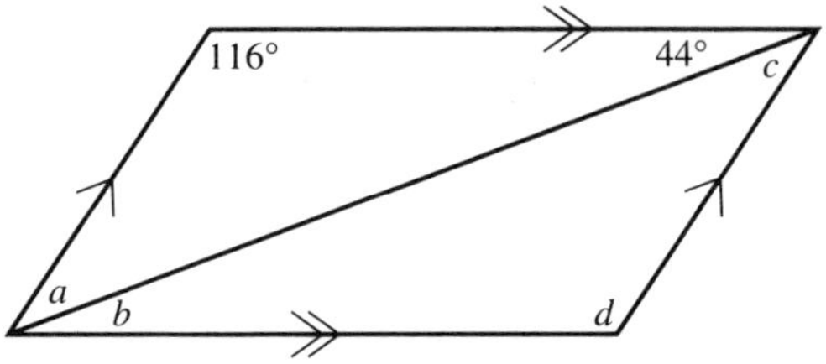

Exercise 13.18

Find the marked angles.

1

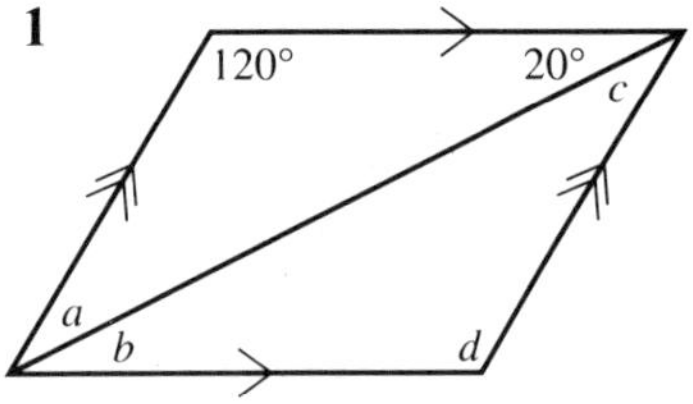

2

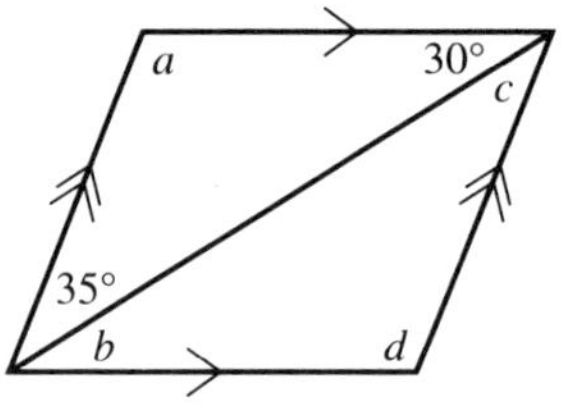

3

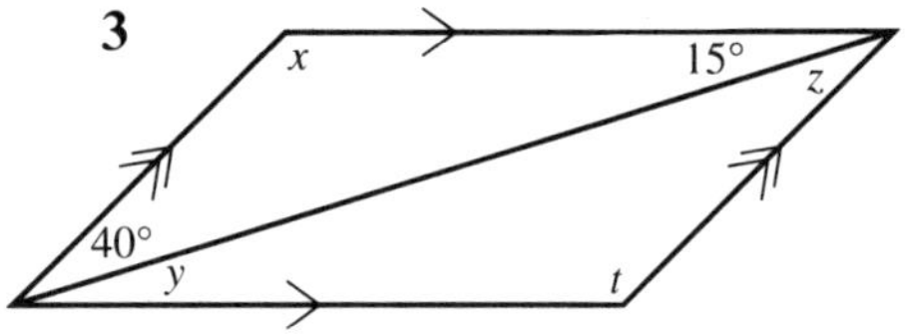

4

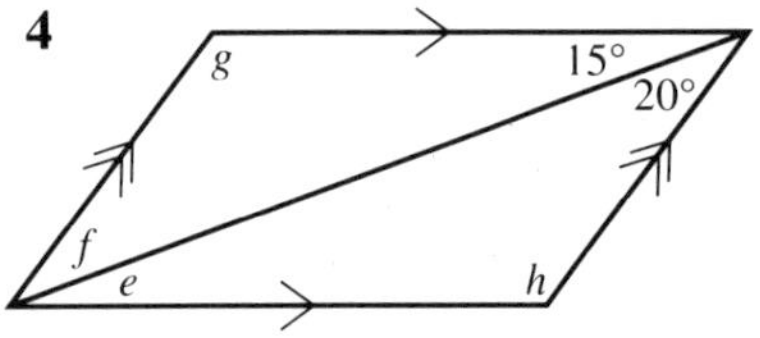

5

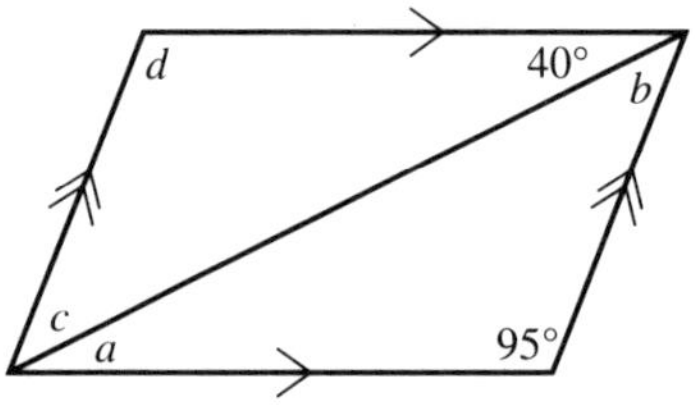

6

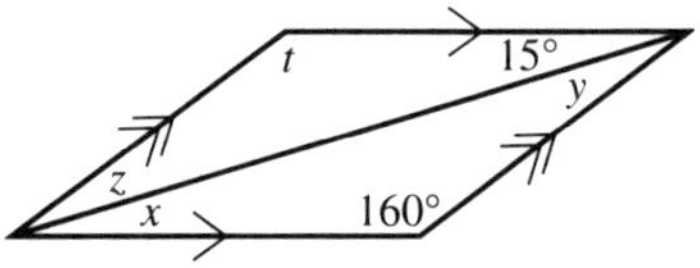

13.4 Knowing and using angle and symmetry properties of quadrilaterals and other polygons

Reminder 1

Read over the properties of quadrilaterals given in Section 12.4 on pp. 186–7.

Example 1

Find the sides and angles:

a DC
b BC
c OA
d $A\widehat{D}C$
e $B\widehat{C}D$

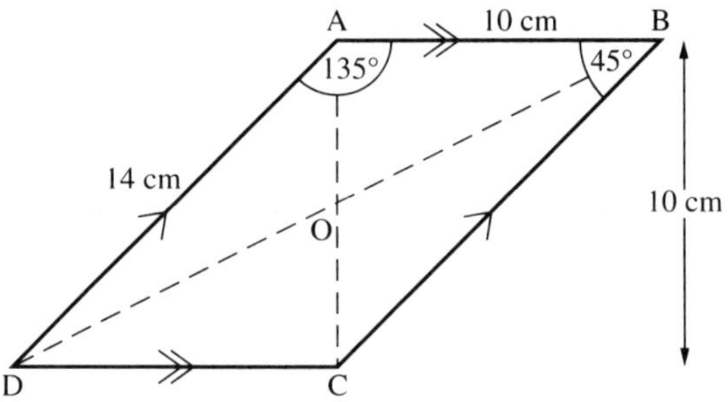

a DC = 10 cm (equal to AB)
b BC = 14 cm (equal to AD)
c OA = 5 cm (the diagonals of a parallelogram bisect each other)
d $A\widehat{D}C = 45°$ (equal to $A\widehat{B}C$)
e $B\widehat{C}D = 135°$ (equal to $B\widehat{A}D$)

Exercise 13.19

For all questions find the sides and angles listed.

1 a AB = ?
b BC = ?
c OA = ?
d $A\widehat{B}C$ = ?
e $B\widehat{C}D$ = ?

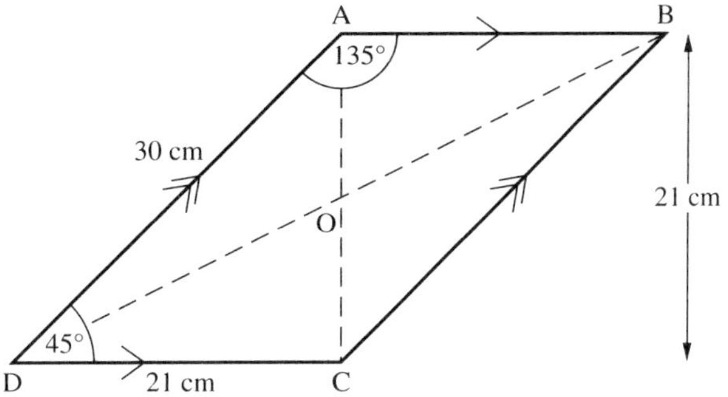

2 a MN = ?
b KN = ?
c OM = ?
d $K\widehat{N}M$ = ?
e $L\widehat{K}N$ = ?

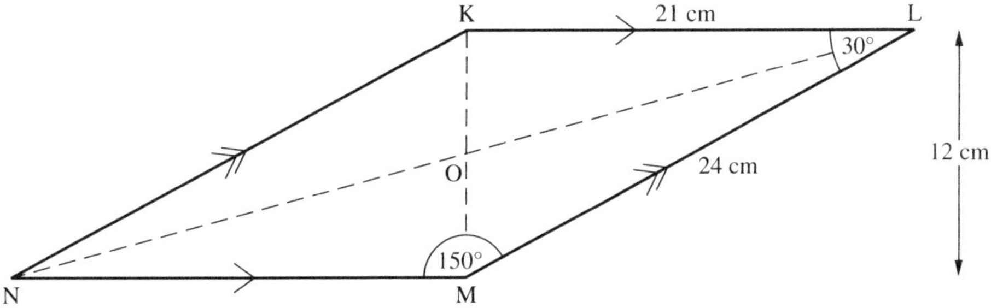

3 a XY = ?
b WX = ?
c WY = ? (given that XZ = 31 cm)
d $X\widehat{Y}Z$ = ?
e $X\widehat{W}Z$ = ?

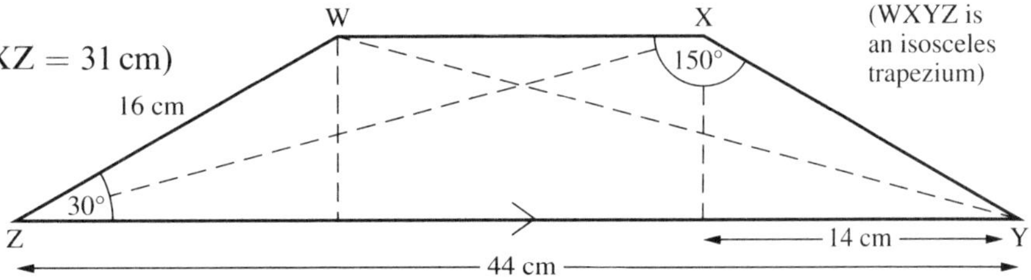

4 a AD = ?
b DC = ?
c OB = ?
d $C\widehat{A}D$ = ?
e $A\widehat{D}C$ = ?

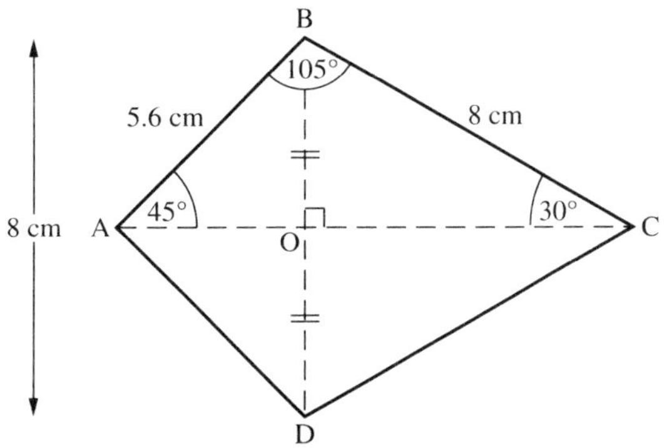

5 a KL = ?
b LM = ?
c ON = ?
d $M\widehat{K}L$ = ?
e $K\widehat{M}L$ = ?
f $K\widehat{L}M$ = ?

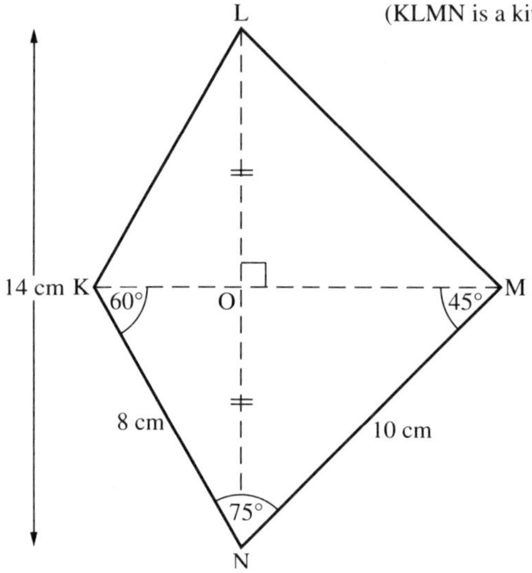

6 a WX = ?
b ZY = ?
c WZ = ?
d XY = ?
e WY = ?
f XZ = ?

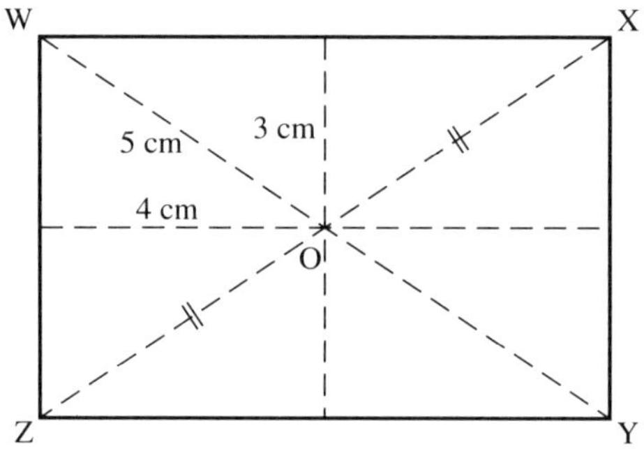

Reminder 2 Symmetry of regular polygons

A regular polygon has all its sides of the same length and all its angles of the same size.

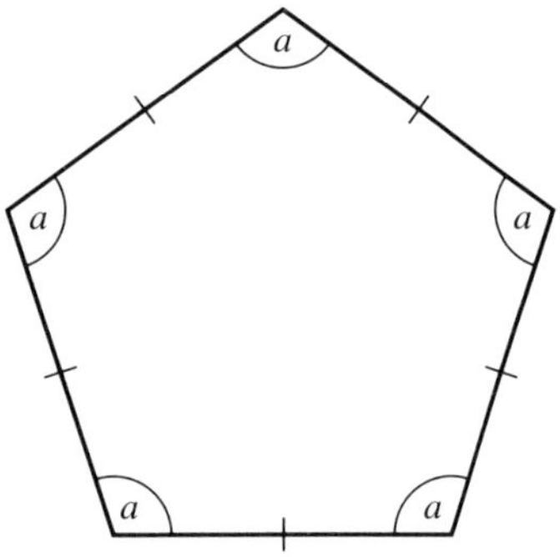

Regular pentagon (5 sides)

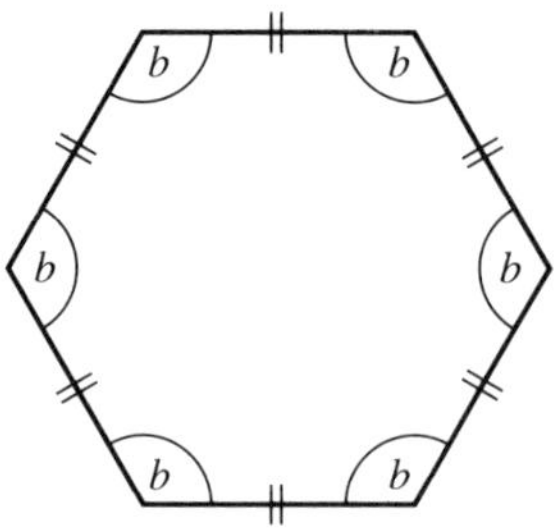

Regular hexagon (6 sides)

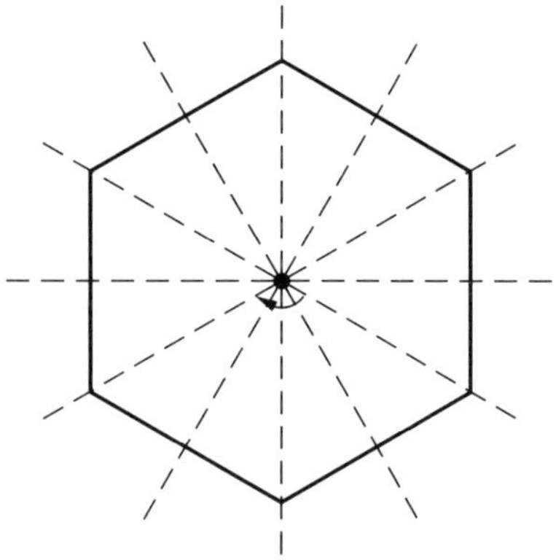

A regular hexagon (six sides) has six lines of symmetry and rotational symmetry of order six.

Exercise 13.20

Ask your teacher for a copy of each of the following regular polygons.
Draw the axes of symmetry on each of your copies.
For each polygon state the number of axes of symmetry and the order of rotational symmetry.

1 Equilateral triangle

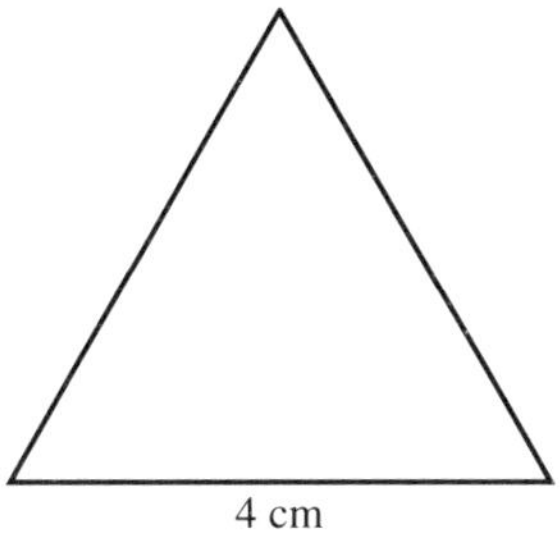

2 Square

3 Regular pentagon

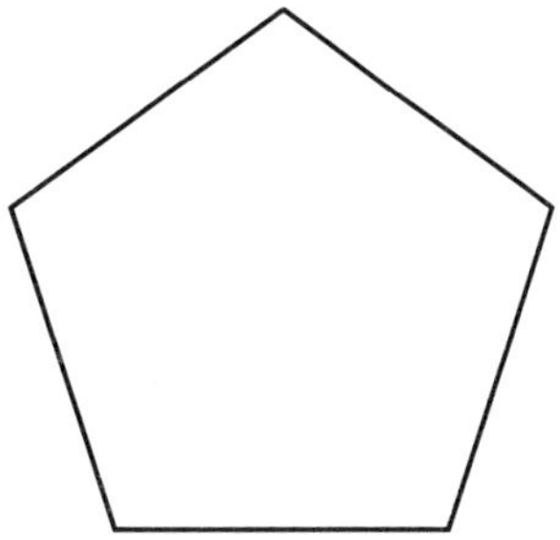

4 Regular heptagon

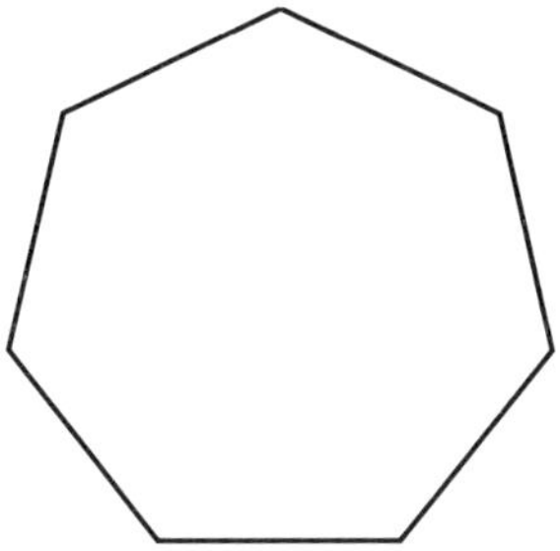

5 Regular octagon

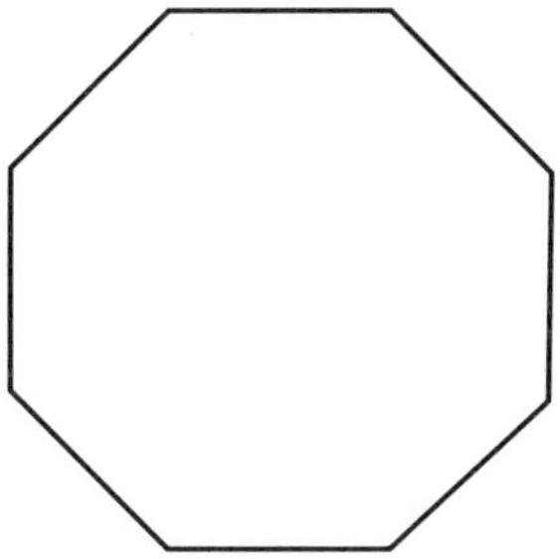

6 State the relationship between the number of sides, number of axes of symmetry and order of rotational symmetry for any regular polygon.

Reminder 3

If you walk around the edge of a regular polygon, you will turn through an angle of 360°, so each of the exterior angles is 360° ÷ number of sides.

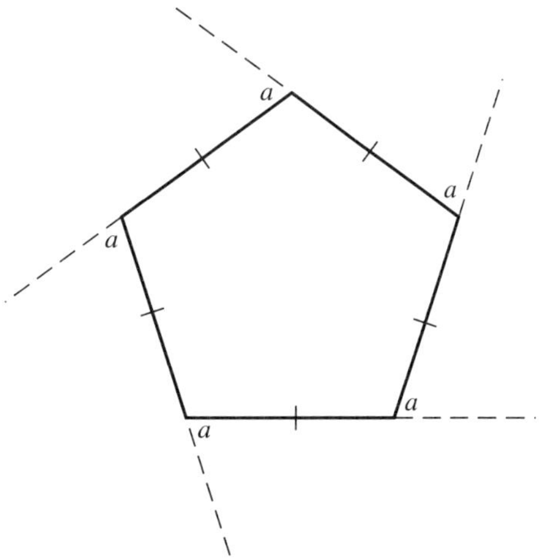

Example 2

Find **a** the exterior angle and **b** the interior angle of a regular pentagon.

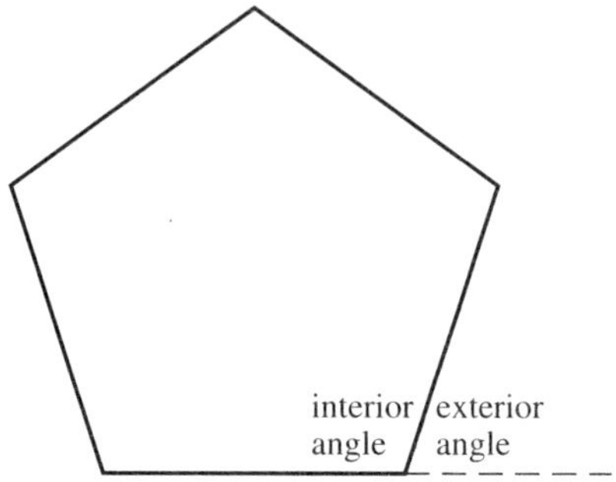

a Exterior angle $= 360° \div$ number of sides
$= 360° \div 5$
$= 72°$

b Interior angle $= 180° -$ exterior angle
$= 180° - 72°$
$= 108°$

Example 3

Find **a** the exterior angle and **b** the number of sides of a regular polygon whose interior angle is 120°.

a Exterior angle $= 180° -$ interior angle $= 180° - 120° = 60°$
b Number of sides $= 360° \div$ exterior angle $= 360° \div 60° = 6$

Exercise 13.21

Find **a** the exterior angle and **b** the interior angle of a regular polygon which has:

1 8 sides **2** 15 sides **3** 24 sides **4** 36 sides

Find **a** the exterior angle and **b** the number of sides of a polygon with an interior angle of:

5 144° **6** 160° **7** 171° **8** 176°

Reminder 4 Angle sum for a polygon

Any polygon can be divided up into a number of triangles.
The regular hexagon shown can be divided up into 4 triangles.
The number of triangles is always 2 less than the number of sides.

Therefore if n is the number of sides the angle sum of a polygon is $(n - 2) \times 180°$
because the angles of any triangle total 180°.

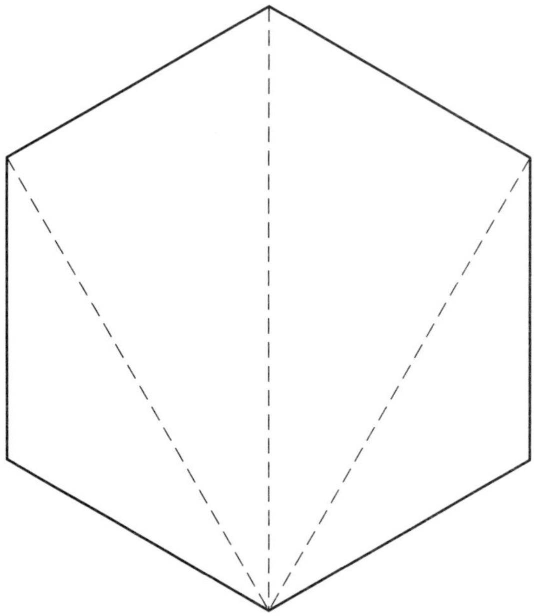

Example 4

Find **a** the angle sum and **b** the interior angle for a regular hexagon.

a Angle sum $= (n-2) \times 180° = (6-2) \times 180° = 4 \times 180° = 720°$
b Interior angle $=$ angle sum $\div n = 720° \div 6 = 120°$

Example 5

Find **a** the number of sides, and **b** the interior angle for a regular pentagon. (Note that the angle sum is 540°.)

a $(n-2) \times 180° = 540°$
so $n - 2 = 540° \div 180°$
or $n - 2 = 3$
therefore $n = 3 + 2$
so $n = 5$

b Interior angle $=$ angle sum $\div n$
$= 540° \div 5$
$= 108°$

Exercise 13.22

Find **a** the angle sum, and **b** the interior angle for a regular polygon with:

1 9 sides **2** 20 sides **3** 30 sides **4** 45 sides

Find **a** the number of sides, **b** the interior angle for a regular polygon with an angle sum of:

5 1800° **6** 10 440° **7** 12 600° **8** 180°

14 Understanding and using measures

14.1 Understanding the relationship between the units of length, mass capacity and time

Units of length

Three important units of length are the metre, centimetre and millimetre.
These three units are connected as follows:

10 millimetres (mm) = 1 centimetre (cm)
100 centimetres (cm) = 1 metre (m)

It therefore follows that 1 metre is equal to 100 centimetres or 1000 millimetres.

Example 1

Change 3 metres to: **a** centimetres **b** millimetres

a 3 metres = (3 × 100) centimetres = 300 centimetres
b 3 metres = (3 × 1000) millimetres = 3000 millimetres

Example 2

Change 1900 centimetres to: **a** metres **b** millimetres

a 1900 centimetres = (1900 ÷ 100) metres = 19 metres
b 1900 centimetres = (1900 × 10) millimetres = 19 000 millimetres

Example 3

Change 40 000 millimetres to **a** metres **b** centimetres

a 40 000 millimetres = (40 000 ÷ 1000) metres = 40 metres
b 40 000 millimetres = (40 000 ÷ 10) centimetres = 4000 centimetres

Exercise 14.1

Copy and complete the following table.

	Metres	Centimetres	Millimetres
1	5	5 × 100 =	5 × 1000 =
2	12		
3	25		
4	2400 ÷ 100 =	2400	2400 × 10 =
5		1600	
6		100	
7	6000 ÷ 1000 =	6000 ÷ 10 =	6000
8			11 000
9			26 000
10			50 000

Units of mass and capacity

Two important units of mass are the gram and the kilogram.
They are connected as follows:

1000 grams (g) = 1 kilogram (kg)

Note: mass is often referred to as 'weight'.

Two important units of capacity are the litre and the millilitre.
They are connected as follows:

1000 millilitres (ml) = 1 litre (l)

A box that measures 1 cm × 1 cm × 1 cm has a capacity of 1 millilitre.

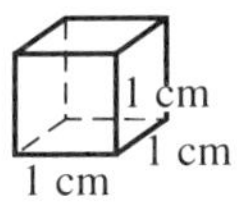

A box that measures 10 cm × 10 cm × 10 cm has a capacity of 1 litre

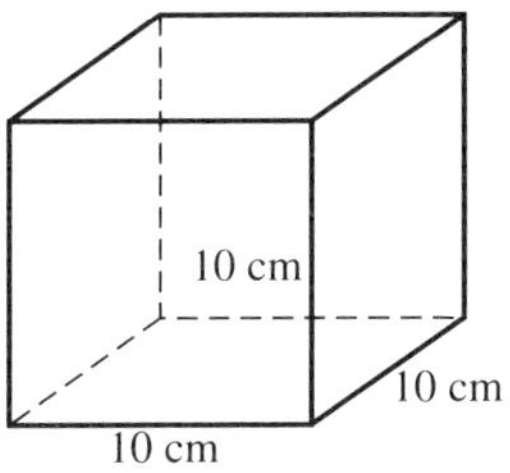

Note also:

a The mass of 1 millilitre of water is 1 gram.
b The mass of 1 litre of water is 1 kilogram.

Example 4

A bucket has a capacity of 5 litres. Find:
a The capacity of the bucket in millilitres.
b The mass of water which fills the bucket in kilograms.
c The mass of water which fills the bucket in grams.

a 5 litres = (5 × 1000) millilitres = 5000 millilitres.
b 5 kilograms, because 1 litre of water has a mass of 1 kilogram.
c 5000 grams, because 1 millilitre of water has a mass of 1 gram.

Exercise 14.2

Copy and complete the table

	Capacity of container (litres)	Capacity of container (millilitres)	Mass of water which fills the container (kilograms)	Mass of water which fills the container (grams)
1	6			
2	12			
3	10			
4		2000		
5		18 000		
6		25 000		
7		32 000		
8			28	
9			36	
10			40	
11				17 000
12				38 000
13				16 000
14				50 000

Units of time

Three important units of time are the hour, second and minute.
These three units are connected as follows:

60 seconds = 1 minute
60 minutes = 1 hour

It therefore follows that 1 hour is equal to 60 × 60 or 3600 seconds.

Example 5

Change 4 hours to **a** minutes and **b** seconds

a 4 hours = (4 × 60) minutes = 240 minutes
b 4 hours = (4 × 60) minutes × 60 seconds = 14 400 seconds

Example 6

Change 720 minutes to **a** hours and **b** seconds

a (720 ÷ 60) hours = 12 hours
b (720 × 60) seconds = 43 200 seconds

Example 7

Change 54 000 seconds to **a** minutes and **b** hours

a 54 000 seconds = (54 000 ÷ 60) minutes = 900 minutes
b 54 000 seconds = (54 000 ÷ 60) minutes ÷ 60 = 15 hours

Exercise 14.3

Copy and complete the table.

	Hours	Minutes	Seconds
1	5		
2	11		
3	10		
4		1020	
5		780	
6		120	
7			57 600
8			144 000

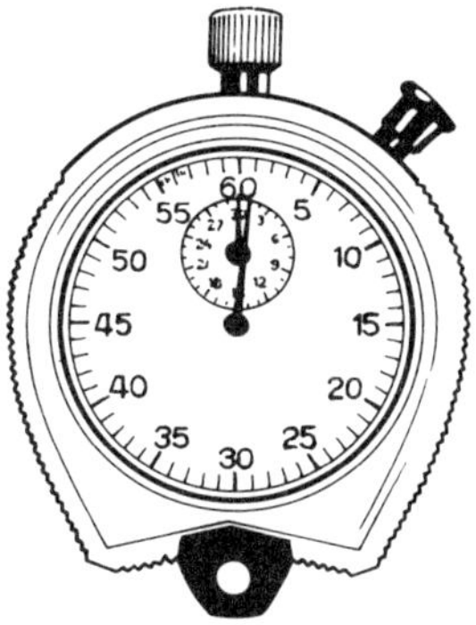

14.2 Using imperial units and knowing their rough metric equivalents

At present we still have two different systems of measurement in common use. Sometimes we use the imperial system and sometimes we use the metric system.
The table over the page shows the most commonly used imperial units and their metric equivalents.

Quantity to be measured	Imperial unit	Accurate conversion to and from metric equivalent	Approximate conversion to and from metric equivalent
Length	1 inch (1″ or 1 in)	1 in = 2.54 cm 1 cm = 0.3937″	1″ = 2.5 (or $2\frac{1}{2}$) cm 1 cm = 0.4″
	1 foot (1′ or 1 ft) (1 ft = 12 in)	1 ft = 0.3048 m 1 m = 3.281 ft	1′ = 0.3 m 1 m = 3.3′
	1 yard (1 yd) (1 yd = 36 in)	1 yd = 0.9144 m 1 m = 1.0936 yd	1 yd = 0.9 m 1 m = 1.1 yd
	1 mile (1 mile = 1760 yd)	1 mile = 1.6093 km 1 km = 0.6214 miles	1 mile = 1.6 km 1 km = 0.625 miles (or $\frac{5}{8}$ mile)
Mass	1 ounce (1 oz)	1 oz = 28.35 g 1 g = 0.0353 oz	1 oz = 28 g 1 g = 0.04 oz (or $\frac{1}{25}$ oz)
	1 pound (1 lb) (1 lb = 16 oz)	1 lb = 0.4536 kg 1 kg = 2.205 lb	1 lb = 0.45 kg 1 kg = 2.2 lb
Capacity	1 pint	1 pint = 0.568 litres 1 litre = 1.761 pints	1 pint = 0.6 litres 1 litre = 1.75 pints (or $1\frac{3}{4}$ pints)
	1 gallon (1 gallon = 8 pints)	1 gallon = 4.544 litres	1 gallon = 4.5 litres (or $4\frac{1}{2}$ litres)
		1 litre = 0.2205 gallons	1 litre = 0.22 gallons

Example 1

a A shelf is 40 inches long.
What is its length in centimetres, given that 1 inch equals 2.5 centimetres?

b A lead for a large dog is 55 centimetres long.
What is its length in inches, given that 1 centimetre equals 0.4 inches?

a 40 inches = $40 \times 2.5 = 100$ centimetres

b 55 centimetres = $55 \times 0.4 = 22$ inches

Exercise 14.4

For Questions 1–6 convert the dimension in inches to centimetres, given that 1 inch equals 2.5 centimetres.

1 A pencil is 6 inches long.

2 A cup is 4 inches high.

3 A spanner is 10 inches long.

4 A screwdriver is 16 inches. long.

5 A saw is 24 inches long.

6 A knife is 14 inches long.

For Questions 7–12 convert the dimension in centimetres to inches, given that 1 centimetre = 0.4 inches.

7 A nail is 5 centimetres long.

8 A glass is 20 centimetres tall.

9 A hammer is 30 centimetres long.

10 A lamp is 45 centimetres high.

11 A gate is 110 centimetres high.

12 A table is 80 centimetres high.

Example 2

a The cliff at Achill Island in Ireland is the tallest in the British Isles and it is 2190 feet high.
Given that 1 foot = 0.3 metres, find its height in metres.

b The tallest of the Egyptian pyramids is 150 metres high.
Given that 1 metre is 3.3 feet, find its height in feet.

a 2190 feet = (2190×0.3) metres = 657 metres

b 150 metres = (150×3.3) feet = 495 feet

Exercise 14.5

For Questions 1–6 convert the dimension in feet to metres, given that 1 foot = 0.3 metres.

1 A telegraph pole is 40 feet tall.

2 A bungalow is 20 feet high.

3 An aeroplane is 60 feet long.

4 Nelson's Column is 180 feet tall.

5 Blackpool Tower is 540 feet tall.

6 The highest mountain in England is Scafell Pike, and its height is 3210 feet above sea level.

For Questions 7–12 convert the dimension in metres to feet, given that 1 metre = 3.3 feet.

7 A street lamp is 10 metres tall.

8 A tree is 30 metres tall.

9 A railway coach is 20 metres long.

10 A ship is 70 metres long. metres tall.

11 The Eiffel Tower is 320 metres tall.

12 The highest mountain in the British Isles is Ben Nevis and its height is 1340 metres above sea level.

Example 3

a The playground at Sally's school is 70 yards long.
Given that 1 yard = 0.9 metres, finds its length in metres.
b The football pitch at Paul's school is 80 metres long.
Given that 1 metre = 1.1 yards, find its length in yards.

a 70 yards = (70×0.9) metres = 63 metres
b 80 metres = $(80 \times 1.1$ yards$)$ = 88 yards

Exercise 14.6

For Questions 1–6 convert the dimension in yards to metres, given that 1 yard = 0.9 metres.

1. The garden at Nathan's house is 50 yards long.
2. A swimming pool is 40 yards long.
3. Larch Avenue is 90 yards long.
4. The platforms at a railway station are 250 yards long.
5. The distance between two mile posts is 1760 yards.
6. Two advance signs before a motorway exit are 900 yards apart.

For Questions 7–11 convert the dimension in metres to yards, given that 1 metre = 1.1 yards.

7. A cotton reel contains 10 metres of cotton thread.
8. The telegraph posts along London Road are 40 metres apart.
9. Two locks on a canal are 360 metres apart.
10. The distance between the underground stations at Marble Arch and Bond Street is 650 metres.
11. The distance between two kilometre posts along a road in France is 1000 metres.

Example 4

a The distance from London to Leeds is 190 miles.
Given that 1 mile = 1.6 kilometres, convert this distance to kilometres.
b The distance from Paris to Lyon is 472 kilometres.
Given that 1 kilometre = 0.625 $\left(\text{or } \frac{5}{8}\right)$ miles, convert this distance to miles.

a 190 miles = (190×1.6) kilometres = 304 kilometres
b 472 kilometres = (472×0.625) miles = 295 miles

Exercise 14.7

For Questions 1–8 convert the distance given from miles to kilometres, given that 1 mile = 1.6 kilometres.

1 London to Edinburgh – 395 miles

2 London to Aberdeen – 525 miles

3 London to Liverpool – 195 miles

4 London to Leicester – 105 miles

5 London to Plymouth – 225 miles

6 London to Weymouth – 130 miles

7 London to Portsmouth – 75 miles

8 London to Brighton – 50 miles

For Questions 9–16 convert the distance given from kilometres to miles, given that 1 kilometre = 0.625 (or $\frac{5}{8}$) miles.

9 Paris to Marseilles – 792 kilometres

10 Paris to Bordeaux – 568 kilometres

11 Paris to Luxembourg – 344 kilometres

12 Paris to Amsterdam – 488 kilometres

13 Rome to Naples – 216 kilometres

14 Paris to Cologne – 464 kilometres

15 Paris to Munich – 824 kilometres

16 Paris to Madrid – 1272 kilometres

Example 5

a Theo has a bag which contains 22 ounces of nuts. Given that 1 ounce = 28 grams, convert this weight to grams.
b A block of ice cream in a cardboard packet weighs 425 grams. Given that 1 gram = 0.04 (or $\frac{1}{25}$) ounces, convert this weight to ounces.

a 22 ounces = (22 × 28) grams = 616 grams
b 425 grams = (425 × 0.04) ounces = 17 ounces

Exercise 14.8

For Questions 1–5 convert the weight to grams, given that 1 ounce = 28 grams.

1 A lollipop has a weight of 2 ounces.

2 A jar contains 10 ounces of coffee.

3 A tub contains 16 ounces of butter.

4 A bag contains 20 ounces of raisins.

5 A packet contains 24 ounces of flour.

For Questions 6–10 convert the weight to ounces, given that 1 gram = 0.04 (or $\frac{1}{25}$) ounces.

6 A tin contains 75 grams shoe polish.

7 A paper bag contains 100 of grams of sweets.

8 A tin contains 200 grams of beans.

9 A tub contains 325 grams of cream.

10 A jar contains 375 grams of jam.

Example 6

a A piano weighs 360 pounds (lb).
Convert this weight to kilograms, given that 1 lb = 0.45 kg.
b A sack of potatoes weighs 30 kg.
Convert this weight to pounds, given that 1 kg = 2.2 lb.

a $360\,\text{lb} = (360 \times 0.45)\,\text{kg} = 162\,\text{kg}$
b $30\,\text{kg} = (30 \times 2.2)\,\text{lb} = 66\,\text{lb}$

Exercise 14.9

In Questions 1–6 the weights of the members of Peter's family and his pets are given in pounds.
Convert each of these weights to kilograms, given that 1 lb = 0.45 kg.

1 Peter's cat – 20 lb

2 Peter's younger brother – 60 lb

3 Peter – 100 lb

4 Peter's older brother – 160 lb

5 Peter's grandmother – 140 lb

6 Peter's grandfather – 200 lb

In Questions 7–12 the weights of the members of Kelly's family and her pets are given in kilograms.
Convert these weights to pounds, given that 1 kg = 2.2 lb

7 Kelly's baby brother – 15 kg

8 Kelly's dog – 25 kg

9 Kelly's older brother – 75 kg

10 Kelly's younger sister – 35 kg

11 Kelly's grandmother – 65 kg

12 Kelly's grandfather – 80 kg

Example 7

a A full carton contains 0.75 pints of orange juice.
Convert its capacity to litres, given that 1 pint = 0.6 litres
b A full bottle contains 0.3 litres of vinegar.
Convert its capacity to pints, given that 1 litre = 1.75 pints

a 0.75 pints = (0.75 × 0.6) litres = 0.45 litres
b 0.3 litres = (0.3 × 1.75) pints = 0.525 pints

Exercise 14.10

For Questions 1 and 2 convert the capacities to litres, given that 1 pint = 0.6 litres

1 Goodgrub's supermarket sells milk in cartons of four sizes: 0.8 pints, 1.5 pints, 2.5 pints and 4.0 pints.

2 Goodgrub's supermarket sells lemonade in bottles of four sizes: 0.4 pints, 0.6 pints, 1.4 pints and 2.4 pints.

For Questions 3 and 4 convert the capacities to pints, given that 1 litre = 1.75 pints

3 The Foodhall supermarket sells orange squash in bottles of four sizes: 0.8 litres, 1.2 litres, 2.0 litres and 2.8 litres.

4 The Foodhall supermarket sells liquid soap in plastic bottles of four sizes: 0.2 litres, 0.5 litres, 1.5 litres and 2.5 litres.

Example 8

a A large drum of paraffin holds 15 gallons.
Convert this capacity to litres, given that 1 gallon = 4.5 litres.
b A lorry has a fuel tank which can hold 90 litres of diesel.
Convert this capacity to gallons, given that 1 litre = 0.22 gallons.

a 15 gallons = (15×4.5) litres = 67.5 litres.
b 90 litres = (90×0.22) gallons = 19.8 gallons.

Exercise 14.11

1 There are 6 cars at Refill's garage. The amount of petrol that each driver buys for his or her car is shown below.
Convert each amount to litres, given that 1 gallon = 4.5 litres.

First car:	4 gallons	Second car:	10 gallons
Third car:	6 gallons	Fourth car:	8 gallons
Fifth car:	12 gallons	Sixth car:	2 gallons

2 There are 6 cars at Fulltank's garage. The amount of petrol that each driver buys for his or her car is shown below.
Convert each amount to gallons, given that 1 litre = 0.22 gallons.

First car:	15 litres	Second car:	20 litres
Third car:	35 litres	Fourth car:	50 litres
Fifth car:	60 litres	Sixth car:	45 litres

3 Domestic oil tanks come in these sizes:

a 750 litres **b** 150 gallons, **c** 160 gallons, **d** 700 litres, **e** 154 gallons, **f** 725 litres **g** 144 gallons, **h** 650 litres, **i** 680 litres.

Convert all the gallons to litres (1 gallon = 4.5 litres) and list the sizes in order, starting with the letter that is the smallest.

14.3 Converting one metric unit to another

Here is a reminder of the metric length table:

10 millimetres (mm) = 1 centimetre (cm)

100 centimetres (or 100 millimetres) = 1 metre (m)

Example 1

Convert 1.527 metres to **a** centimetres and **b** millimetres.

a 1.527 m = (1.527 × 100) cm = 152.7 cm

b 1.527 m = (1.527 × 1000) mm = 1527 mm

Exercise 14.12

Convert each of the following to **a** centimetres and **b** millimetres.

1 14 m	**2** 19 m	**3** 12 m	**4** 8 m
5 3.328 m	**6** 6.745 m	**7** 2.419 m	**8** 8.634 m
9 4.56 m	**10** 7.95 m	**11** 5.2 m	**12** 8.4 m
13 0.819 m	**14** 0.273 m	**15** 0.053 m	**16** 0.036 m

Example 2

Convert 863.2 centimetres to **a** millimetres and **b** metres.

a 863.2 cm = (863.2 × 10) mm = 8632 mm

b 863.2 cm = (863.2 ÷ 100) m = 8.632 m

Exercise 14.13

Convert each of the following to **a** millimetres and **b** metres.

1 284 cm	**2** 659 cm	**3** 670 cm	**4** 940 cm
5 5600 cm	**6** 7400 cm	**7** 634.5 cm	**8** 528.6 cm
9 24.8 cm	**10** 53.6 cm	**11** 72.5 cm	**12** 87.3 cm
13 4.3 cm	**14** 5.7 cm	**15** 9.2 cm	**16** 7.1 cm

Example 3

Convert 4973 millimetres to **a** centimetres and **b** metres.

a 4973 mm = (4973 ÷ 10) cm = 497.3 cm

b 4973 mm = (4973 ÷ 1000) m = 4.973 m

Example 4

Convert 376 millimetres to **a** centimetres and **b** metres.

a 376 mm = (376 ÷ 10) cm = 37.6 cm

b 376 mm = (376 ÷ 1000) m = 0.376 m

Exercise 14.14

Convert each of the following to **a** centimetres and **b** metres.

1 9615 mm	**2** 8976 mm	**3** 3540 mm	**4** 6270 mm
5 4100 mm	**6** 6700 mm	**7** 9000 mm	**8** 4000 mm
9 78 000 mm	**10** 65 000 mm	**11** 92 000 mm	**12** 83 000 mm
13 725 mm	**14** 379 mm	**15** 96 mm	**16** 78 mm

We now need to remind ourselves of the connection between the metric mass units. (Mass is often referred to as 'weight'.)

1000 grams (g) = 1 kilogram (kg)

Example 5

Convert each of the following to grams:
a 8.254 kg **b** 0.0018 kg

a 8.254 kg = (8.254 × 1000) g = 8254 g
b 0.0018 kg = (0.0018 × 1000) g = 1.8 g

Exercise 14.15

Convert each of the following to grams.

1 86 kg	**2** 6.254 kg	**3** 9.386 kg	**4** 7.521 kg
5 3.86 kg	**6** 7.52 kg	**7** 4.37 kg	**8** 0.372 kg
9 0.19 kg	**10** 0.045 kg	**11** 0.082 kg	**12** 0.074 kg
13 0.056 kg	**14** 0.029 kg	**15** 0.0036 kg	**16** 0.0057 kg

Example 6

Convert 932 g to kilograms.

932 g = (932 ÷ 1000) kg = 0.932 kg

Exercise 14.16

Convert each of the following to kilograms.

1 25 000 g	**2** 2475 g	**3** 5286 g	**4** 7394 g
5 8390 g	**6** 4280 g	**7** 2950 g	**8** 526 g
9 170 g	**10** 76 g	**11** 59 g	**12** 43 g
13 28 g	**14** 17 g	**15** 4.8 g	**16** 6.3 g

We now need to remind ourselves of the connection between the metric capacity units.

1000 millimetres (ml) = 1 litre (l)

Example 7

Convert 4.5 litres to millimetres.

4.5 l = (4.5 × 1000) ml = 4500 ml

Exercise 14.17

Convert each of the following to millilitres.

1 5 litres	**2** 60 litres	**3** 6.4 litres	**4** 3.7 litres
5 0.6 litres	**6** 0.3 litres	**7** 0.05 litres	**8** 0.08 litres
9 0.003 litres	**10** 0.009 litres		

Example 8

Convert 2800 millilitres to litres.

2800 ml = (2800 ÷ 1000) litres = 2.8 litres

Exercise 14.18

Convert each of the following to litres.

1 6000 ml	**2** 30 000 ml	**3** 9500 ml	**4** 1400 ml
5 800 ml	**6** 500 ml	**7** 90 ml	**8** 40 ml
9 7 ml	**10** 5 ml		

15 Perimeter, area and volume

15.1 Finding perimeters of simple shapes

The perimeter of a shape is the total distance around the outside of the shape.

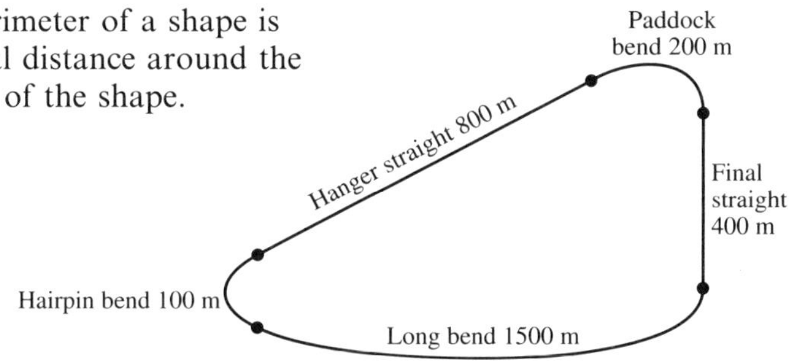

> **Example 1**
>
> Find the perimeter of the race track illustrated above.
>
> The perimeter is: $(1500 + 100 + 800 + 200 + 400)\,\text{m} = 3000\,\text{m}$

Exercise 15.1

1 Find the perimeter of this car park.

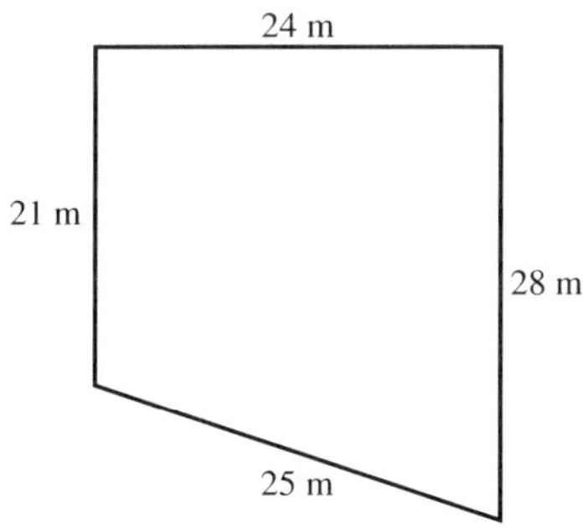

2. Find the perimeter of this bus station.

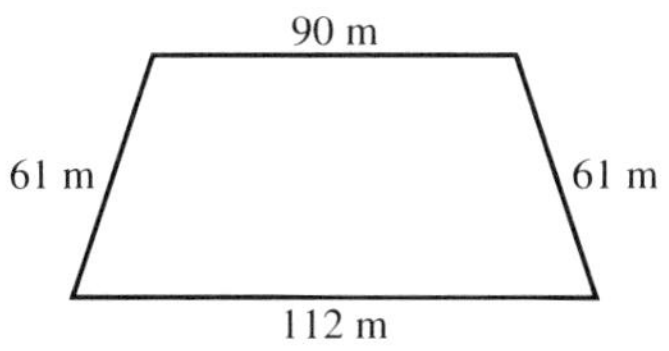

3 Find the perimeter of this village green.

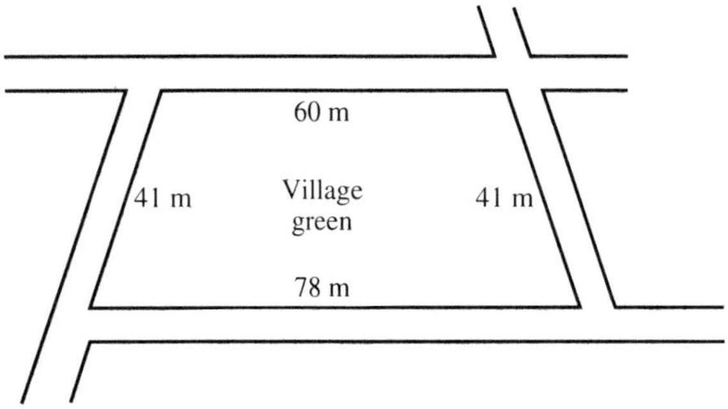

4 Find the perimeter of this recreation ground.

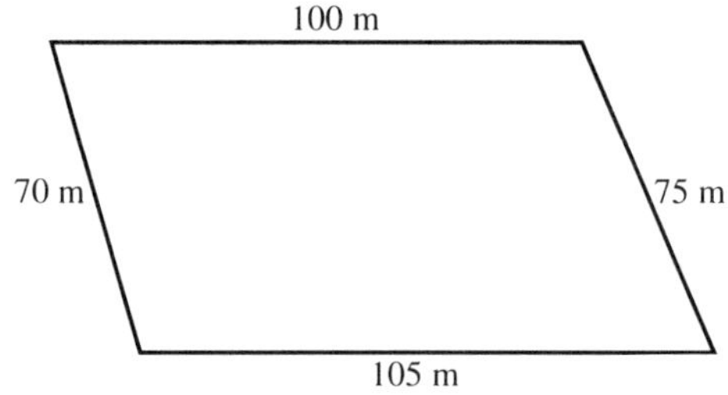

5. Find the perimeter of Wood Common.

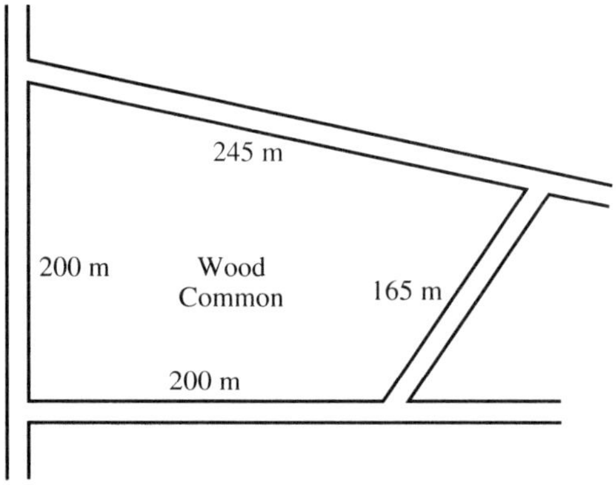

For regular shapes there is often a quick way of finding the perimeter.

Example 2

Find the perimeter of:
a a square of side length 5 cm and **b** an equilateral triangle of side length 5 cm.

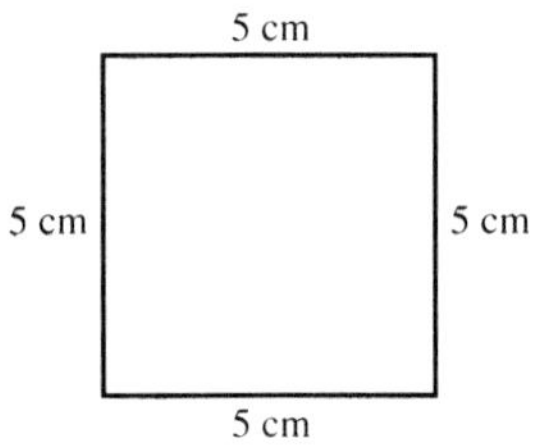

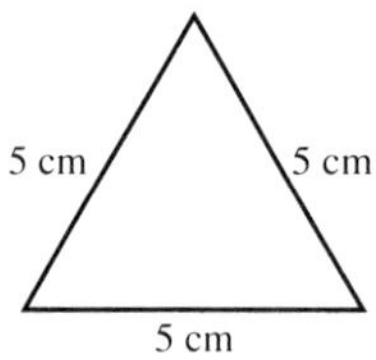

a Perimeter = (4×5) cm = 20 cm **b** Perimeter = (3×5) cm = 15 cm

Exercise 15.2

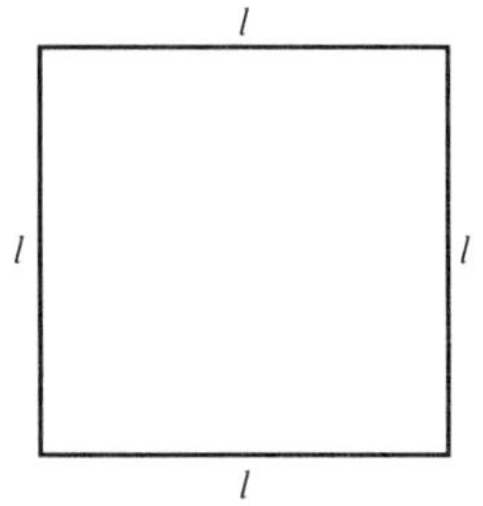

1 Find the perimeter of the square if the side length l is:

a 6 cm **b** 9 cm **c** 15 cm
d 24 mm **e** 65 mm **f** 90 mm

2 Find the perimeter of the equilateral triangle if the side length l is
a 8 cm **b** 6 cm **c** 20 cm
d 45 mm **e** 72 mm **f** 96 mm

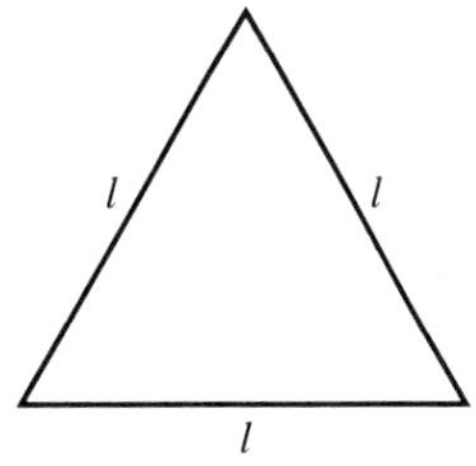

3 Find the perimeter of the regular hexagon if the side length l is
a 12 cm **b** 18 cm **c** 25 cm
d 42 mm **e** 75 mm **f** 60 mm

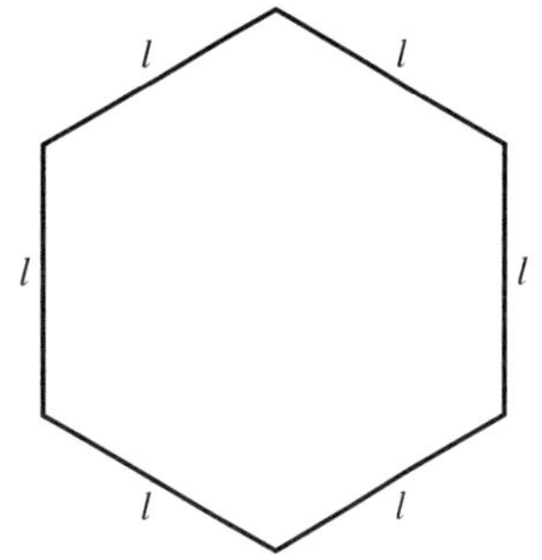

Example 3

Find the perimeter of this rectangle.

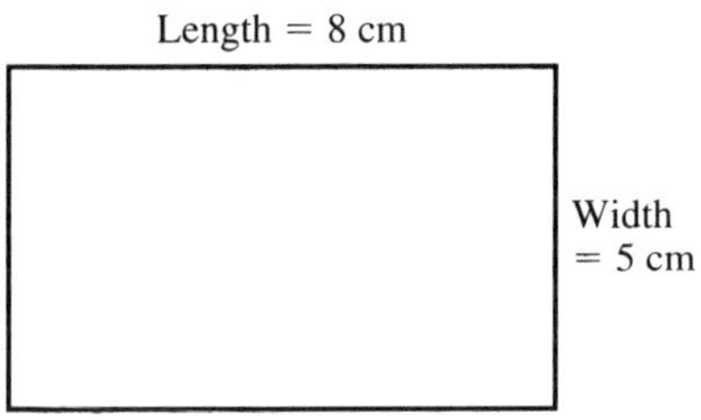

Perimeter $= (2 \times 8) + (2 \times 5)$ cm $= (16 + 10)$ cm $= 26$ cm

Exercise 15.3

For Questions 1–4 find which rectangle, **a**, **b** or **c** has a different perimeter from the other two.

1 **a** Length = 25 cm Width = 15 cm **b** Length = 28 cm Width = 12 cm **c** Length = 21 cm Width = 17 cm

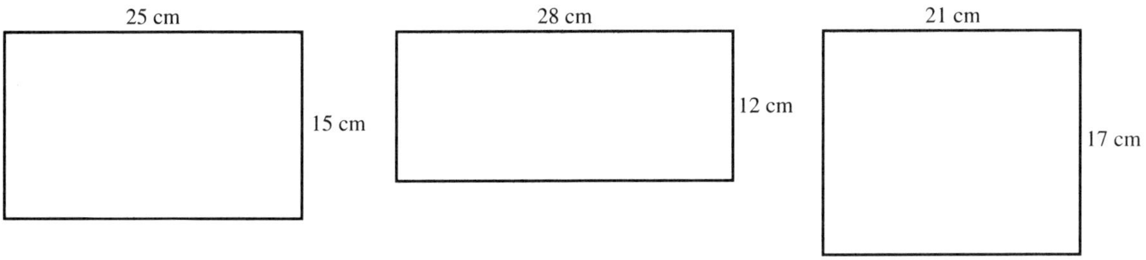

2 **a** Length = 96 cm
Width = 26 cm
b Length = 75 cm
Width = 45 cm
c Length = 83 cm
Width = 37 cm

3 **a** Length = 96 mm
Width = 34 mm
b Length = 72 mm
Width = 56 mm
c Length = 101 mm
Width = 29 mm

4 **a** Length = 105 mm
Width = 63 mm
b Length = 122 mm
Width = 46 mm
c Length = 93 mm
Width = 77 mm

For Questions 5 and 6 measure the side lengths of the rectangle and find its perimeter. *Note:* each square is a $\frac{1}{2}$cm square.

5

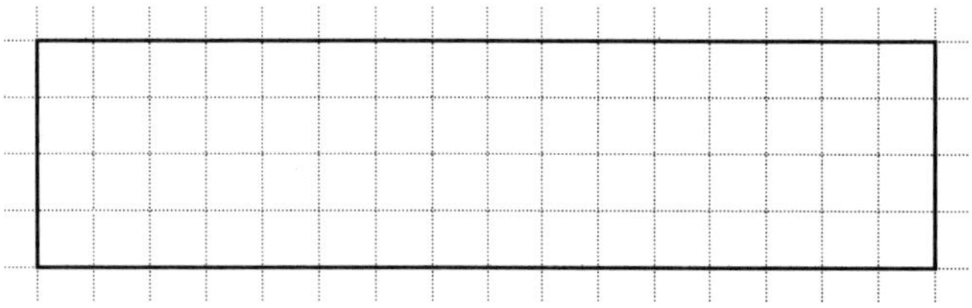

6

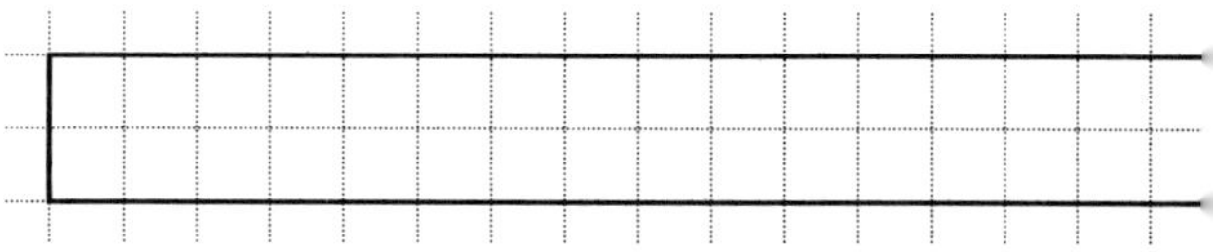

For Questions 7 and 8 measure the side lengths of the triangle and find its perimeter. *Note:* each square is a $\frac{1}{2}$cm square.

7

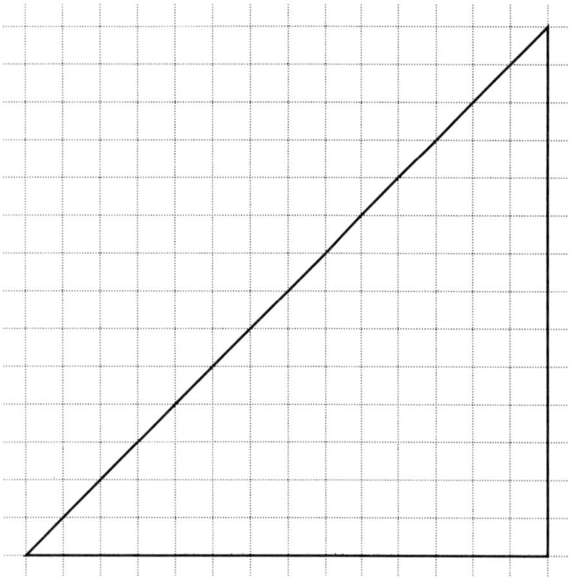

8

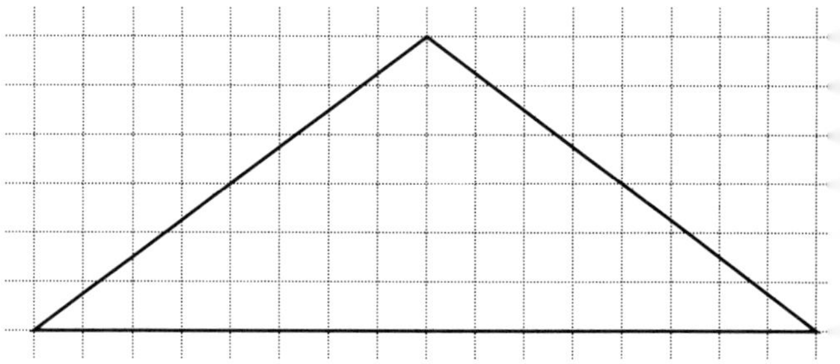

For Questions 9 and 10 measure the side lengths of the parallelogram and find its perimeter. *Note:* each square is a $\frac{1}{2}$cm square.

9

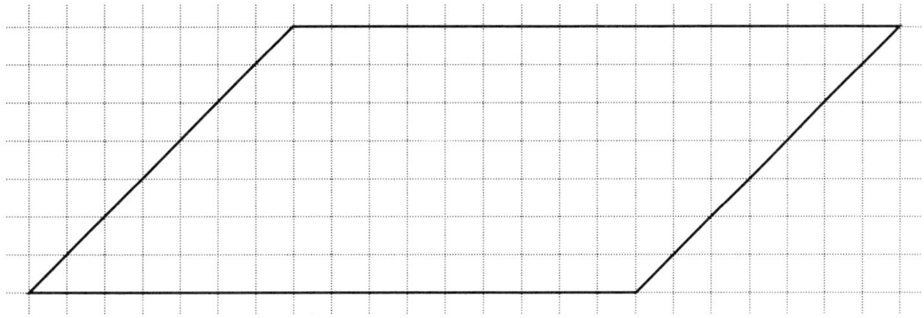

10

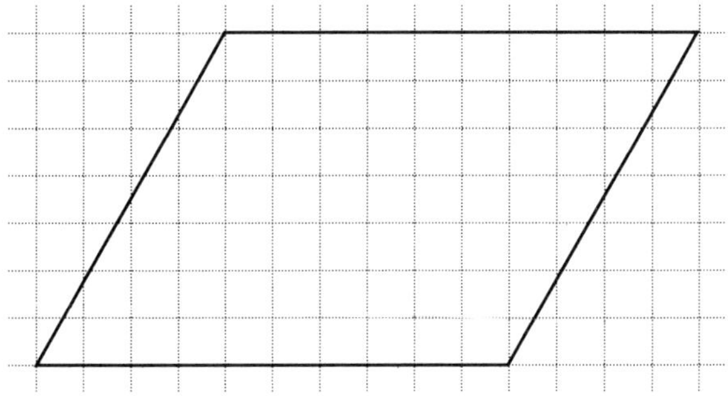

For Questions 11 and 12 measure the side lengths of the trapezium and find its perimeter. *Note:* each square is a $\frac{1}{2}$cm square.

11

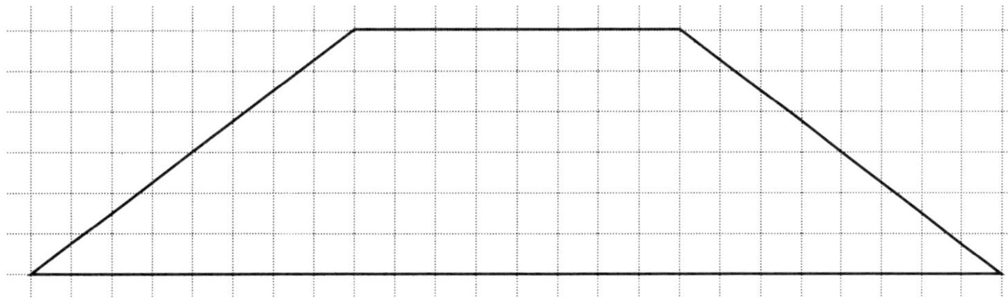

12

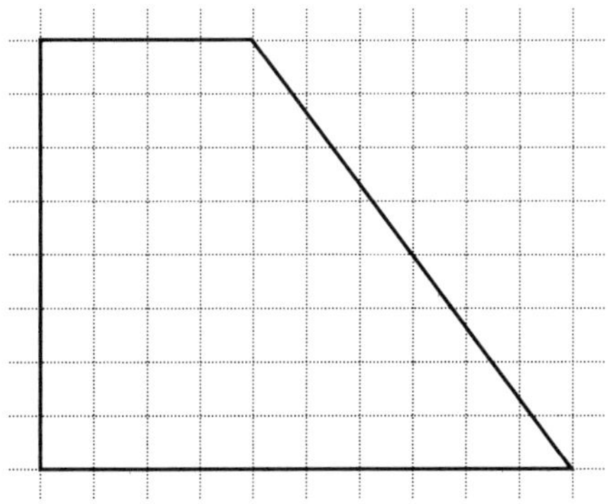

Example 4

The perimeter of each shape is 24 cm.
Find the length of the unknown side.

a

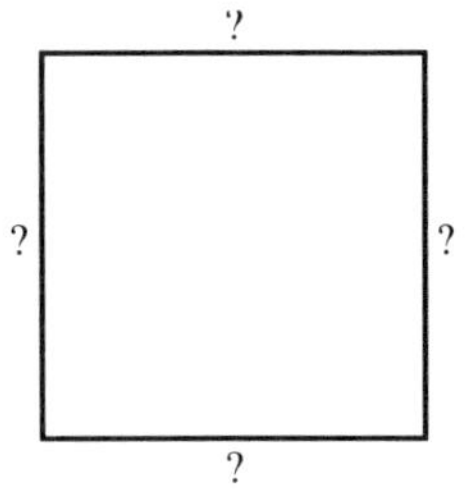

b

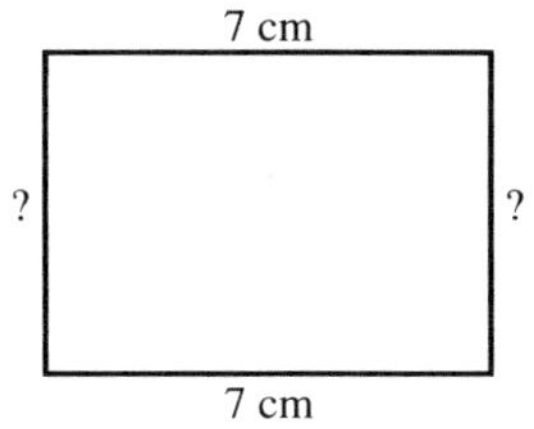

c

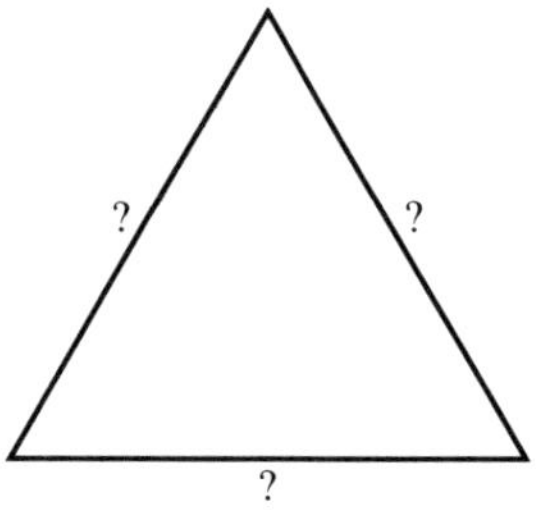

d

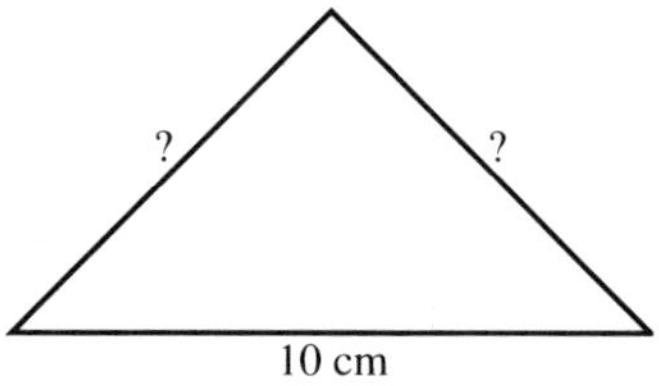

a The length of each side of the square is: $(24 \div 4)\,\text{cm} = 6\,\text{cm}$

b The length of the short side of the rectangle is: $\dfrac{(24 - 14)}{2}\,\text{cm} = 5\,\text{cm}$

c The length of each side of the equilateral triangle is: $(24 \div 3)\,\text{cm} = 8\,\text{cm}$

d The length of each of the equal sides of the isosceles triangle is: $\dfrac{(24 - 10)}{2}\,\text{cm} = 7\,\text{cm}$

Exercise 15.4

1 Find the side length of the square if the perimeter is:
a 32 cm **b** 48 cm **c** 100 cm
d 180 cm **e** 272 mm **f** 348 mm

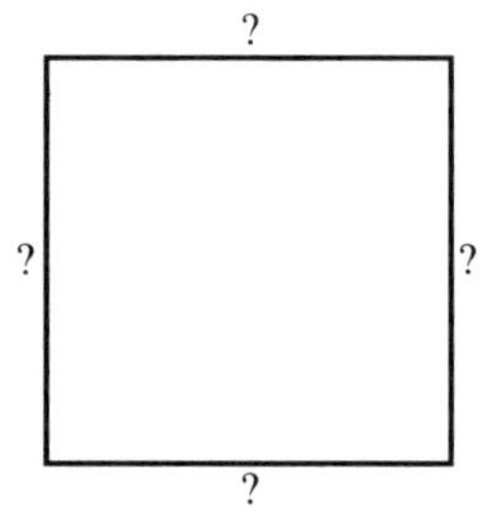

2 Find the width of the rectangle if the perimeter and length are:
a Perimeter = 54 cm, length = 15 cm
b Perimeter = 82 cm, length = 24 cm
c Perimeter = 114 cm, length = 32 cm

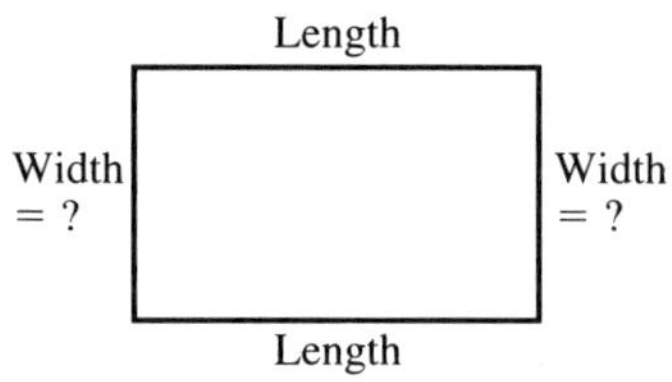

3 Find the length of the rectangle if the perimeter and width are:
a Perimeter = 82 cm, width = 13 cm
b Perimeter = 114 cm, width = 18 cm
c Perimeter = 150 cm, width = 23 cm

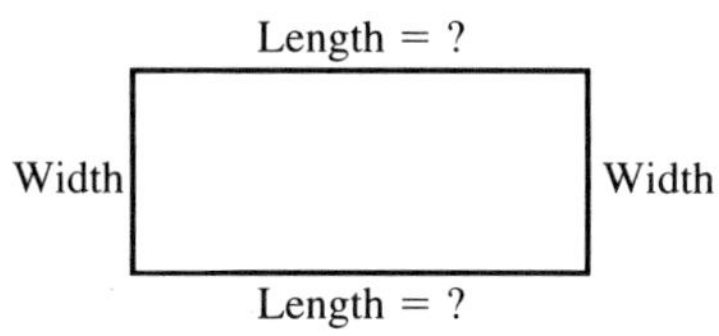

4 Find the side length of the equilateral triangle if the perimeter is:
a 36 cm **b** 27 cm **c** 45 cm

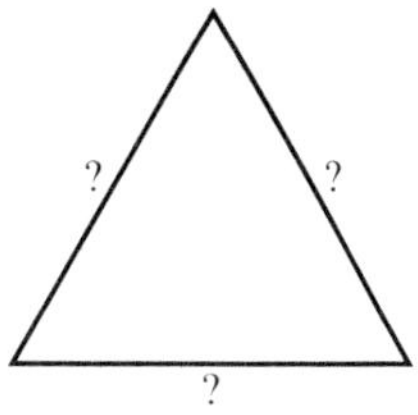

5 Find the length of each of the equal sides of the isosceles triangle if the perimeter and unequal side are:
a Perimeter = 17 cm, unequal side = 5 cm
b Perimeter = 25 cm, unequal side = 7 cm
c Perimeter = 42 cm, unequal side = 12 cm

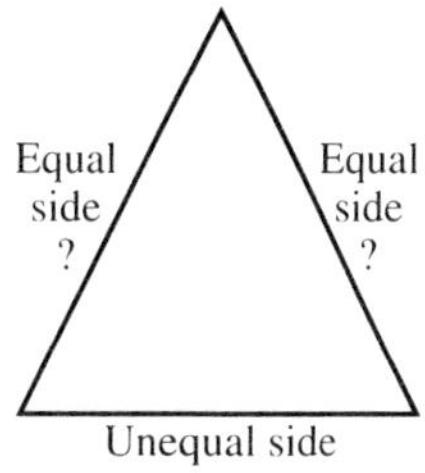

Example 5

A rectangle has a perimeter of 20 cm.
If the side dimensions are a whole number of centimetres, find the five possible pairs of values for the length and width.

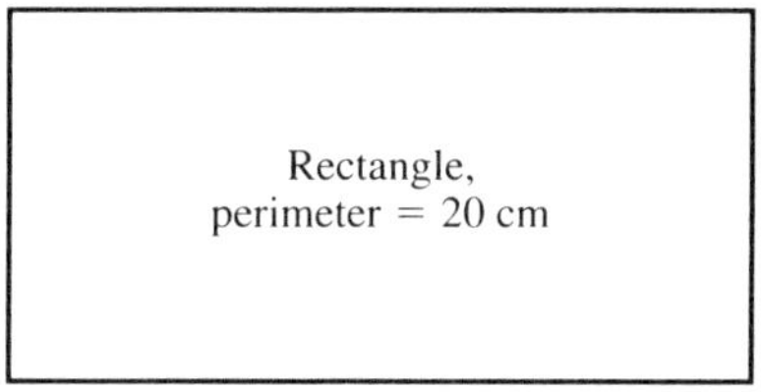

The length and width must total 10 cm, therefore the five possible values are:
9 cm and 1 cm, 8 cm and 2 cm, 7 cm and 3 cm, 6 cm and 4 cm and 5 cm and 5 cm.

Exercise 15.5

1 A rectangle has a perimeter of 18 cm.

If the side dimensions are a whole number of centimetres, find the four possible pairs of values for the length and width.

2 A rectangle has a perimeter of 14 cm.
If the side dimensions are a whole number of centimetres, find the three possible pairs of values for the length and width.

3 A parallelogram has a perimeter of 16 cm.
If the side lengths are a whole number of centimetres, find the four possible pairs of values for the side length.

4 A parallelogram has a perimeter of 12 cm.
If the side lengths are a whole number of centimetres, find the three possible pairs of values for the side length.

Example 6

An isosceles triangle has a perimeter of 7 cm.
If the sides are a whole number of centimetres, find the two possible sets of values for the side lengths.

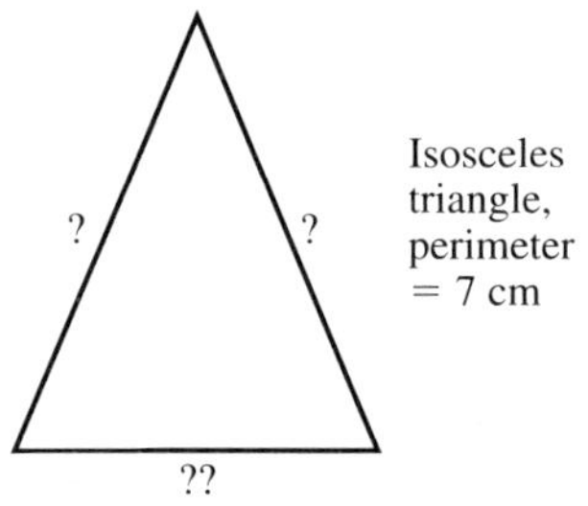

Try 1 cm for the length of the unequal side:

The length of the equal sides is therefore $\frac{(7-1)}{2}$ cm $= \frac{6}{2}$ cm
$= 3$ cm

Therefore one set of values is 1 cm, 3 cm and 3 cm.

The length of the unequal side cannot be 2 cm or any other even number of centimetres because when the length of the unequal side is subtracted from 7 cm he result must be an even number.

Try 3 cm for the length of the unequal side:

The length of the equal sides is therefore $\frac{(7-3)}{2}$ cm $= \frac{4}{2}$ cm
$= 2$ cm

Therefore one set of values is 3 cm, 2 cm and 2 cm.

(*Note:* the length of the unequal side cannot be 5 cm as a triangle cannot be formed.)

Exercise 15.6

1 An isosceles triangle has a perimeter of 13 cm.
If the side lengths are a whole number of centimetres, find the three possible sets of values for the side lengths.

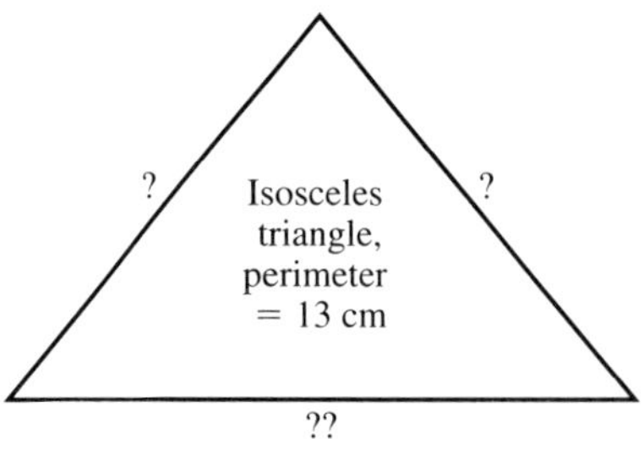

2 An isosceles triangle has a perimeter of 11 cm.
If the side lengths are a whole number of centimetres, find the three possible sets of values for the side lengths.

3 An isosceles triangle has a perimeter of 18 cm.
If the side lengths are a whole number of centimetres, find the four possible sets of values for the side lengths.

4 An isosceles triangle has a perimeter of 8 cm.
If the side lengths are a whole number of centimetres, find the only set of values for the side lengths.

15.2 Finding areas by counting squares and volumes by counting cubes

Example 1

Find the area of each shape.
Both shapes are drawn on a grid of 1 cm squares.

a

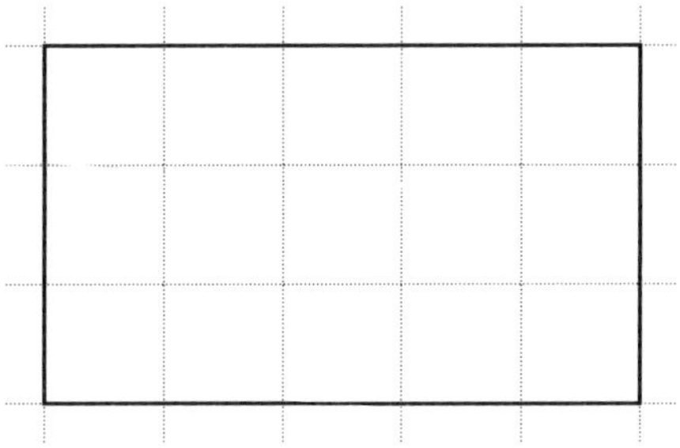

b

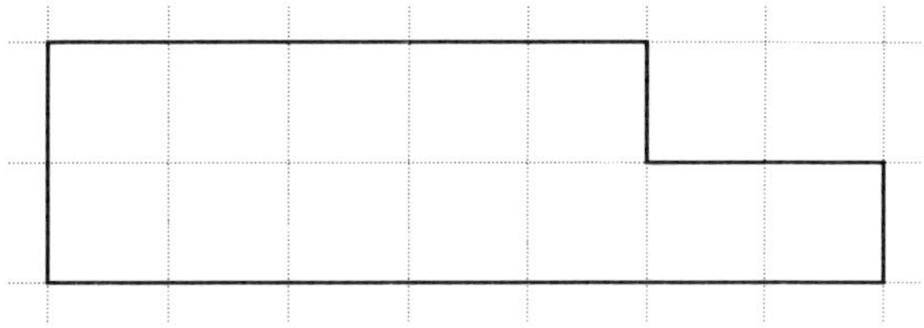

Shape **a** has an area of 15 cm^2 (15 squares)   Shape **b** has an area of 12 cm^2 (12 squares)

Exercise 15.7

Find the area of each of these shapes. Assume each square is a 1 cm square, even though some are a little smaller.

1

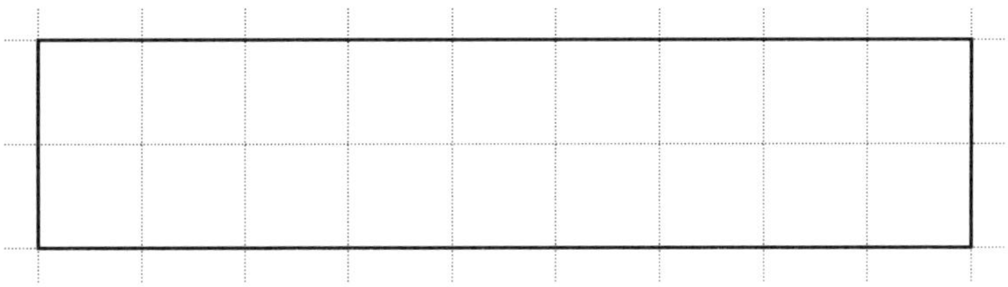

2

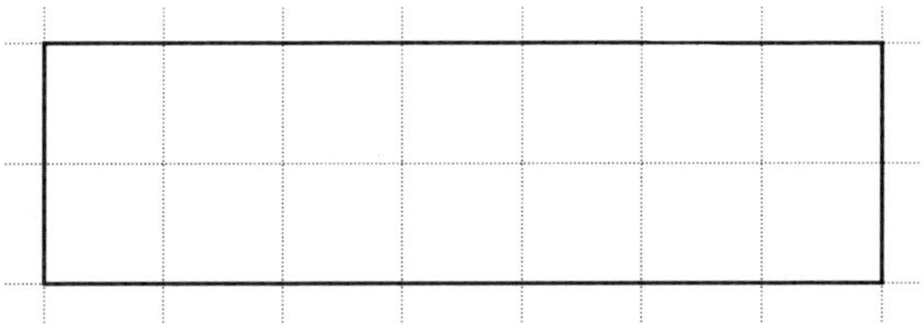

3

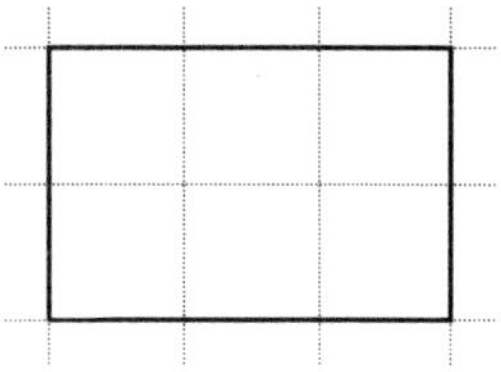

4

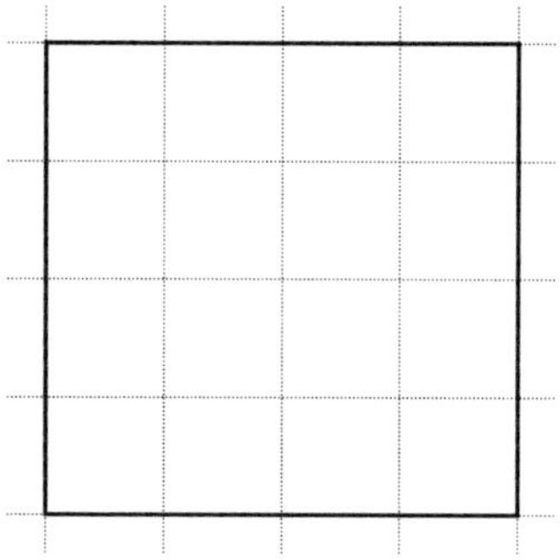

5

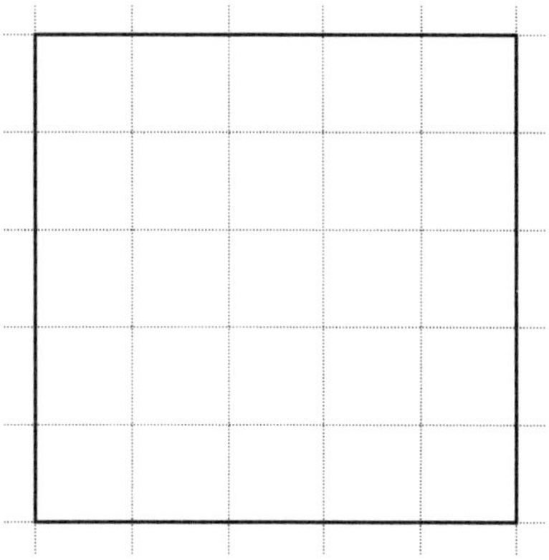

6

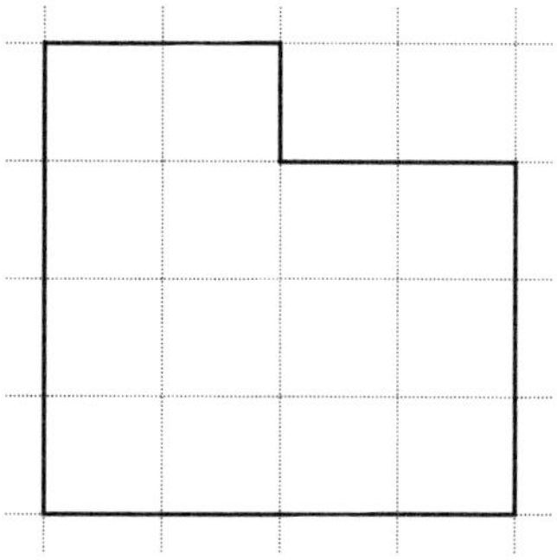

7

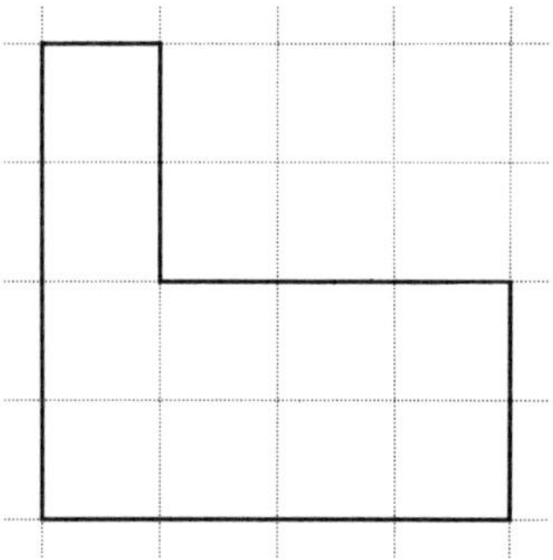

8

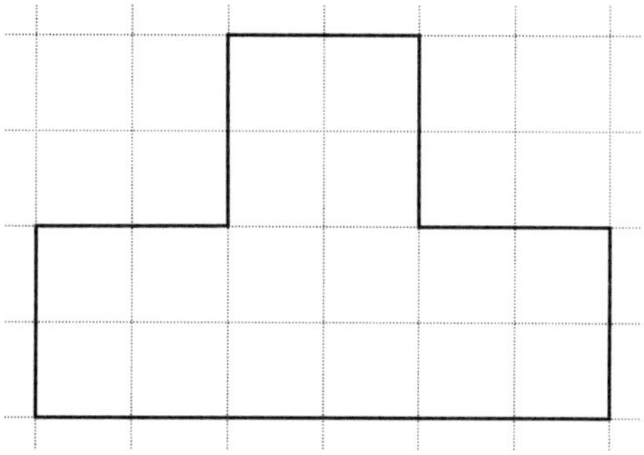

9

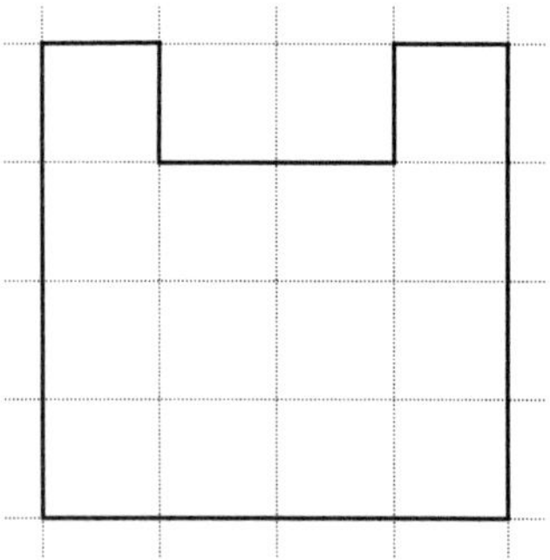

10

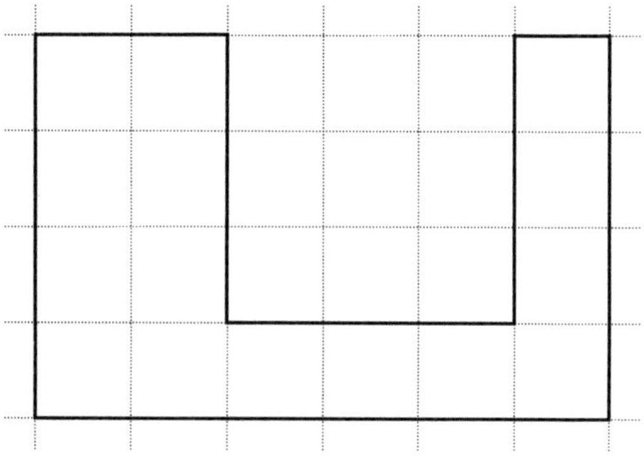

Example 2

Find the area of each triangle.
Both triangles are drawn on a grid of 1 cm squares.

a

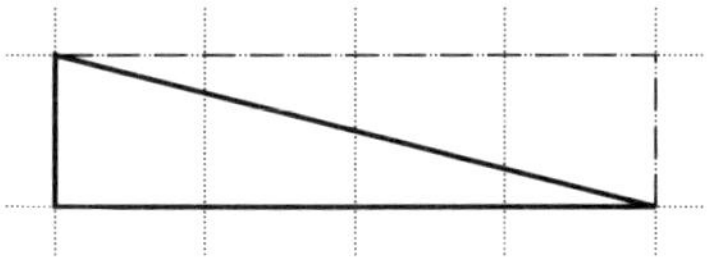

b

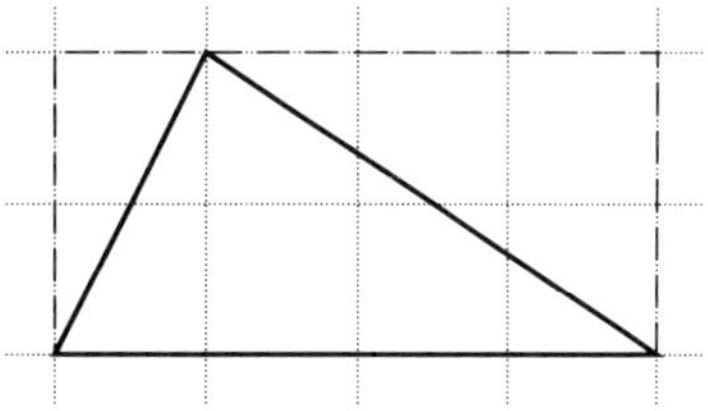

Triangle **a** has an area of 2 cm^2
(one half of a rectangle of area 4 cm^2)

Triangle **b** has an area of 4 cm^2
(one half of a rectangle of area 8 cm^2)

Exercise 15.8

Find the area of each of these triangles. Assume, as before, that each square is a 1 cm square.

1

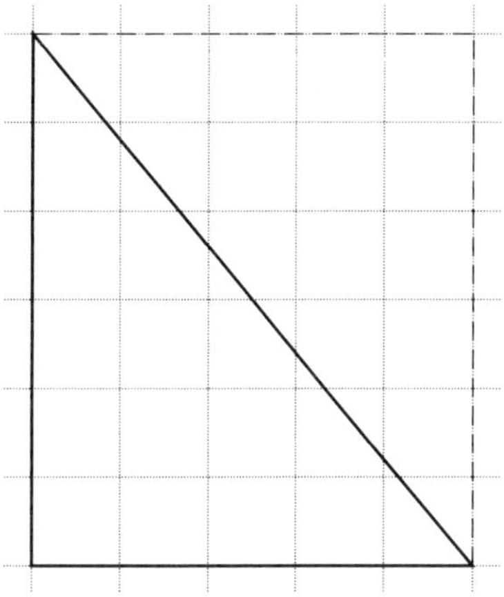

2

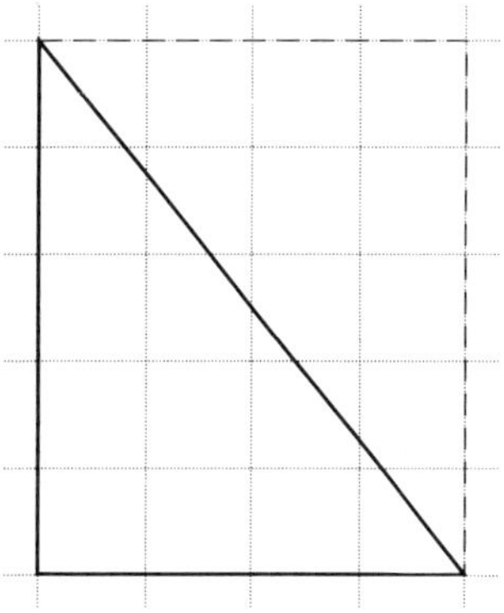

3

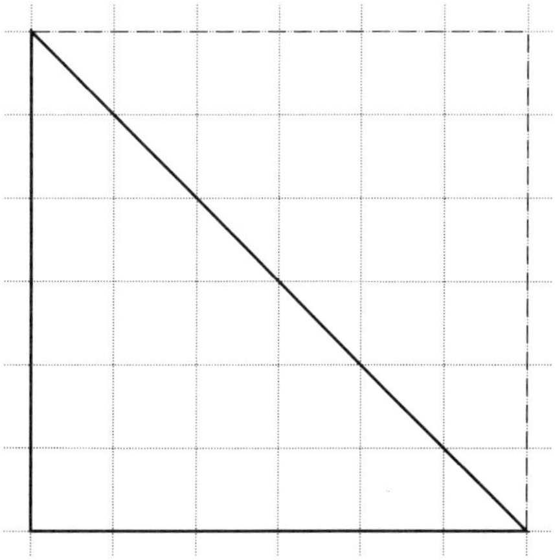

4

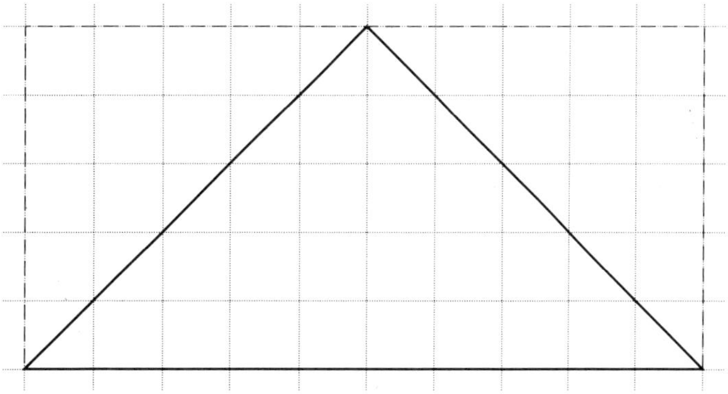

5

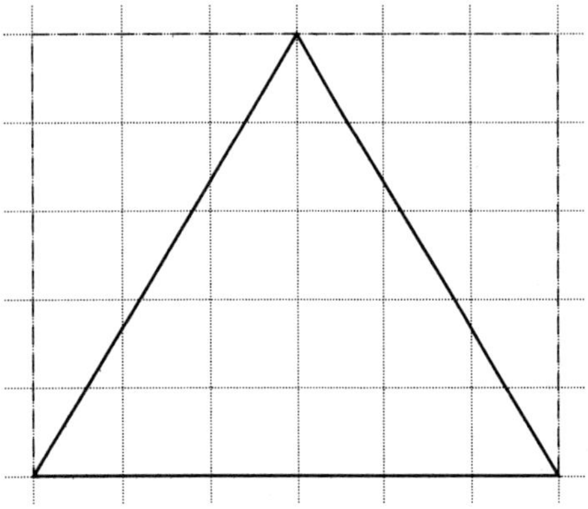

6

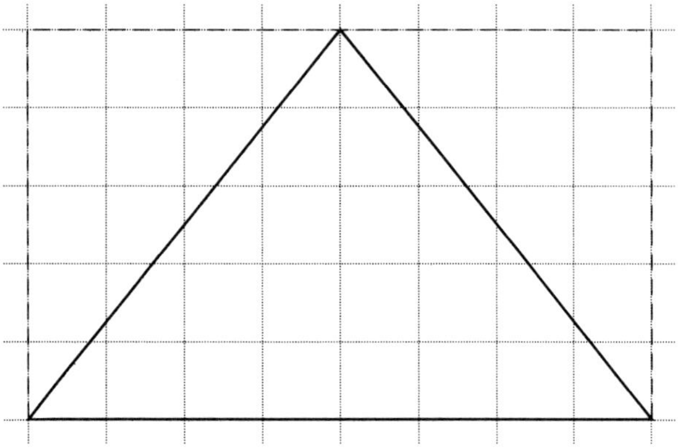

7

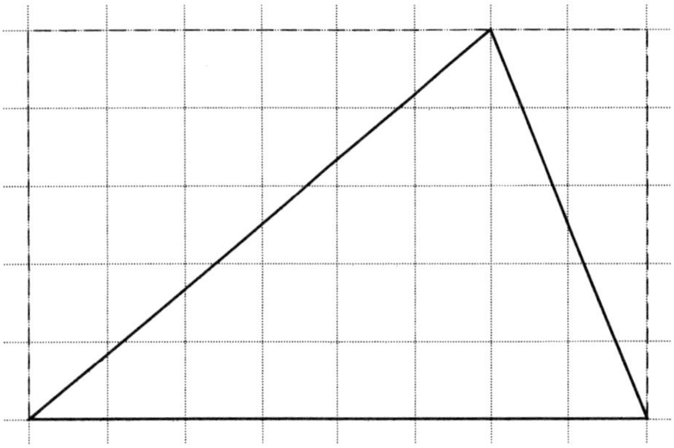

8

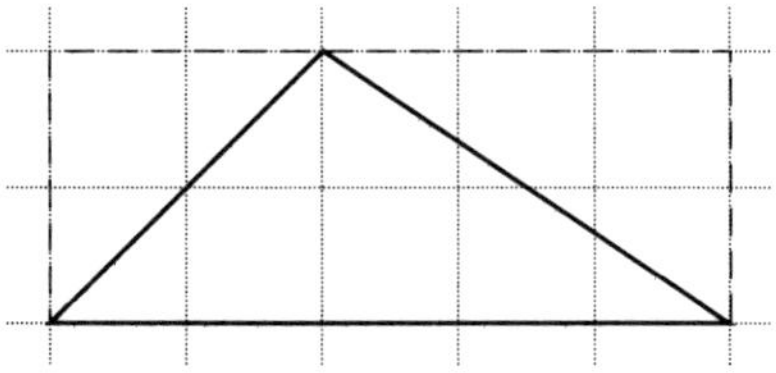

9

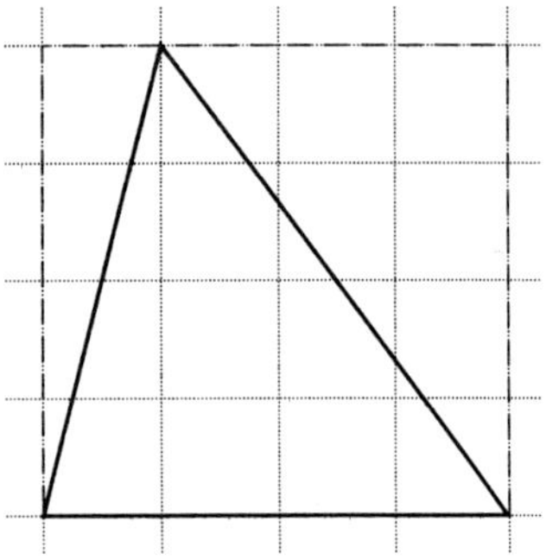

10

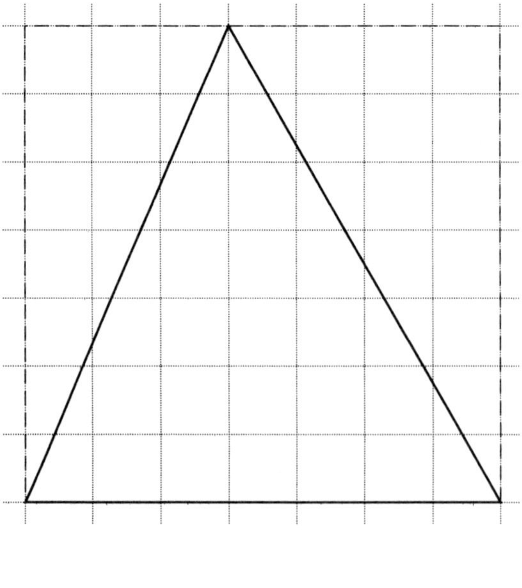

Example 3

Find the surface area of this leaf, which is drawn on a grid of 1 cm squares.
Each whole square is marked with a cross and each half square is marked with a dot.

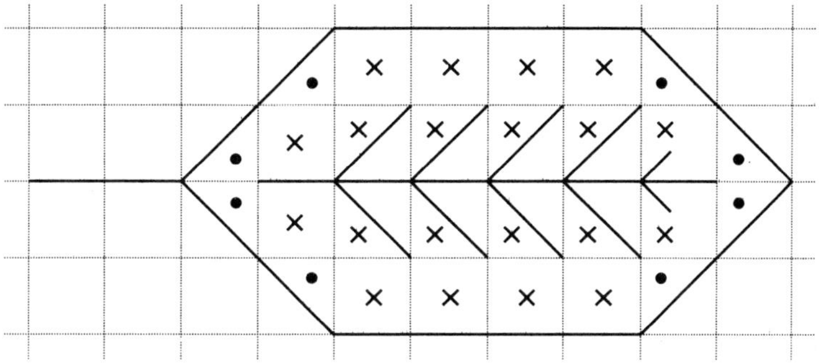

There are 20 crosses and 8 dots.
The surface area is: $(20 + (\frac{1}{2} \times 8))\,\text{cm}^2 = (20 + 4)\,\text{cm}^2 = 24\,\text{cm}^2$

Exercise 15.9

Find the surface area of each of these shapes. Assume that each one is drawn on 1 cm squares.

1 Small leaf

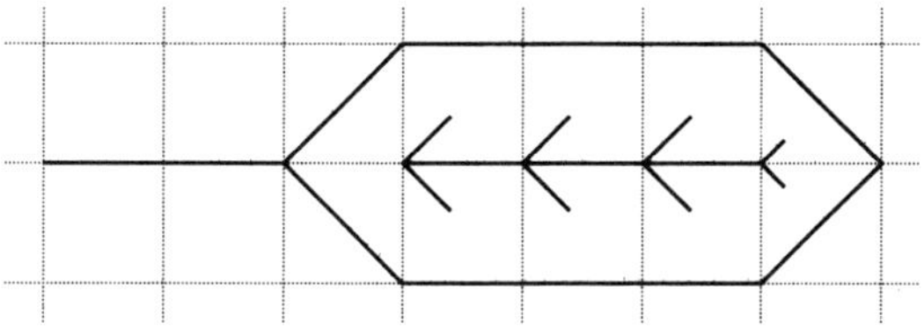

2 Knife

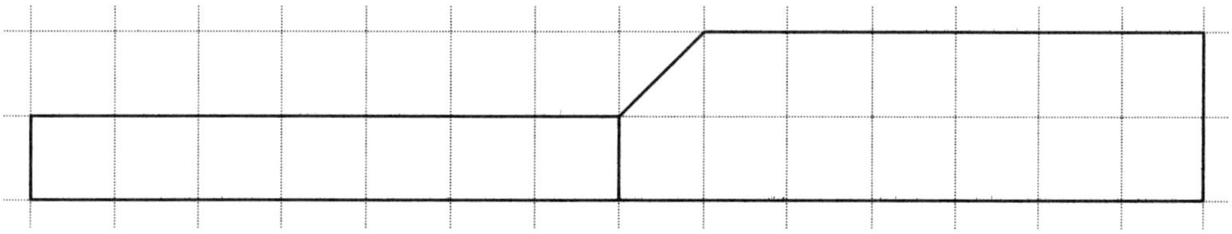

3 Mini TV screen

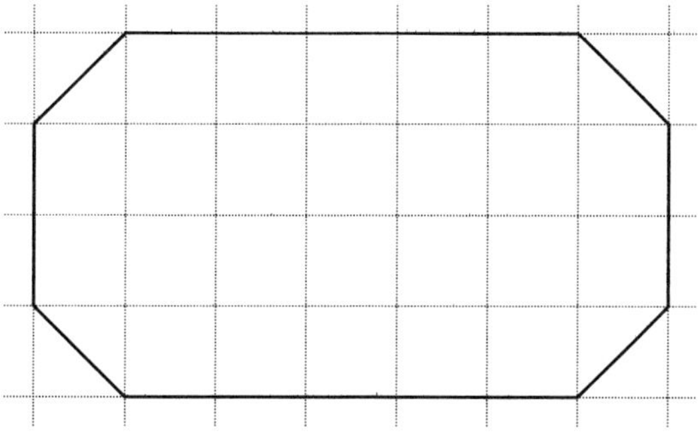

4 Small set square

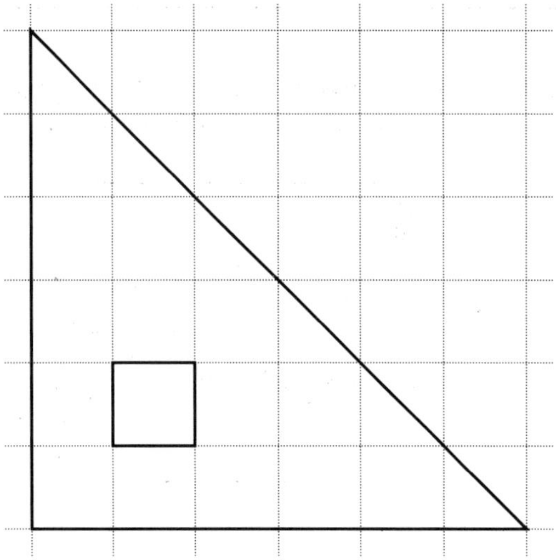

5 Large set square

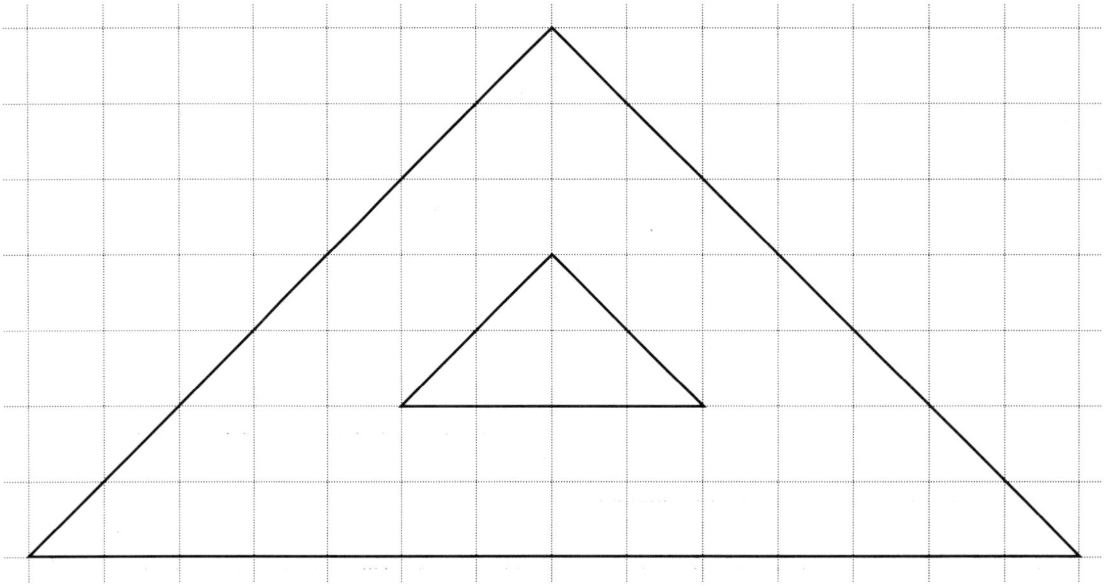

6 Key

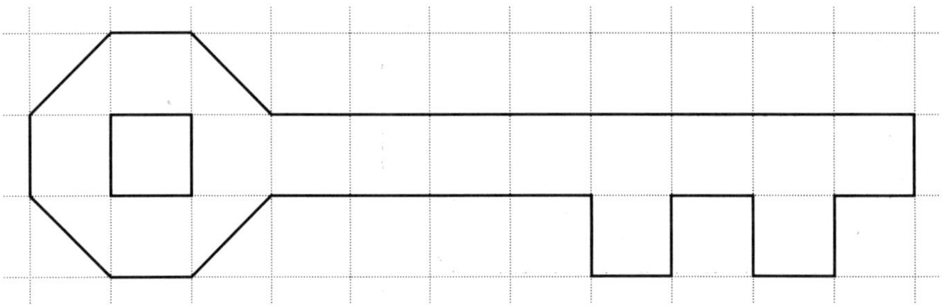

Example 4

Find the volume of each shape illustrated.
Both shapes are made with 1 cm cubes.

a

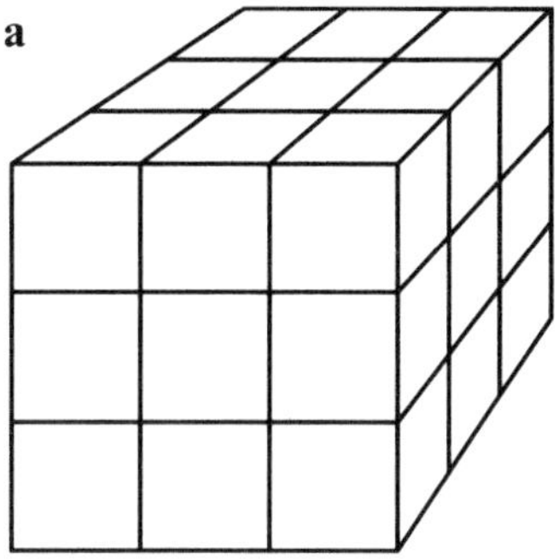

b

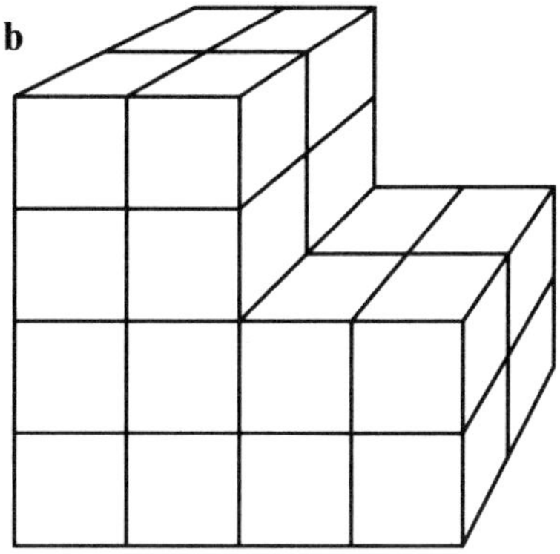

Shape **a** has a volume of $27\,cm^3$ (it is a $3 \times 3 \times 3$ cube)

Shape **b** has a volume of $24\,cm^3$ (it is a $4 \times 2 \times 2$ cuboid and a $2 \times 2 \times 2$ cube)

Exercise 15.10

Find the volume of each of these shapes made from 1 cm cubes.

1

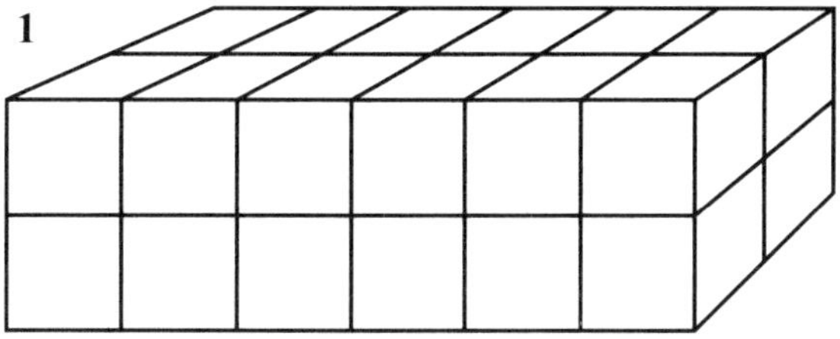

2

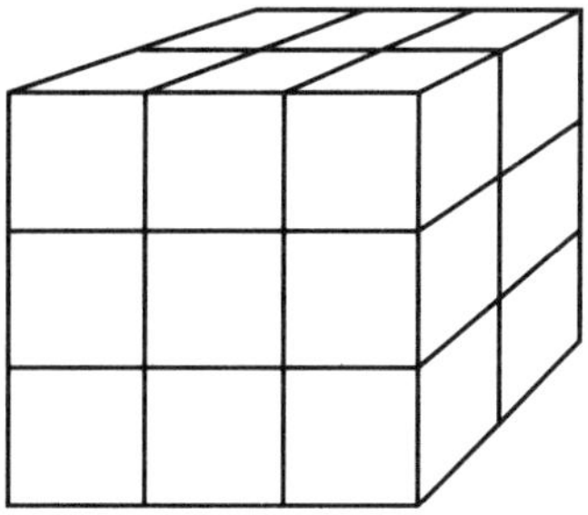

3

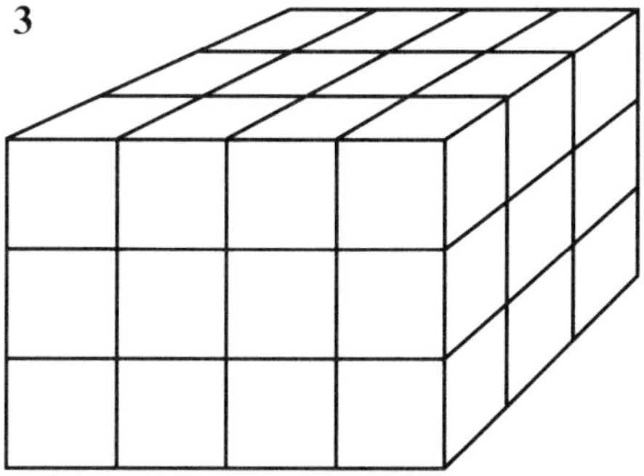

4

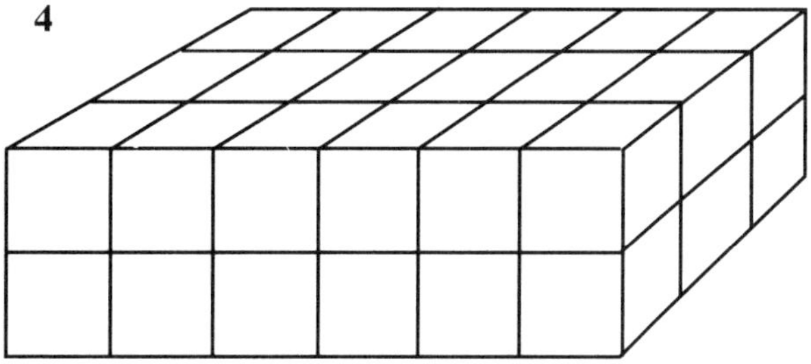

5

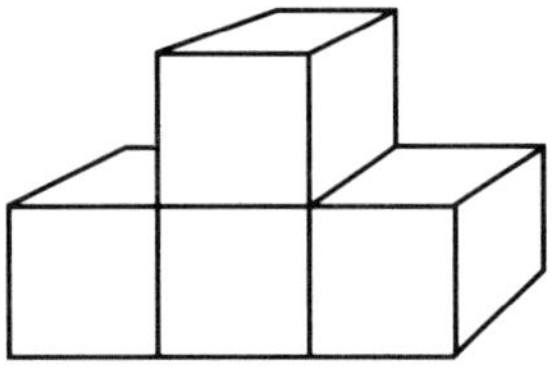

6

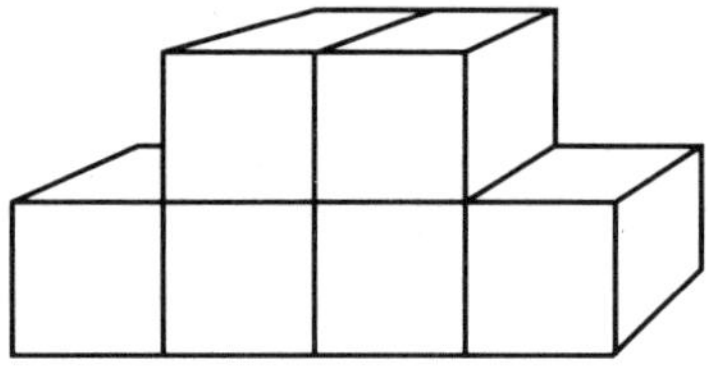

7

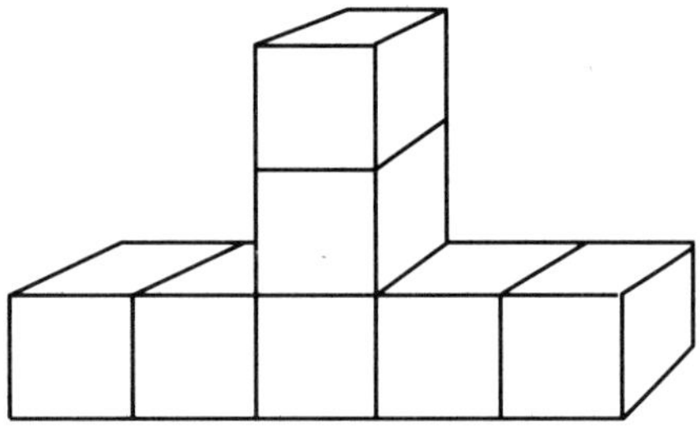

8

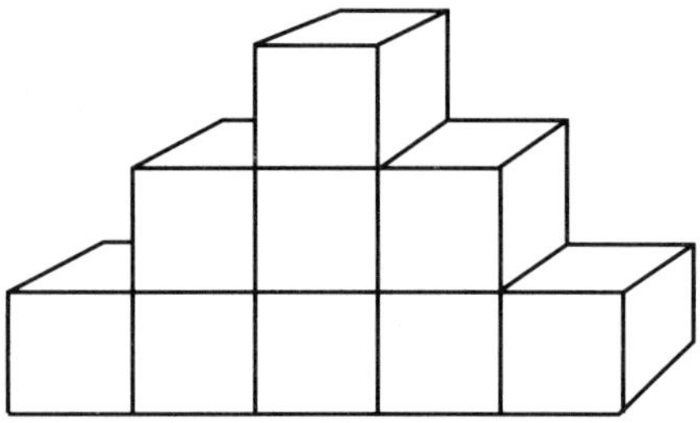

Example 5

Find the volume of a cuboid whose edge lengths are 5 cm, 4 cm and 2.5 cm.

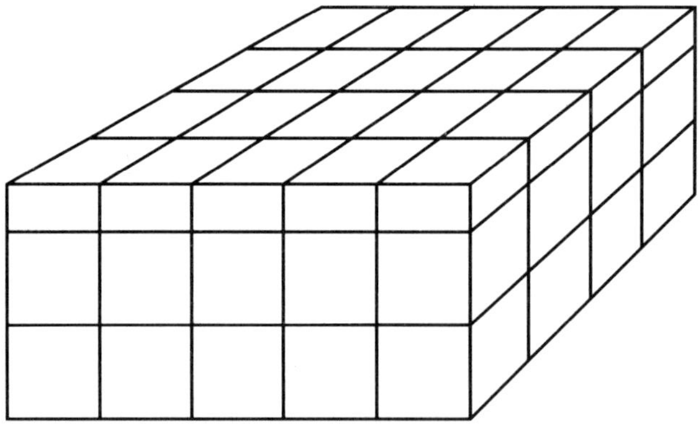

The cuboid has two layers with $20\,\text{cm}^3$ in each and one layer with $20 \times \frac{1}{2}\,\text{cm}^3$ in it.

The volume is: $(20 + 20 + 10)\,\text{cm}^3 = 50\,\text{cm}^3$

Exercise 15.11

Find the volume of each of these cuboids made from 1 cm cubes and half cubes.

1

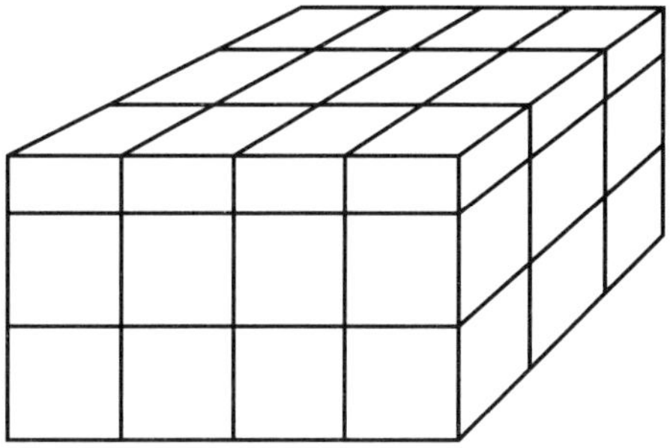

2

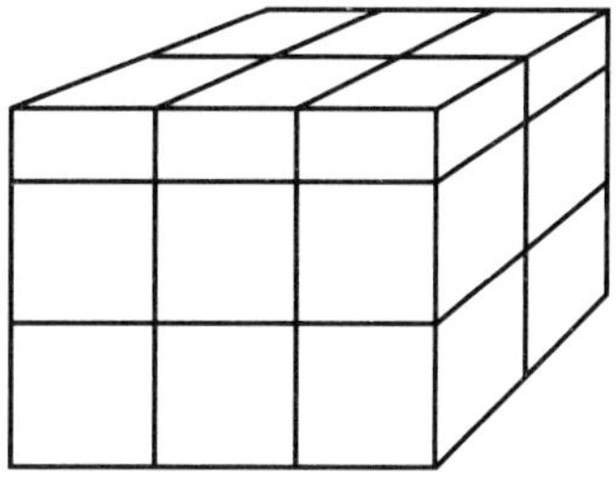

3

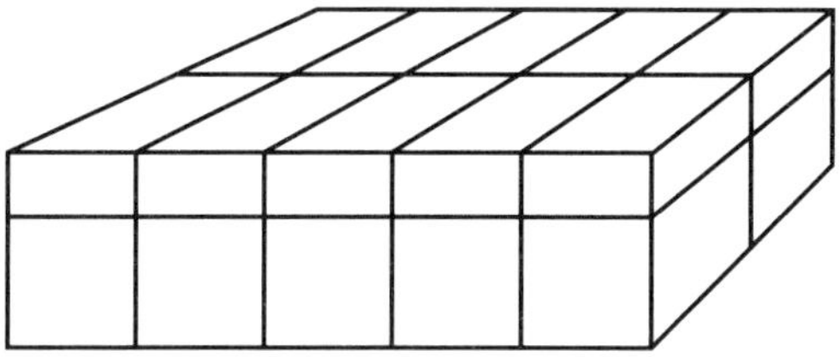

4

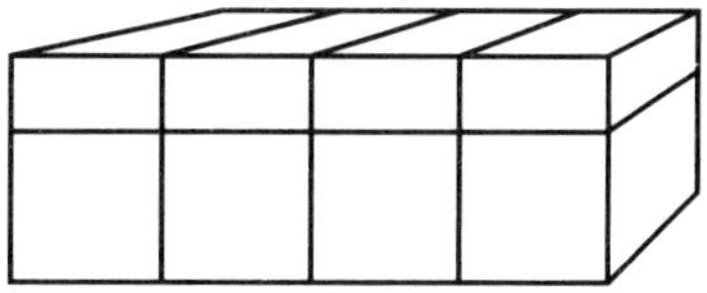

5

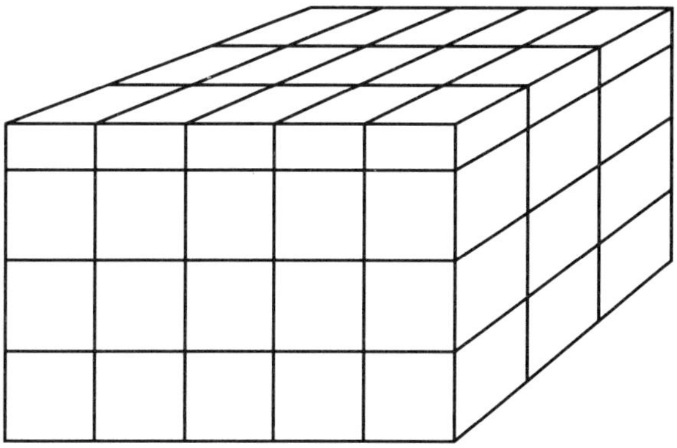

15.3 Finding areas of plane figures using appropriate formulae

Reminder 1

The area of a square of side length a cm is $(a \times a)$ cm^2 or a^2 cm^2.

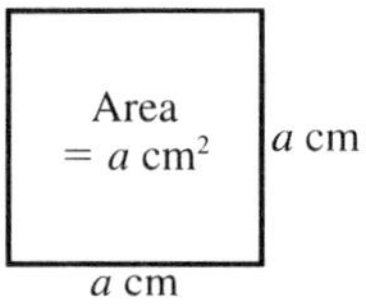

Example 1

Find the area of a square of side length 60 mm.

Area = (60×60) mm^2 = 3600 mm^2

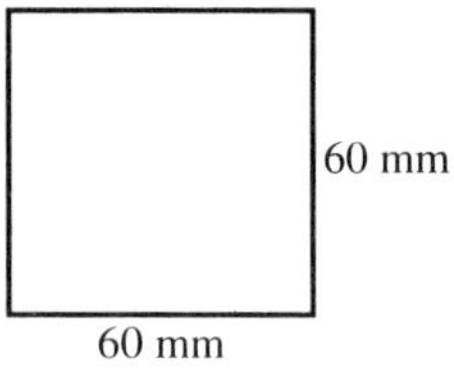

Exercise 15.12

Copy and complete this table.

Side length of square	Area of square
1 5 cm	
2 12 cm	
3 30 mm	
4 1.4 m	
5 2.4 m	

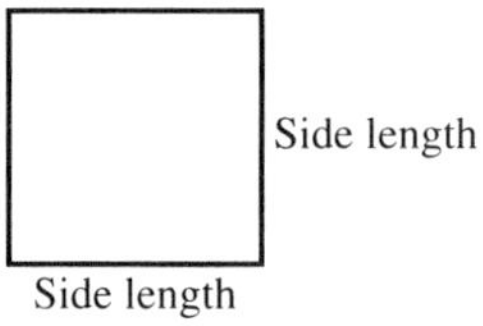

Example 2

Find the side length of a square of area 2.56 m^2

Side length = square root of 2.56 m^2 (or $\sqrt{(2.56)}$ m)

= 1.6 m

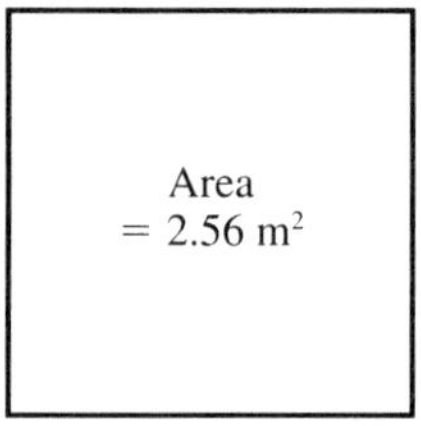

Exercise 15.13

Copy and complete this table.

Area of square	Side length of square
1 36 cm^2	
2 225 cm^2	
3 4900 mm^2	
4 1.69 m^2	
5 2.89 m^2	

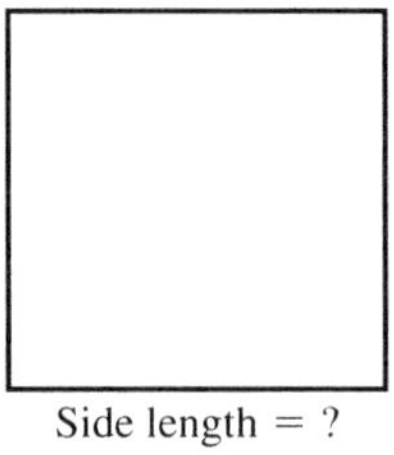

Reminder 2

The area of a rectangle of length a cm and width b cm is $(a \times b)\,\text{cm}^2$ or $ab\,\text{cm}^2$

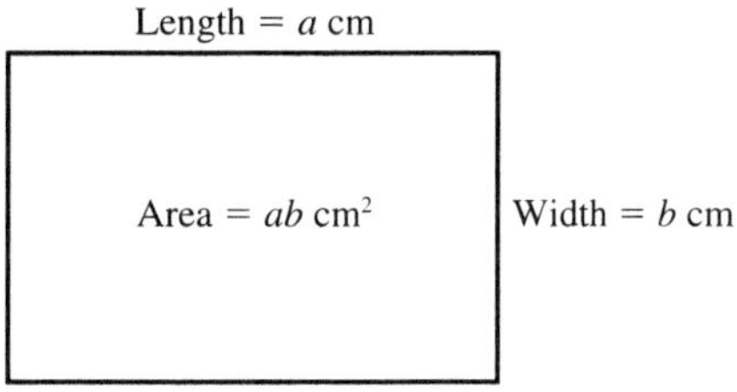

Example 3

Find the area of each of the three rectangles below.
State which one has a different area from the other two.

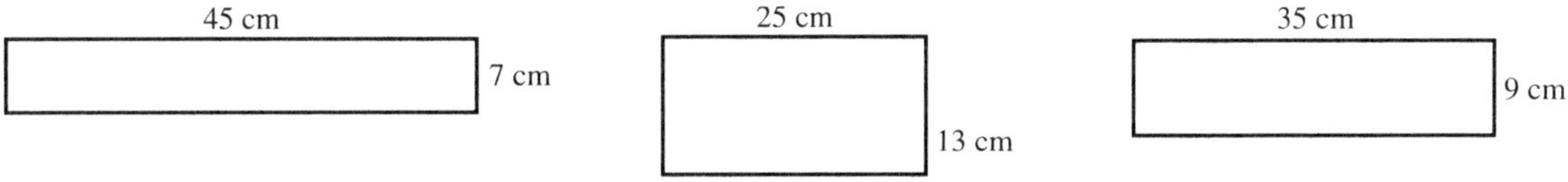

a Length = 45 cm
width = 7 cm

Area = $(45 \times 7)\,\text{cm}^2$
$= 315\,\text{cm}^2$

b Length = 25 cm
width = 13 cm

Area = $(25 \times 13)\,\text{cm}^2$
$= 325\,\text{cm}^2$

c Length = 35 cm
width = 9 cm

Area = $(35 \times 9)\,\text{cm}^2$
$= 315\,\text{cm}^2$

Therefore **b** has a different area from the other two.

Exercise 15.14

Copy and complete this table. For each question state which rectangle has a different area from the other two.

	Length	Width	Area
1 a	24 cm	6 cm	
b	18 cm	8 cm	
c	35 cm	4 cm	
2 a	75 mm	36 mm	
b	60 mm	45 mm	
c	76 mm	35 mm	
3 a	3.5 m	1.2 m	
b	3.2 m	1.5 m	
c	3.84 m	1.25 m	
4 a	4.48 m	1.25 m	
b	2.4 m	2.25 m	
c	3.2 m	1.75 m	
5 a	6 m	2.25 m	
b	11 m	1.5 m	
c	7.5 m	2.2 m	

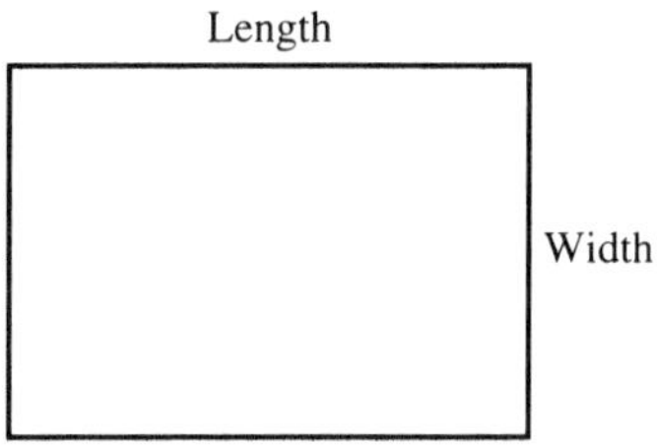

Reminder 3

The area of a triangle with base *l* 1 cm, length *b cm and vertical height h* cm is $(\frac{1}{2} \times b \times h)\,\text{cm}^2$ or $\frac{1}{2}bh\,\text{cm}^2$.

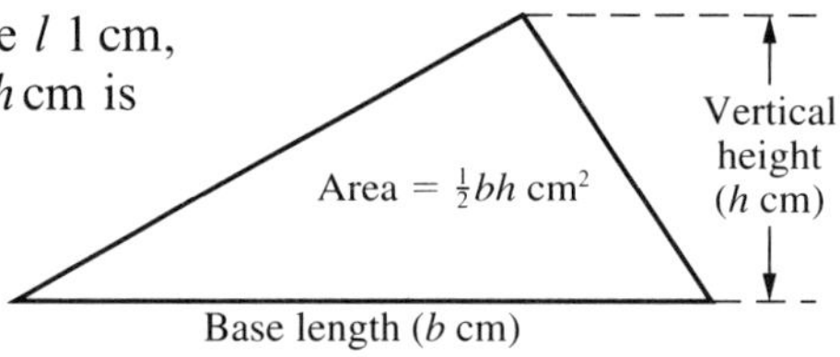

> **Example 4**
>
> Find the area of triangle with a base length of 1.6 m and a vertical height of 3.5 m
>
> $\text{Area} = (\frac{1}{2} \times 1.6 \times 3.5)\,\text{m}^2$
>
> $= 2.8\,\text{m}^2$
>
> 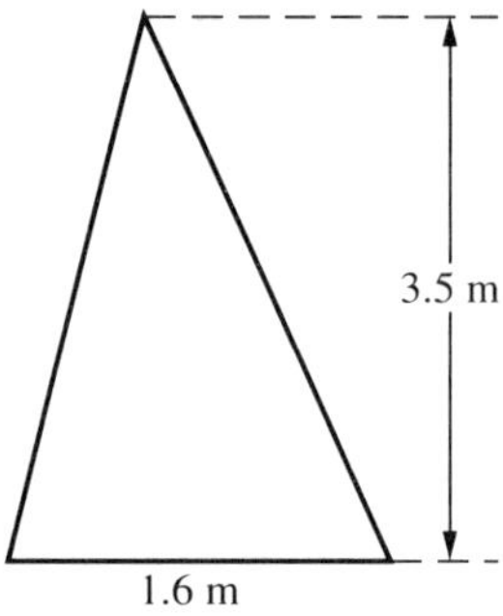
>

Exercise 15.15

For Questions 1–6 find the area of the triangle if:

1 $b = 18\,\text{cm}$ $h = 25\,\text{cm}$
2 $b = 24\,\text{cm}$ $h = 45\,\text{cm}$
3 $b = 80\,\text{mm}$ $h = 16\,\text{mm}$
4 $b = 70\,\text{mm}$ $h = 48\,\text{mm}$
5 $b = 1.2\,\text{m}$ $h = 2.5\,\text{m}$
6 $b = 1.6\,\text{m}$ $h = 1.5\,\text{m}$

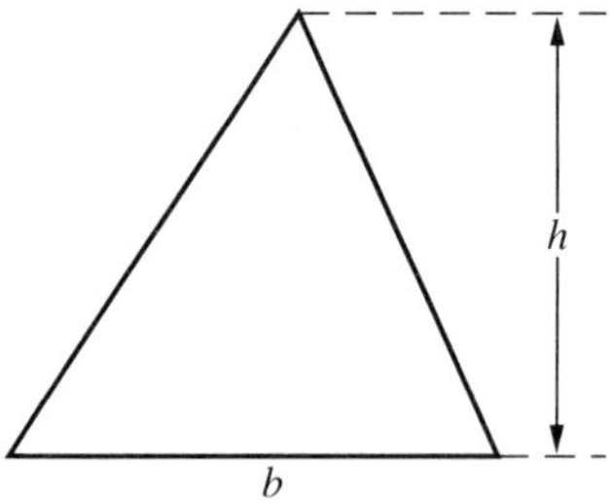

For Questions 7–10 find the area of the right-angled triangle if:

7 AB = 2.4 m BC = 1.5 m
8 AB = 3.2 m BC = 1.25 m
9 AB = 18 cm BC = 35 cm
10 AB = 45 mm BC = 60 mm

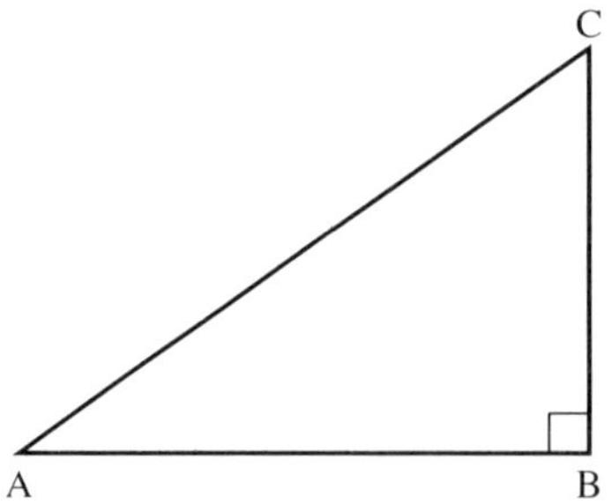

Reminder 4

The area of a parallelogram of base length b cm and vertical height h cm is $(b \times h)$ cm^2 or bh cm^2

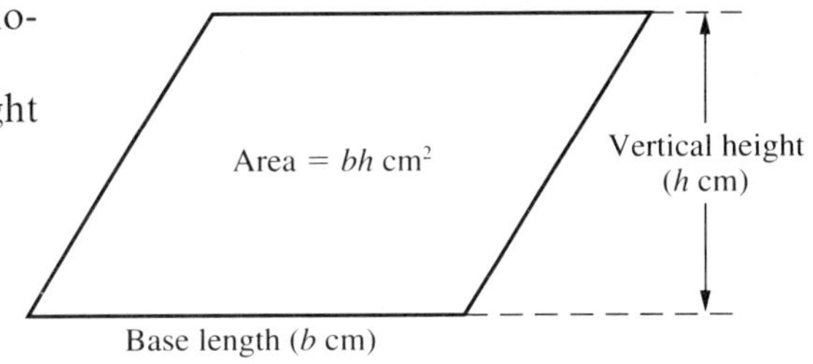

Example 5

Find the area of a parallelogram with a base length of 2.5 m and a vertical height of 1.4 m.

Area = (2.5×1.4) m^2 = 3.5 m^2

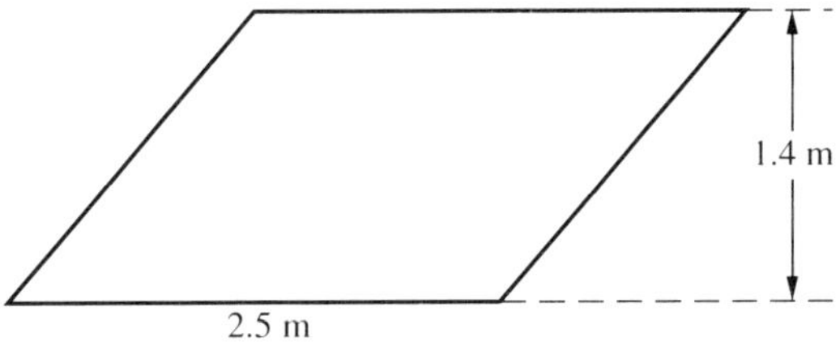

Exercise 15.16

For Questions 1–4 find the area of the parallelogram if:

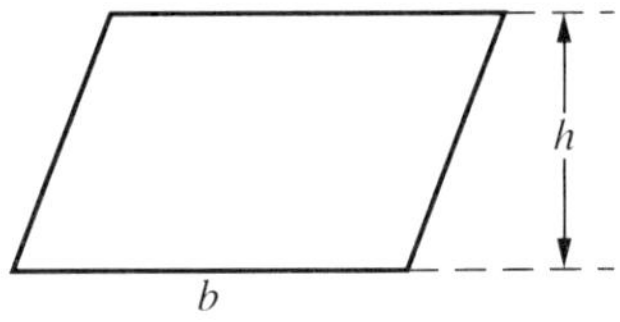

1 $b = 4.5$ m $h = 1.2$ m
2 $b = 7.5$ m $h = 1.8$ m
3 $b = 10.5$ m $h = 2.4$ m
4 $b = 18$ cm $h = 5$ cm

For Questions 5–8 find the area of the parallelogram if:

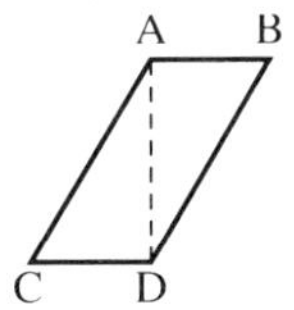

5 AB = 44 cm AD = 25 cm
6 AB = 35 cm AD = 22 cm
7 AB = 28 cm AD = 15 cm
8 AB = 75 mm AD = 42 mm

Reminder 5

The area of a trapezium with parallel sides of length a cm and b cm and a height of h cm is $\frac{1}{2}(a+b)h$ cm^2.

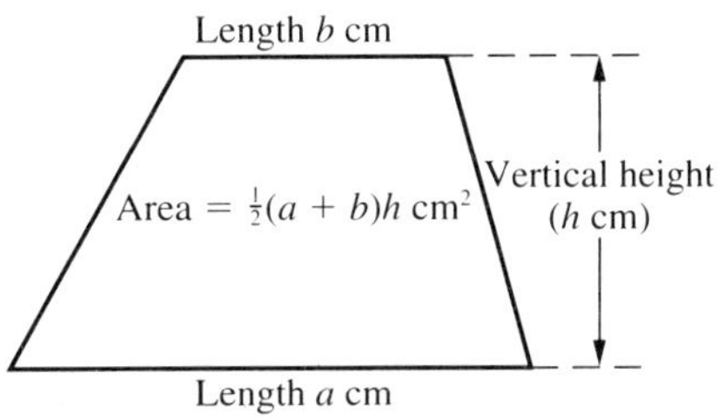

Example 6

Find the area of a trapezium with parallel sides of lengths 27 cm and 18 cm and a height of 12 cm.

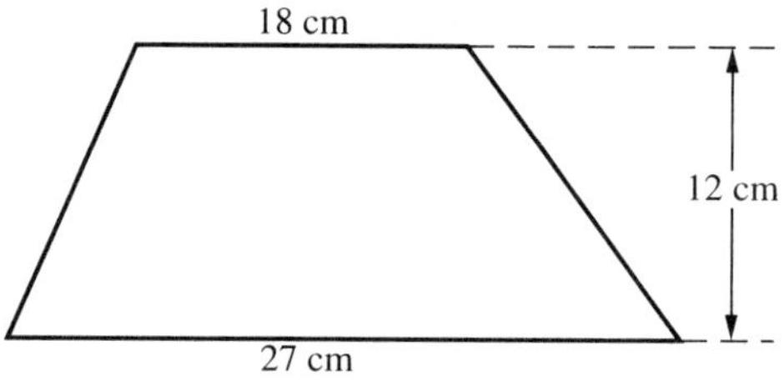

$$\text{Area} = \tfrac{1}{2} \times (27 + 18) \times 12\,\text{cm}^2$$
$$= \tfrac{1}{2} \times 45 \times 12\,\text{cm}^2$$
$$= 270\,\text{cm}^2$$

Exercise 15.17

For all questions find the area of the trapezium if:

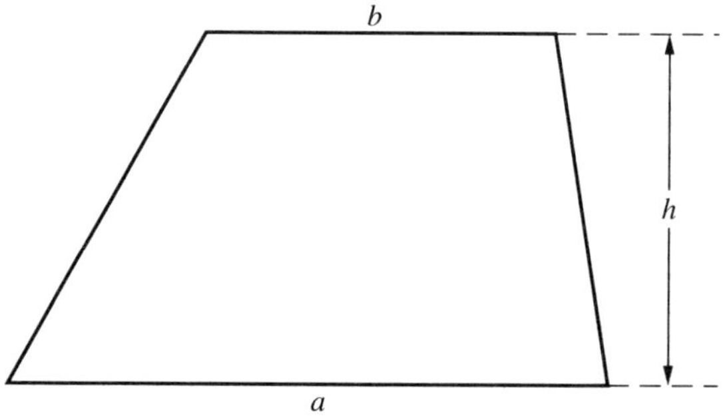

1 $a = 24\,\text{cm}$, $b = 21\,\text{cm}$ and $h = 16\,\text{cm}$
2 $a = 39\,\text{cm}$, $b = 16\,\text{cm}$ and $h = 24\,\text{cm}$
3 $a = 98\,\text{mm}$, $b = 42\,\text{mm}$ and $h = 54\,\text{mm}$
4 $a = 114\,\text{mm}$, $b = 96\,\text{mm}$ and $h = 72\,\text{mm}$
5 $a = 1.6\,\text{m}$, $b = 1.2\,\text{m}$ and $h = 1.5\,\text{m}$
6 $a = 5.8\,\text{m}$, $b = 3.4\,\text{m}$ and $h = 4.5\,\text{m}$

Example 7

Find the area of the floor in Daniel's bedroom.

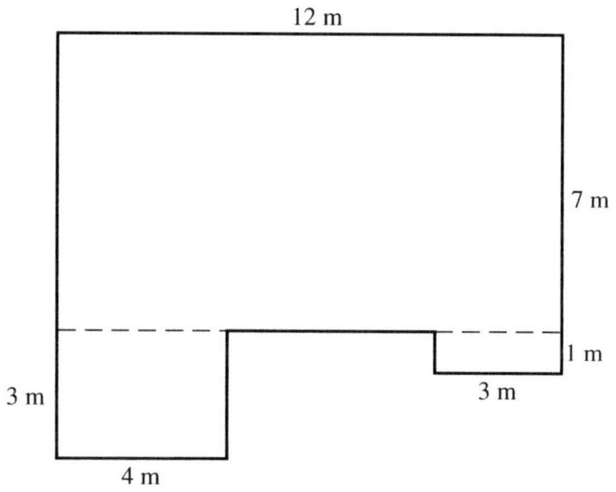

The floor can be divided into three rectangles.
The areas of these are:

$$(12 \times 7)\,\text{m}^2 = 84\,\text{m}^2$$
$$(4 \times 3)\,\text{m}^2 = 12\,\text{m}^2$$
$$(3 \times 1)\,\text{m}^2 = 3\,\text{m}^2$$

Therefore the total area is $(84 + 12 + 3)\,\text{m}^2 = 99\,\text{m}^2$

Example 8

Find the area of the end wall of this bus station.

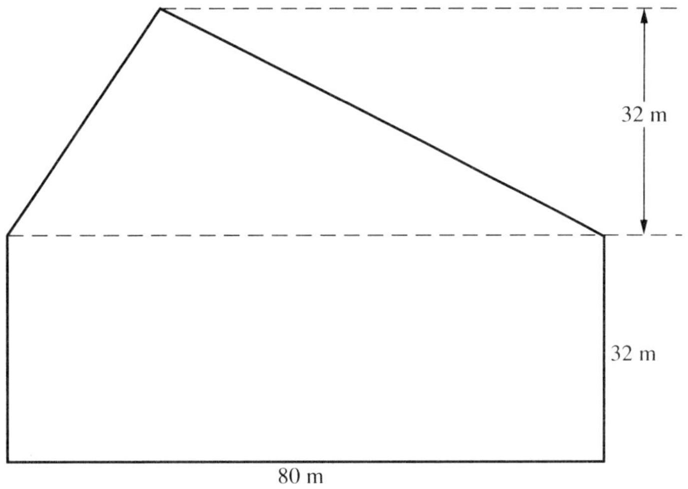

The area of the rectangular part is:

$(80 \times 32)\,\text{m}^2 = 2560\,\text{m}^2$

The area of the triangular part is:

$(\frac{1}{2} \times 32 \times 80)\,\text{m}^2 = 1280\,\text{m}^2$

Therefore the total area is:

$(2560 + 1280)\,\text{m}^2 = 3840\,\text{m}^2$

Exercise 15.18

For all questions find the area.

1 The playground at Lucy's school

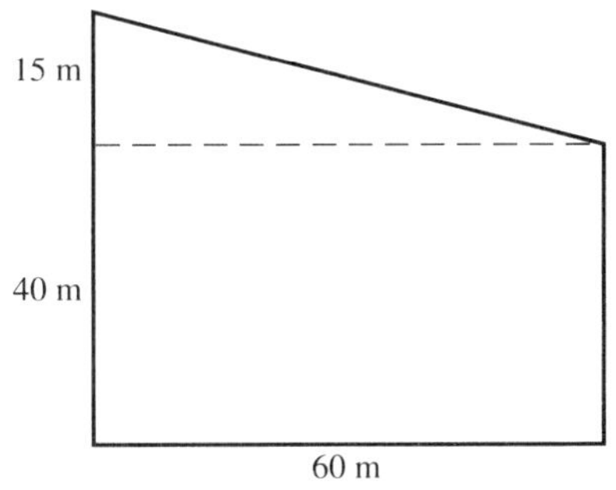

2 The car park

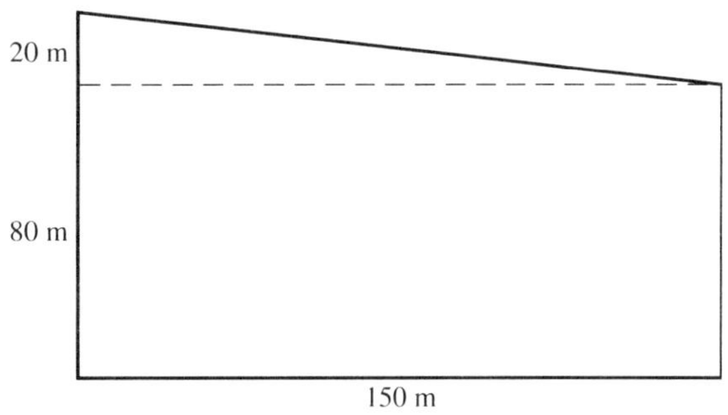

3 The end of a dolls's house

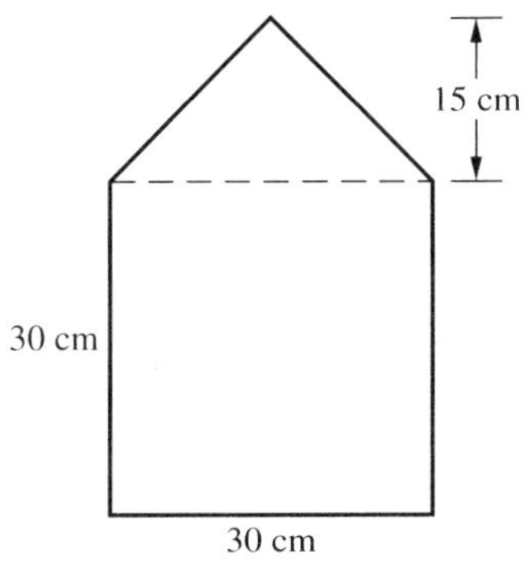

4 A sign board

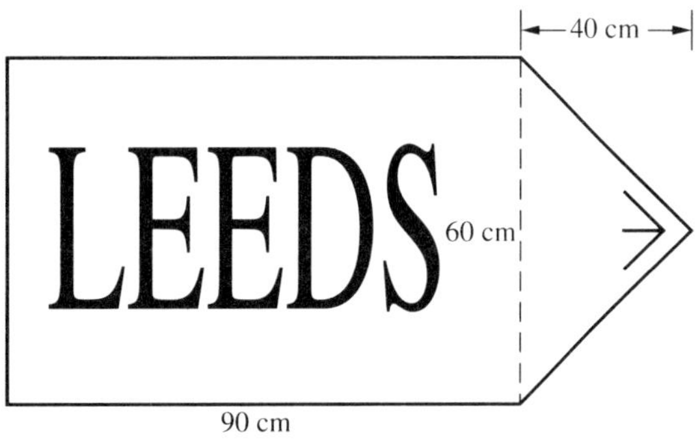

5 A cardboard divider in a file

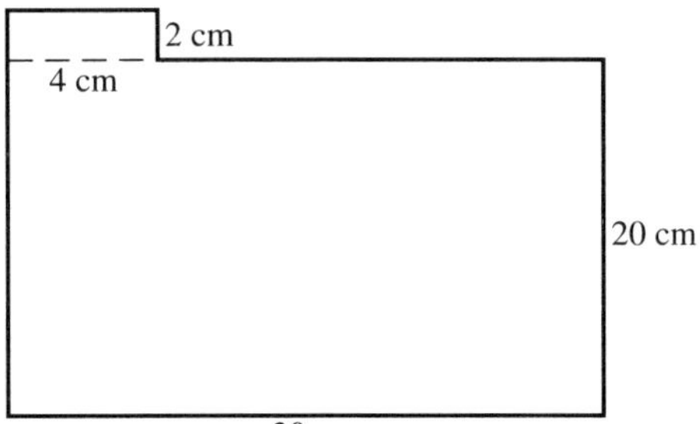

6 The playground at Amarjit's school

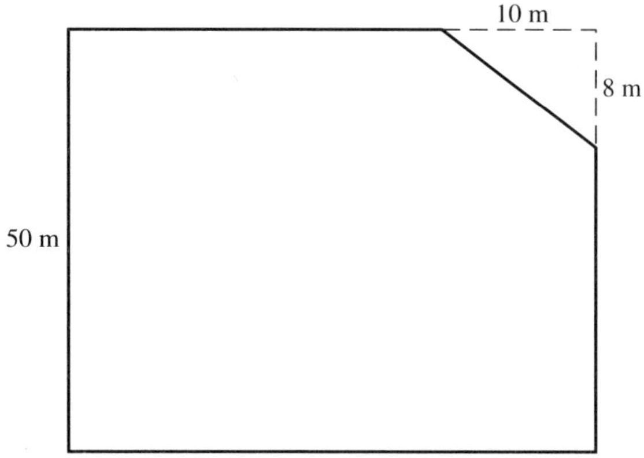

7 **a** The stage in the hall at Jolene's school
b The floor space in the hall at Jolene's school

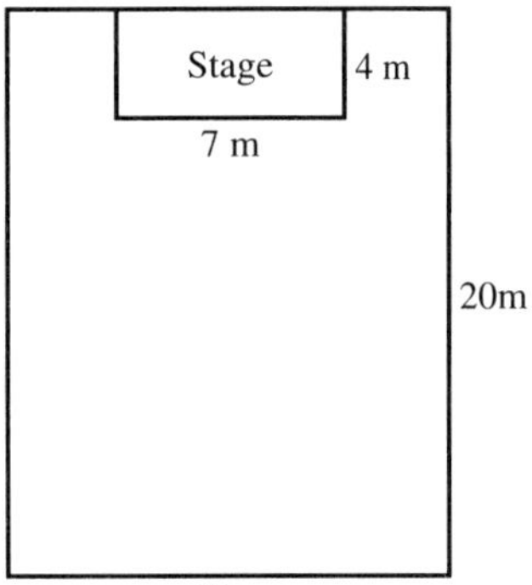

8 The picture frame Joshua has made

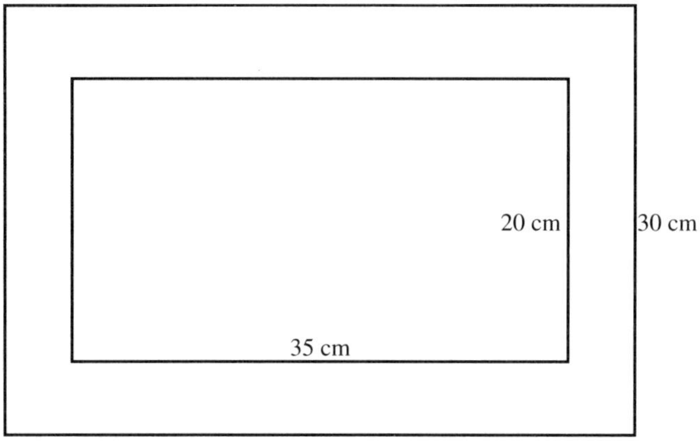

9 The floor space of a room which has a bay window

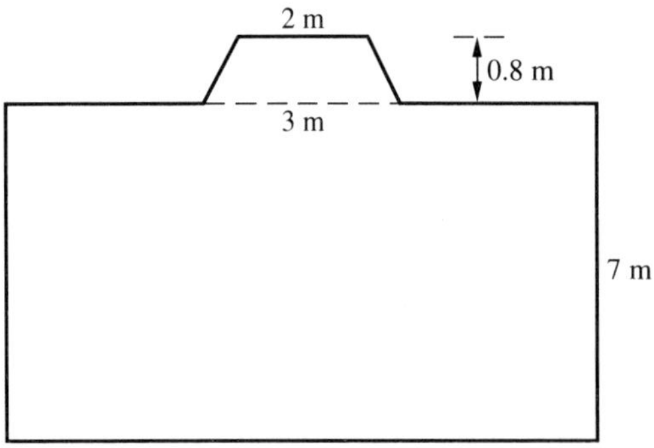

10 **a** One of these magnets
b The wooden separator between the magnets
c One of the keepers at either end
d All five pieces together
e The top surface area of the rectangular box that the five pieces fit into.

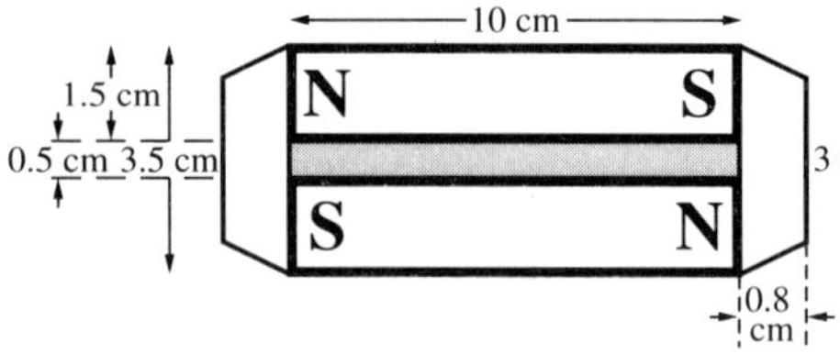

15.4 Finding volumes of simple solids using appropriate formulae

Reminder 1

The volume of a cube of side length a cm is a^3 cm^3.

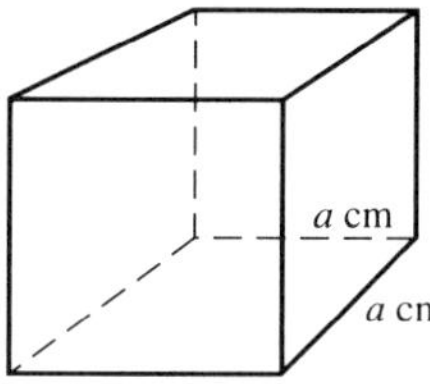

Example 1

Find the volume of a cube whose side length is 9 cm.

Volume = $(9 \times 9 \times 9)$ cm^3 = 729 cm^3.

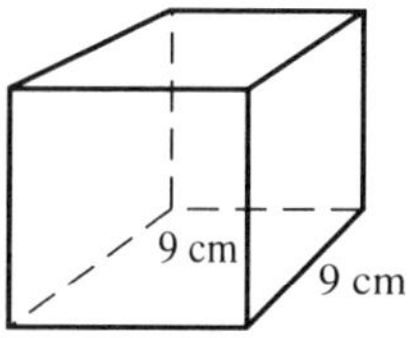

Exercise 15.19

Copy and complete the table.

Side length of square	Volume of cube
1 8 cm	
2 12 cm	
3 50 mm	
4 25 mm	

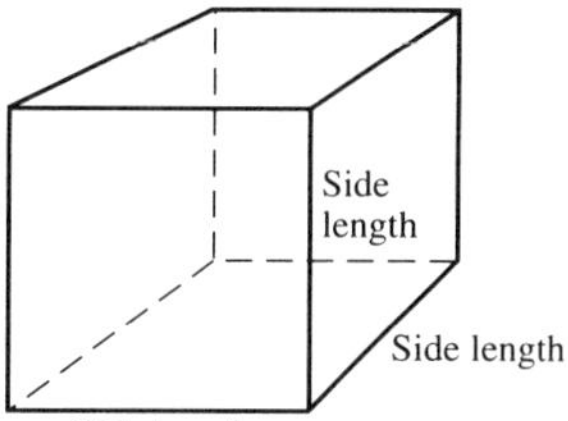

Example 2

Find the side length of a cube whose volume is 125 cm^3

Side length

= cube root of 125 cm^3

= $\sqrt[3]{125}$ cm

= 5 cm

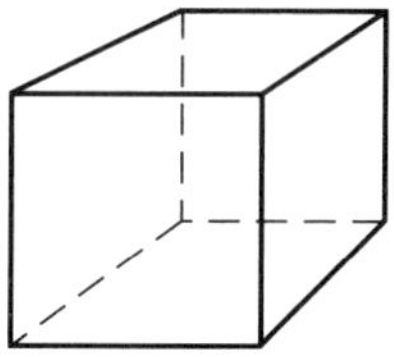

Exercise 15.20

Copy and complete the table.

Volume of cube	Side length of cube
1 27 cm^3	
2 64 cm^3	
3 1000 cm^3	
4 8000 mm^3	
5 27 000 mm^3	

Reminder 2

The volume of a cuboid of length l cm, width w cm and height h cm is lwh cm^3.

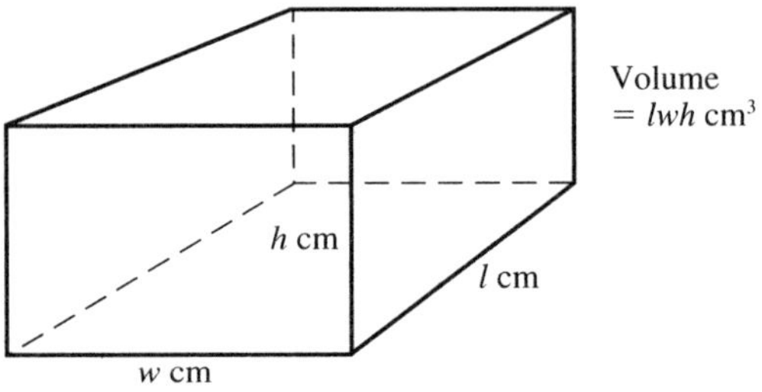

Example 3

Find the volume of a cuboid of length 4 m, width 2.5 m and height 1.05 m.

Volume $= (4 \times 2.5 \times 1.05)\,\text{m}^3 = 10.5\,\text{m}^3$

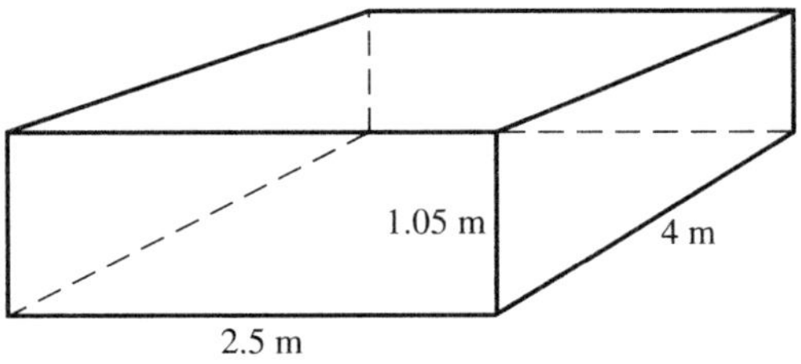

Exercise 15.21

For Questions 1–3 copy and complete the table. State which cuboid has a different volume from the other two.

	Length	Width	Height	Volume
1 a	13 cm	11 cm	2 cm	
b	12 cm	8 cm	3 cm	
c	16 cm	9 cm	2 cm	
2 a	28 mm	7 mm	5 mm	
b	24 mm	20 mm	2 mm	
c	30 mm	8 mm	4 mm	
3 a	2.0 m	1.5 m	0.8 m	
b	4.0 m	1.75 mm	0.4 m	
c	1.6 m	1.25 mm	1.2 m	

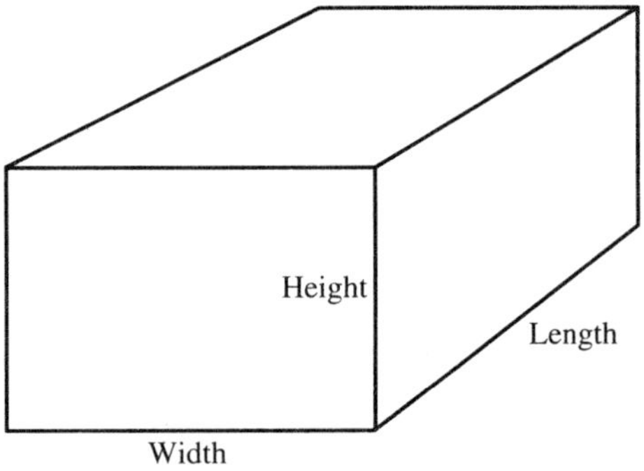

Find the volume of:

4 a match box

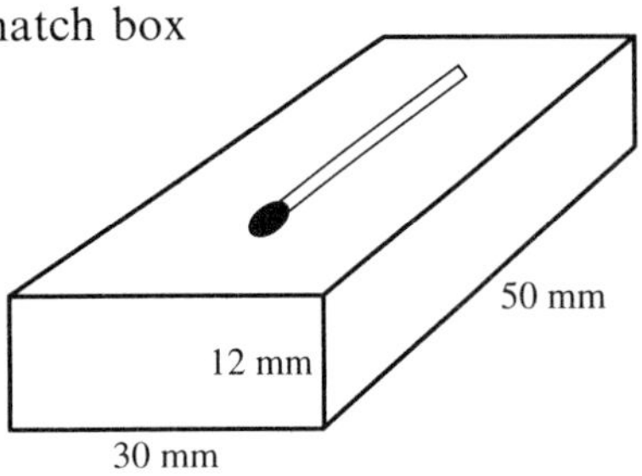

5 a brick

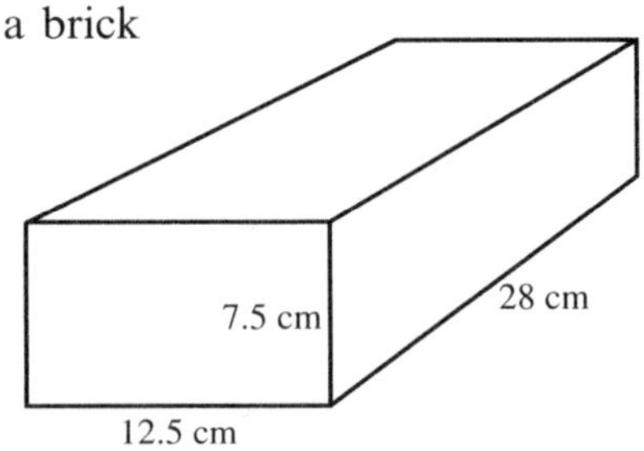

Example 4

Find the volume of this solid.

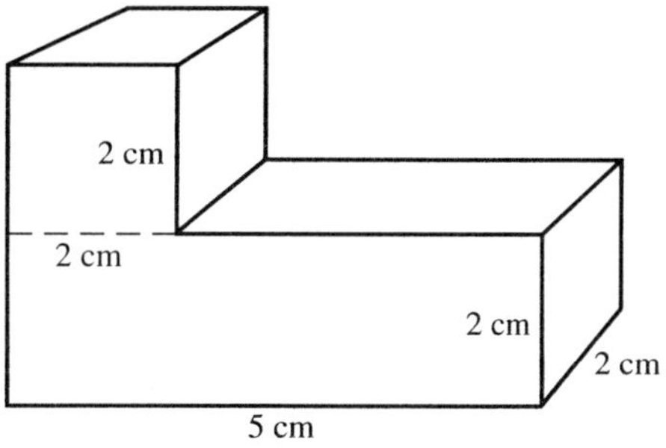

The solid can be divided into a cube and cuboid.

Volume of cube $= (2 \times 2 \times 2)\,\text{cm}^3 = 8\,\text{cm}^3$
Volume of cuboid $= (5 \times 2 \times 2)\,\text{cm}^3 = 20\,\text{cm}^3$

so Volume of solid $= 8\,\text{cm}^3 + 20\,\text{cm}^3 = 28\,\text{cm}^3$

Exercise 15.22

Find the volume of:

1

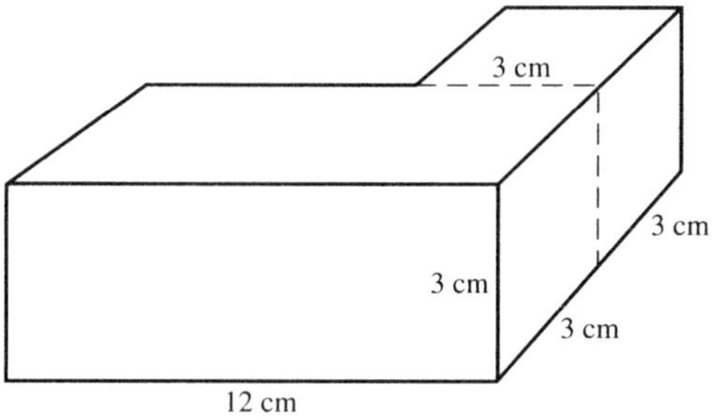

2

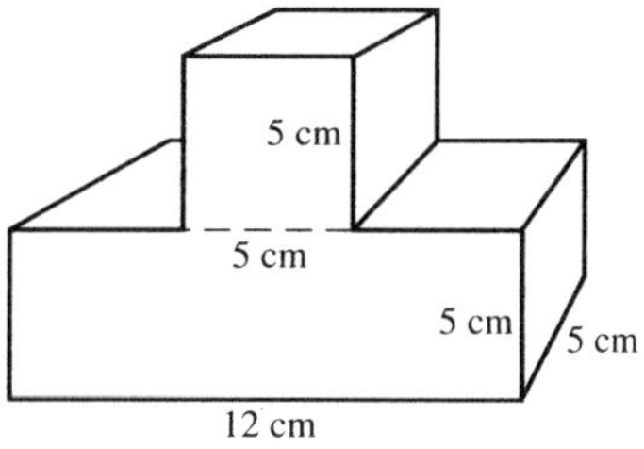

3

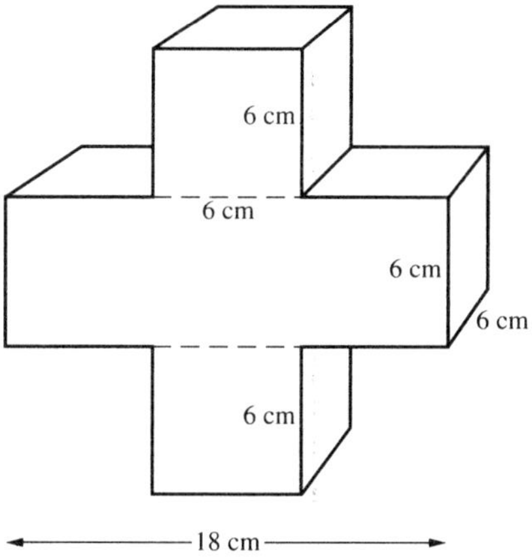

4

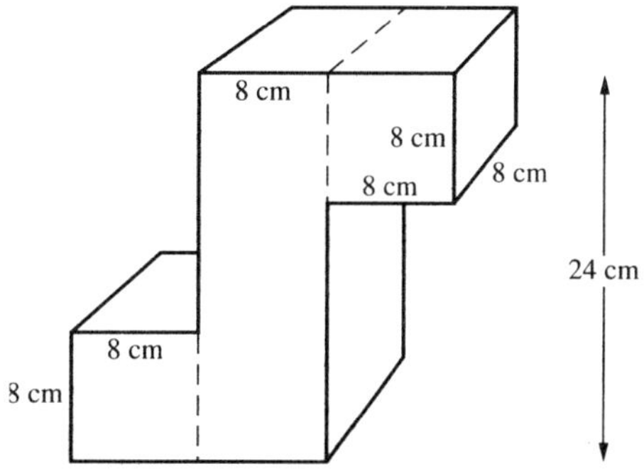

5

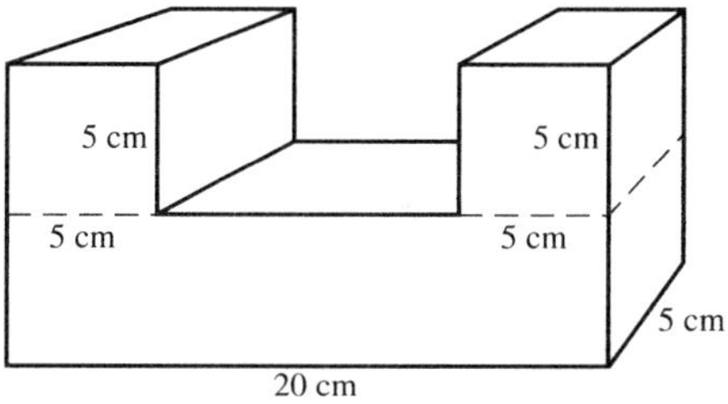

16 Circumference and area of circles

16.1 Finding the circumferences of circles

Reminder 1

The circumference (or perimeter) of a circle of diameter D cm is πD cm, where $\pi = 3.14$ (correct to two decimal places).

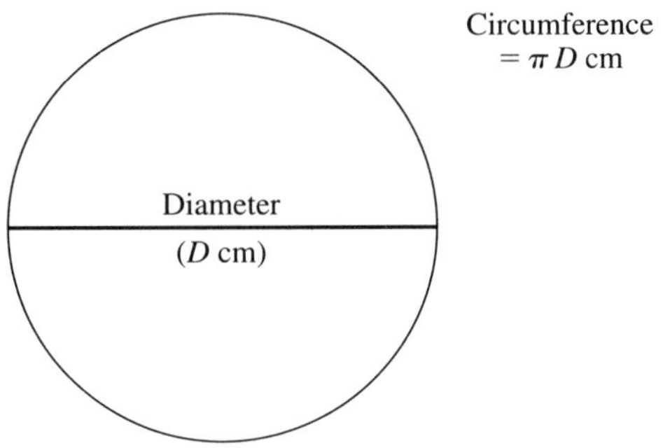

Reminder 2

The circumference of a circle of radius r cm is $2\pi r$ cm.

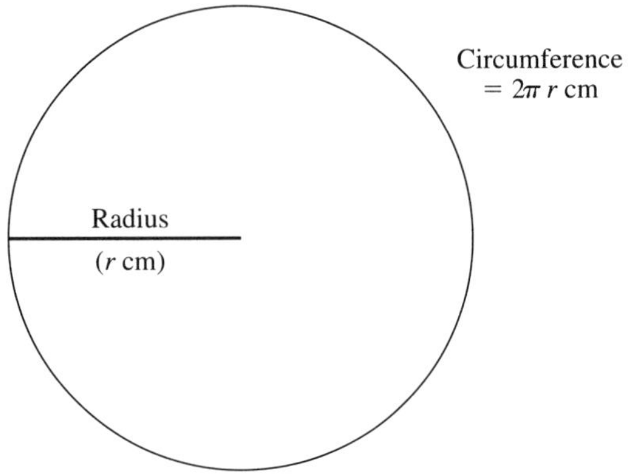

Example 1

a Find the circumference of a circle of diameter 0.5 m (take $\pi = 3.14$).

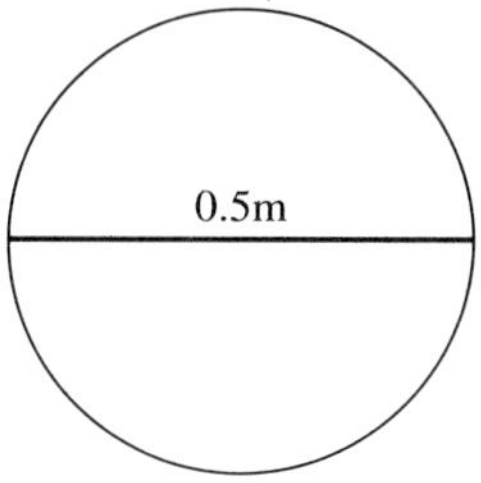

b Find the circumference of a circle of radius 5 cm ($\pi = 3.14$).

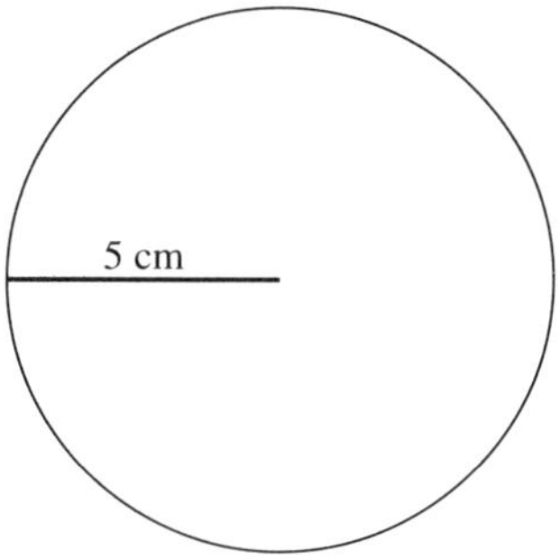

a Circumference $= \pi \times 0.5\,\text{m}$
$= (3.14 \times 0.5)\,\text{m}$
$= 1.57\,\text{m}$

b Circumference $= 2\pi \times 5\,\text{cm}$
$= (2 \times 3.14 \times 5)\,\text{cm}$
$= 31.4\,\text{cm}$

Exercise 16.1

Find the circumference of the circle for each of the following cases. Take $\pi = 3.14$

1 Diameter = 5 cm **2** Diameter = 4 cm **3** Diameter = 15 cm

4 Diameter = 35 cm **5** Diameter = 4.5 cm **6** Diameter = 7.5 cm

7 Radius = 3 cm **8** Radius = 10 cm **9** Radius = 25 cm

10 Radius = 22.5 mm **11** Radius = 12.5 mm **12** Radius = 0.75 m

Example 2

Find **a** the diameter and **b** the radius of a circle with circumference 94.2 cm. Take $\pi = 3.14$

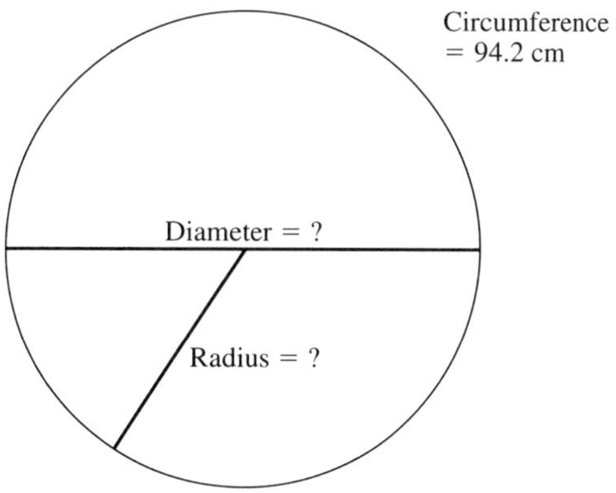

a Diameter $= (94.2 \div \pi)\,\text{cm}$
$= (94.2 \div 3.14)\,\text{cm}$
$= 30\,\text{cm}$

b Radius = Diameter $\div\, 2$
$= (30 \div 2)\,\text{cm}$
$= 15\,\text{cm}$

Exercise 16.2

For Questions 1–6 find the diameter of the circle. Take $\pi = 3.14$

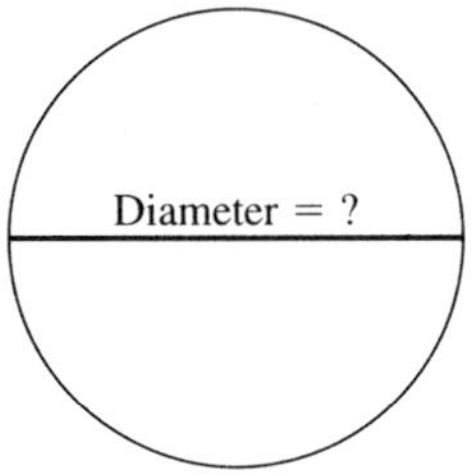

1 Circumference = 25.12 cm **2** Circumference = 28.26 cm

3 Circumference = 31.4 cm **4** Circumference = 204.1 cm

5 Circumference = 235.5 cm **6** Circumference = 172.7 cm

For Questions 7–12 find **a** the diameter and **b** the radius of the circle.
Take $\pi = 3.14$

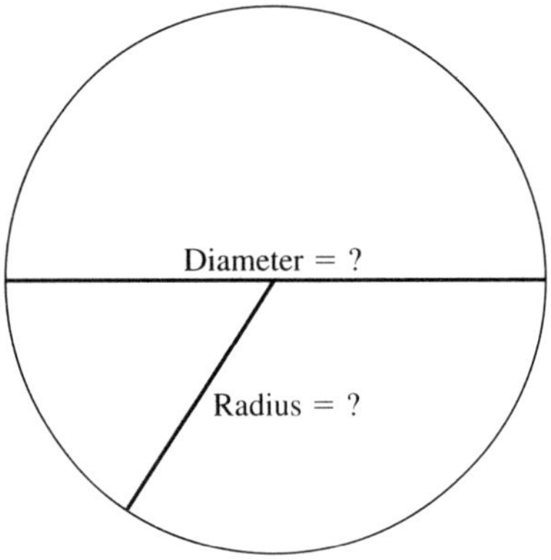

7 Circumference = 21.98 mm **8** Circumference = 9.42 mm

9 Circumference = 34.54 mm **10** Circumference = 502.4 mm

11 Circumference = 565.2 mm **12** Circumference = 659.4 mm

Example 3

A birthday cake has a diameter of 40 cm.

What length of ribbon is required to go round it, assuming it has a 1 cm overlap? Take $\pi = 3.14$

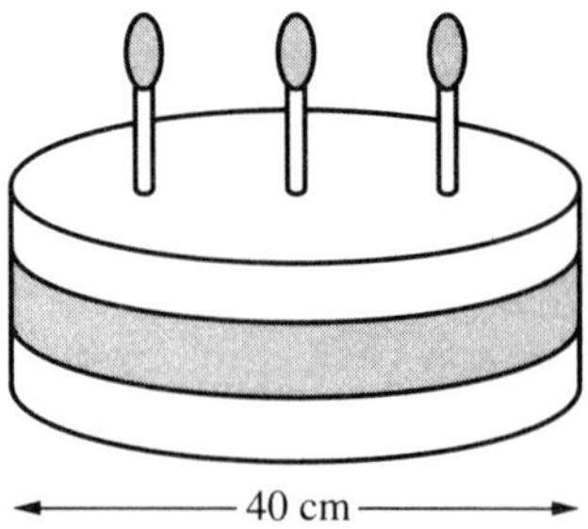

Circumference of cake $= (3.14 \times 40)$ cm
$= 125.6$ cm
So the length of ribbon = Circumference + overlap
$= (125.6 + 1)$ cm
$= 126.6$ cm

Exercise 16.3

Take $\pi = 3.14$

1 An electromagnet is made from an iron cylinder of diameter 1.5 cm and some wire.
Find:
a the circumference of the cylinder and
b the length of wire used
if it is coiled round 150 times.

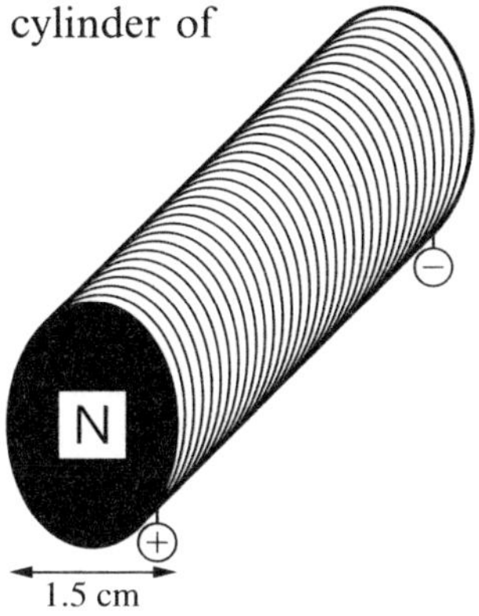

2 The windlass illustrated is used to raise water from a well.
If the drum has a diameter of 15 cm, find:
a the circumference of the drum and
b the length of rope on the drum
when the bucket is fully raised if the handle has to be turned 20 times.

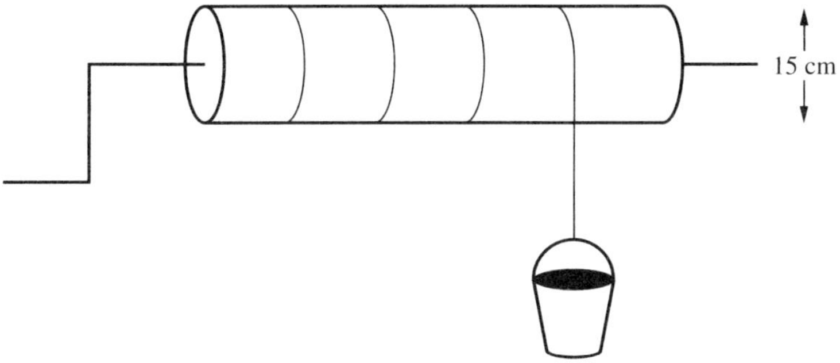

3 The inside of a biscuit tin is to be lined with thick paper.
If the diameter of the tin is 25 cm and there is to be 1 cm overlap, what length of paper is required?

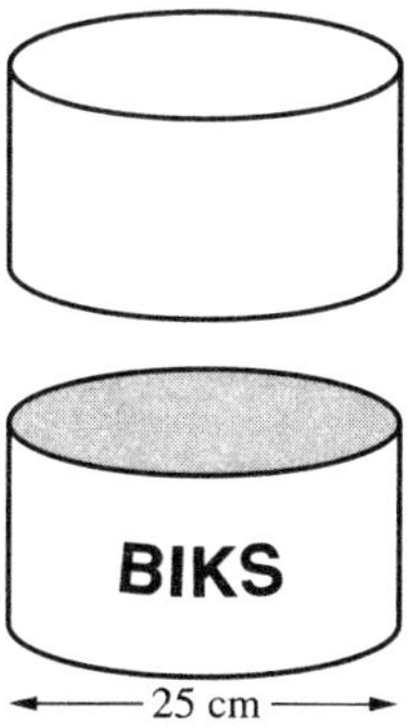

4 Jonathan has a marble of diameter 1.25 cm.
Find
a the circumference of the marble and
b the number of times it rotates while it rolls 235.5 cm across a floor before stopping.

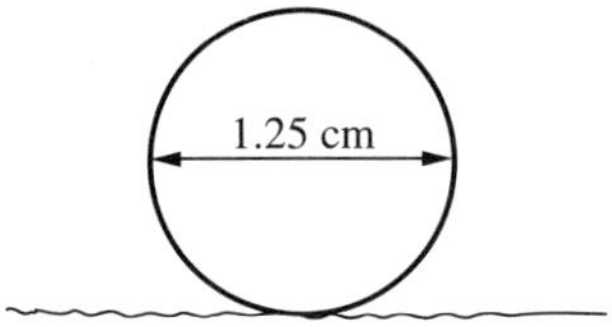

5 A garden roller has a diameter of 40 cm.
Find
a the circumference of the roller and
b the number of times it rotates while rolling a lawn of length 94.2 m (9420 cm).

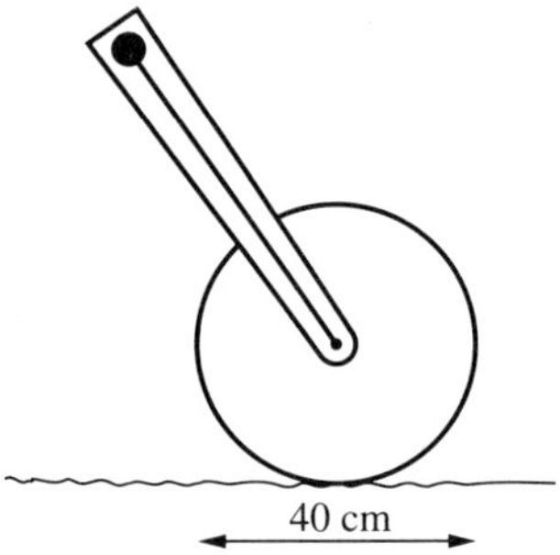

6 The wheels of Melanie's bicycle have a diameter of 0.6 m.
Find
a the circumference of the wheels and
b the number of times they rotate while she rides to see her grandmother who lives 14.13 kilometres (14 130 m) away.

Example 4

Find the perimeter of the athletics track. Take $\pi = 3.14$

The perimeter is equal to the sum of the circumference of the circle and the lengths of the two straight sides.

Therefore the perimeter $= (3.14 \times 60 + 55.8 + 55.8)$ m
$= (188.4 + 55.8 + 55.8)$ m
$= 300$ m

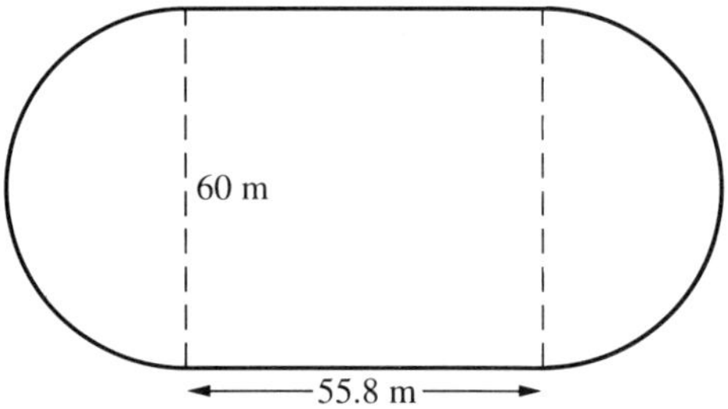

Exercise 16.4

For all questions find the perimeter. Take $\pi = 3.14$

1 The athletics track

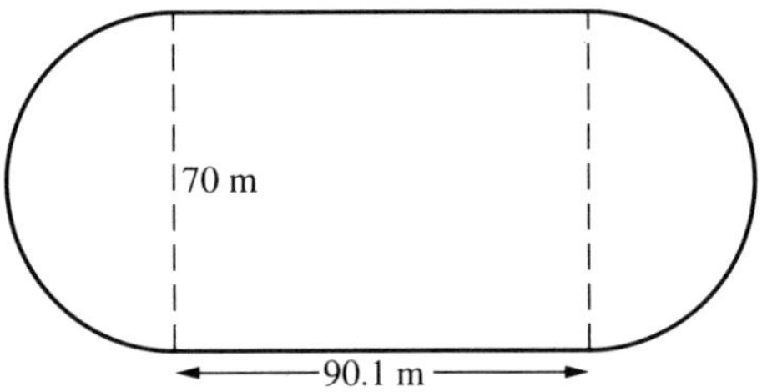

2 The ice rink

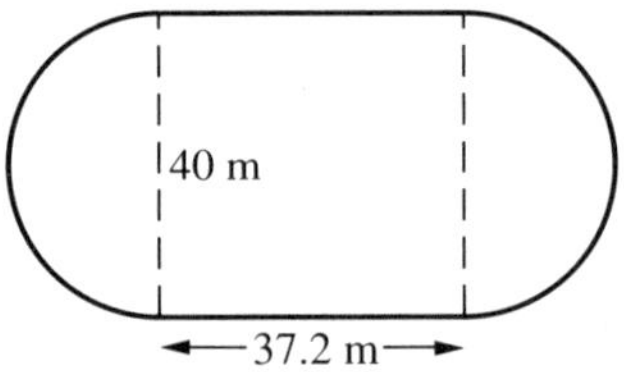

3 The oval swimming pool

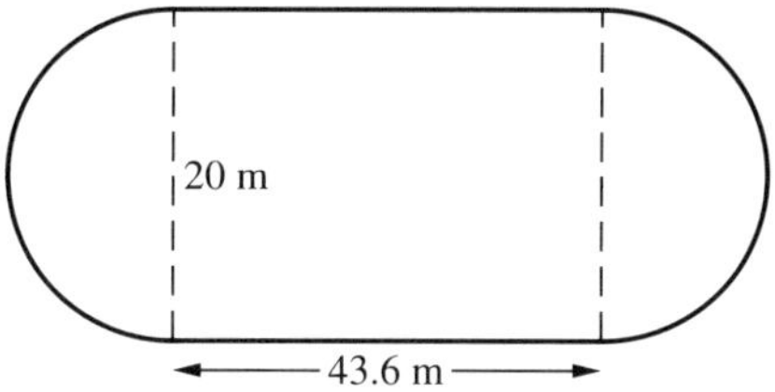

4 The dodgem car circuit at a fairground

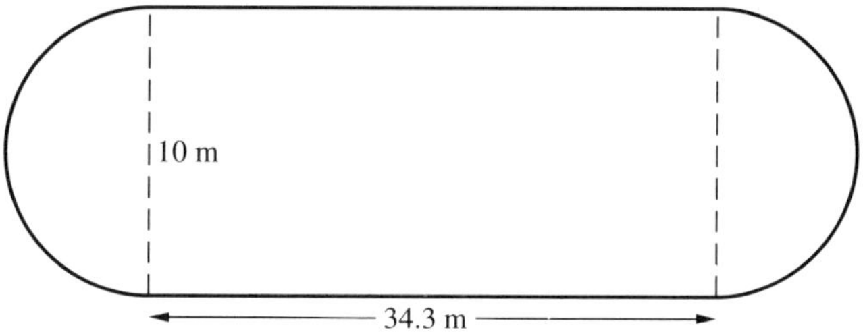

16.2 Finding areas of circles using the formula

Reminder

The area of a circle of radius r cm is πr^2 cm^2 (where $\pi = 3.14$ correct to two decimal places).

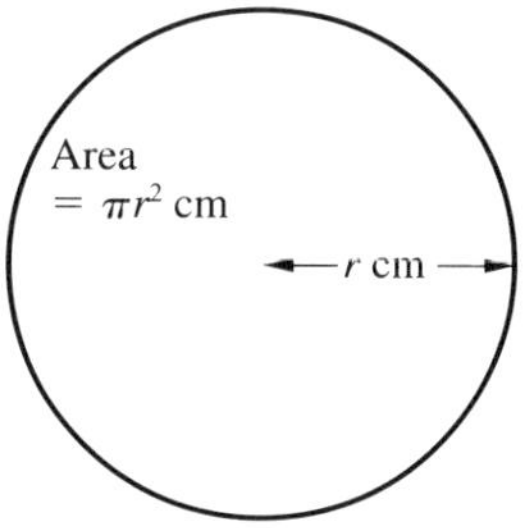

Example 1

Find the area of a circle of radius 10 cm.
Take $\pi = 3.14$

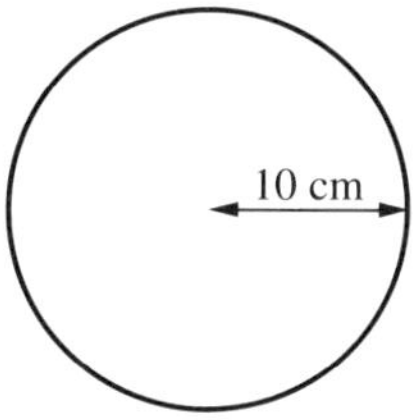

$$\begin{aligned}\text{Area} &= (\pi \times 10 \times 10)\,\text{cm}^2\\ &= (3.14 \times 10 \times 10)\,\text{cm}^2\\ &= 314\,\text{cm}^2\end{aligned}$$

Exercise 16.5

Find the area of these circles. Take $\pi = 3.14$

1 Radius = 5 cm **2** Radius = 3 cm **3** Radius = 15 cm

4 Radius = 20 cm **5** Radius = 40 mm **6** Radius = 35 mm

7 Diameter = 120 mm **8** Diameter = 90 mm **9** Diameter = 110 m

10 Diameter = 11 m **11** Diameter = 5 m **12** Diameter = 2 m

Example 2

Find **a** the radius and **b** the diameter of a circle of area $50.24\,\text{cm}^2$. Take $\pi = 3.14$

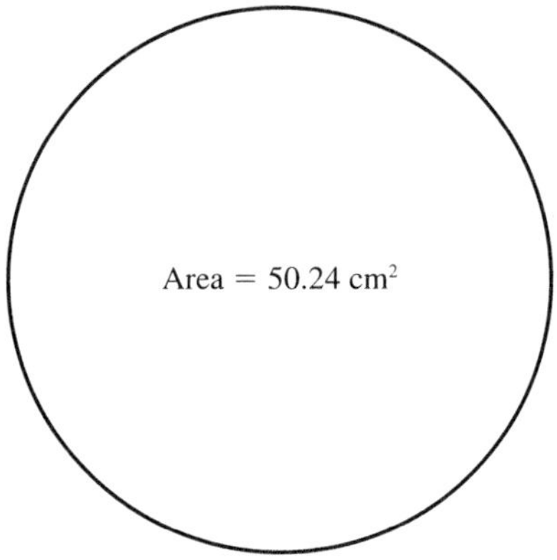

a $\pi r^2 = 50.24\,\text{cm}^2$

so $3.14r^2 = 50.24\,\text{cm}^2$

so $r^2 = (50.24 \div 3.14)\,\text{cm}^2$

or $r^2 = 16\,\text{cm}^2$

and so $r = \sqrt{16}\,\text{cm}$

or $r = 4\,\text{cm}$

b Diameter $= (2 \times 4)\,\text{cm} = 8\,\text{cm}$

Exercise 16.6

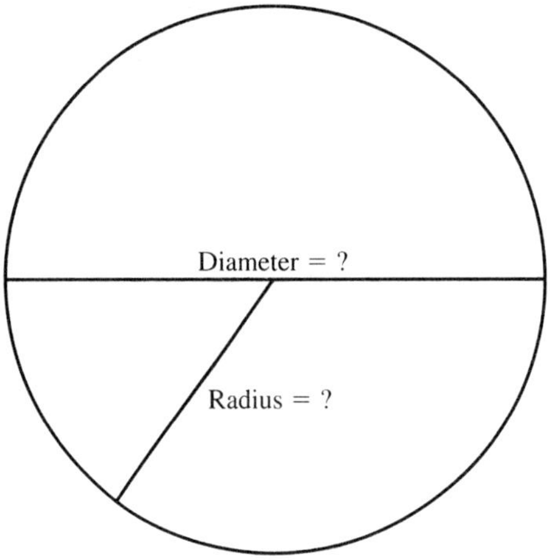

Find **a** the radius and **b** the diameter of these circles. Take $\pi = 3.14$

1 Area $= 1962.5\,\text{mm}^2$ **2** Area $= 2826\,\text{mm}^2$ **3** Area $= 7850\,\text{mm}^2$

4 Area $= 12.56\,\text{cm}^2$ **5** Area $= 63.585\,\text{cm}^2$ **6** Area $= 176.625\,\text{cm}^2$

Example 3

Find the area of **a** the circle, **b** the rectangle and **c** the shaded part. Take $\pi = 3.14$

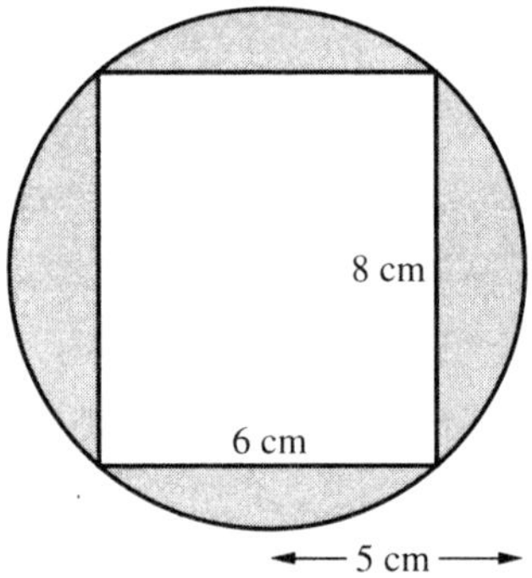

a Area of circle $= (\pi \times 5 \times 5)\,\text{cm}^2$

$= (3.14 \times 5 \times 5)\,\text{cm}^2$

$= 78.5\,\text{cm}^2$

b Area of rectangle $= (8 \times 6)\,\text{cm}^2 = 48\,\text{cm}^2$

c Area of shaded part $= (78.5 - 48)\,\text{cm}^2 = 30.5\,\text{cm}^2$

Exercise 16.7

Take $\pi = 3.14$

1 Find the area of
a the square
b the circle
c the shaded part.

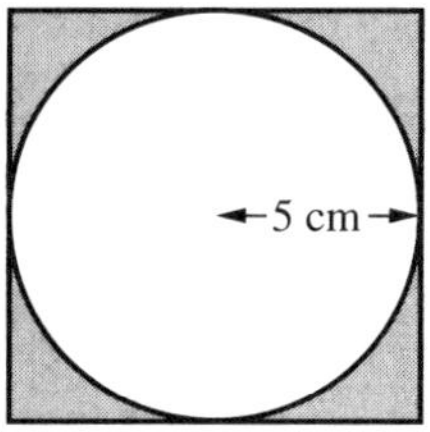

2 Find the area of
a the circle
b the square
c the shaded part.

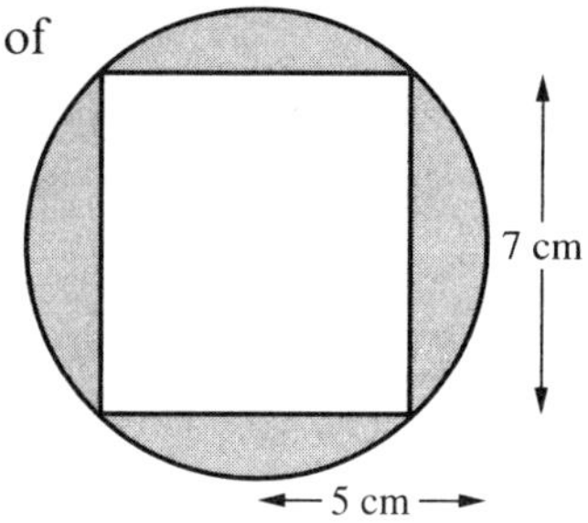

3 Find the area of
a the circle
b the triangle
c the shaded part.

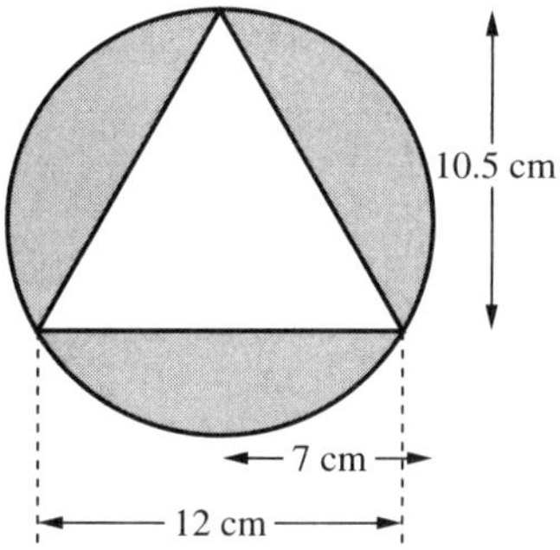

4 Find the area of
a the triangle
b the circle
c the shaded part.

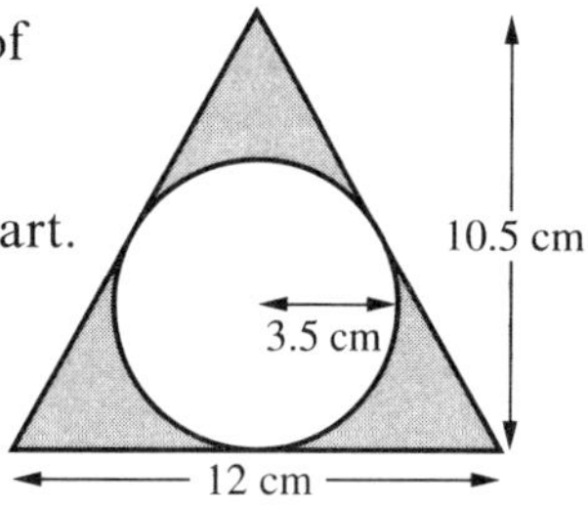

5 Find the area of the washer.

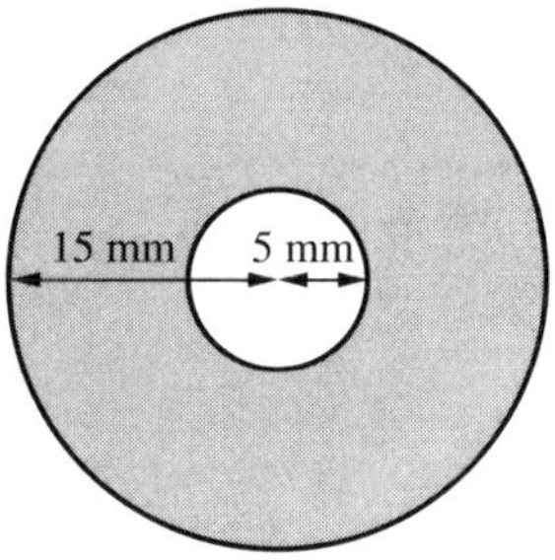

6 Find the area of the frame around the clock face.

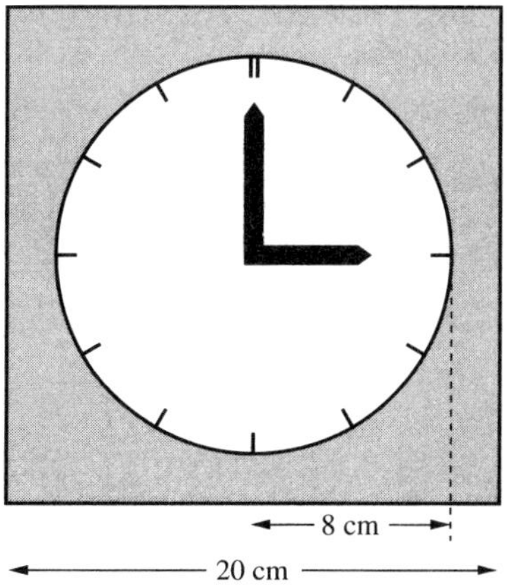

7 Find the area of the frame around the calendar.

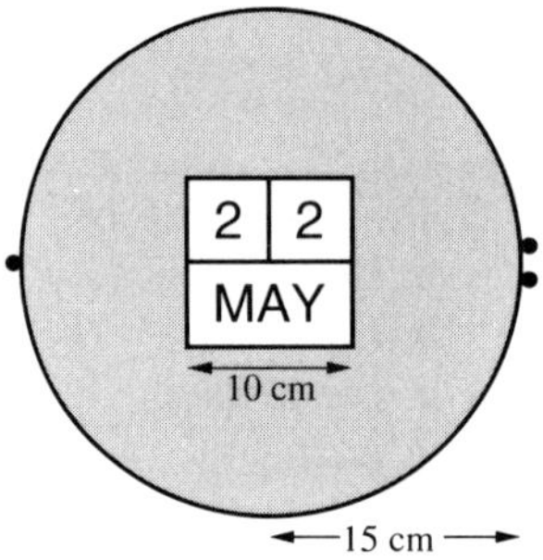

8 Find the area of the set square.

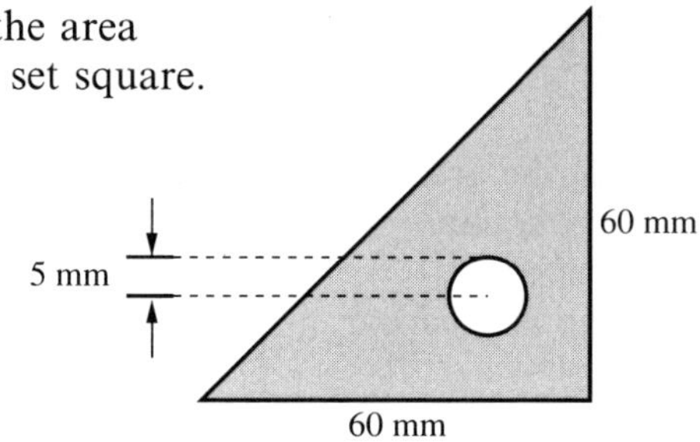

9 Find the area of each of these semicircles.

a Radius = 10 cm
b Radius = 20 cm
c Radius = 5 cm
d Radius = 40 mm
e Radius = 60 mm

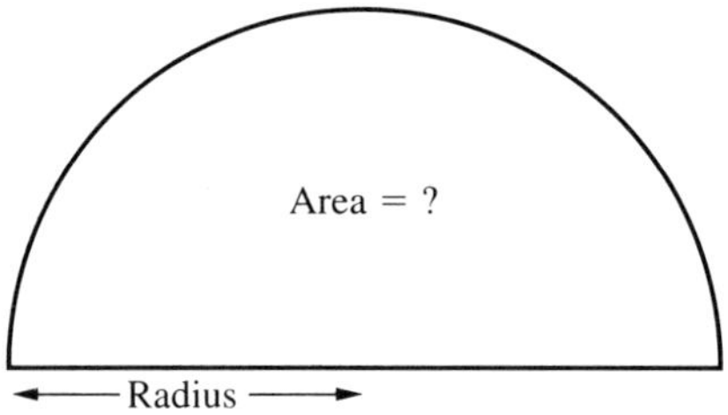

10 Find the area of

a the triangle
b the semicircle
c the shaded part.

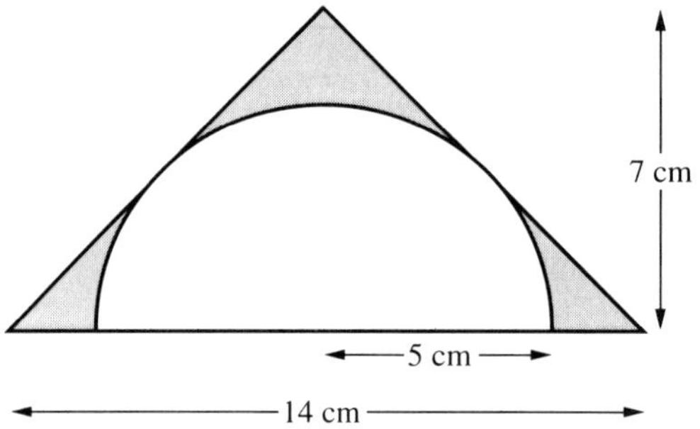

11 Find the area of

a the rectangle
b the semicircle
c the shaded part.

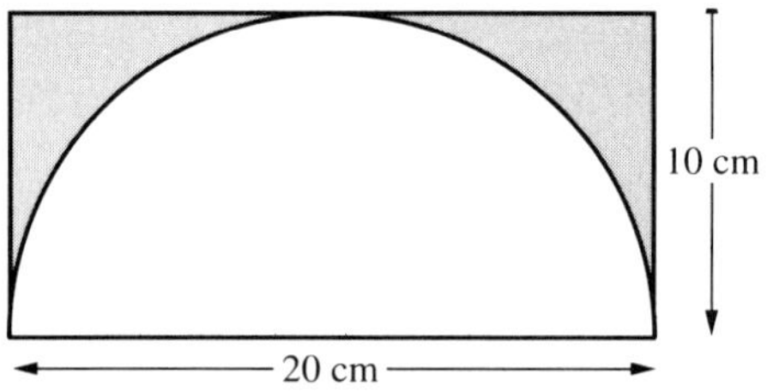

12 Find the area of the upright surface of the bridge, which is shaded in the diagram.

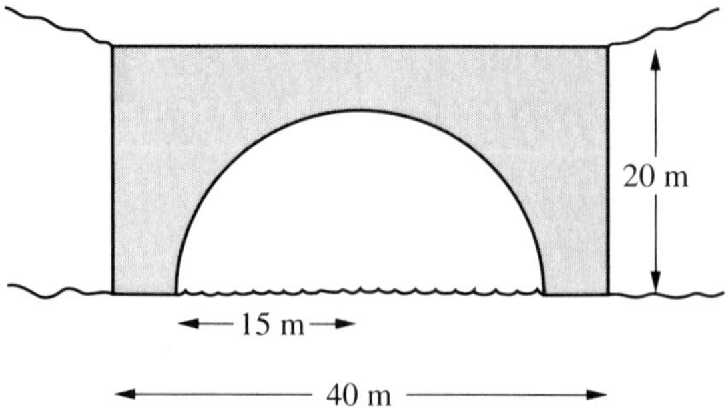

Example 4

A circle has a radius of 3 cm.
Find the side length of the square which has the same area.

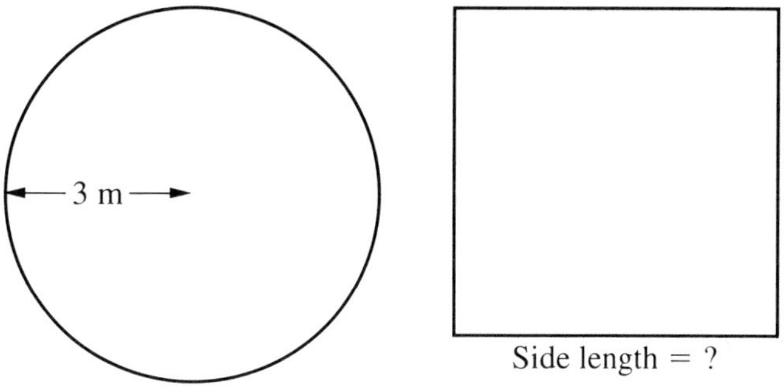

Let the side length of the square be l.
As both figures have the same area:

$$l^2 = \pi r^2$$

or $l = (\sqrt{\pi})\,r$

Now $\sqrt{3.14} = 1.772\,004\ldots$, so we use a very close approximation of 1.77

Hence $l = (1.77 \times 3)\,\text{cm} = 5.31\,\text{cm}$

Exercise 16.8

For Questions 1–4 find the side length of the square which has the same area as the circle.

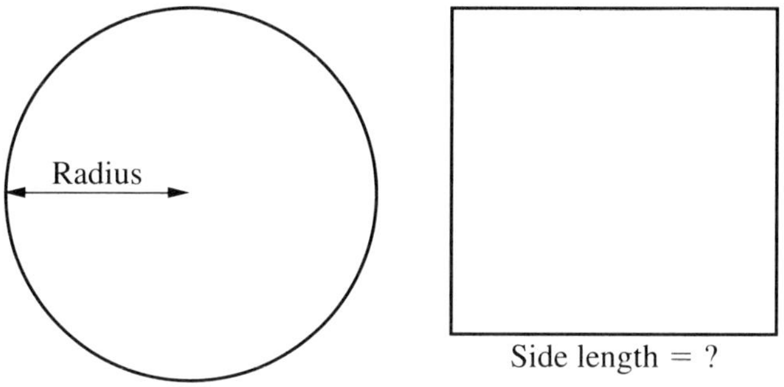

1 Radius = 4 cm **2** Radius = 6 cm

3 Radius = 60 cm **4** Radius = 1.2 m

5 A glassworks makes two mirrors of the same surface area. One is circular and the other is square.
If the radius of the circular one is 0.2 m, what is the side length of the square one?

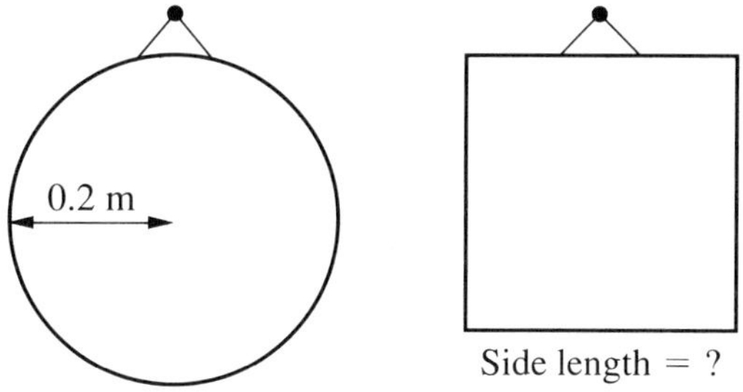

6 Wasim's mother has two trays, one circular and one square.
If the radius of the circular one is 0.3 m, find the side length of the square one if its surface area is the same.

Example 5

A square has side length 12.5 cm. Find the radius of the circle which has the same area.

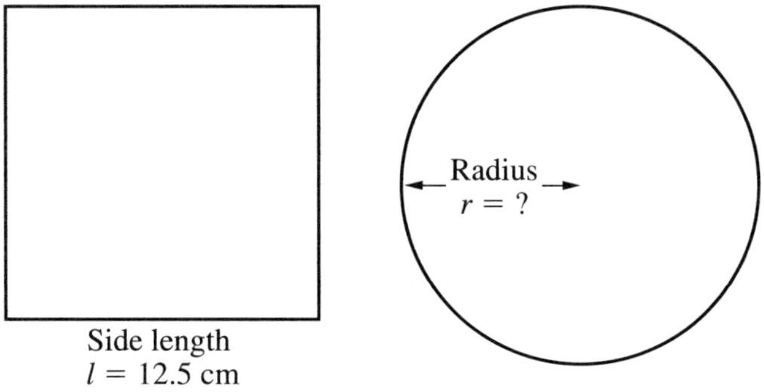

As both figures have the same area
$\pi r^2 = l^2$
or $\sqrt{\pi} r = l$
so $r = l \div (\sqrt{\pi})$

Now $\sqrt{3.14} = 1.772\,004\ldots$, and
$1 \div \sqrt{3.14}$ is very nearly 0.564
So we can make a very close approximation by multiplying l by 0.564.

Hence $r = (12.5 \times 0.564)$ cm = 7.05 cm

Exercise 16.9

For Questions 1–6 find the radius of the circle which has the same area as the square whose side length is given.

1 Side length = 5 cm **2** Side length = 15 cm

3 Side length = 50 cm **4** Side length = 40 mm

5 Side length = 70 mm **6** Side length = 1.25 m

7 At the Manor House there are two ornamental ponds, one square and one circular.
If the side length of the square one is 10 m, what is the radius of the circular one if it has the same surface area?

8 Amy's mother has two breadboards, one square and one circular.
If the side length of the square one is 25 cm, what is the radius of the circular one if it has the same surface area?

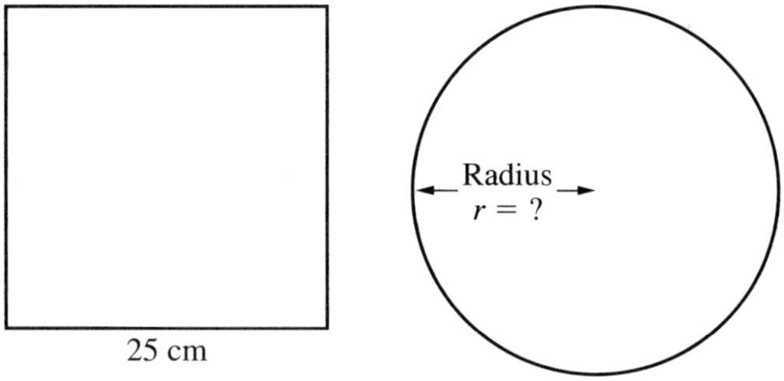

17 Specifying location using coordinates

17.1 Specifying location using ordered pairs

Example 1

a What shape is located in the square described by (i) B5 (ii) E4?
b Give the location of (i) the circle (ii) the star.

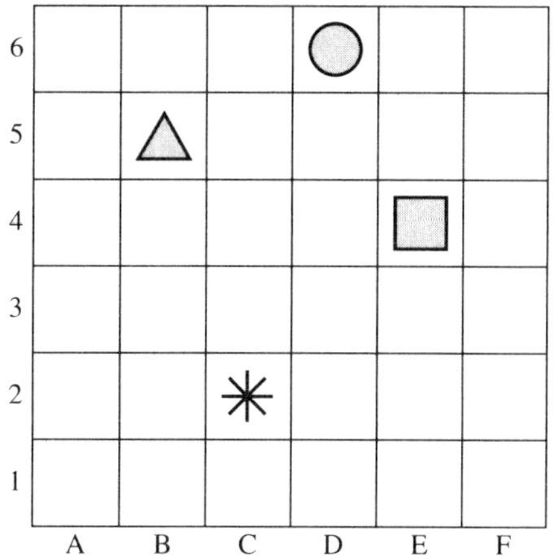

a (i) The triangle is in the square B5.
(ii) The square is in the square E4.
b (i) The location of the circle is D6.
(ii) The location of the star is C2.

Exercise 17.1

1 There are six kinds of shapes in this square array: circles, squares, triangles, rectangles, stars and arrows.

Copy and complete the table.

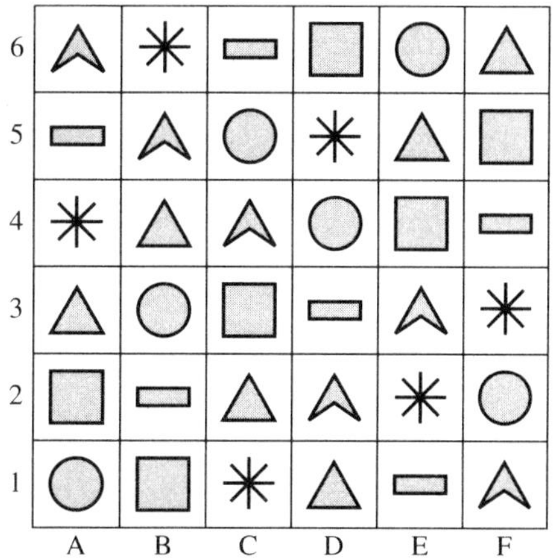

Square	Shape in square	Square	Shape in square	Square	Shape in square	Square	Shape in square	Square	Shape in square	Square	Shape in square
A1		B1		C1		D1		E1		F1	
A2		B2		C2		D2		E2		F2	
A3		B3		C3		D3		E3		F3	
A4		B4		C4		D4		E4		F4	
A5		B5		C5		D5		E5		F5	
A6		B6		C6		D6		E6		F6	

2 Give the locations of the squares on the grid that contain:
a circles
b squares
c triangles
d rectangles
e stars
f arrows

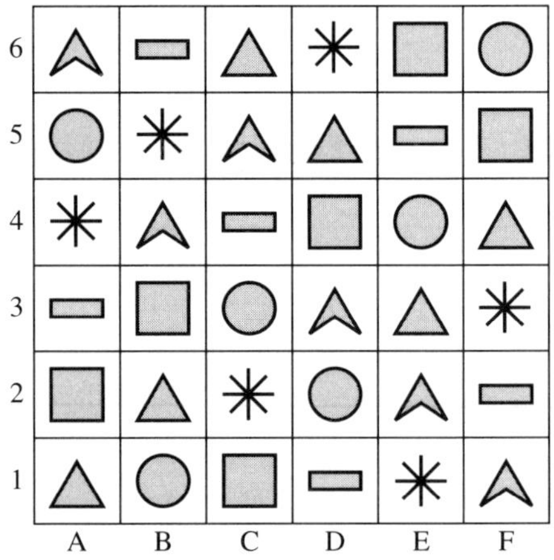

17.2 *Use of coordinates in the first quadrant*

We can use a pair of numbers called coordinates to represent a particular point on a grid.

For example, the point P is represented by (3, 6).

The first coordinate, 3, tells us how far to go to the right.
The second coordinate, 6, tells us how far to go up.

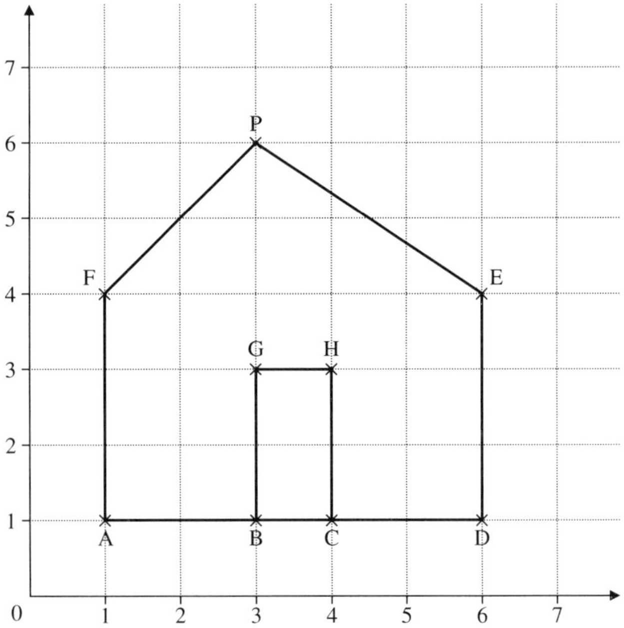

The point (0, 0) is called the origin. This is zero in both directions.

Example 1

From the grid above, write down the coordinates of the points B, C, D, E, F, G and H.

B is 3 across and 1 up, therefore B is (3, 1).
C is 4 across and 1 up, therefore C is (4, 1).
D is 6 across and 1 up, therefore D is (6, 1).
E is 6 across and 4 up, therefore E is (6, 4).
F is 1 across and 4 up, therefore F is (1, 4).
G is 3 across and 3 up, therefore G is (3, 3).
H is 4 across and 3 up, therefore H is (4, 3).

Note that (4, 1) and (1, 4) represent different points, so the order of the numbers is important.

Exercise 17.2

1 State the coordinates of each of the points A to Z on this grid.

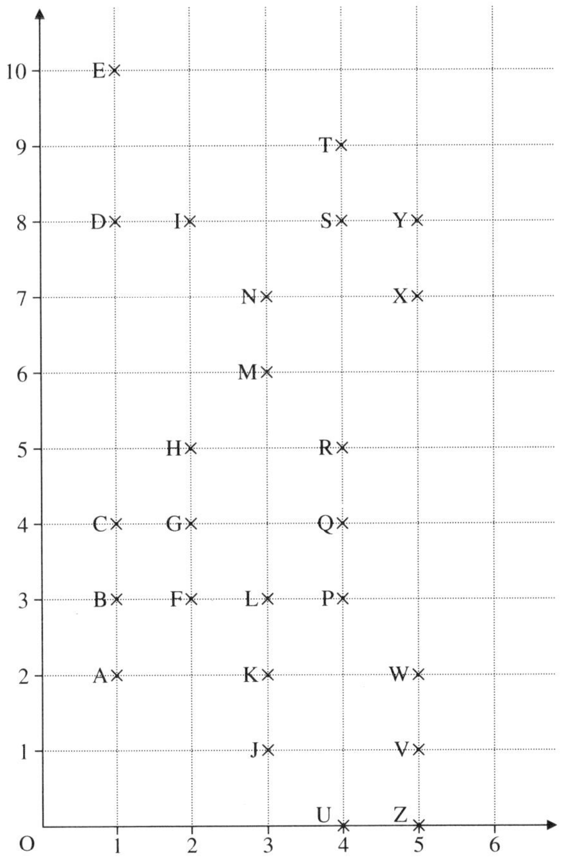

2 **a** State the coordinates of each of the four corners of the square ABCD.

b State the coordinates of each of the four corners of the square PQRS.

c State the coordinates of each of the four corners of the rectangle LMNP.

d State the coordinates of each of the four corners of the rectangle UVXW.

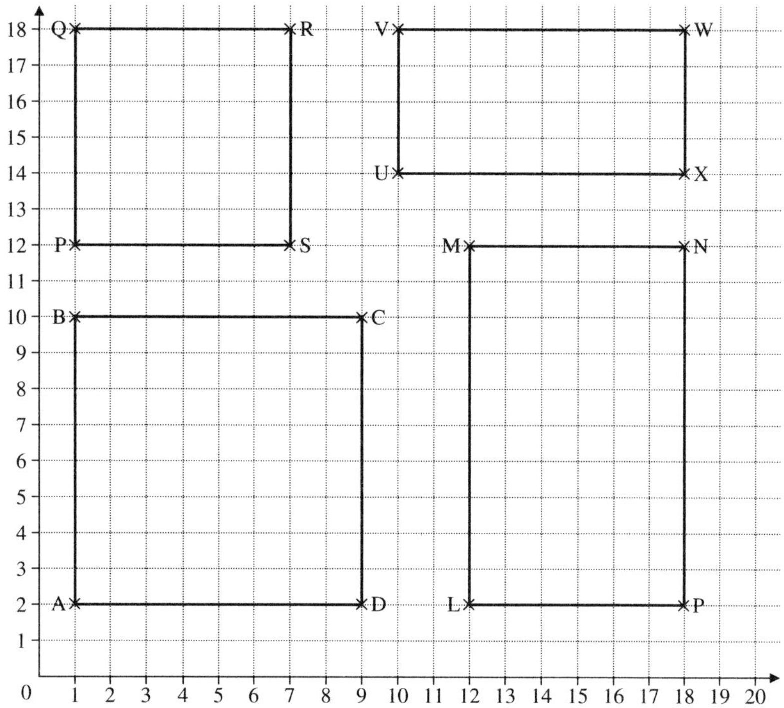

3 State the coordinates of the four corners of:

a the square ABCD

b the square KLMN

c the square UVWX

d the square PQRS.

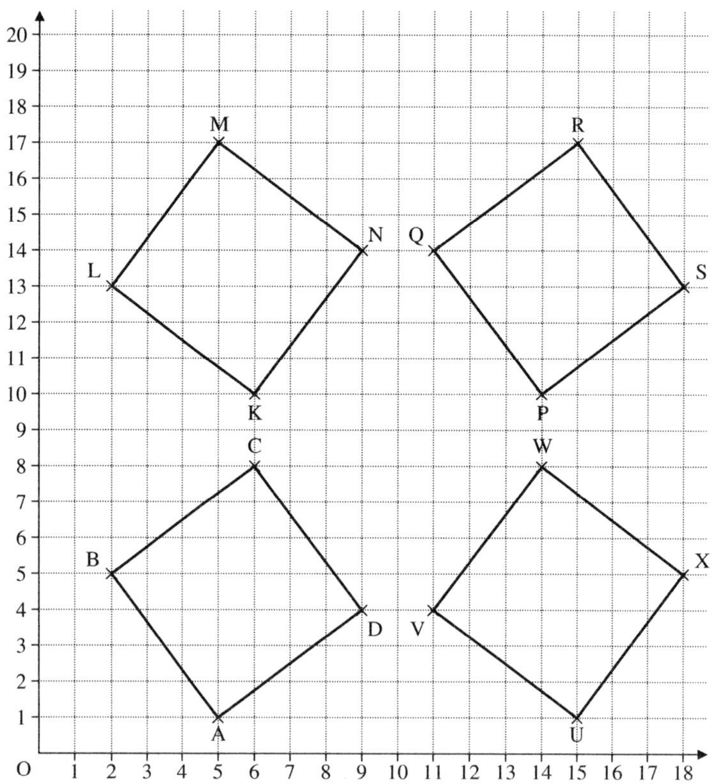

Example 2

Copy this grid and plot these points on the grid.
A (1, 3), B (5, 1), C (3, 3), and D (5, 5).

Join A to B, B to C, C to D, D to A.

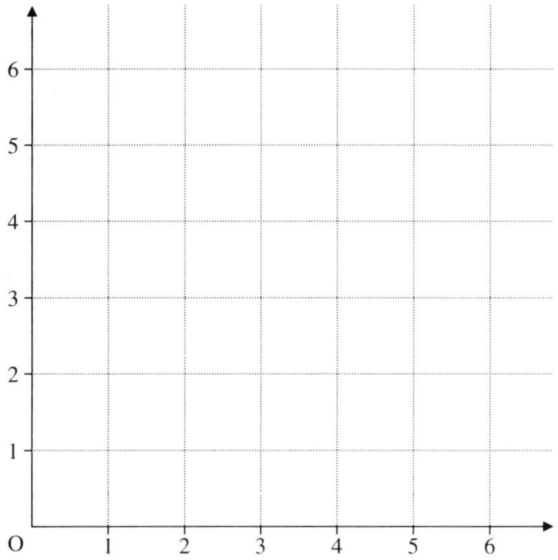

Name the shape that you have made.

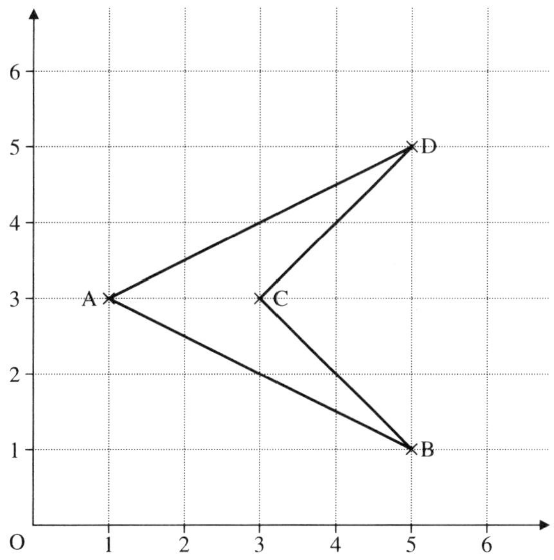

The shape is an arrowhead.

Exercise 17.3

1 Copy the grid and plot these points on it.
A (1, 5), B (1, 6), C (1, 7), D (1, 9), E (1, 1), F (1, 0), G (2, 1), H (2, 2), I (2, 6), K (2, 10), L (2, 0), M (3, 4), N (3, 5), P (3, 8), Q (3, 9), R (3, 0), S (4, 1), T (4, 2), U (4, 7), V (4, 10), W (5, 3), X (5, 4), Y (5, 9), Z (5, 10),

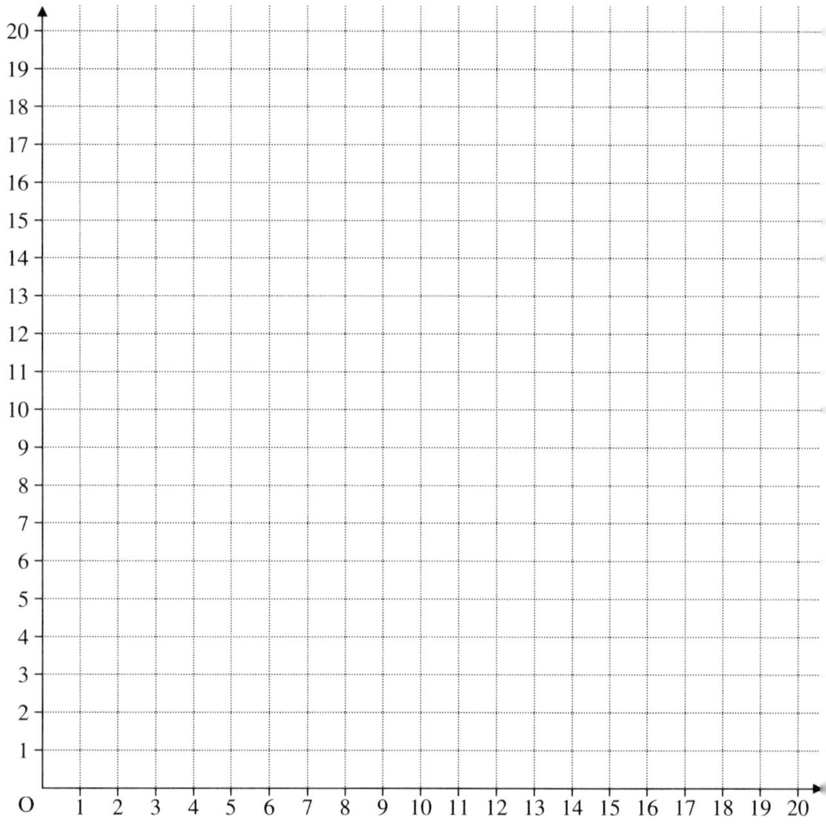

Copy the grid from Question 1 for these questions.

2 Plot each of these points on a grid.
Join them in alphabetical order and describe the shape you have made.
A (6, 16) B (14, 16) C (18, 9) D (14, 2) E (6, 2) F (2, 9)

3 Plot each of these points on a grid.
Join them in alphabetical order and describe the shape that you have made.
A (1, 6) B (1, 13) C (6, 18) D (13, 18) E (18, 13)
F (18, 6) G (13, 1) H (6, 1)

4 Plot each of these points on a grid.
Join them in alphabetical order and describe the shape that you have made.
A (1, 1) B (3, 7) C (1, 13) D (7, 11) E (13, 13) F (11, 7)
G (13, 1) H (7, 3)

Example 3

a Copy the grid below and use this table on the right to complete it. Draw a line through your grid points.

Number of pencils bought	0	1	2	5
Cost	0 p	5 p	10 p	25 p

b Find from your line the cost of (i) 3 pencils (ii) 4 pencils.

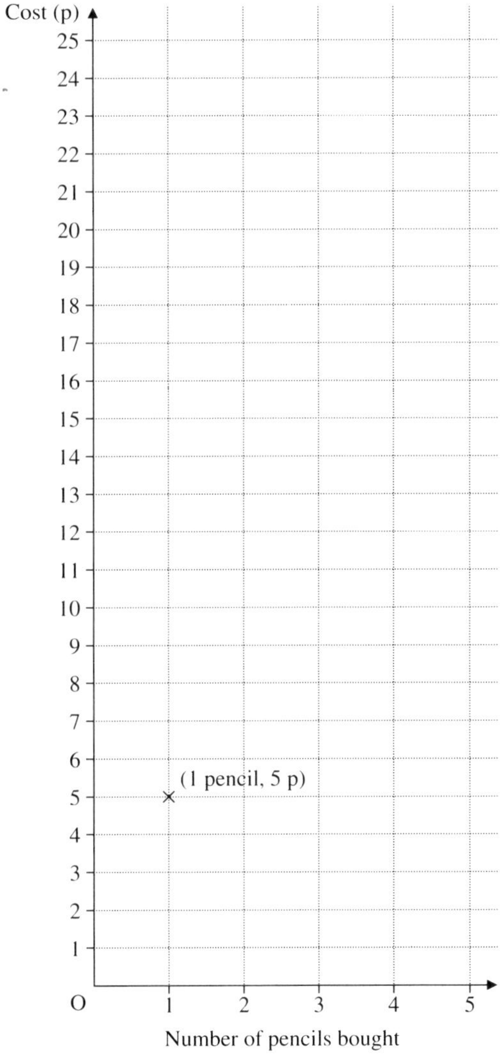

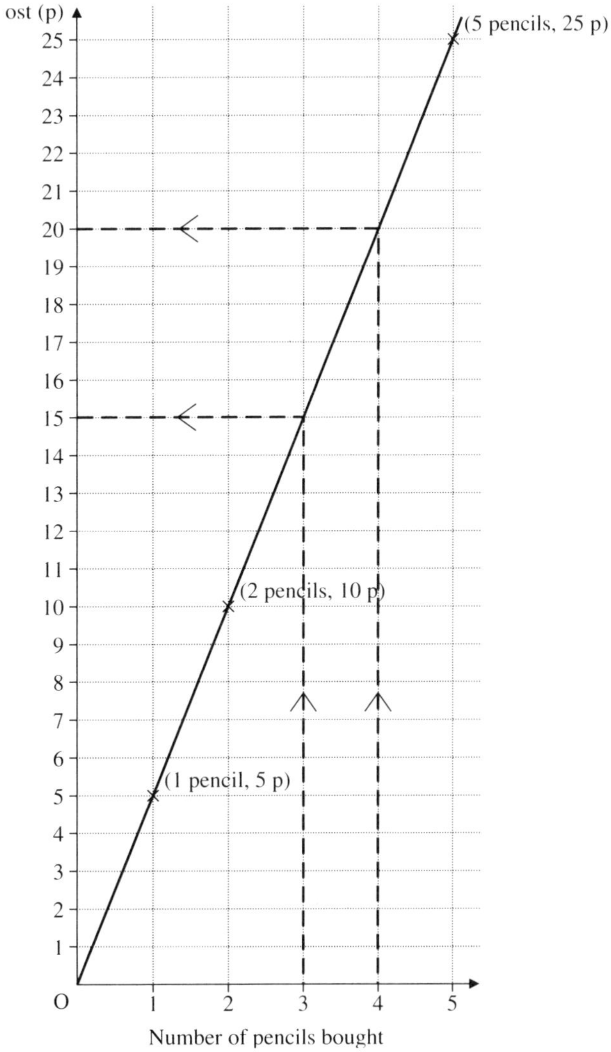

b (i) Cost of 3 pencils = 15 p
(ii) Cost of 4 pencils = 20 p

Exercise 17.4

1 Copy the grid and use the table to complete it. Draw a line through your grid points.

Number of marbles bought	Cost
0	0 p
1	4 p
2	8 p
3	12 p
4	16 p
5	20 p

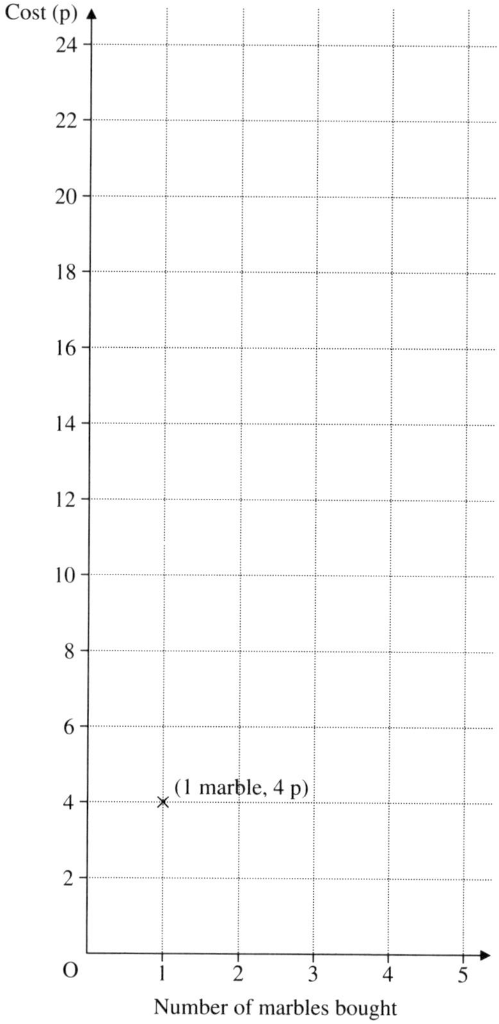

2 a Copy the grid at the top of the next page and use the table to complete it. Draw a line through your grid points.

Number of fuses bought	0	1	2	3	4	6	8	10
Cost	0 p	2 p	4 p	6 p	8 p	12 p	16 p	20 p

b Find from your line the cost of:
(i) 5 fuses
(ii) 7 fuses
(iii) 9 fuses.

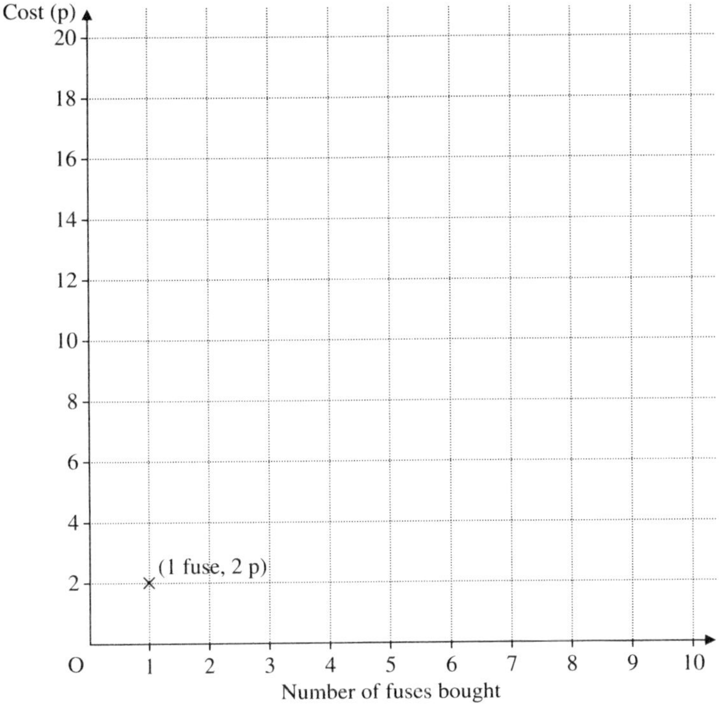

3 a Copy the grid and use the table to complete it. Draw a line through your grid points.

Number of litres of turpentine bought	Cost
0	£0
1	£2
2	£4
5	£10
7	£14
9	£18
10	£20

b Find from your line the cost of:
(i) 3 litres
(ii) 4 litres
(iii) 6 litres
(iv) 8 litres.

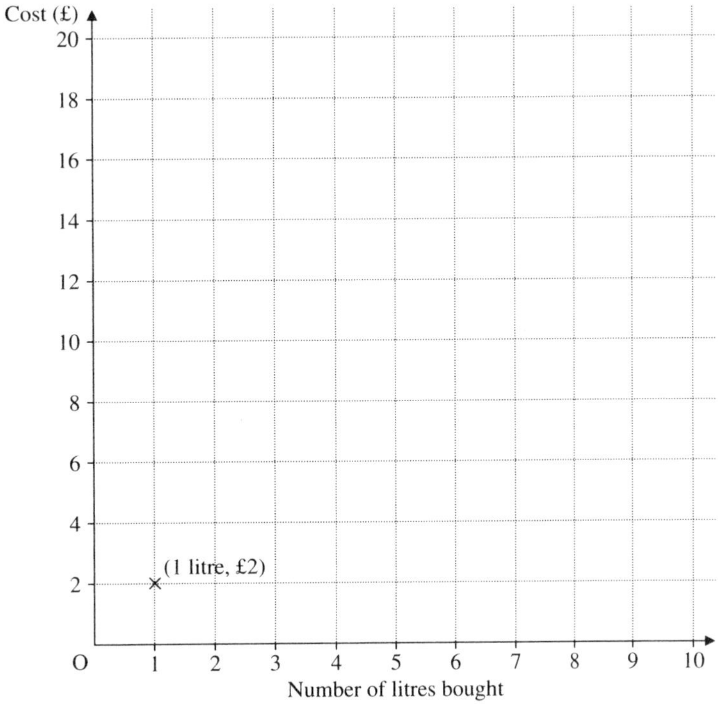

4 a A soap bottle measured 3 cm across when it was first blown. Its size increased as shown below (before it burst).

Time after bubble was blown	0 seconds	1 second	2 seconds	5 seconds	7 seconds
Size of bubble	3 cm	4 cm	5 cm	8 cm	10 cm

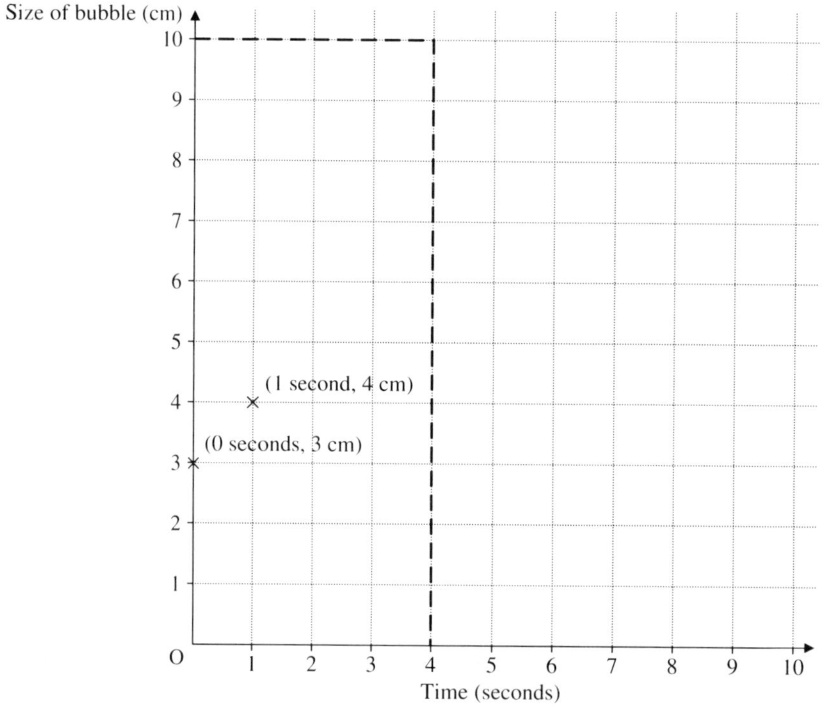

Copy the grid and use the table to complete it.
Draw a line through your grid points

b Use your line to find the size of the bubble after
(i) 3 seconds
(ii) 4 seconds
(iii) 6 seconds.

17.3 Understanding and using coordinates in all four quadrants

In Section 17.2 you used coordinates to describe points on a grid. In that section both coordinates were positive numbers. We will now extend the ideas to include negative coordinates.

Points to the left of the origin 0 have a negative first coordinate.
Points below the origin 0 have a negative second coordinate.

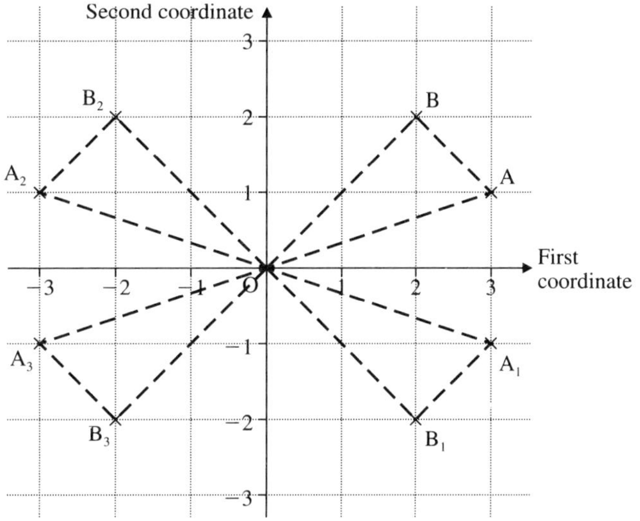

> **Example 1**
>
> Look at the grid.
> Write down the coordinates of:
>
> **a** A and B **b** A_1 and B_1
> **c** A_2 and B_2 **d** A_3 and B_3
>
> **a** A is (3, 1); B is (2, 2)
> **b** A_1 is (3, −1); B_1 is (2, −2)
> **c** A_2 is (−3, 1); B_2 is (−2, 2)
> **d** A_3 is (−3, −1); B_3 is (−2, −2)

Exercise 17.5

1. Write down the coordinates of all the points from A to Z.
2. Write down the coordinate of all the points from A_1 to Z_1.
3. Write down the coordinates of all the points from A_2 to V_2
4. Write down the coordinates of all the points from A_3 to V_3.

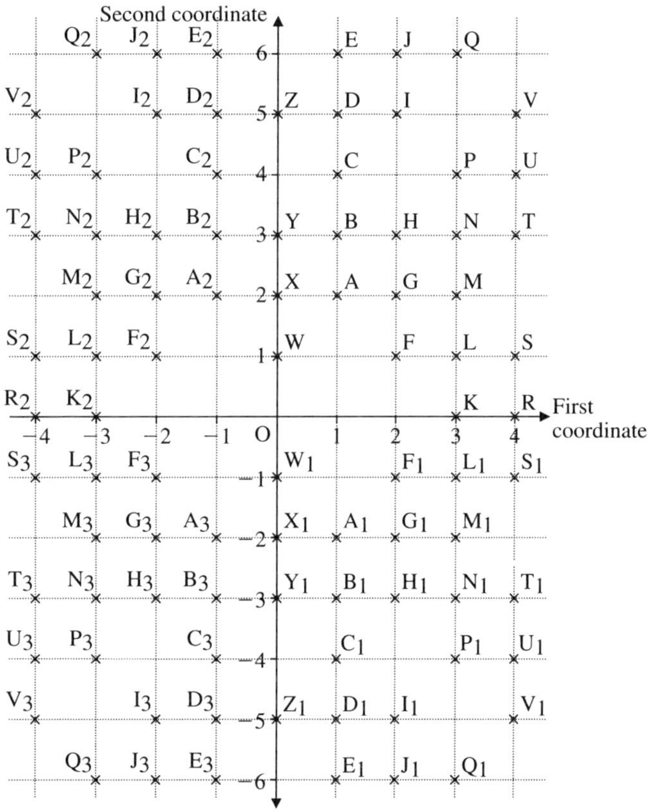

Example 2

ABCDEF is one half of an arrow. The other half is formed by a reflection in the line AF.

Copy the grid and draw in the reflection.

Write down the coordinates of:

a A, B, C, D, E and F

b B_1, C_1, D_1 and E_1
which are the reflections of B, C, D and E.

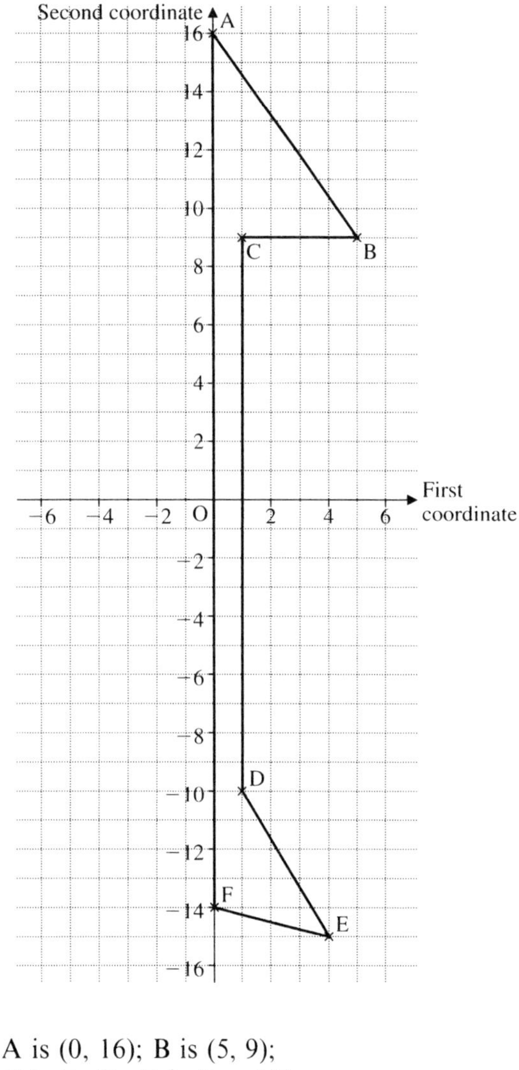

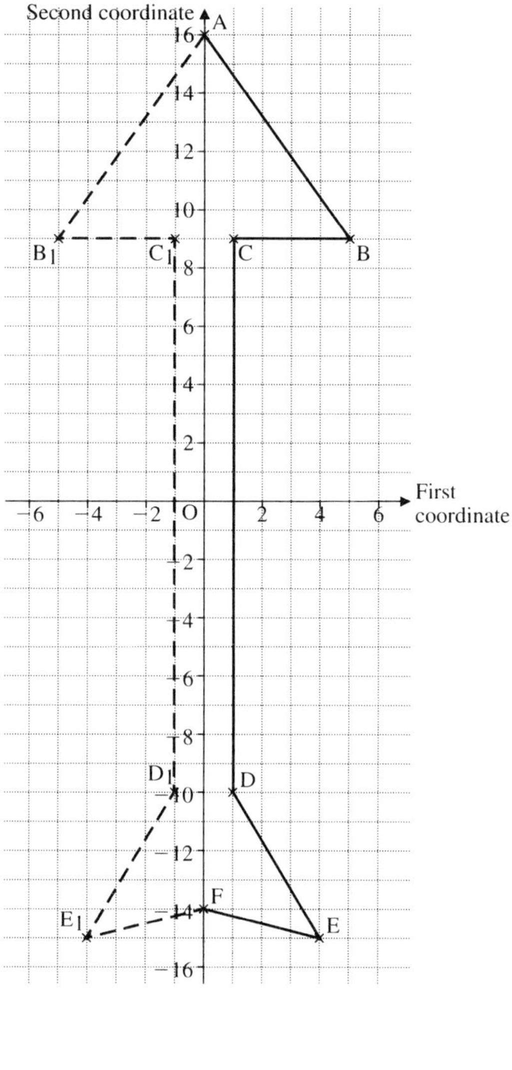

a A is (0, 16); B is (5, 9);
C is (1, 9); D is (1, −10);
E is (4, −15); F is (0,−14)

b B_1 is (−5, 9); C_1 is (−1, 9);
D_1 is (−1, −10); E_1 is (−4, −15)

Exercise 17.6

For each case one half of a figure is drawn.
The other half being formed by a reflection as in Example 2.

For each question copy the grid, draw in the reflection and then write down the coordinates of:

a All the given lettered points.
b All the image points formed.

1

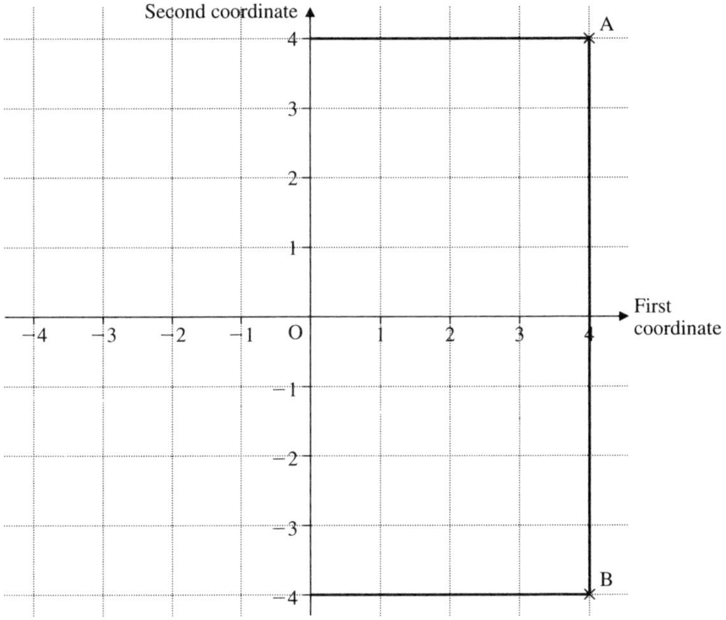

2

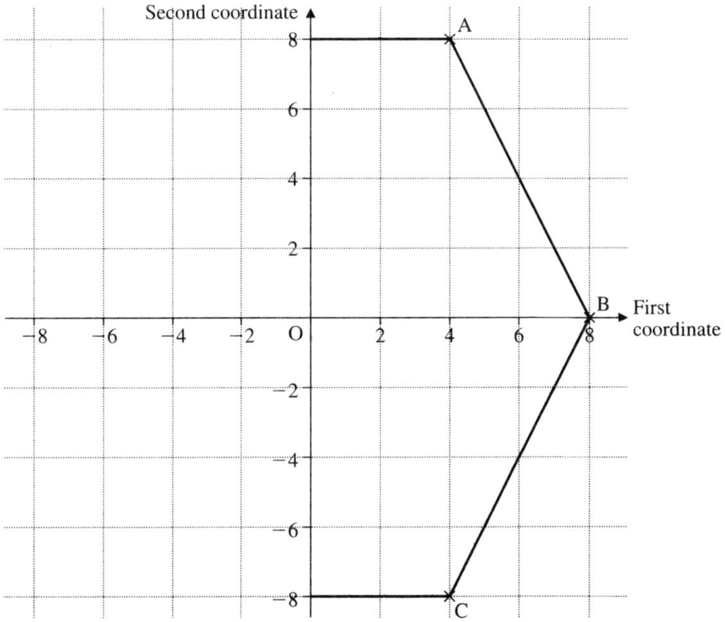

3

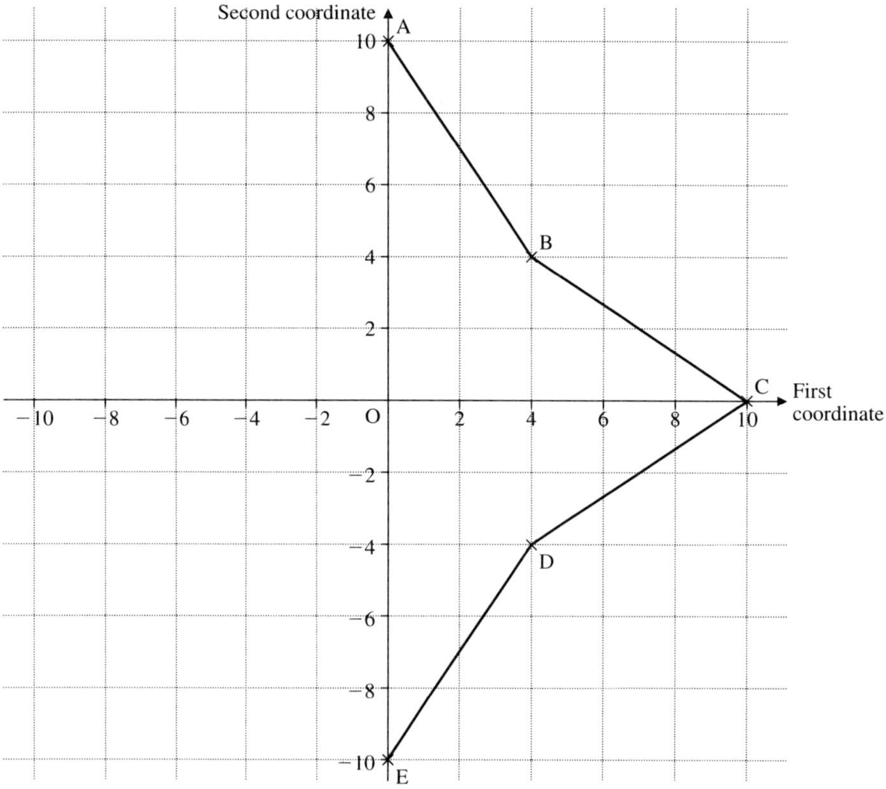
Second coordinate
First coordinate
A
B
C
D
E
O
−10
−8
−6
−4
−2
2
4
6
8
10

4

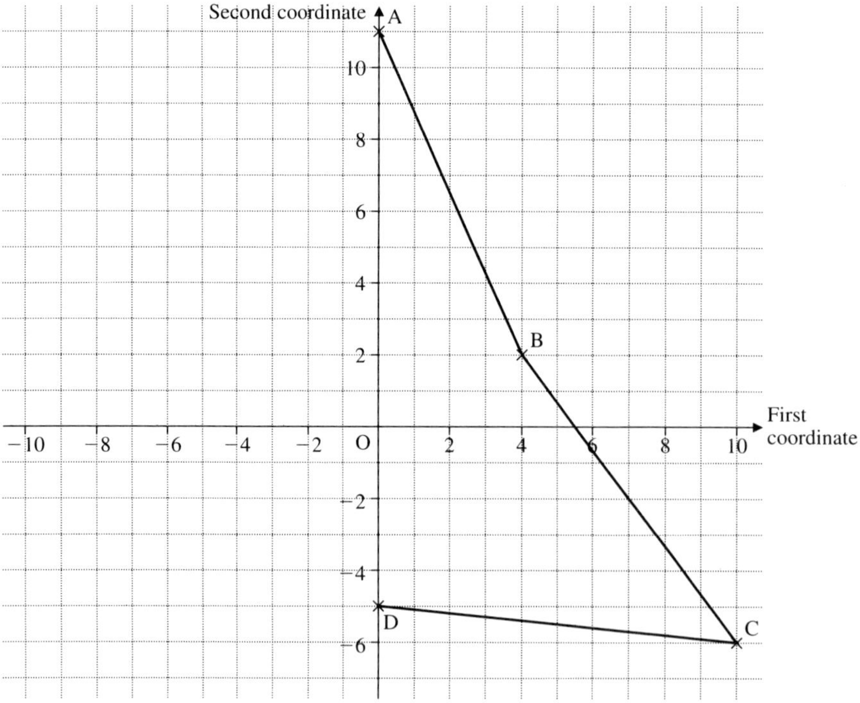
Second coordinate
First coordinate
A
B
C
D
O
−10
−8
−6
−4
−2
2
4
6
8
10

Example 3

Plot the coordinates on a grid.

Join them in order.

What shape have you made?

(13, 1), (19, 2), (19, −2), (13, −1), (−9, −1), (−15, 0), (−9, 1)

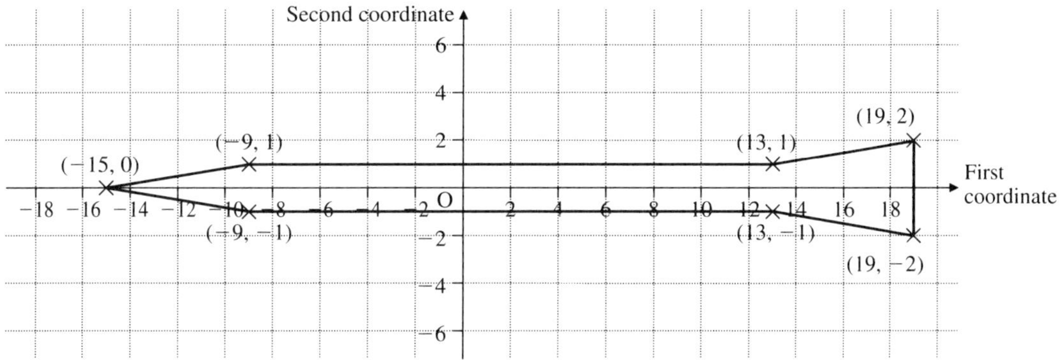

The shape could be a nail or a rocket.

Exercise 17.7

For each question plot the coordinates on a grid.
Join them in order and name the shape that you have made. (Use $\frac{1}{2}$ cm squared paper for all questions.)

In Questions 1 and 2 mark the horizontal axis from −20 to +20 and the vertical axis from −6 to +6.

1 (7, 1), (8, 3), (18, 3), (20, 2), (20, −2), (18, −3), (8, −3), (7, −1), (−13, −1), (−14, −2), (−15, −2),(−15, 2), (−14, 2), (−13, 1) and (7, 1)

2 (10, 1), (15, 4), (14, 0), (15, −4), (10, −1), (−9, −1), (−9, −5), (−16, 0), (−9, 5), (−9, 1) and (10, 1)

In Questions 3 and 4 mark both axes from −20 to +20.

3 (7, 12), (12, 7), (12, 0), (7, −5), (0, −5), (−5, 0), (−5, 7), (0, 12) and (7, 12)

4 (10, 10), (14, 0), (10, −10), (0, −14), (−10, −10), (−14, 0), (−10, 10), (0, 14) and (10, 10)

17.4 Specifying location by means of coordinates in all four quadrants

Example 1

Write down the coordinates of each of the points A, B, C and D on the grid.

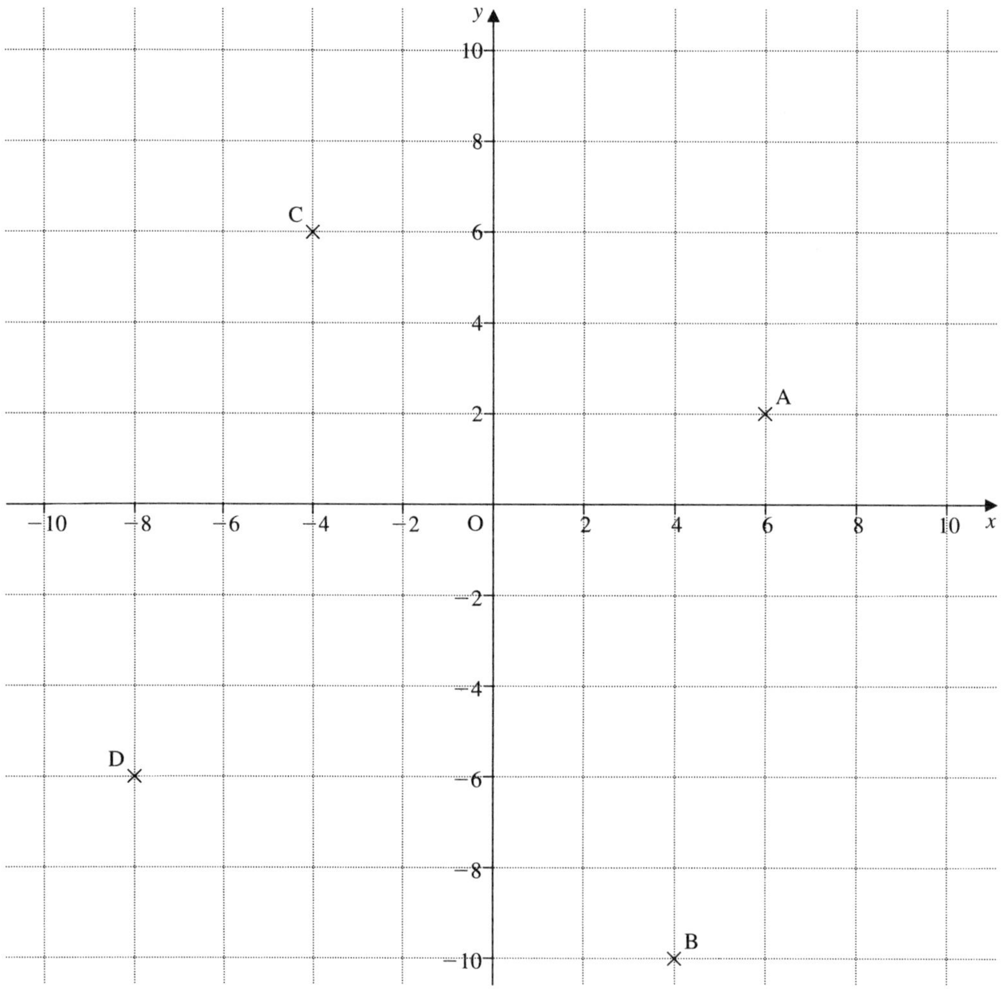

A is (6, 2)
B is (4, −10)
C is (−4, 6)
D is (−8, −6)

Example 2

Mark each of these sets of points on a grid.

Join the points together in order and state the shape that you have made.

a (6, 2) (11, 7) (6, 12) and (1, 7)
b (6, −3), (10, −10) and (2, −10)
c (−12, 5), (−2, 5) and (−7, 10)
d (−6, −4), (−2, −4), (−2, −10) and (−6, −10)

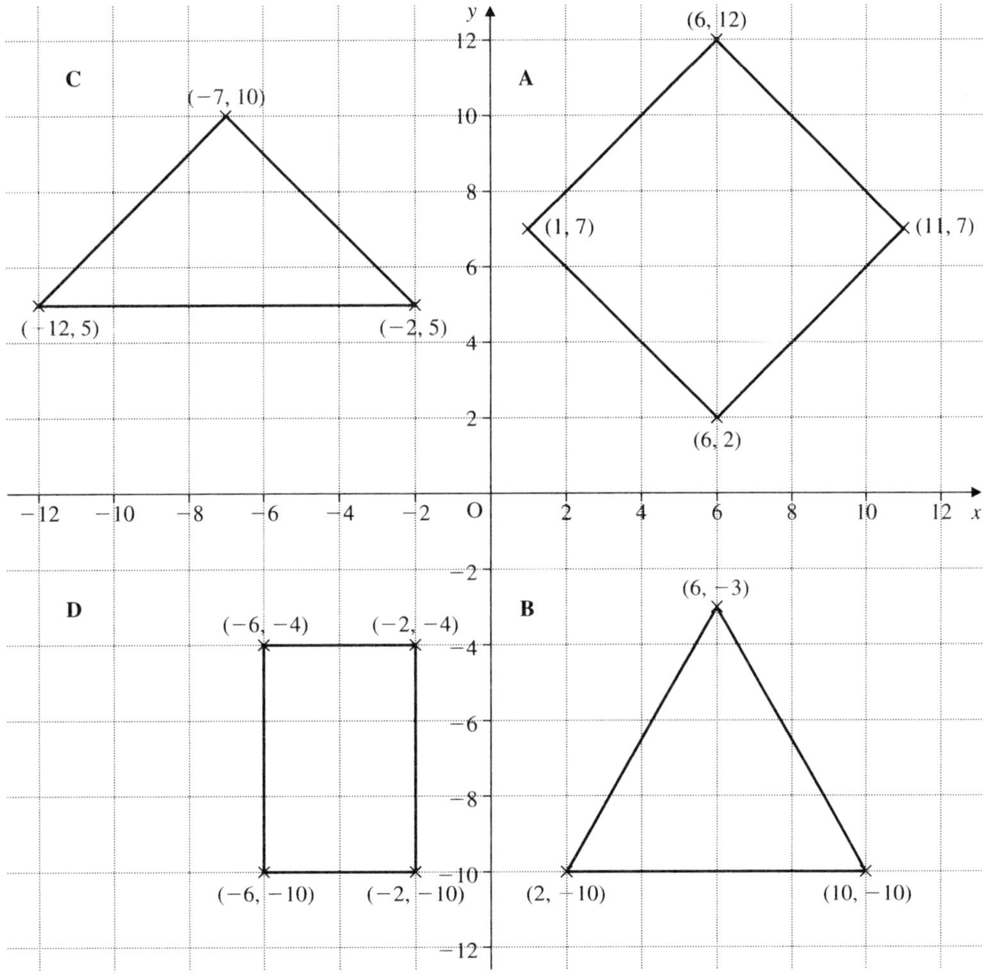

a Square
b Equilateral triangle
c Isosceles right-angled triangle
d Rectangle

Exercise 17.8

1 Write down the coordinates of each of these points on the grid:

a A, A_1, A_2 and A_3
b B, B_1, B_2 and B_3
c C, C_1, C_2 and C_3
d D, D_1, D_2 and D_3
e E, E_1, E_2 and E_3
f F, F_1, F_2 and F_3

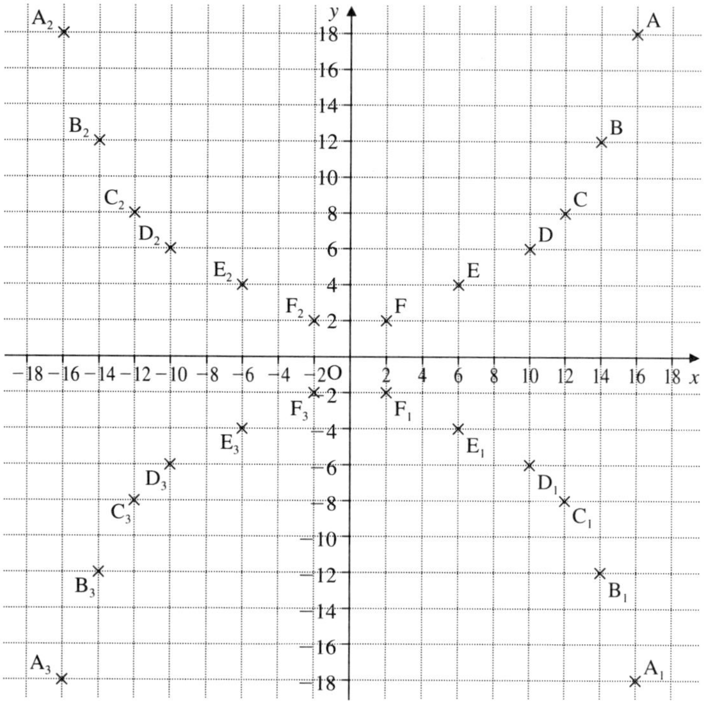

2 Write down the coordinates ofeach of the following points on the grid.

a the four corners of rectangle A
b the four corners of rectangle B
c the four corners of trapezium C
d the four corners of trapezium D
e the four corners of rhombus E
f the four corners of rhombus F
g the four corners of square G
h the four corners of square H

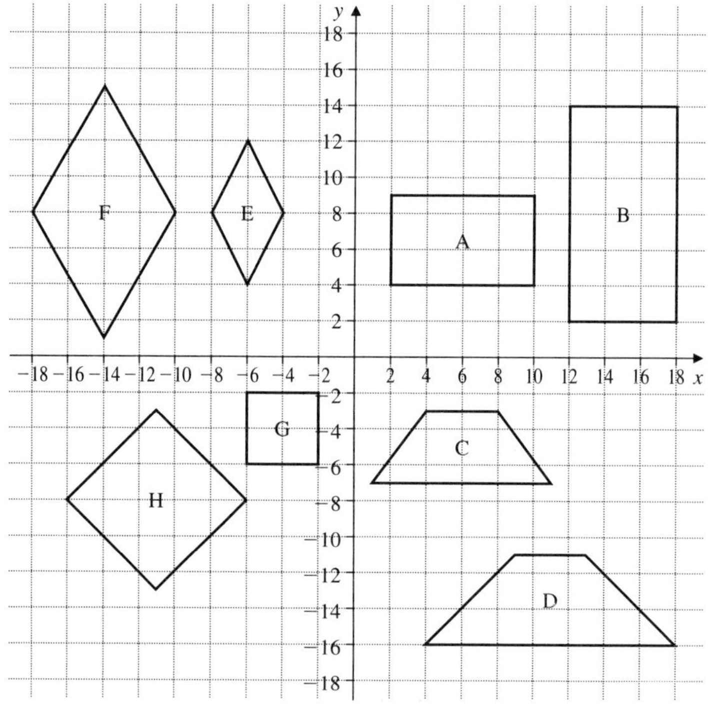

For Questions 3 and 4 draw a pair of axes on $\frac{1}{2}$cm squared paper and label both from −20 to +20.

3 Plot each set of points on the same grid.
Join the points in order and describe the shapes you have made.

a A(18, 16) A_1(18, −16) A_2(−18, −16) A_3(−18, 16) A
b B(16, 12) B_1(16, −12) B_2(−16, −12) B_3(−16, 12) B
c C(14, 10) C_1(14, −10) C_2(−14, −10) C_3(−14, 10) C
d D(12, 6) D_1(12, −6) D_2(−12, −6) D_3(−12, 6) D
e E(4, 2) E_1(4, −2) E_2(−4, −2) E_3(−4, 2) E

4 Plot each set of points on the same grid.
Join the points together in order and name the shape that you have made.

a (1, 2), (9, 2), (5, 5), (1, 2)
b (11, 1), (19, 1), (15, 8), (11, 1)
c (2, −1), (10, −5), (2, −5), (2, −1)
d (3, −11), (15, −6), (15, −11), (3, −11)

Example 3

Mark the two points on a grid.
Find the coordinates of two other points in the same quadrant which will make a square with the given ones.
Points: (5, 5) and (5, 9)
(*Note:* there are three possibilities.)

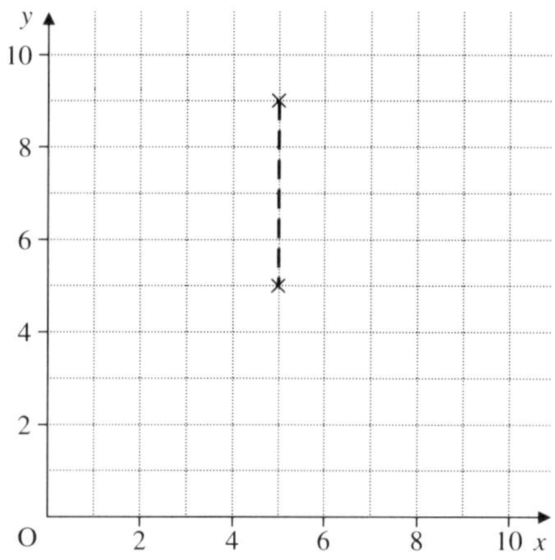

The three possibilities are:
a (1, 5) and (1, 9)
b (9, 5) and (9, 9)
c (3, 7) and (7, 7)

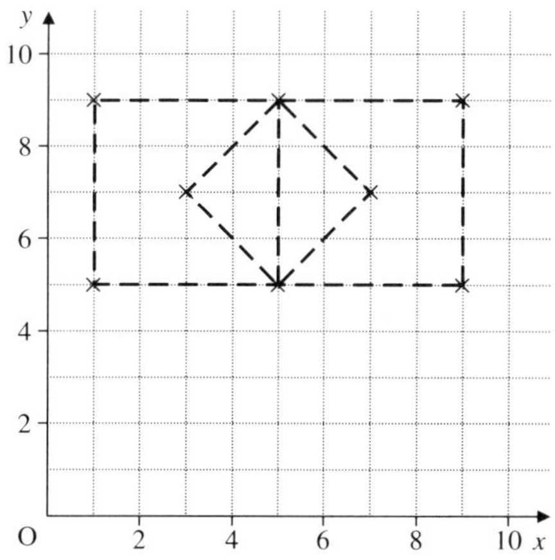

Exercise 17.9

For all four parts of each question, mark the two given points on a grid.
Find the coordinates of two other points in the same quadrant that will make a square with the given ones.

1 a (7, 4) and (7, 10) (3 possibilities)
b (7, −4) and (7, −10) (3 possibilities)
c (−7, 4) and (−7, 10) (3 possibilities)
d (−7, −4) and (−7, −10) (3 possibilities)

2 a (4, 7) and (10, 7) (3 possibilities)
b (4, −7) and (10, −7) (3 possibilities)
c (−4, 7) and (−10, 7) (3 possibilities)
d (−4, −7) and (−10, −7) (3 possibilities)

3 a (7, 14) and (14, 7) (3 possibilities)
b (7, −14) and (14, −7) (3 possibilities)
c (−7, 14) and (−14, 7) (3 possibilities)
d (−7, −14) and (−14, −7) (3 possibilities)

17.5 Specifying location by means of grid references and by angle and distance

On a map we use grid references to specify a square.
The location of Churchville is given by 0604. The '06' is the position on the Easterly grid line. The '04' is the position on the northerly grid line.
All locations in the shaded square are given by 0604.

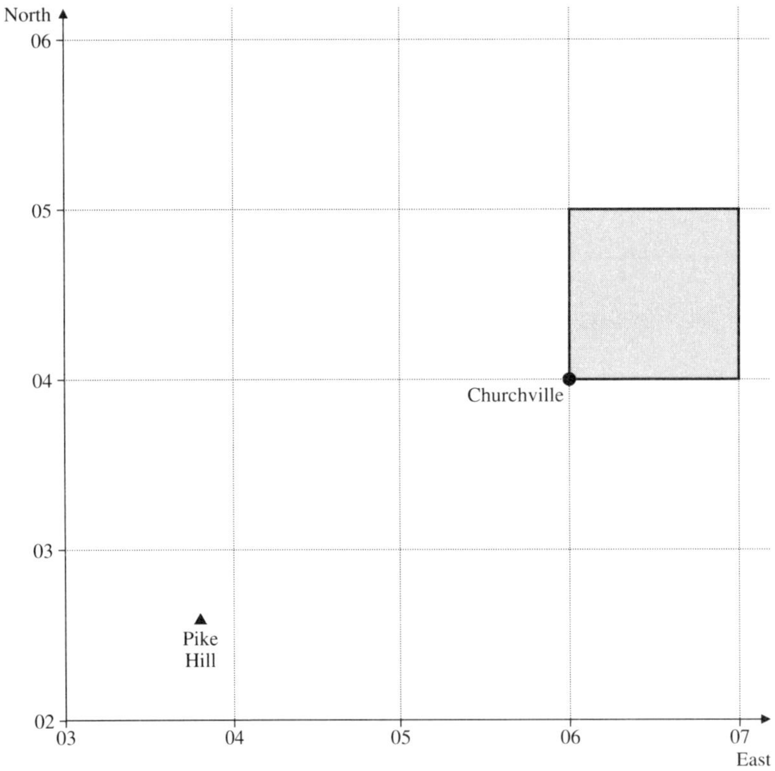

Pike Hill is in the square 0302.
In order to give locations more accurately, each large square on a map is divided into smaller ones. The accurate location of Pike Hill is 038026.

The first three figures refer to the Easterly grid line and the last three figures refer to the Northerly grid line.

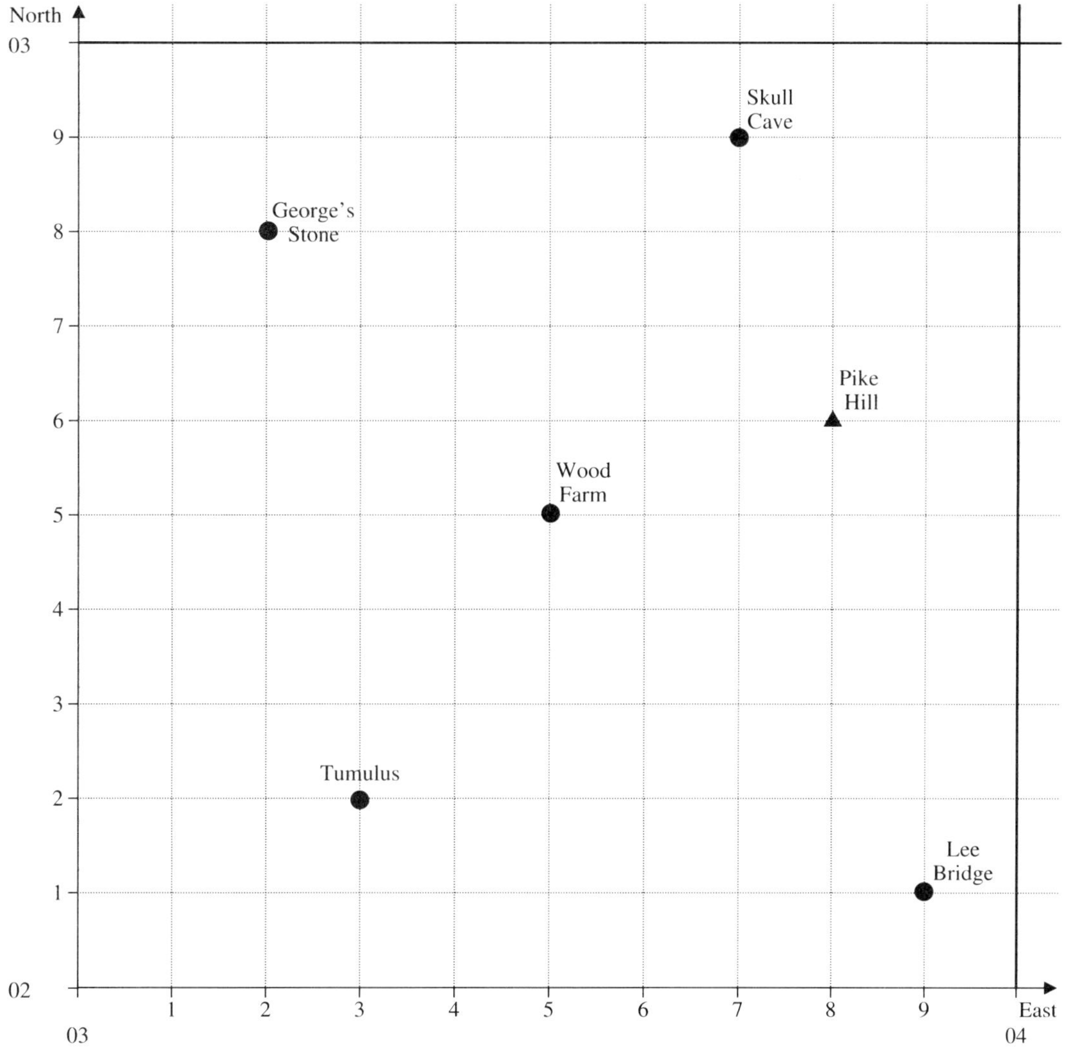

Here are some more locations to check out:

George's Stone is at 032028
Tumulus is at 033022
Wood Farm is at 035025
Skull Cave is at 037029
Lee Bridge is at 039021.

Exercise 17.10

1 The map shows the east of Kent.
Give the grid reference of the squares in which these places are located:

a Hythe
b Bossingham
c Whitstable
d Folkestone
e Aylesham
f Reculver
g Eastry
h Minster
i Sellindge
j Chartham
k Herne Bay
l Selstead
m Wingham
n Dover
o Deal
p Ramsgate
q Lyminge
r Canterbury
s Sandgate
t Shepherdswell
u Littlebourne
v Ashley
w Sandwich
x Margate

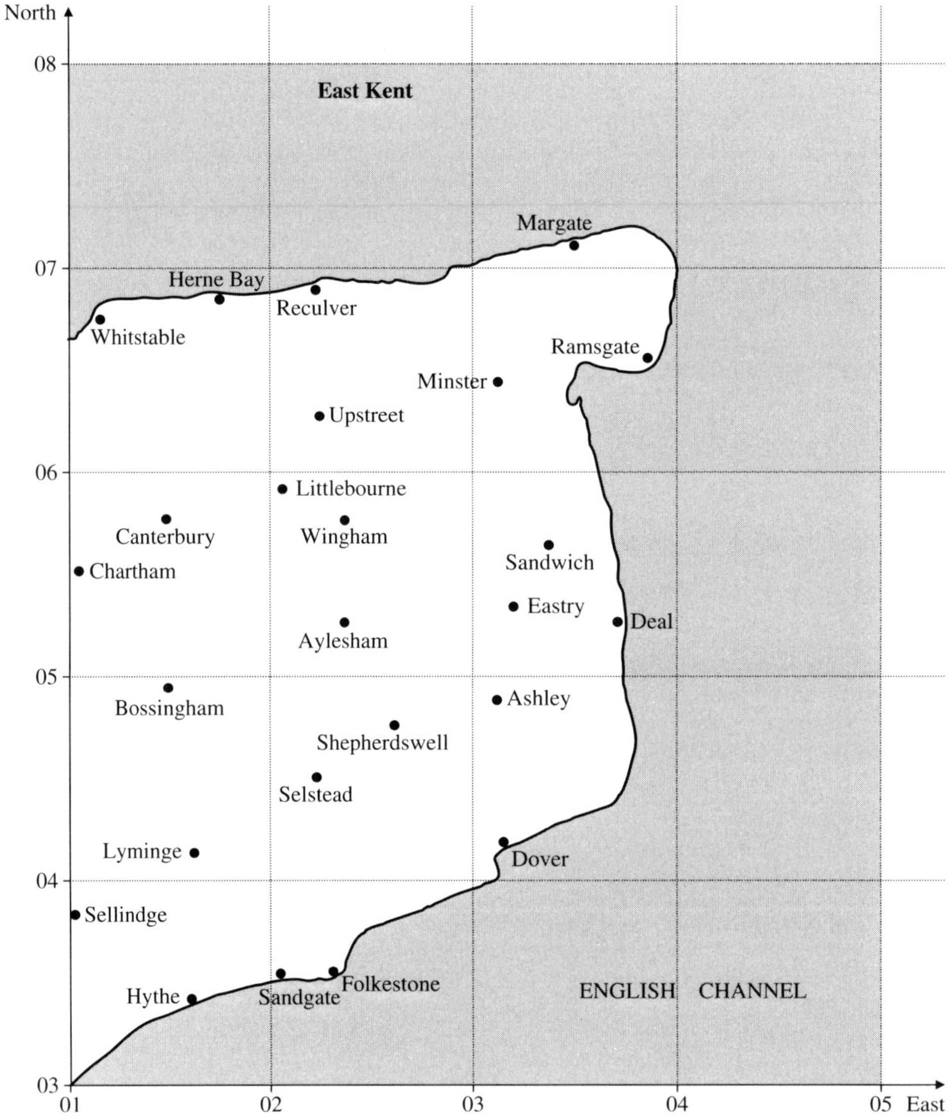

2 The map shows part of the south west of Scotland.
Give the names of the places or hills which are located in these grid squares:

a 0005	**b** 0105	**c** 0205
d 0006	**e** 0106	**f** 0206
g 0007	**h** 0107	**i** 0207

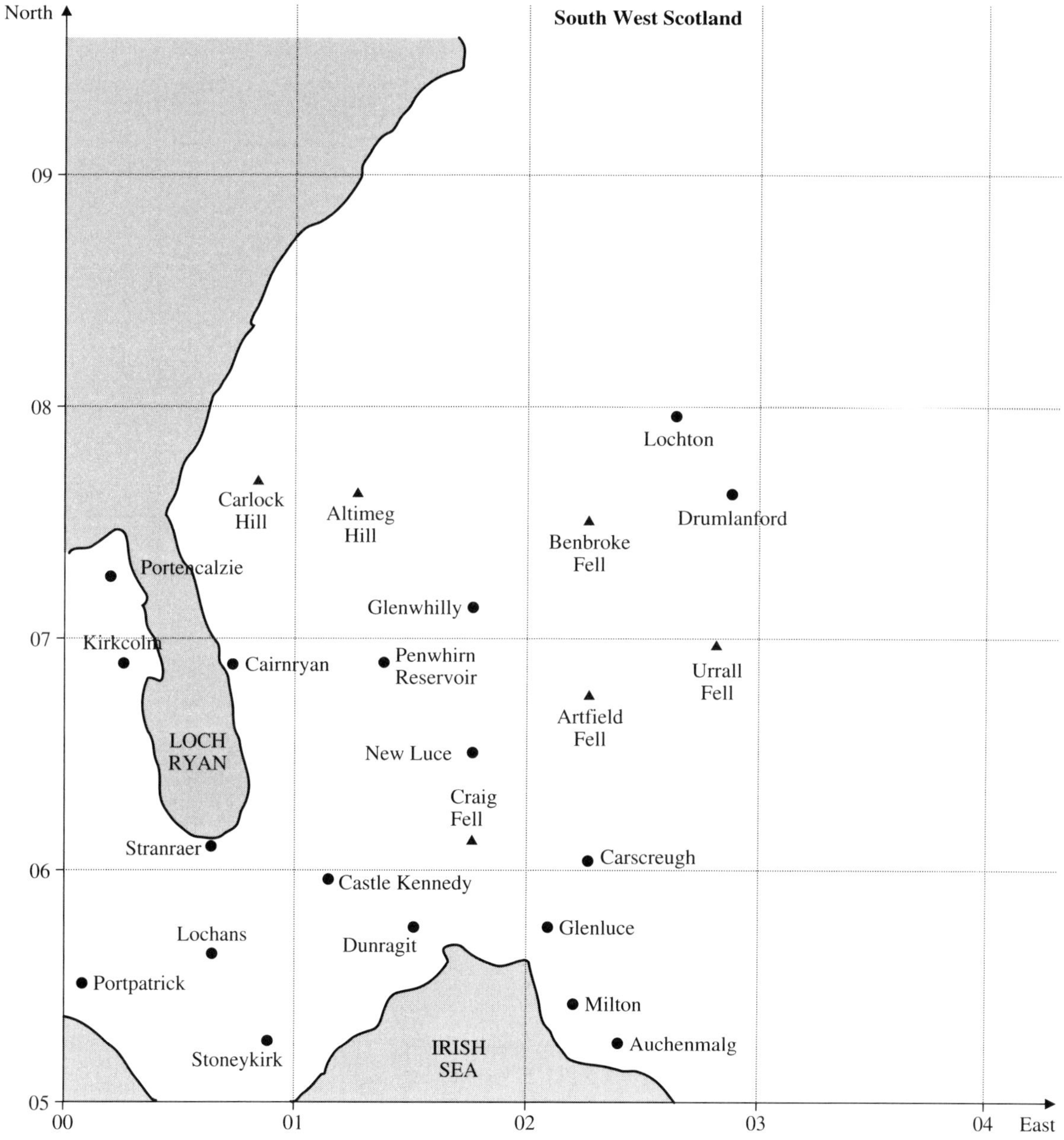

3 The map shows two areas to the north and east of Leeds.
Give the place which is located at each of these map references.

(i) **a** 031032
b 031036
c 031038
d 031039
e 031043
f 031046
g 032034
h 032036
i 032045
j 032047
k 033033
l 033035
m 033043
n 033049
o 034031
p 034036
q 034037
r 034047
s 035033
t 035036
u 035045

(ii) **a** 036034
b 036037
c 036043
d 037034
e 037036
f 037044
g 037048
h 038030
i 038040
j 038041
k 038046
l 039032
m 039047
n 040030
o 040033
p 040038
q 040048
r 040050
s 030033
t 030035
u 030037
v 030045
w 030047
x 030050

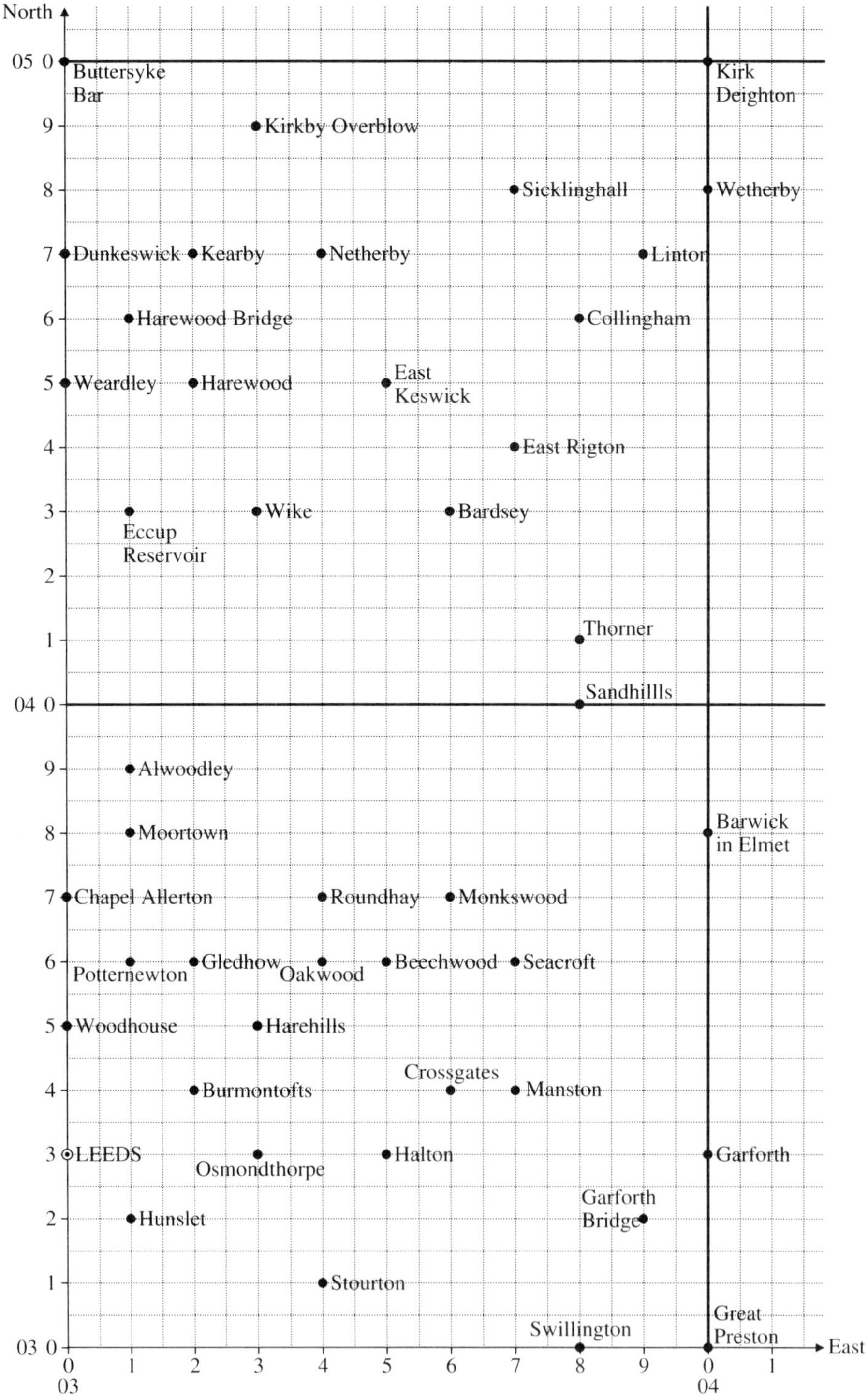

An alternative way of specifying the location of a point is to use a distance and direction.

Example 1

a What is located at each of the following positions from the town centre?
(i) 2 km north (ii) 3 km south east (iii) 4 km west (iv) 2 km north west

b Give the distance and direction of each of the following from the town centre.
(i) cinema (ii) airport (iii) tennis courts (iv) theatre

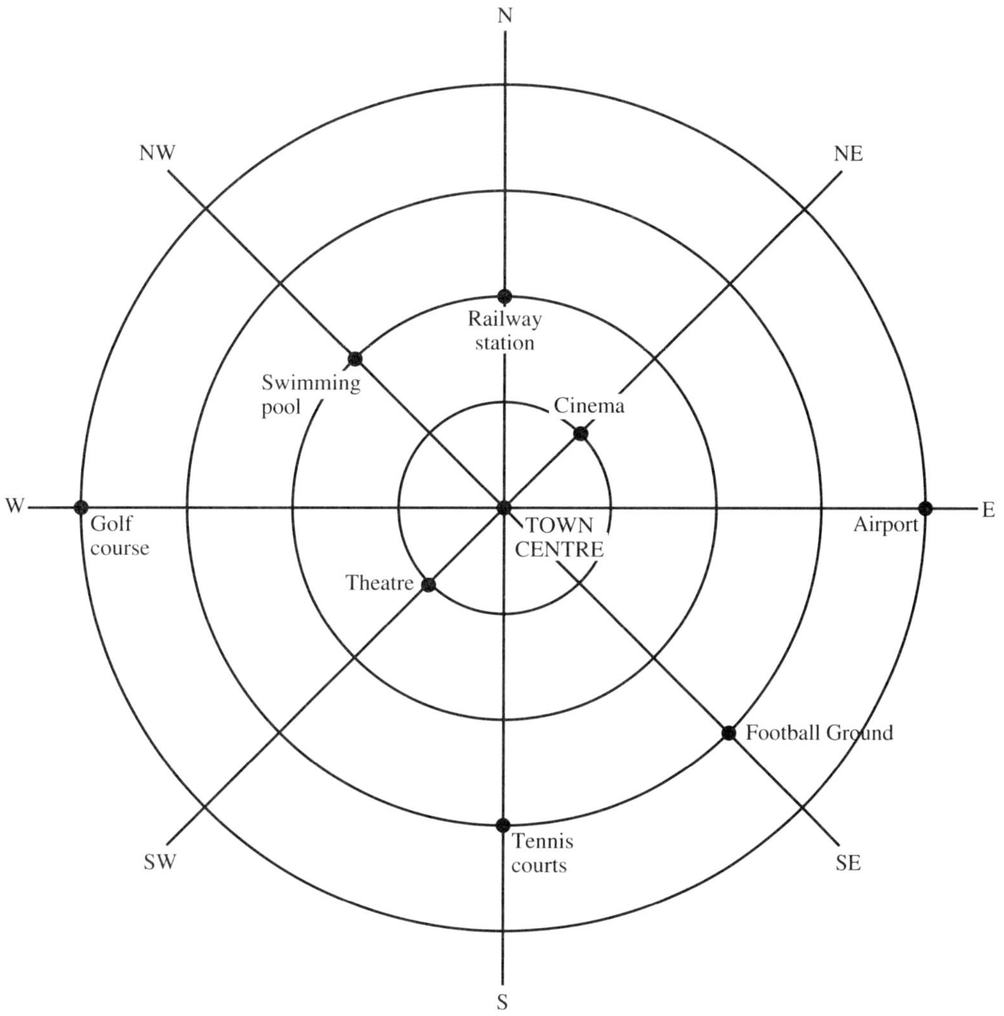

a (i) Railway station (ii) Football ground (iii) Golf course (iv) Swimming pool

b (i) 1 km north east (ii) 4 km east (iii) 3 km south (iv) 1 km south west

Exercise 17.11

1 The map shows the area around Birmingham.
State the name of the place at each of the following locations from the city centre:

a 10 km north	**b** 20 km north	**c** 30 km north
d 10 km east	**e** 20 km east	**f** 30 km east
g 10 km south	**h** 20 km south	**i** 30 km south
j 10 km west	**k** 20 km west	**l** 30 km west
m 10 km north east	**n** 20 km north east	**o** 30 km north east
p 10 km south east	**q** 20 km south east	**r** 30 km south east
s 10 km south west	**t** 20 km south west	**u** 30 km south west
v 10 km north west	**w** 20 km north west	**x** 30 km north west

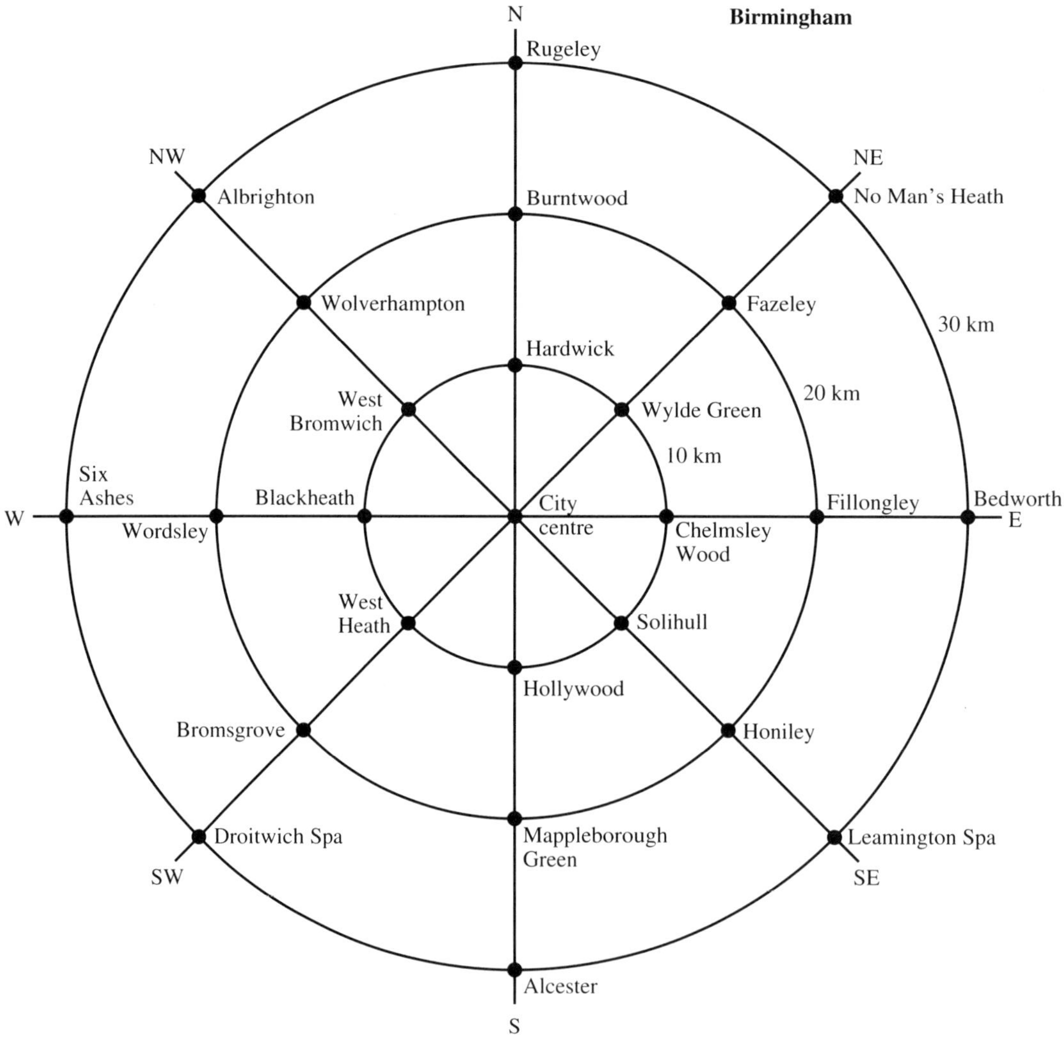

2 The map shows the area around Guildford.
Give the distance and direction from the city centre for each of the 24 places:

a Woking	**b** Virginia Water	**c** Langley
d Ockham	**e** Esher	**f** Putney
g Abinger Hammer	**h** Brockham	**i** Nutfield
j Pitch Hill	**k** Kingsfold	**l** Handcross
m Hascombe	**n** Plaistow	**o** Fittleworth
p Milford	**q** Hindhead	**r** Liss
s Ash Green	**t** Crondall	**u** Upton Grey
v Mytchett	**w** Finchampstead	**x** Shinfield

Guilford

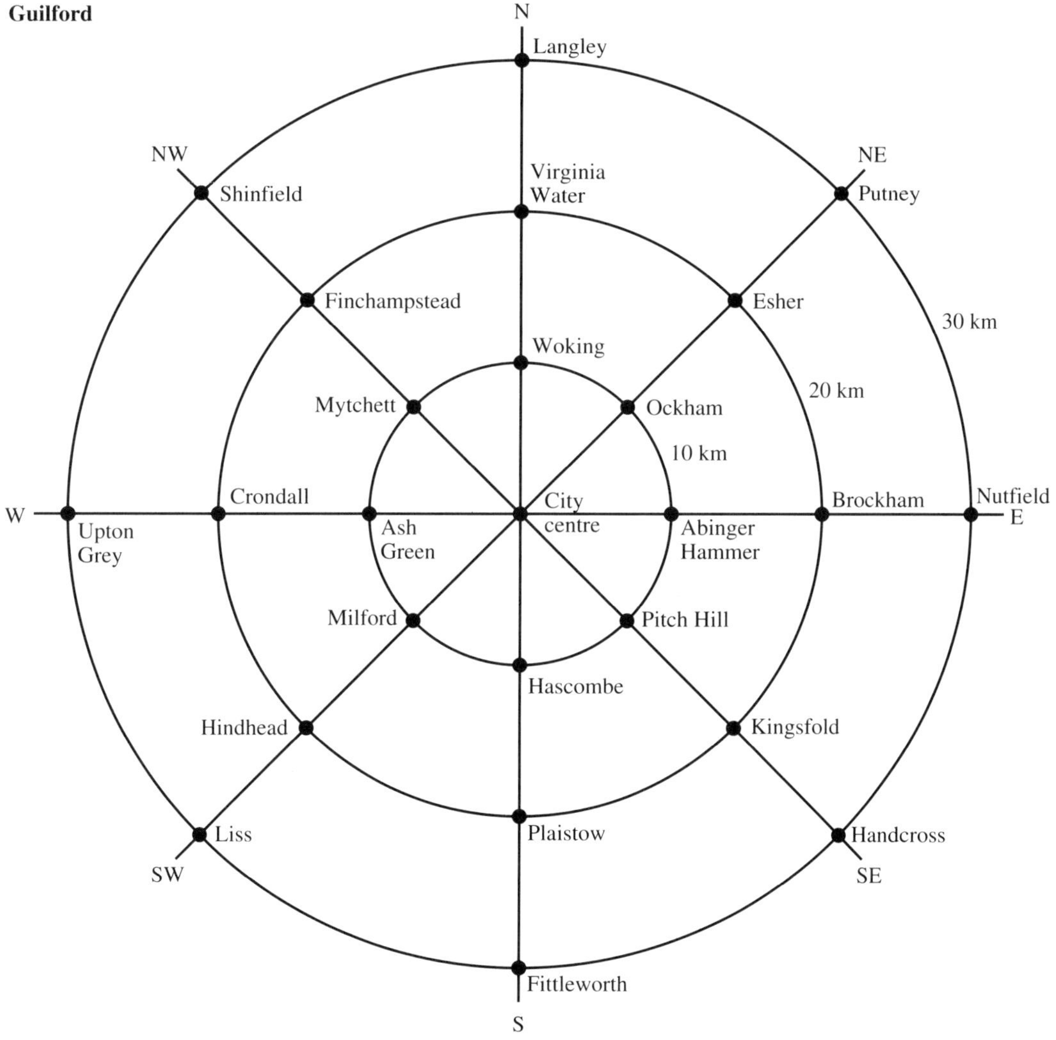

17.6 Understanding and using bearings to define directions

In the previous section we described a position by using the points of the compass (e.g. 3 km NE). When we want to do this more accurately we need to use a bearing.
A bearing is a distance and a direction measured in degrees.

Example 1

Use a bearing to describe the position of:
a Portsmouth
b Dover
c Calais
from Cherbourg.

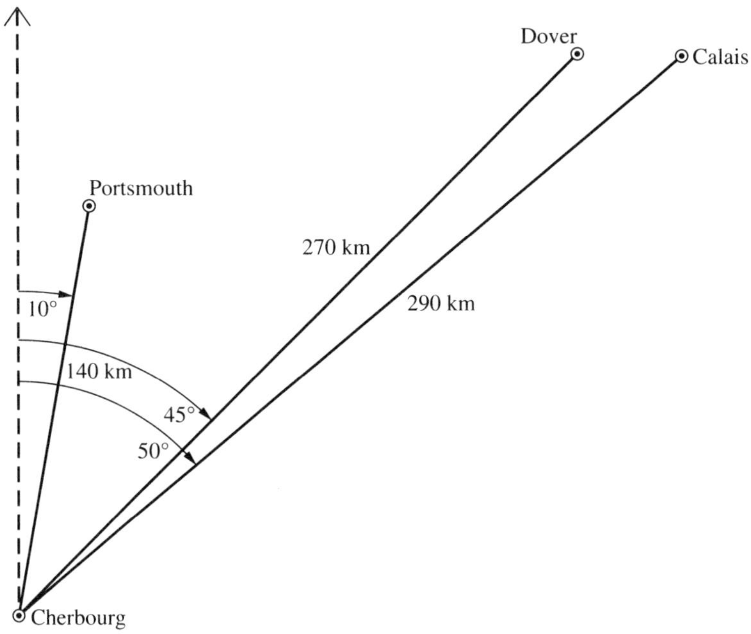

a Portsmouth is 140 km on a bearing of 010°.
b Dover is 270 km on a bearing of 045°.
c Calais is 290 km on a bearing of 050°.

(Note that bearing angles are always given as three-figure numbers.)

Example 2

Use a bearing to describe the position of each of the points A to J.

A 30 km, 030°
B 10 km, 060°
C 30 km, 120°
D 20 km, 150°
E 10 km, 180°
F 20 km, 240°
G 30 km, 270°
H 30 km, 300°
I 10 km, 330°
J 10 km, 000°

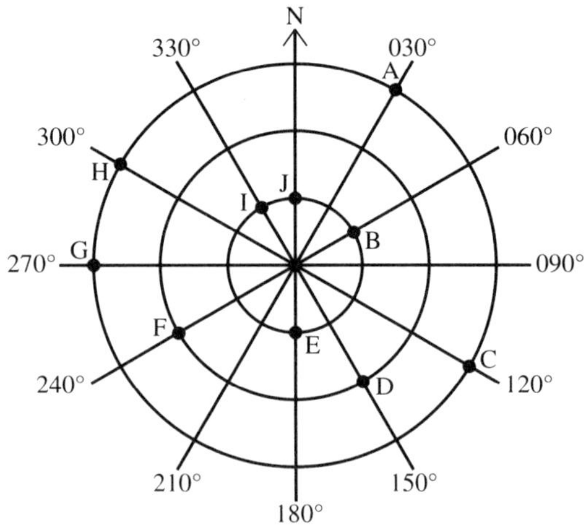

Exercise 17.12

1 Use a bearing to describe the positions of points A to Y.

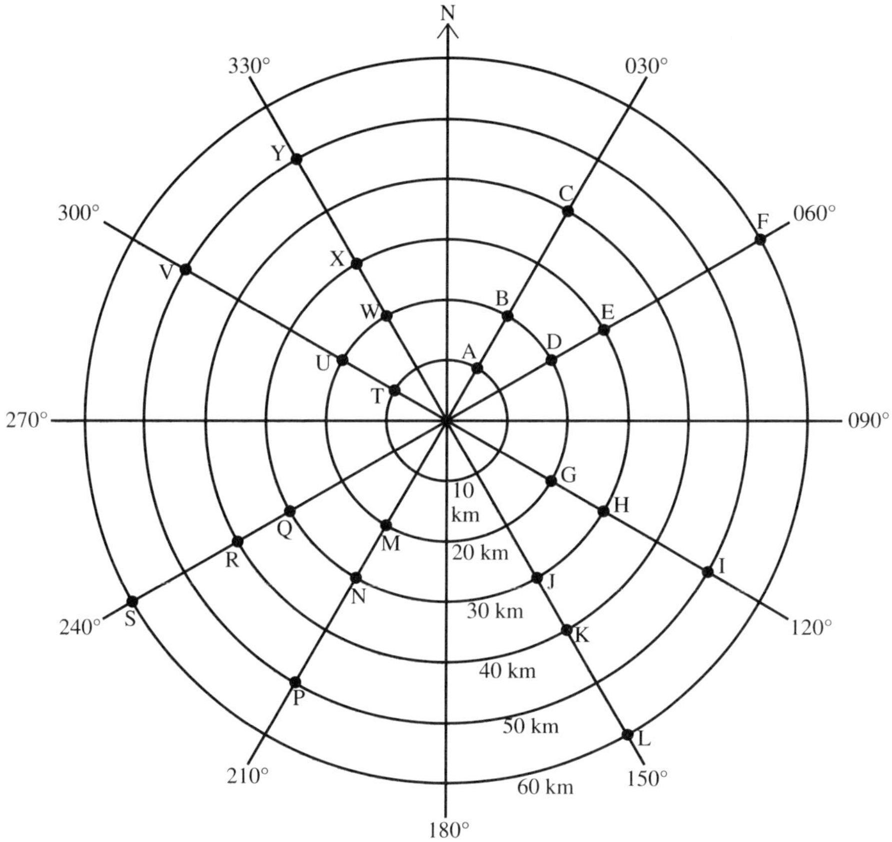

Example 3

Copy the diagram and mark on it each of these points.

A 20 km, 060°
B 10 km, 120°
C 30 km, 210°
D 20 km, 330°

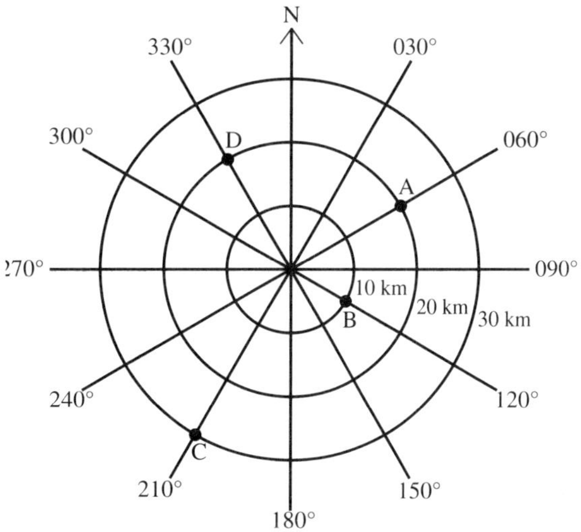

Exercise 17.13

1 Ask your teacher for a photocopy of this grid.
(Note to teacher. Permission is granted for you to photocopy this particular page.)

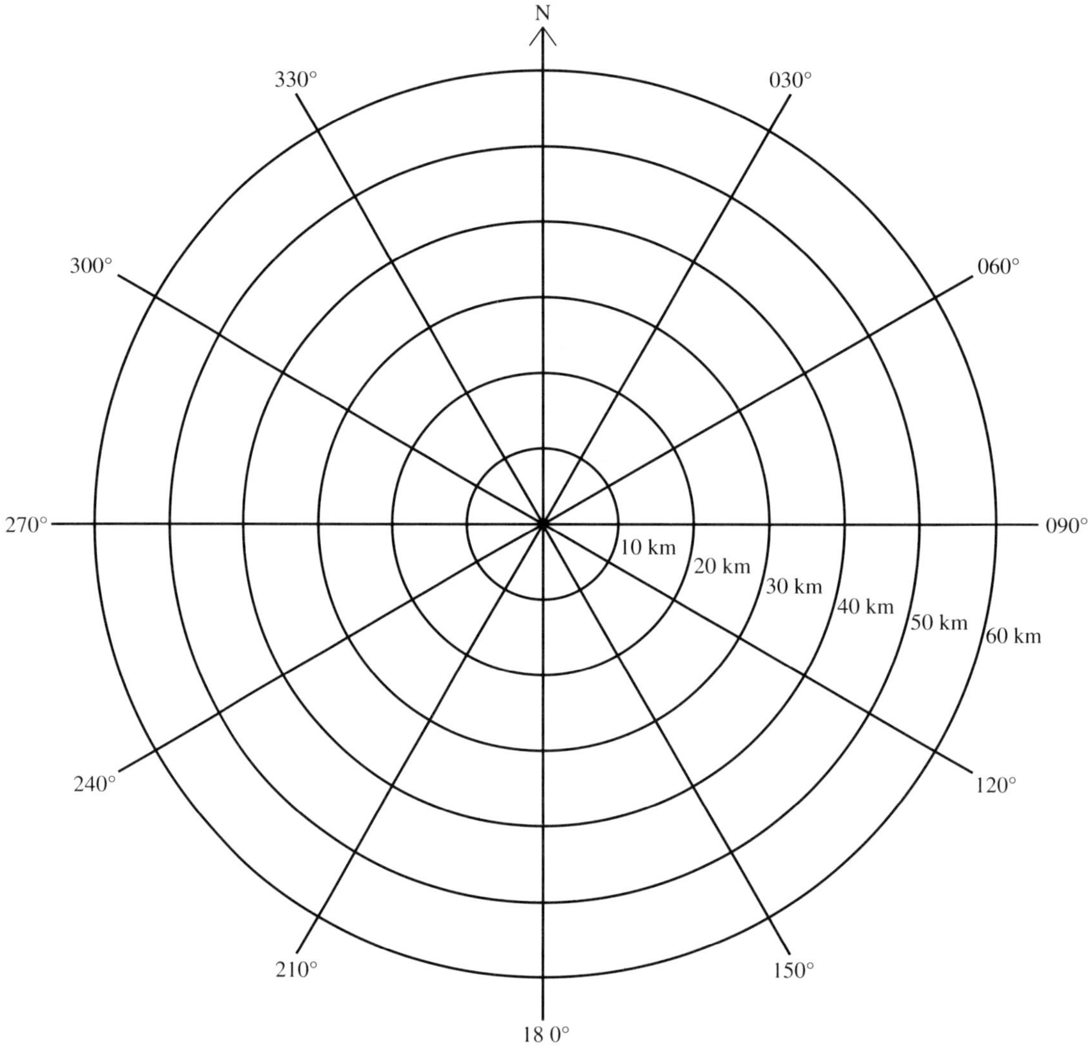

Mark on your grid the points A to Y with bearings

A 30 km, 030°	B 50 km, 030°	C 60 km, 030°
D 10 km, 060°	E 40 km, 060°	F 50 km, 060°
G 30 km, 120°	H 40 km, 120°	I 60 km, 120°
J 10 km, 150°	K 20 km, 150°	L 50 km, 150°
M 10 km, 210°	N 40 km, 210°	P 60 km, 210°
Q 10 km, 240°	R 20 km, 240°	S 50 km, 240°
T 30 km, 300°	U 40 km, 300°	V 60 km, 300°
W 10 km, 300°	X 20 km, 330°	Y 60 km, 330°

Example 4

The cartesian coordinates of four points are given below.
Mark each point on a grid and measure the bearing for each point.
Measure distance with a ruler and angle with a protractor. (Use a scale of 1 cm = 5 km)

A (12 km E, 21 km N) B (28 km E, −16 km N)
C (−16 km E, −28 km N) D (−35 km E, 20 km N)

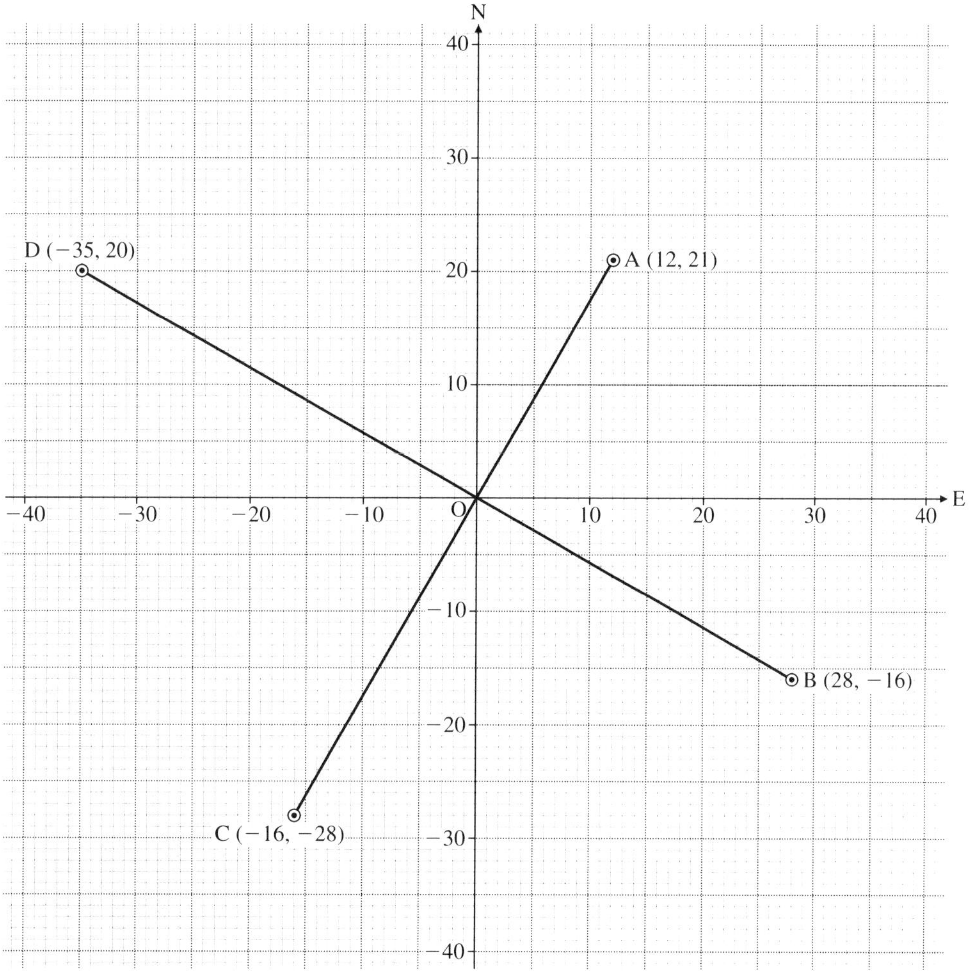

The distance OA is 4.8 cm or 24 km, and OA makes a clockwise angle of 30° with the north. Therefore the bearing of A is 24 km, 030°.

The distance OB is 6.4 cm or 32 km, and OB makes a clockwise angle of 120° with the north. Therefore the bearing of B is 32 km, 120°.

The distance OC is 6.4 cm or 32 km, and OC makes a clockwise angle of 210° with the north. Therefore the bearing of C is 32 km, 210°.

The distance OD is 8 cm or 40 km and OD makes a clockwise angle of 300° with the north. Therefore the bearing of D is 40 km, 300°.

Exercise 17.14

1 The cartesian coordinates of points A to Y are given.
Mark each point on a grid and measure the bearing for each point.
(Use a scale of 1 cm = 5 km)

A (20 km E, 35 km N) B (28 km E, 16 km N)
C (30 km E, 30 km N) D (24 km E, 42 km N)
E (20 km E, 20 km N) F (42 km E, 24 km N)
G (35 km E, −20 km N) H (24 km E, −42 km N)
I (25 km E, −25 km N) J (15 km E, −15 km N)
K (12 km E, −21 km N) L (21 km E, −12 km N)
M (−24 km E, −42 km N) N (−42 km E, −24 km N)
P (−20 km E, −20 km N) Q (−35 km E, −20 km N)
R (−40 km E, −40 km N) S (−20 km E, −35 km N)
T (−16 km E, 28 km N) U (−21 km E, 12 km N)
V (−24 km E, 42 km N) W (−35 km E, 35 km N)
X (−28 km E, 16 km N) Y (−15 km E, 15 km N)

2 The bearings of points A to Y are given.
Mark each point on a grid, using a ruler and protractor.
State the cartesian coordinates of each point.
(Use a scale of 1 cm = 5 km)

A 40 km, 060° B 40 km, 045° C 28 km, 030°
D 24 km, 060° E 42 km, 045° F 16 km, 030°
G 24 km, 120° H 32 km, 150° I 48 km, 120°
J 50 km, 135° K 30 km, 135° L 40 km, 150°
M 32 km, 210° N 24 km, 240° P 40 km, 225°
Q 21 km, 225° R 24 km, 210° S 32 km, 240°
T 40 km, 330° U 35 km, 315° V 40 km, 300°
W 56 km, 315° X 48 km, 300° Y 24 km, 330°

3 Copy the grid illustrated using a scale of 1 cm = 5 m.

a Find the cartesian coordinates for each of the following boats from the bearings given:

(i) 31m, 065° (ii) 62m, 025°
(iii) 51m, 045° (iv) 31m, 155°
(v) 50m, 160° (vi) 34m, 135°
(vii) 31m, 335° (viii) 51m, 315°

b Find the bearing for each of the following boats from the cartesian coordinates given:

(i) (13,28) (ii) (17,47)
(iii) (24,24) (iv) (28, −13)
(v) (26, −56) (vi) (−28, 13)
(vii) (−17, 47) (viii) (−24, 24)

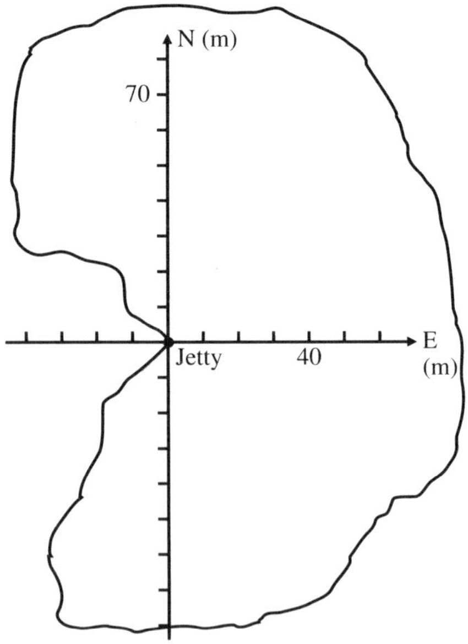

Exercise 17.15

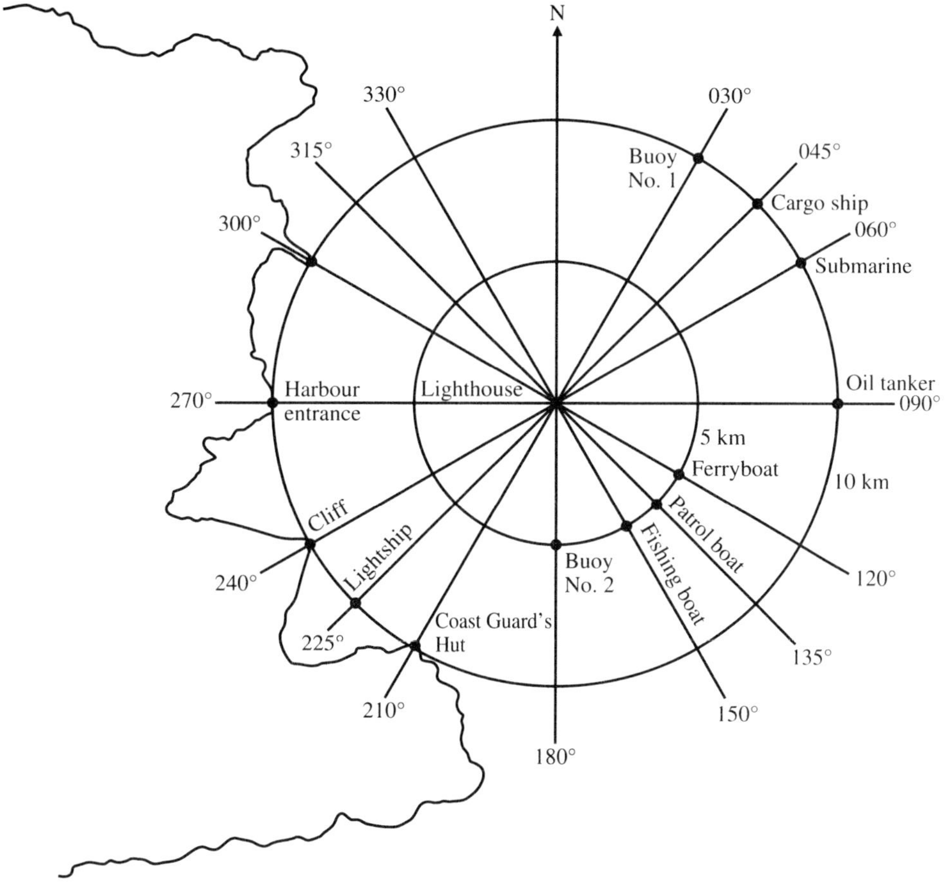

Look at the map above.
Give the bearing of each of the following from the lighthouse.

1 Buoy No. 1 **2** Cargo ship **3** Submarine
4 Oil tanker **5** Ferry boat **6** Patrol boat
7 Fishing boat **8** Buoy No. 2 **9** Coastguard's hut
10 Lightship **11** Cliff **12** Harbour entrance

18 Enlarging shapes

18.1 Enlarging a shape by a whole number scale factor

Example 1

A triangle ABC has its vertices at (1, 1), (1, 3) and (3, 1) respectively.

Enlarge this triangle by using a scale factor of 2 with the origin as the centre of enlargement.

Write down the coordinates of the vertices of the enlargement (A_1, B_1 and C_1)

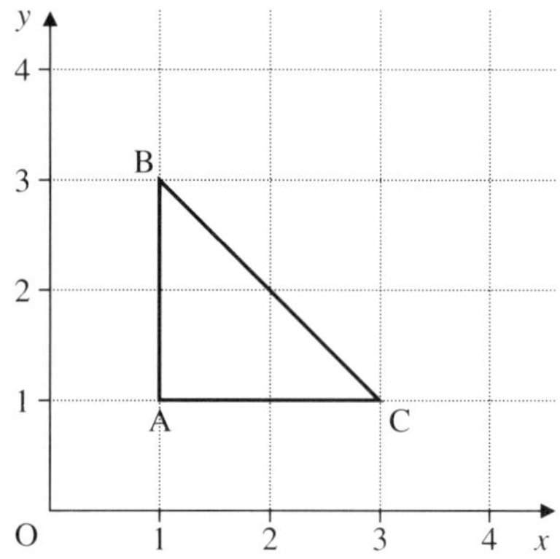

Join OA, OB and OC.
Extend the lines to A_1, B_1 and C_1 respectively such that $OA_1 = 2OA$, $OB_1 = 2OB$ and $OC_1 = 2OC$

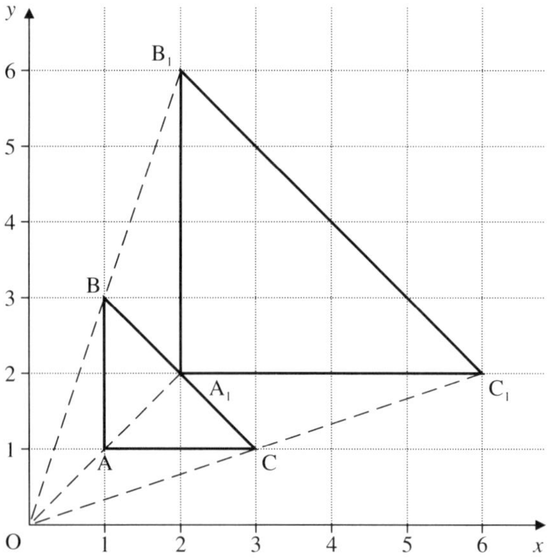

The coordinates of the vertices of the enlargement are:
A_1 (2, 2) B_1 (2, 6) C_1 (6, 2)
Note that for the above example the centre of enlargement is outside the figure.

Exercise 18.1

For each case enlarge the given triangle by using the origin as the centre of enlargement. For each question you will need a grid with both axes labelled from 0–10.

Write down the coordinates of the vertices of each enlargement.

	Coordinates of vertices	Scale factor		Coordinates of vertices	Scale factor
1	A(1, 1), B(1, 3), C(3, 1)	3	**4**	X(1, 2), Y(2, 2), Z(2, 1)	**a** 3
2	P(1, 1), Q(1, 2), R(2, 1)	**a** 2			**b** 5
		b 4	**5**	X(1, 2), Y(2, 2), Z(2, 1)	**a** 2
3	P(1, 1), Q(1, 2), R(2, 1)	**a** 3			**b** 4
		b 5	**6**	L(1, 3), M(3, 3), N(3, 1)	**a** 2
					b 3

In Example 1 the centre of enlargement was outside the figure, but the centre of enlargement can be on the figure.

Example 2

A triangle ABC has its vertices at (0, 0), (3, 2) and (4, 0) respectively.

Enlarge this triangle by using a scale factor of 3, with the origin as the centre of the enlargement.

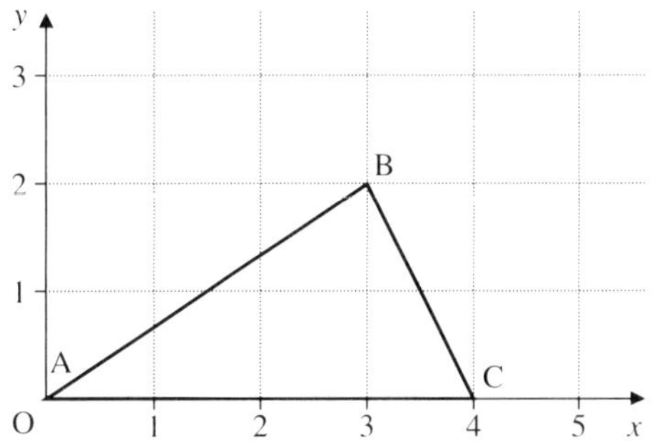

Write down the coordinates of the vertices of the enlargement, A_1, B_1 and C_1.

Extend AB and AC to B_1 and C_1, respectively such that $AB_1 = 3AB$ and $AC_1 = 3AC$.

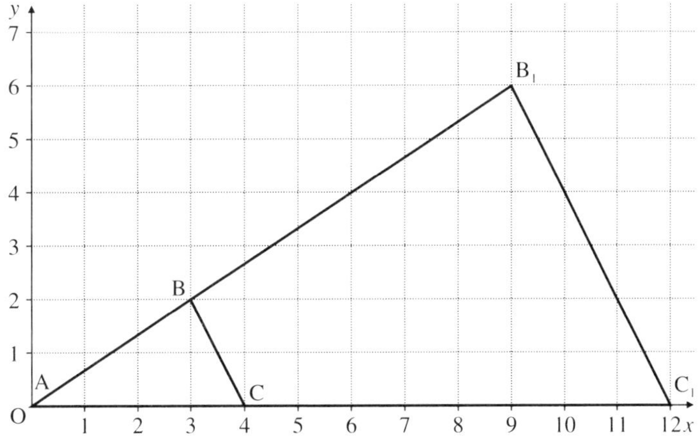

The coordinates of the vertices of the enlargement are:
A_1 (0, 0) (same as A)
B_1 (9, 6)
C_1 (12, 0)

Exercise 18.2

For each case enlarge the given triangle by using the origin as the centre of enlargement.
For each question you will need a grid with both axes labelled from 0–12.

Write down the coordinates of the vertices of each enlargement.

	Coordinates of vertices	Scale factor		Coordinates of vertices	Scale factor
1	A (0, 0), B (3, 2), C (4, 0)	2	**3**	L (0, 0), M (2, 2), N (3, 0)	**a** 2
2	P (0, 0), Q (1, 2), R (4, 0)	**a** 2			**b** 3
		b 3			**c** 4

For the enlargements dealt with so far, the centre of enlargement has either been outside the figure or on the edge of the figure.
In Example 3 the centre of enlargement is inside the figure.

Example 3

A square ABCD has its vertices at (5, 5), (5, 7), (7, 7) and (7, 5) respectively.

Enlarge this square by using a scale factor of 3 with the point (6, 6) as the centre of enlargement.

Write down the coordinates of the vertices of the enlargement, A_1, B_1, C_1 and D_1

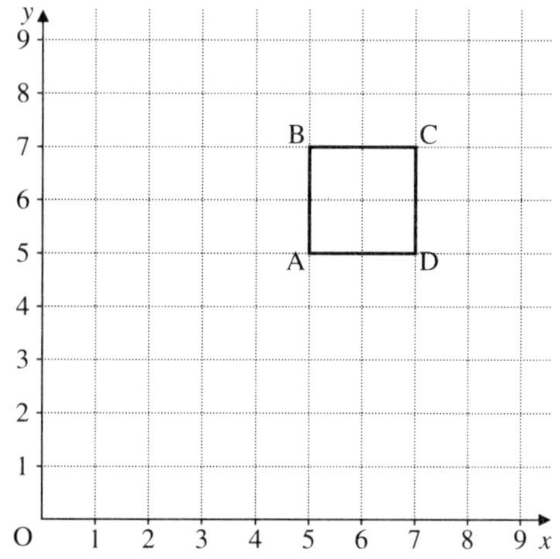

Join OA, OB, OC and OD. Extend the lines A, B, C and D respectively such that $AO_1 = 3OA$, $OB_1 = 3OB$, $OC_1 = 3OC$ and $OD_1 = 3OD$.

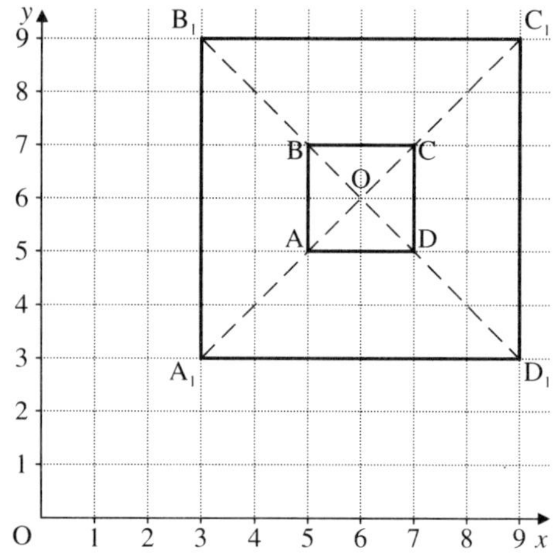

The coordinates of the vertices of the enlargement are:
A_1 (3, 3), B_1 (3, 9), C_1 (9, 9), D_1 (9, 3)

Exercise 18.3

For each case enlarge the given four-sided figure.
For each question you will need a grid with both axes labelled from 0 to 12.

Write down the coordinates of the vertices of each enlargement.

	Coordinates of vertices	Scale factor	Centre of enlargement
1	A (5, 5), B (5, 7), C (7, 7), D (7, 5)	**a** 2	(6, 6)
		b 4	(6, 6)
2	P (4, 5), Q (4, 7), R (8, 7), S (8, 5)	2	(6, 6)
3	A (5, 4), X (5, 8), Y (7, 8), Z (7, 4)	2	(6, 6)

Example 4

Enlarge the triangle OAB so that it fits exactly into the rectangle.

State the scale factor of the enlargement.

Extend OA to A_1 and OB to B_1, then join A, B.

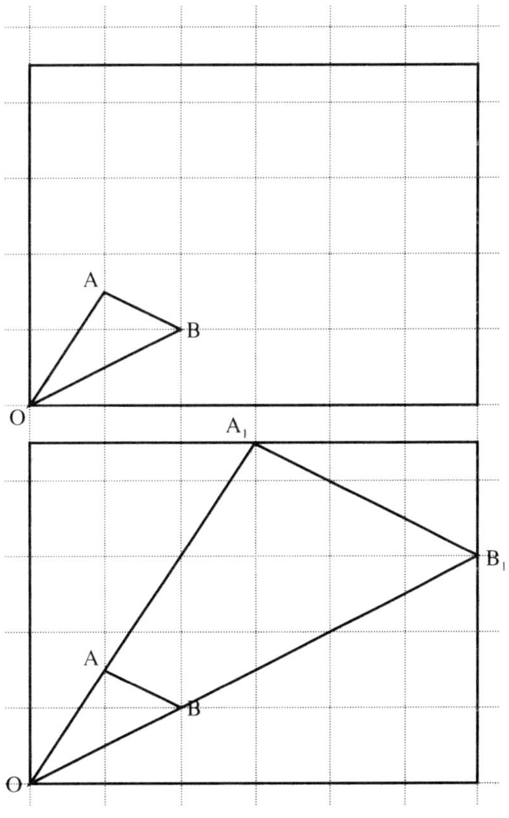

The scale factor of the enlargement is 3 because
$AO_1 = 3OA$,
$OB_1 = 3OB$

Exercise 18.4

Copy the triangles and rectangles on to squared paper. Enlarge each triangle so that it fits exactly into the rectangle.
State the scale factor of the enlargement.

1

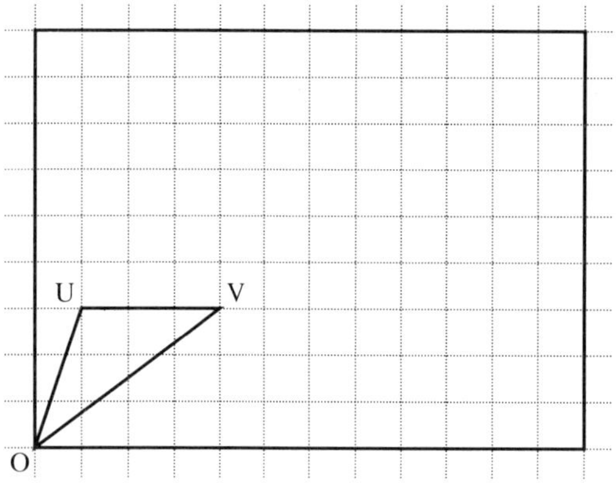

2

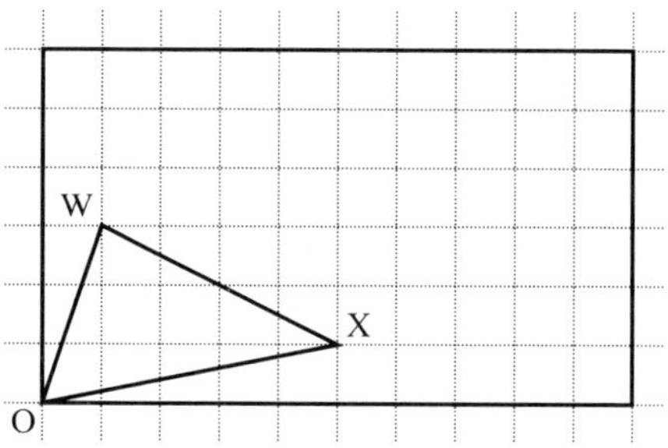

3

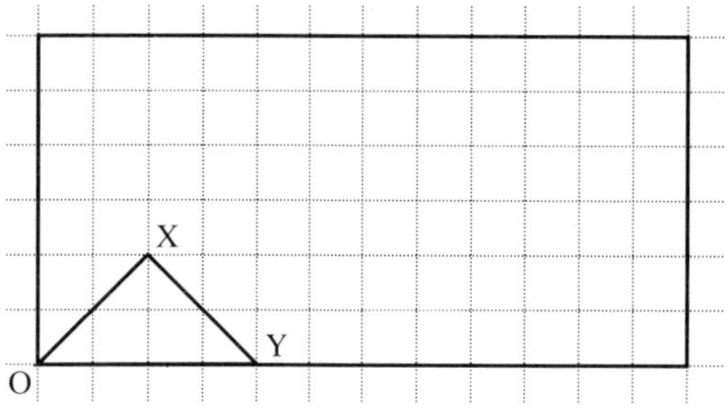

4

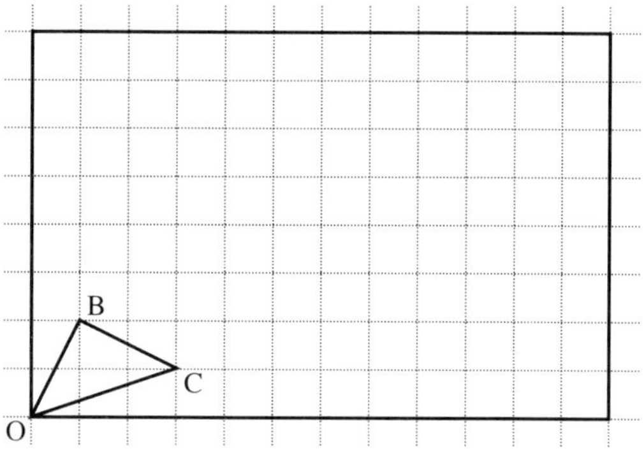

5

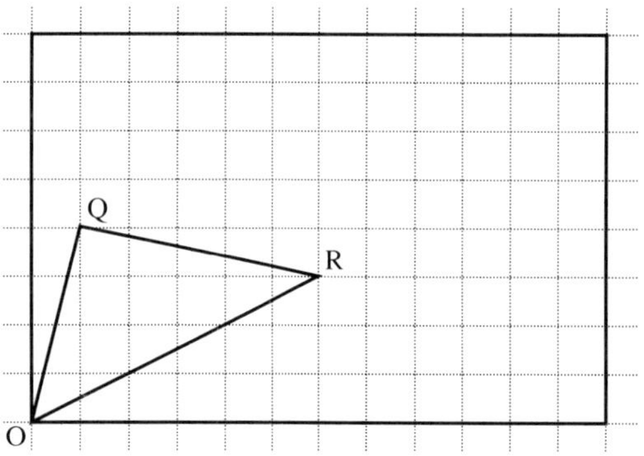

6

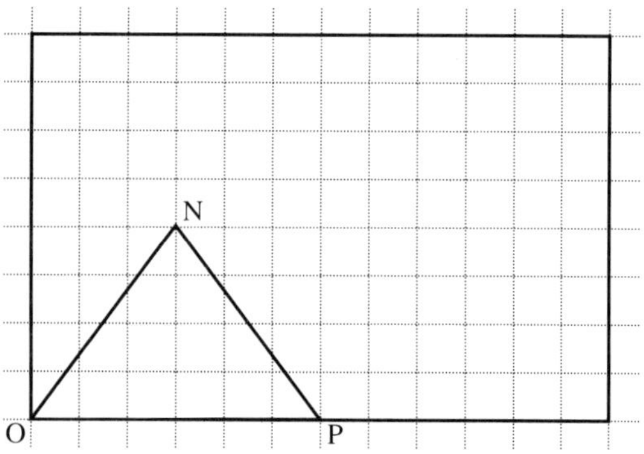

SECTION D: HANDLING DATA

19 Collecting, representing and interpreting data

19.1 Collecting, grouping and ordering discrete data using tallying methods and creating a frequency table for grouped data

Example 1

Fifteen children were asked their favourite milk drink from:
tea, coffee, chocolate, Horlicks, hot milk, Bournvita

Their answers were:

Anne: tea	George: tea	Azil: hot milk
Adam: Horlicks	Miranda: coffee	Melissa: Bournvita
David: coffee	Razi: chocolate	Paul: coffee
Emma: coffee	Salik: hot milk	Satwinder: chocolate
Fran: hot milk	Zorba: coffee	Yusuf: coffee

a Draw a tally chart to show this information.
b Show this information on a barchart.

a

Drink	Tally	Total
Bournvita	|	1
chocolate	||	2
coffee	~~||||~~ |	6
Horlicks	|	1
hot milk	|||	3
tea	||	2
	Total answers	15

Note that ~~||||~~ means 5

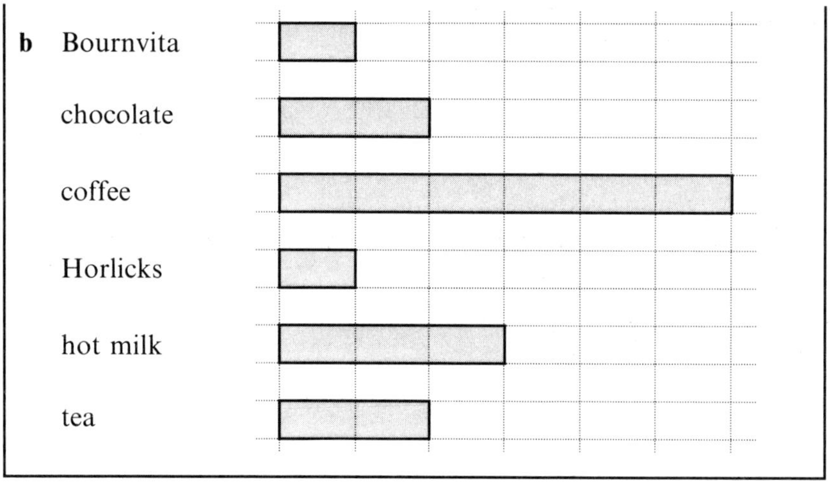

Exercise 19.1

For each question show the information **a** on a tally chart and **b** on a bar chart.

1 Twenty children in London were asked which football team they supported.
Their answers were:

George: Arsenal
Peter: West Ham
William: Spurs
David: Arsenal
Joanne: Chelsea
Robert: Arsenal
Andrew: Spurs
Richard: Chelsea
Kelly: Queen's Park Rangers
Sophie: Arsenal
Luke: Spurs
Steven: Chelsea
Melanie: Spurs
Anna: Queen's Park Rangers
Sunil: Arsenal
Sanjay: Chelsea
Rahul: West Ham
Manjula: Spurs
Daniel: Spurs
Thomas: West Ham

2 All twenty children in Class 2A bought a can of pop during one morning break.
The flavours they bought were:

Saraya: apple
Ranjit: lime
Abbie: lemon
Jennifer: raspberry
Candace: orange
Shani: lemon
Polly: lime
Tanya: cola
Natalie: lemon
Rebecca: raspberry
Asiti: orange
Dean: lime
David: orange
George: apple
Joshua: lemon
Tyrone: raspberry
Peter: orange
Patrick: lime
Andrew: cola
James: orange

3 All fifteen children in Class 2B bought a packet of crisps during one morning break. The flavours they bought were:

Kate: salt and vinegar
Chantelle: cheese and onion
Amy: plain
Laura: Worcester sauce
Aisha: salt and vinegar
Suzanne: cheese and onion
Lucy: plain
Herjinder: salt and vinegar
Michael: plain
George: cheese and onion
Carl: prawn cocktail
David: Worcester sauce
William: salt and vinegar
Pritesh: cheese and onion
Robert: salt and vinegar

4 All twenty-five girls in Year 11 were asked to choose their summer sports option. Their choices were as follows:

Nicola: swimming
Jaswinder: tennis
Jodie: golf
Nina: golf
Laura: swimming
Shani: athletics
Rani: athletics
Janis: swimming
Kayleigh: swimming
Marie: tennis
Drashma: athletics
Katie: athletics
Frances: tennis
Natalie: golf
Sian: athletics
Becky: tennis
Samantha: tennis
Amanda: swimming
Mandeep: tennis
Annie: rounders
Vicky: rounders
Deborah: rounders
Kirsty: athletics
Joanne: rounders
Rosie: tennis

Example 2

The fifteen children in Example 1, who were all born in 1980 were asked their month and date of birth. Their answers were:

Anne: 12 July
Miranda: 15 October
Paul: 12 December
Fran: 14 July
George: 12 January
Melissa: 4 July
Emma: 9 June
Zorba: 12 March
Azil: 15 November
David: 7 January
Salik: 9 April
Yusuf: 14 January
Adam: 15 May
Razi: 17 December
Satwinder: 1 February

a Write their names in order of age, starting with the eldest.
b Draw a tally chart to show how many were born in each month.
c Draw a tally chart to show how many were born on a particular date in the month.

a David, George, Yusuf, Satwinder, Zorba, Salik, Adam, Emma, Melissa, Anne, Fran, Miranda, Azil, Paul, Razi.

b

Month	Tally	Totals
January	|||	3
February	|	1
March	|	1
April	|	1
May	|	1
June	|	1
July	|||	3
August		0
September		0
October	|	1
November	|	1
December	||	2
	Total tally:	15

c

Date of month	Tally	Totals
1	|	1
4	|	1
7	|	1
9	||	2
12	||||	4
14	||	2
15	|||	3
17	|	1
	Total tally:	15

Exercise 19.2

1 The children in Class 7A were asked the month and day of their birth.
Their answers were:

Pupil	Month	Day	Pupil	Month	Day
George	June	Monday	James	October	Saturday
Peter	January	Wednesday	Jotinder	November	Wednesday
David	September	Friday	Richard	April	Monday
Rahul	March	Tuesday	Daniel	December	Sunday
Paul	July	Sunday	William	February	Monday

Pupil	Month	Day	Pupil	Month	Day
Jandeep	January	Wednesday	Laura	September	Sunday
Rita	June	Thursday	Chenise	May	Saturday
Daisy	August	Sunday	Saadiya	August	Thursday
Sally	December	Wednesday	Keeley	April	Tuesday
Nicola	March	Monday	Joanna	November	Friday

a Draw a tally chart to show how many children had birthdays in each month.
b Draw a tally chart to show how many children had birthdays on each day of the week.

2 On an island there are five counties and fifteen market towns.
The table shows the county that each market town is in and the market day for each town.

Town	County	Market day
Batchton	South	Tuesday
Colestown	North	Monday
Dunford	East	Thursday
Entonville	Midland	Wednesday
Garton	East	Thursday
Luntonbridge	Midland	Friday
Milltown	West	Saturday
Broomford	East	Friday
Restonville	Midland	Wednesday
St John's	South	Saturday
Swaytown	Midland	Thursday
Wentley	North	Monday
Wallingbridge	South	Saturday
Waltonville	West	Tuesday
Oldcastle	East	Wednesday

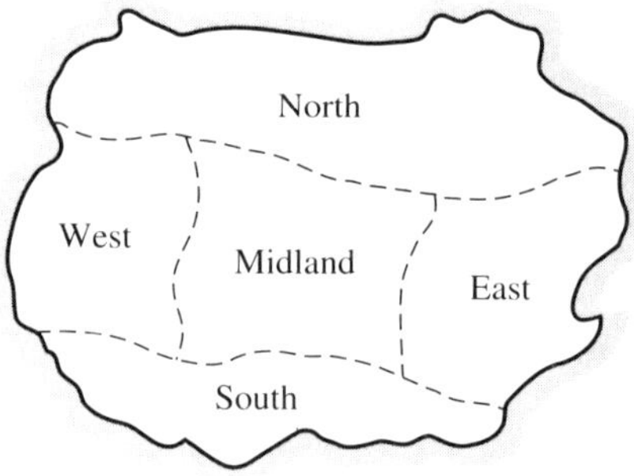

a Draw a tally chart to show how many market towns there are in each of the counties.
b Draw a tally chart to show how many towns have their market day on each of the six weekdays.

Example 3

The heights of the children in Example 1 are:

Anne: 127 cm	George: 111 cm	Azil: 119 cm
Adam: 136 cm	Miranda: 125 cm	Melissa: 120 cm
David: 158 cm	Razi: 116 cm	Paul: 145 cm
Emma: 147 cm	Salik: 148 cm	Satwinder: 128 cm
Fran: 142 cm	Zorba: 151 cm	Yusuf: 149 cm

a Write their names in order of height, starting with the shortest.

b Draw a tally chart to show how many are in each of these height groups:
110–119 cm 120–129 cm 130–139 cm
140–149 cm 150–159 cm

c Show this information on a frequency chart.

a George, Razi, Azil, Melissa, Miranda, Anne, Satwinder, Adam, Fran, Paul, Emma, Salik, Yusuf, Zorba, David.

b

Height	Tally	Frequency
110–119 cm	|||	3
120–129 cm	||||	4
130–139 cm	|	1
140–149 cm	~~||||~~	5
150–159 cm	||	2
	Total:	15

c

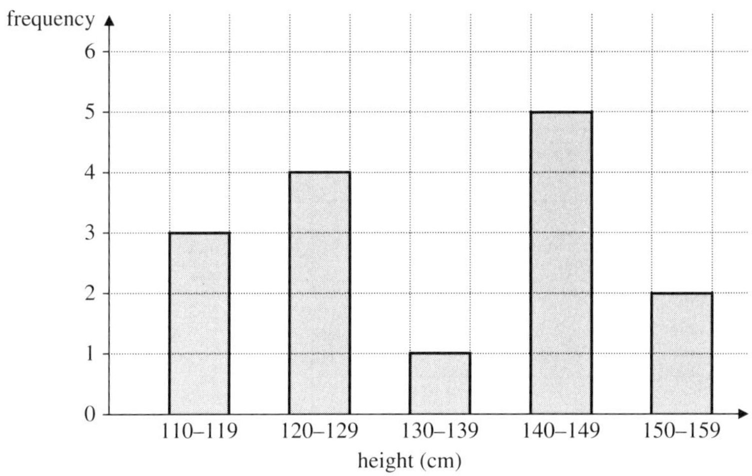

Exercise 19.3

1 The table shows how many caravans were parked on a holiday site over a four-week period.

Day	Number	Day	Number	Day	Number	Day	Number
Monday	11	Monday	17	Monday	13	Monday	15
Tuesday	5	Tuesday	7	Tuesday	6	Tuesday	8
Wednesday	9	Wednesday	12	Wednesday	14	Wednesday	28
Thursday	18	Thursday	24	Thursday	16	Thursday	29
Friday	38	Friday	35	Friday	22	Friday	33
Saturday	32	Saturday	27	Saturday	37	Saturday	23
Sunday	26	Sunday	28	Sunday	21	Sunday	35

a Copy and complete this tally chart.
b Show this information on a frequency chart.

Number of caravans	Tally	Number of days (frequency)
0–9		
10–19		
20–29		
30–39		

2 The table shows the best performance of each of the members of Class 5A at high jump.

Name	Height cleared	Name	Height cleared	Name	Height cleared
James	105 cm	Robert	123 cm	Joanna	126 cm
Peter	156 cm	Drew	163 cm	Alix	108 cm
Linford	132 cm	David	119 cm	Lucy	122 cm
Nishil	112 cm	Jolene	137 cm	Michelle	148 cm
Michael	134 cm	Melanie	142 cm	Vepula	125 cm
Richard	116 cm	Gaynor	138 cm	Yogita	145 cm
Mark	135 cm	Amy	152 cm		

a Copy and complete this tally chart.
b Show this information on a frequency chart.

Height cleared	Tally	Number of pupils (frequency)
100–109 cm		
110–119 cm		
120–129 cm		
130–139 cm		
140–149 cm		
150–159 cm		
160–169 cm		

19.2 Constructing and interpreting bar-line and line graphs and frequency diagrams with suitable class intervals for discrete variables

Example 1

A patient in hospital had his temperature taken at hourly intervals from 6 a.m. to 12 noon. The readings were:

Time	6 a.m.	7 a.m.	8 a.m.	9 a.m.	10 a.m.	11 a.m.	12 noon
Temperature	35.7 °C	35.3 °C	35.4 °C	36.0 °C	36.4 °C	37.0 °C	36.4 °C

a Draw a bar-line graph for this information.
b Draw a line graph for this information.
c At what time was the temperature highest?
d At what time was the temperature lowest?
e Estimate the patient's temperature at:
(i) 6.30 a.m. (ii) 9.30 a.m. (iii) 10.30 a.m.
f What is the median of the seven temperatures in the table?

a

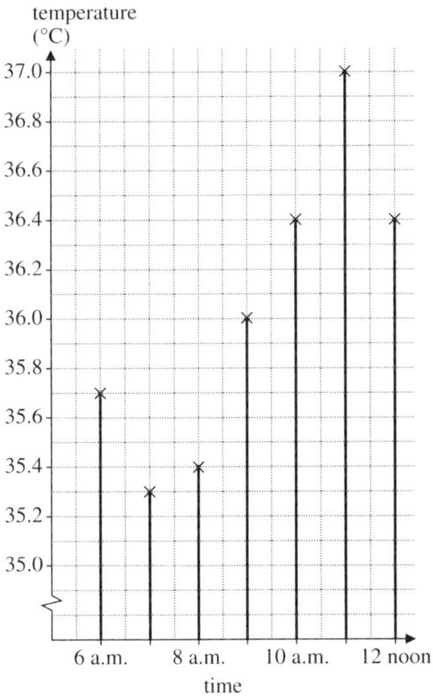

b

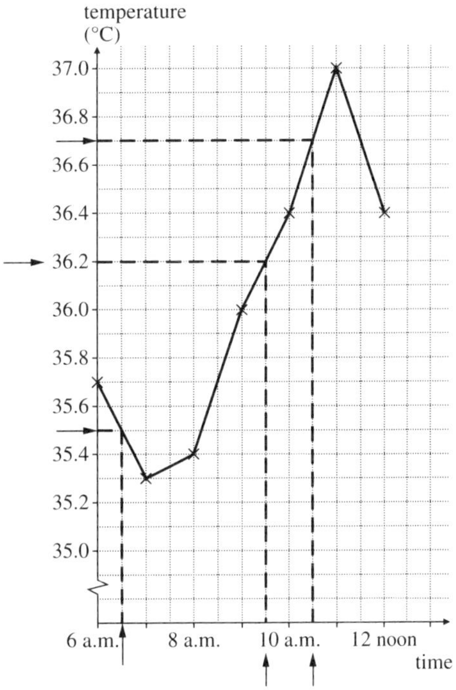

c 11.00 a.m. (37.0 °C) **d** 7 a.m. (35.3 °C)
e (i) 35.5 °C (ii) 36.2 °C (iii) 36.7°C (see graph)
The temperatures in ascending order are:
f 35.3 °C, 35.4 °C, 35.7 °C, 36.0 °C, 36.4 °C, 36.4 °C, 37.0 °C
The median temperature is therefore 36.0 °C

Exercise 19.4

1 a John was born in 1984. The bar-line graph shows his height at two-year intervals. Copy the table and use the bar-line graph to complete it.

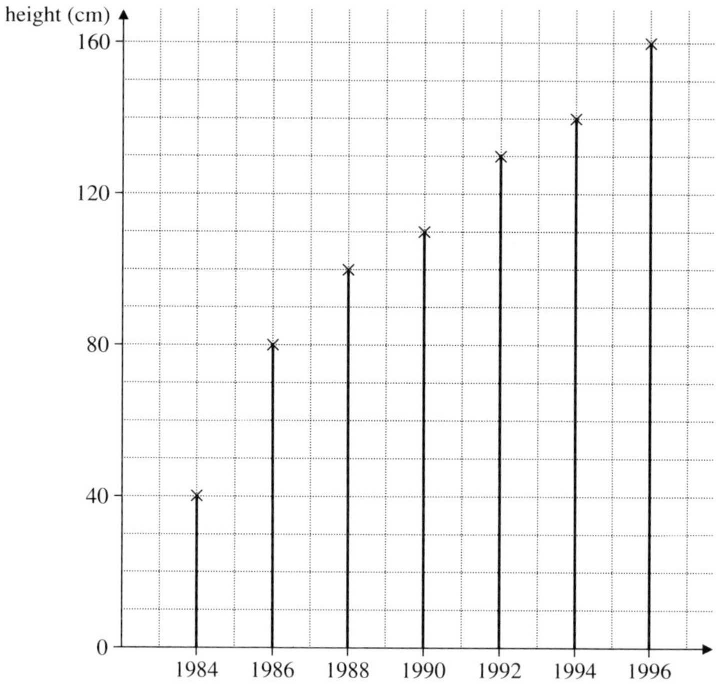

Year	Height (cm)
1984	
1986	
1988	
1990	
1992	
1994	
1996	

b Use your table to draw a line graph on a copy of the grid illustrated. (Your teacher has permission to photocopy it.)

c From your line graph find John's height in:
(i) 1985 (ii) 1987
(iii) 1991 (iv) 1995

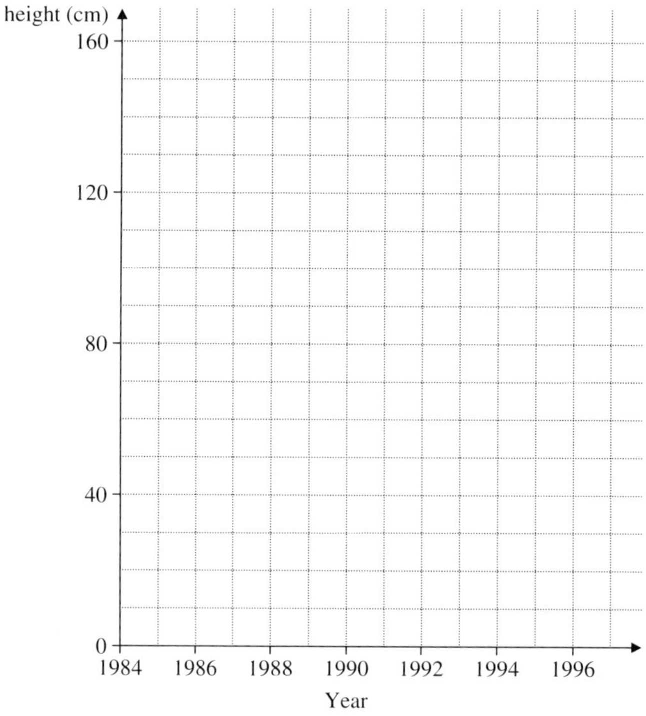

2 Some men are digging a trench. The lengths they have dug out at certain times are shown on the bar-line graph.

a Copy the table and use the bar-line graph to complete it.

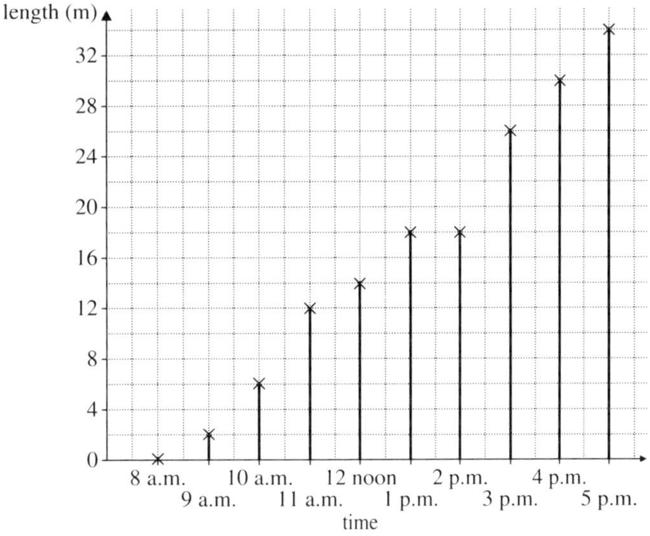

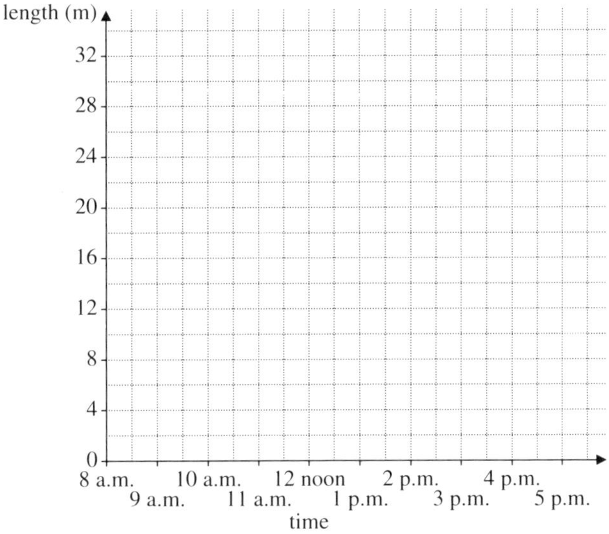

Time	8 a.m.	9 a.m.	10 a.m.	11 a.m.	12 noon	1 p.m.	2 p.m.	3 p.m.	4 p.m.	5 p.m.
Length (m)										

b Use your table to draw a line graph on a copy of the grid illustrated.

c Find from your graph the length of the trench at:
(i) 9.30 a.m. (ii) 12.30 p.m. (iii) 2.30 p.m.
(iv) 3.30 p.m. (v) 4.30 p.m.

d Between which times did the men take a break? Explain your answer.

3 a An immersion heater is turned on at 8 a.m. and turned off at 3 p.m. The bar-line graph shows the temperature of the water at hourly intervals.

Copy the table and use the bar-line graph to complete it.

Time	8 a.m.	9 a.m.	10 a.m.	11 a.m.	12 noon	1 p.m.	2 p.m.	3 p.m.	4 p.m.	5 p.m.
Temperature °C										

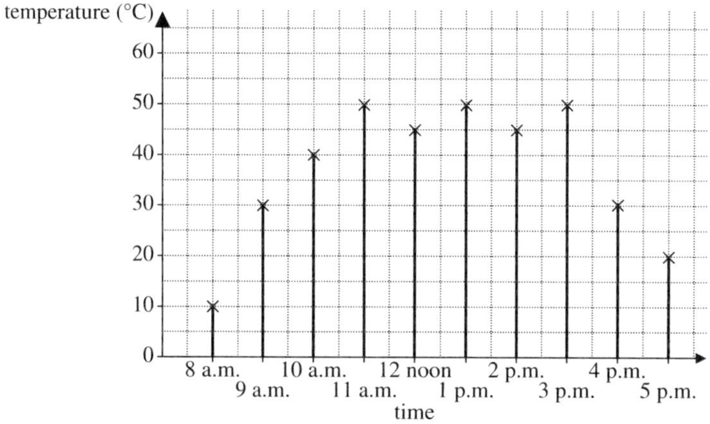

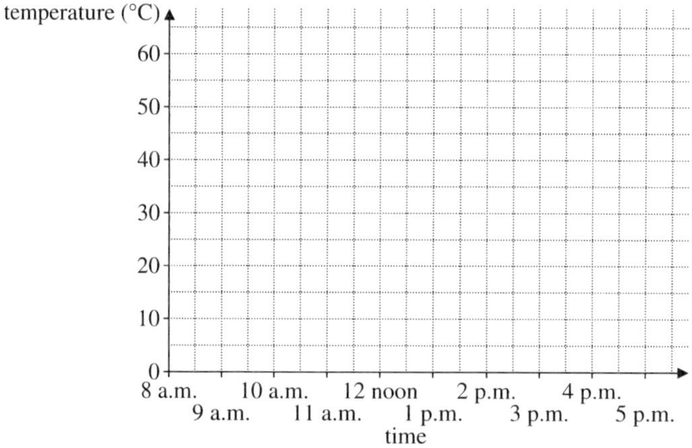

b Use your table to draw a line graph on a copy of the grid illustrated. (Your teacher has permission to photocopy this.)

c Find from your graph the temperature of the water at:
(i) 8.30 a.m. (ii) 9.30 a.m. (iii) 10.30 a.m.
(iv) 3.30 p.m. (v) 4.30 p.m.

d Explain why the temperature oscillates between 11 a.m. and 3 p.m.

4 On a very wet day rain from a roof gutter ran into a barrel. The depth of the water in the barrel at hourly intervals is given in the table.

Time	8 a.m.	9 a.m.	10 a.m.	11 a.m.	12 a.m.	1 p.m.	2 p.m.	3 p.m.	4 p.m.	5 p.m.
Depth (cm)	0	2	6	12	14	18	18	18	26	30

a Draw a bar-line graph of this information on a copy of the grid illustrated.

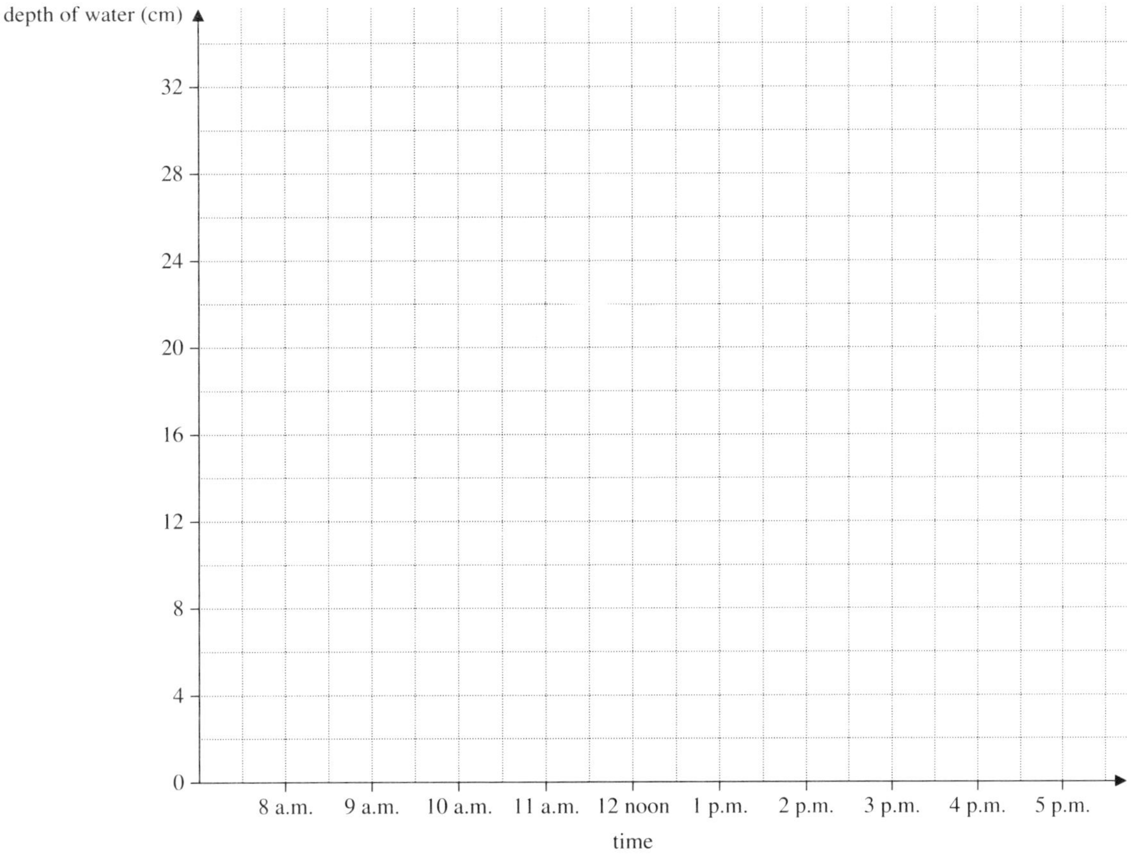

b Draw a line graph of the information on another copy of the grid illustrated. (Your teacher has permission to photocopy this.)

c Find from your line graph the depth of the water:
(i) 9.30 a.m. (ii) 12.30 p.m. (iii) 3.30 p.m. (iv) 4.30 p.m.

d Between which times did it not rain?

5 Pritesh records the speed of his father' car at ten-second intervals.

Time (s)	0	10	20	30	40	50	60	70	80	90
Speed ($km\,h^{-1}$)	0	20	60	80	120	120	120	80	60	0

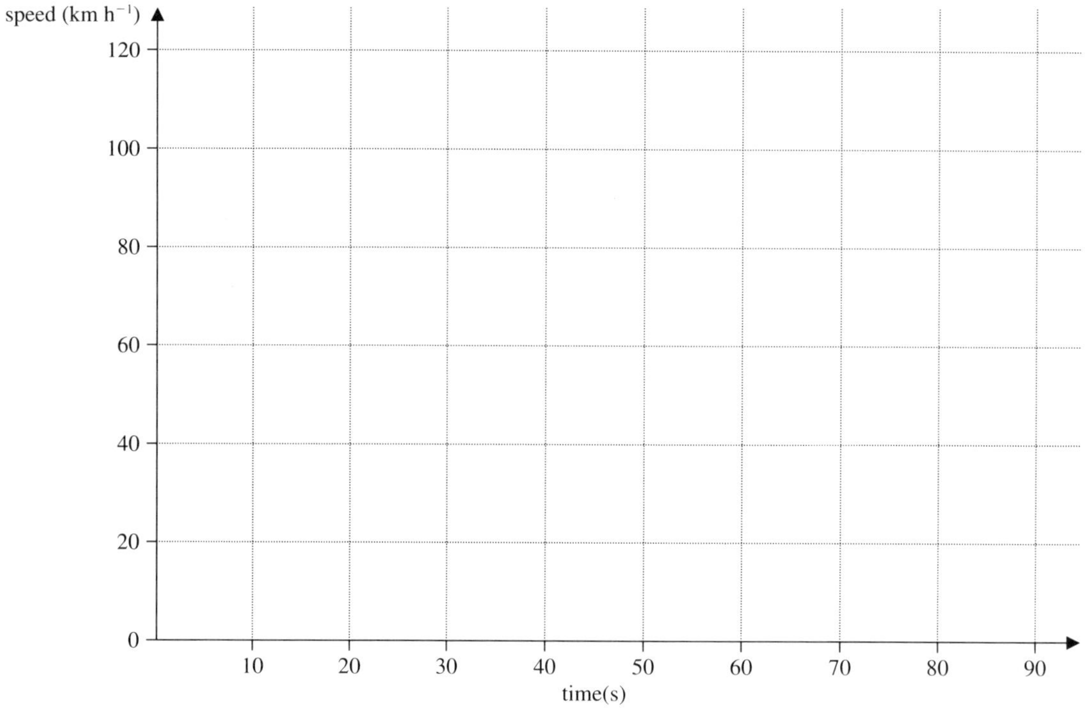

a Draw a bar-line graph of this information on a copy of the grid illustrated.

b Draw a line graph of the information on a second copy of the grid illustrated.

c Find the speed of the car after:
(i) 5 seconds
(ii) 15 seconds
(iii) 25 seconds
(iv) 35 seconds
(v) 65 seconds
(vi) 75 seconds
(vii) 85 seconds

d What is the highest speed reached by the car?

e For how long does the car travel at its highest speed?

6 The table shows the height of a viaduct above ground level at various distances from one end.

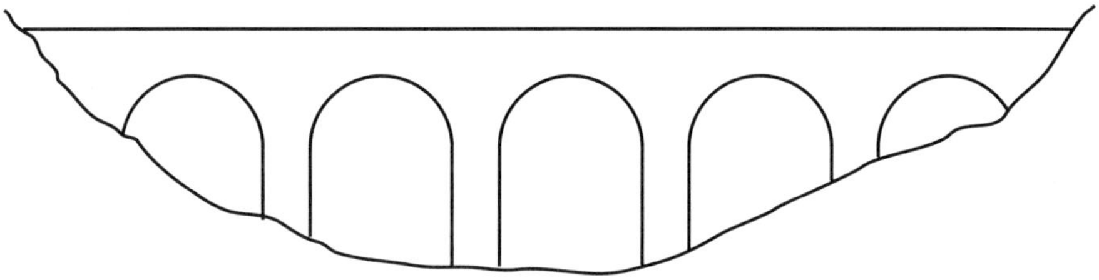

Distance from end (m)	0	20	40	60	80	100	120	140	160
Height above ground (m)	0	20	40	50	60	50	50	10	0

a Draw a line graph of this information on a copy of the grid illustrated.

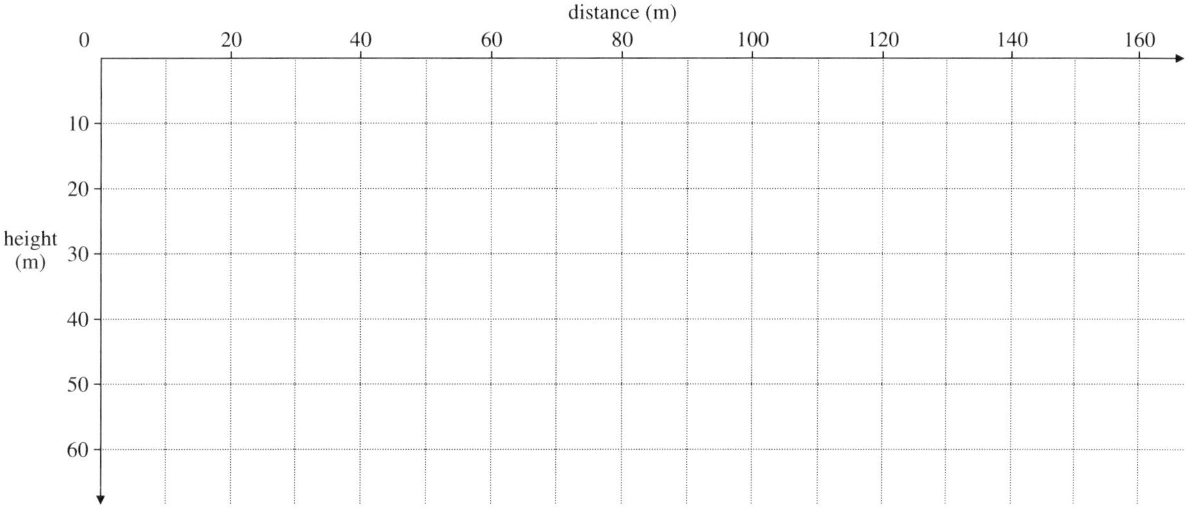

b Find the height of the viaduct at each of the following distances from the end: (i) 10 m (ii) 50 m (iii) 125 m (iv) 135 m

c Jodie walks across the viaduct. She is frightened of heights and must not look down from a height greater than 30 m.
How far is she from the end when she can:
(i) no longer look down, (ii) start looking down again?

Example 2

The tally chart shows the number of cars passing my window between each of the given times:

9–10 a.m.	200	10–11 a.m.	100	11–12 noon	50	12–1 p.m.	100
1–2 p.m.	50	2–3 p.m.	150	3–4 p.m.	250		

(i) Draw a frequency diagram to show this information.
(ii) During which time interval does the greatest number of cars pass?
(iii) Suggest reasons why a large number pass between 9 and 10 a.m. and between 3 and 4 p.m.
(iv) Suggest reasons why the number of cars passing increases between 12 noon and 1 p.m.

(i)

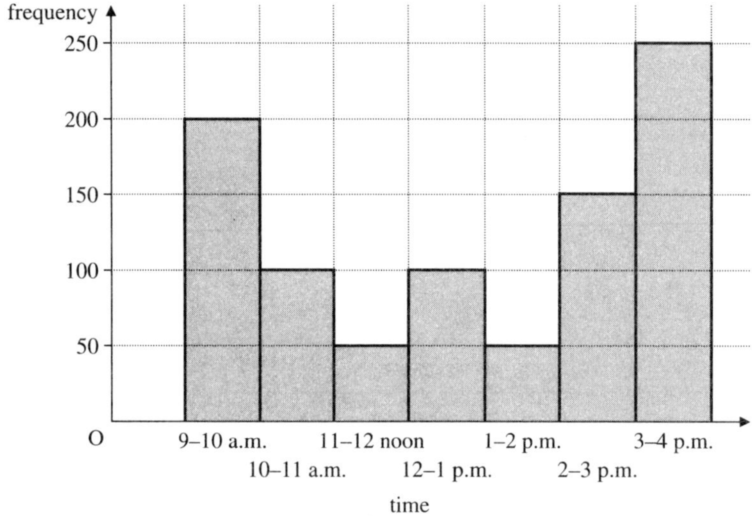

(ii) 3–4 p.m. (250 cars)
(iii) People going to or coming from work and people taking children to and from school.
(iv) People going out or home for lunch.

Exercise 19.5

1 Nathan records the number of Intercity trains that depart from a busy station during hourly intervals:

8–9 a.m.	12	1–2 p.m.	7
9–10 a.m.	8	2–3 p.m.	6
10–11 a.m.	6	3–4 p.m.	5
11–12 noon	5	4–5 p.m.	9
12–1 p.m.	4	5–6 p.m.	11

a Draw a frequency diagram to show this information.
b During which interval did the largest number of trains depart?
c Suggest reasons why a large number of trains departed during the intervals 8–9 a.m. and 5–6 p.m.

2 A cafe is open from 8 a.m. to 6 p.m. The numbers of customers served during each of the hourly intervals on a certain day were recorded.

8–9 a.m.	30	1–2 p.m.	50
9–10 a.m.	15	2–3 p.m.	25
10–11 a.m.	10	3–4 p.m.	35
11–12 noon	20	4–5 p.m.	40
12–1 p.m.	40	5–6 p.m.	45

a Draw a frequency diagram to show this information.
b During which interval did the cafe serve the largest number of customers?
c Suggest reasons why a large number of customers were served in the intervals 8–9 a.m., 12 noon–2 p.m. and 4–6 p.m.

3 An ice-cream seller works from 1 p.m. to 8 p.m. He records how many ice-creams he sells during each of the hourly intervals.

1–2 p.m.	80	5–6 p.m.	50
2–3 p.m.	100	6–7 p.m.	60
3–4 p.m.	20	7–8 p.m.	40
4–5 p.m.	30		

a Draw a frequency diagram to show this information.
b During which hourly interval did he sell the largest number of ice-creams?
c Suggest why his sales dropped after 3 p.m.

4 The table shows the readings on the electricity meter in Sadie's house at hourly intervals on a certain day.

Interval	Initial reading	Units consumed
8–9 a.m.	325.0	1.5
9–10 a.m.	326.5	0.5
10–11 a.m.	327.0	0.4
11–12 noon	327.4	
12–1 p.m.	327.8	
1–2 p.m.	328.4	
2–3 p.m.	329.0	
3–4 p.m.	329.3	
4–5 p.m.	330.1	
5–6 p.m.	331.1	
6–7 p.m.	333.1	
7–8 p.m.	335.4	
8–9 p.m.	337.6	
9–10 p.m.	339.4	
10–11 p.m.	341.2	

a Copy and complete the table.
b Draw a frequency diagram showing the units consumed.
c During which interval were most units consumed?
Can you suggest why?

19.3 Designing and using an observation sheet to collect data; collating and analysing results

Example 1

Traffic surveys are often carried out at junctions and roundabouts to collect data so that the design can be improved to minimise delays. Car manufacturers also use information on make of car, body shape (saloon, hatchback or estate), colour, two- or four-door model, and number of occupants.

Melissa carried out a traffic survey. Here is her observation sheet.

Make of car	Body shape (S, H or E)	Colour	Doors	Number of occupants
1 Rover	S	red	4	1
2 Cavalier	H	red	4	1
3 Ford	E	blue	4	5
4 Volvo	E	blue	4	1
5 Rover	H	green	4	1
6 Rover	H	blue	4	1
7 Cavalier	S	red	4	4
8 Rover	S	green	2	1
9 Cavalier	S	white	2	1
10 Ford	H	white	2	2

(Note that she made 10 entries in her table.)

a Use the details from Melissa's survey to complete each of the following frequency tables.

b Only three cars had more than one occupant.
Suggest a reason for each of the three cars.

Make	Cavalier	Ford	Rover	Volvo
Frequency				

Body shape	saloon	hatchback	estate
Frequency			

Colour	blue	green	red	white
Frequency				

Number of doors	2-door	4-door
Frequency		

Number of occupants	1	2	3	4	5
Frequency					

a

Make	Cavalier	Ford	Rover	Volvo
Frequency	3	2	4	1

Body shape	saloon	hatchback	estate
Frequency	4	4	2

Colour	blue	green	red	white
Frequency	3	2	3	2

Number of doors	2-door	4-door
Frequency	3	7

Number of occupants	1	2	3	4	5
Frequency	7	1	0	1	1

b Car 3 has five occupants. It is probably a family.
Car 7 has four occupants. It could be a family, or four business people or two couples.
Car 10 has two occupants. It could be two friends or two work colleagues.

Exercise 19.6

1 Eighteen children on a school trip went into a cafe for lunch. Their teacher made out an observation sheet about the food and drink they chose.

Pupil	First course	Second course	Drink
Peter	fish and chips	ice cream	tea
Joel	pie and chips	sponge pudding	lemonade
David	pie and chips	ice cream	orange squash
James	baked potato	fruit salad	coffee
Jasbinder	pie and chips	ice cream	orange squash
Martin	baked potato	fruit salad	tea
Saleem	baked potato	sponge pudding	orange squash
Andrew	baked potato	ice cream	lemonade
George	fish and chips	sponge pudding	orange squash
Becki	pie and chips	sponge pudding	orange squash
Lucy	pie and chips	fruit salad	tea
Kelly	baked potato	ice cream	coffee
Amarjit	baked potato	sponge pudding	orange squash
Tanya	pie and chips	fruit salad	lemonade
Kate	pie and chips	ice cream	tea
Sally	baked potato	fruit salad	orange squash
Serena	fish and chips	ice cream	lemonade
Jandeep	pie and chips	fruit salad	tea

Copy and complete these frequency tables.

First course	fish and chips	pie and chips	baked potato
Frequency			

Second course	ice cream	fruit salad	sponge pudding
Frequency			

Drink	tea	coffee	lemonade	orange squash
Frequency				

2 Nadim is collecting train numbers near a railway junction where he can see trains going North, South and East. He sees 12 trains between 8 a.m. and 9 a.m. and makes out this observation sheet below.

Time of train	Type of train	Direction train is travelling
8.05 a.m.	Intercity	south
8.10 a.m.	local passenger	south
8.15 a.m.	Intercity	north
8.20 a.m.	local passenger	south
8.25 a.m.	local passenger	east
8.30 a.m.	Intercity	north
8.32 a.m.	local passenger	south
8.40 a.m.	freight	north
8.45 a.m.	local passenger	north
8.48 a.m.	local passenger	east
8.50 a.m.	freight	south
8.56 a.m.	Intercity	east

a Copy and complete these frequency tables.

	local passenger	Intercity	freight
Type of train			
Frequency			

Direction	north	south	east
Frequency			

b Suggest why more local passenger trains were travelling south than in any other direction.

c Suggest why more Intercity rains were travelling north than in any other direction.

19.4 Collecting, ordering and grouping data using equal class intervals and creating frequency tables

The heights (correct to the nearest cm) and weights (correct to the nearest kg) of 20 children in a class were collected.

	Height (cm)	Weight (kg)		Height (cm)	Weight (cm)
1	146	42	11	138	37
2	152	46	12	142	46
3	137	39	13	149	42
4	142	41	14	153	45
5	159	51	15	158	51
6	143	47	16	142	47
7	154	43	17	147	43
8	155	48	18	135	38
9	161	53	19	146	45
10	129	35	20	158	53

Example 1

Write **a** the heights and **b** the weights as lists in ascending order.

a 129, 135, 137, 138, 142, 142, 142, 143, 146, 146, 147, 149, 152, 153, 154, 155, 158, 158, 159 and 161 cm.

b 35, 37, 38, 39, 41, 42, 42, 43, 43, 45, 45, 46, 46, 47, 47, 48, 51, 51, 53 and 53 kg.

Example 2

Using a class interval of **a** 10 cm for the heights and **b** 10 kg for the weights, show the above information on frequency tables.

a

Height	120–129 cm	130–139 cm	140–149 cm	150–159 cm	160–169 cm
Frequency	1	3	8	7	1

b

Weight	30–39 kg	40–49 kg	50–59 kg
Frequency	4	12	4

Example 3

Using a class interval of **a** 5 cm for the heights and **b** 5 kg for the weights, show the above information on frequency tables.

a

Height	125–129 cm	130–134 cm	135–139 cm	140–144 cm	145–149 cm	150–154 cm	155–159 cm	160–164 cm
Frequency	1	0	3	4	4	3	4	1

b

Weight	35–39 kg	40–44 kg	45–49 kg	50–54 kg
Frequency	4	5	7	4

Exercise 19.7

1 Eighteen boys entered a weight-lifting contest. Their best lifts were:

George	33 kg	Sunil	38 kg	Peter	44 kg
Chris	36 kg	Robert	42 kg	James	37 kg
Clifford	41 kg	Roger	36 kg	Andrew	48 kg
Thomas	31 kg	Fred	43 kg	Paul	39 kg
Rahul	47 kg	Michael	32 kg	Simon	34 kg
Nishil	37 kg	Dean	46 kg	Craig	41 kg

a Write the weights as a list in ascending order.
b Copy and complete this frequency table.

Weight	30–34 kg	35–39 kg	40–44 kg	45–49 kg
Frequency				

2 The times for the twenty children in Class 3C to travel to school one Monday morning were:

Chantelle	7 min	Joanne	8 min	Edward	19 min
Melanie	11 min	Daisy	16 min	Richard	15 min
Polly	22 min	Elizabeth	12 min	Russell	11 min
Sandeep	16 min	Thomas	17 min	Nathan	21 min
Nina	13 min	David	6 min	Luke	9 min
Samantha	14 min	Sean	18 min	Sanjay	20 min
Natalie	18 min	Daniel	17 min		

a Write down these times as a list in ascending order.
b Copy and complete this frequency table.

Time	5–9 min	10–14 min	15–19 min	20–24 min
Frequency				

3 One day fifteen Intercity trains arrive at a station, as follows.

Time due	Time arrived	Minutes late
9.10 a.m.	9.17 a.m.	
10.20 a.m.	10.21 a.m.	
11.10 a.m.	11.16 a.m.	
12.30 p.m.	12.33 p.m.	
1.40 p.m.	1.51 p.m.	
2.50 p.m.	2.51 p.m.	
3.10 p.m.	3.18 p.m.	
3.30 p.m.	3.40 p.m.	
4.20 p.m.	4.23 p.m.	
5.10 p.m.	5.16 p.m.	
5.30 p.m.	5.46 p.m.	
6.20 p.m.	6.20 p.m.	
6.50 p.m.	6.59 p.m.	
7.40 p.m.	7.40 p.m.	
8.30 p.m.	8.43 p.m.	

a Copy and complete the table.
b Write down the lateness in ascending order.
c Copy and complete this frequency table.

Lateness	0–4 min	5–9 min	10–14 min	15–19 min
Frequency				

4 Fifteen girls took part in a high jump contest.
Their best clearances are:

Yasmin	95 cm	Trisha	136 cm	Elizabeth	98 cm
Abbie	108 cm	Sonya	105 cm	Charlotte	143 cm
Carolyn	115 cm	Vanesa	118 cm	Sadie	125 cm
Mandeep	122 cm	Janice	132 cm	Rosie	116 cm
Vepula	128 cm	Kayleigh	112 cm	Alex	102 cm

a Write down the heights in ascending order.
b Copy and complete this frequency table.

Height	90–99 cm	100–109 cm	110–119 cm	120–129 cm	130–139 cm	140–149 cm
Frequency						

5 The twenty-four children in Class 5C had tests in Mathematics and English.
Their results were:

	Mathematics	English
Pranav	55	68
Jack	75	80
Patrick	64	57
Rajesh	24	38
Sanjay	45	28
Tyrone	13	43
Wayne	54	49
Richard	48	32
George	32	55
Marcus	90	59
Asif	49	42
Alex	81	71
Jennifer	35	25
Anya	65	44
Helena	85	64
Eva	56	45
Shelley	39	39
Jade	72	75
Diana	27	36
Amy	53	51
Drashma	8	35
Keeley	51	56
Suzanne	68	67
Samantha	42	48

a Write down the Mathematics marks in ascending order.
b Copy and complete this table.

Mark	0–9	10–19	20–29	30–39	40–49	50–59	60–69	70–79	80–89	90–99
Frequency										

c Write down the English marks in ascending order.
d Copy and complete this table.

Mark	0–9	10–19	20–29	30–39	40–49	50–59	60–69	70–79	80–89	90–99
Frequency										

e Do you notice any difference between the two frequency tables?
Can you suggest reasons for any differences?

19.5 Constructing and interpreting pie charts from a collection of data with a few variables

Example 1

In Example 1 of Section 19.3 we looked at the results of a traffic survey. Here is a first frequency table from that example.

Make	Cavalier	Ford	Rover	Volvo
Frequency	3	2	4	1

Construct a pie chart to show this information.

As there are 10 cars altogether and 360° in a circle, we use 36° to represent each car.

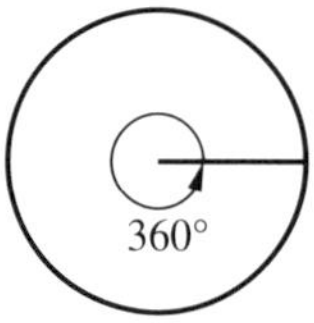

Make	Frequency	Angle in pie chart
Cavalier	3	$3 \times 36^\circ = 108^\circ$
Ford	2	$2 \times 36^\circ = 72^\circ$
Rover	4	$4 \times 36^\circ = 144^\circ$
Volvo	1	$1 \times 36^\circ = 36^\circ$
Check	10	360°

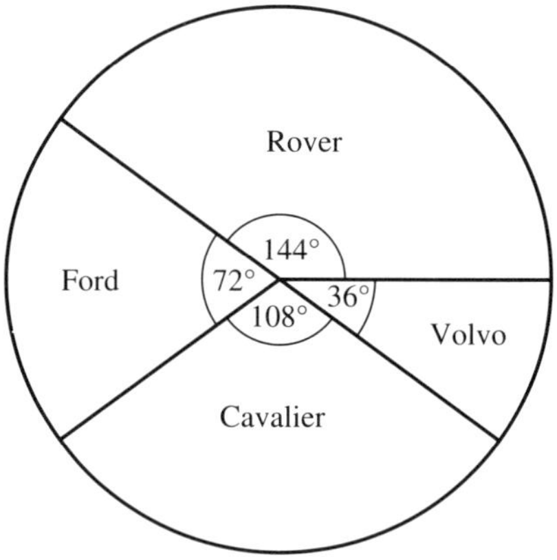

From the pie chart it is easy to see that Rover cars were the most common and that Volvos were the least common.

Exercise 19.8

1 Construct a pie chart for each of these frequency tables from Exercise 19.6.

First course	fish and chips	pie and chips	baked potato
Frequency	3	8	7

Second course	ice cream	fruit salad	sponge pudding
Frequency	7	6	5

Drink	tea	coffee	lemonade	orange squash
Frequency	5	2	4	7

2 Construct a pie chart for each of these frequency tables from Exercise 19.6.

Type of train	local passenger	Intercity	freight
Frequency	6	4	2

Direction	north	south	east
Frequency	4	5	3

3 Draw a pie chart for each of these frequency tables from Example 1, Section 19.3.

Body shape	saloon	hatchback	estate
Frequency	4	4	2

Colour	blue	green	red	white
Frequency	3	2	3	2

4 Draw a pie chart for each of these frequency tables from Example 1, Section 19.3.
(Take care when constructing an angle larger than 180°.)

Number of doors	2-door	4-door
Frequency	3	7

Number of occupants	1	2	3	4	5
Frequency	7	1	0	1	1

Example 2

In Example 2 of Section 19.4 we looked at the heights of 20 children.
Here is a frequency table from this example.

Height	120–129 cm	130–139 cm	140–149 cm	150–159 cm	160–169 cm
Frequency	1	3	8	7	1

Construct a pie chart to show this information.

As there are 20 pupils altogether and 360° in a circle, we can use 18° to represent each pupil.

Height	Frequency	Angle in pie chart
120–129 cm	1	$1 \times 18° = 18°$
130–139 cm	3	$3 \times 18° = 54°$
140–149 cm	8	$8 \times 18° = 144°$
150–159 cm	7	$7 \times 18° = 126°$
160–169 cm	1	$1 \times 18° = 18°$
Check	20	360°

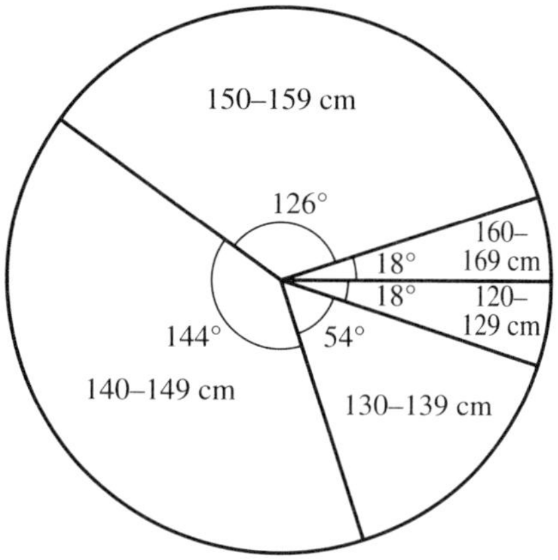

From the pie chart it can be seen that most children were in the interval 140–149 cm.

Exercise 19.9

Draw a pie chart for each of these frequency tables from Exercise 19.7.

1

Weight	30–34 kg	35–39 kg	40–44 kg	45–49 kg
Frequency	4	6	5	3

2

Time	5–9 min	10–14 min	15–19 min	20–24 min
Frequency	4	5	8	3

3

Lateness	0–4 min	5–9 min	10–14 min	15–19 min
Frequency	6	4	4	1

4

Height	90–99 cm	100–109 cm	110–119 cm	120–129 cm	130–139 cm	140–149 cm
Frequency	2	3	4	3	2	1

5

Mathematics mark	0–9	10–19	20–29	30–39	40–49	50–59	60–69	70–79	80–89	90–99
Frequency	1	1	2	3	4	5	3	2	2	1

English mark	0–9	10–19	20–29	30–39	40–49	50–59	60–69	70–79	80–89	90–99
Frequency	0	0	2	5	6	5	3	2	1	0

6 Draw a pie chart for each of these frequency tables from Example 3, Section 19.4.

Height	125–129 cm	130–134 cm	135–139 cm	140–144 cm	145–149 cm	150–154 cm	155–159 cm	160–164 cm
Frequency	1	0	3	4	4	3	4	1

Weight	35–39 kg	40–44 kg	45–49 kg	50–54 kg
Frequency	4	5	7	4

Example 3

The pie chart shows how the 20 pupils in Class 3B travel to school. From the chart find:

a the number who travel by each method

b the percentage who travel by each method.

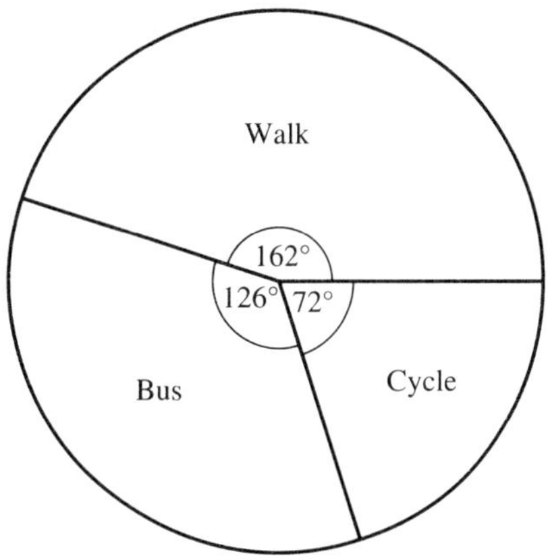

a The number who walk $= \dfrac{162^\circ}{360^\circ} \times 20 = 9$

The number who travel by bus $= \dfrac{126^\circ}{360^\circ} \times 20 = 7$

The number who cycle $= \dfrac{72^\circ}{360^\circ} \times 20 = 4$

b The percentage who walk $= \dfrac{162^\circ}{360^\circ} \times 100\% = 45\%$

The percentage who travel by bus $= \dfrac{126^\circ}{360^\circ} \times 100\% = 35\%$

The percentage who cycle $= \dfrac{72^\circ}{360^\circ} \times 100\% = 20\%$

Exercise 19.10

1 The pie chart shows how all 180 boys in Year 5 chose their sports option.
From the chart find the number who chose each of the five sports.

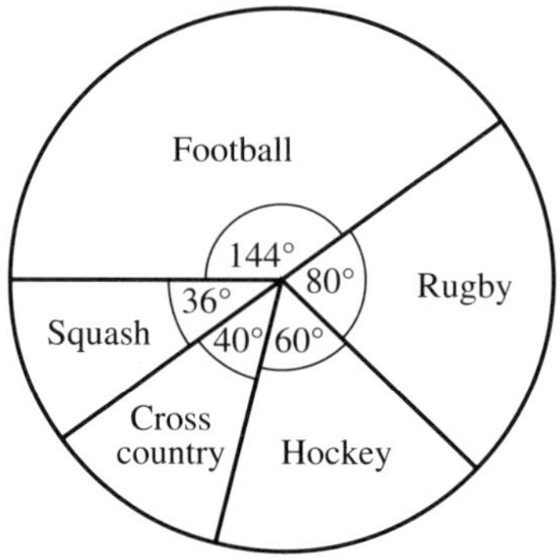

2 300 passengers have boarded a train at Waterloo Station in London. The pie chart shows how many are travelling to each of the five destinations.
From the chart find out how many are travelling to each of the destinations.

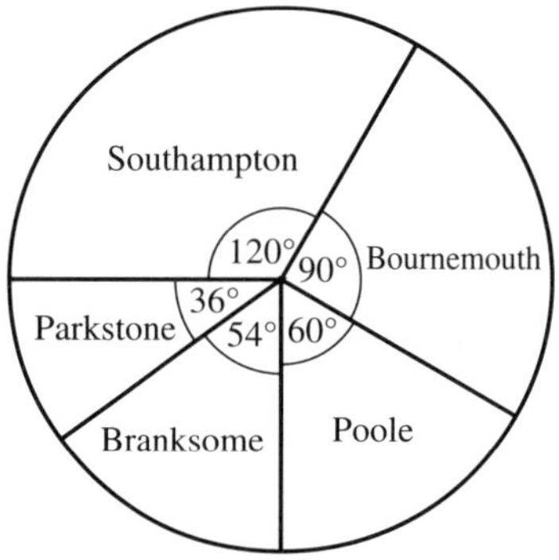

3 One day 150 packets of crisps are dispensed by a vending machine. The pie chart shows how many of each flavour.
From the chart find the number for each flavour.
Find also the percentage of the total for each flavour.

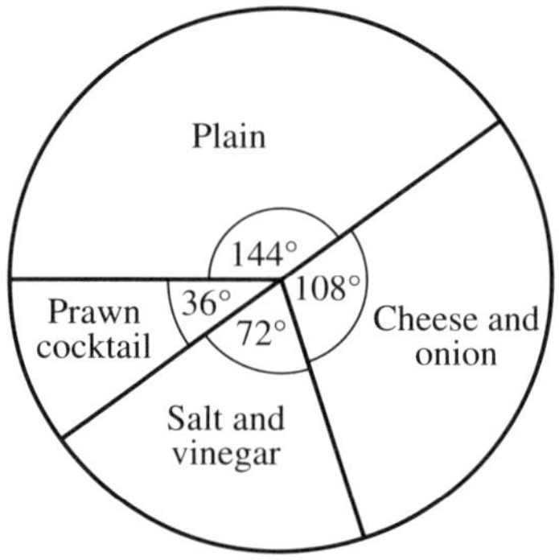

4 One week a car dealer sold 80 cars. The pie chart shows how many on each day.
From the chart find the number sold on each day.
Find also the percentage of the total for each day.

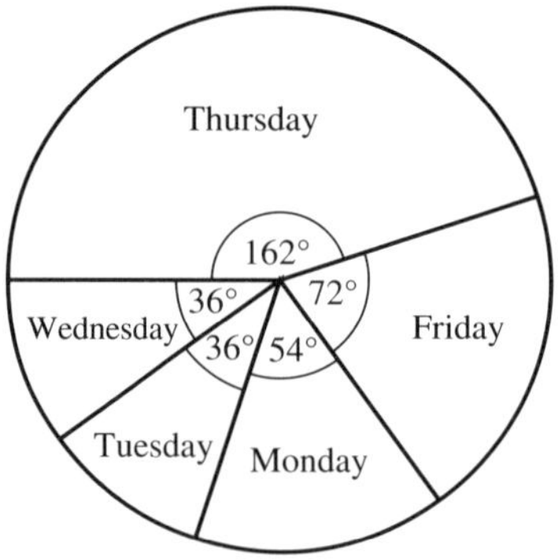

Example 4

The crowd at a football match consisted of:
45% men; 25% boys; 20% women and 10% girls.
Show these details on a pie chart.

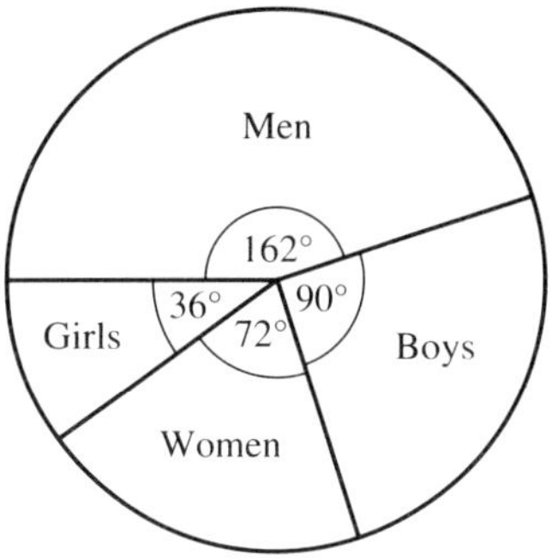

The sector angles are:

Men 45% of $360° = 162°$
Boys 25% of $360° = 90°$
Women 20% of $360° = 72°$
Girls 10% of $360° = 36°$

The pie chart can therefore be drawn as shown.
(*Note:* the sector angles must total 360°.)

Exercise 19.11

1 On an island the population distribution is:
40% live in North County; 25% live in South County;
20% live in East County; 15% live in West County.
Show these details on a pie chart.

2 One year a census was taken to find out how many people went to Europe by the four different modes.
40% went through the Chunnel; 15% went by air;
35% went by boat; 10% went by hovercraft.
Show these details on a pie chart.

3 A certain car is available in five different colours.
One year the sales were:
40% black; 25% white; 15% red; 10% green; 10% blue.
Show these details on a pie chart.

Example 5

480 people who live in a Scottish village are eligible to vote at an election.

120 voted Labour; 160 voted Liberal; 120 voted Conservative;
80 voted Scottish Nationalist.
Show these details on a pie chart.

Sector angle for Labour and Conservative $= \dfrac{120}{480} \times 360° = 90°$

Sector angle for Liberal $= \dfrac{160}{480} \times 360° = 120°$

Sector angle for Scottish Nationalist $= \dfrac{80}{480} \times 360° = 60°$

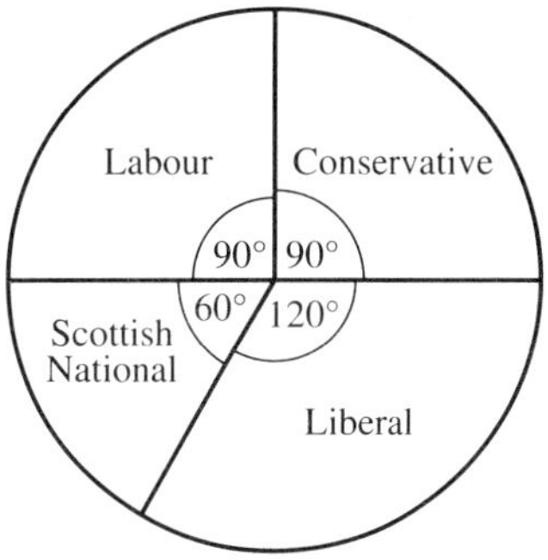

The pie chart can therefore be drawn as shown.

Exercise 19.12

For all questions show the details on a pie chart.

1 120 girls choose from four summer sports options.
40 choose tennis; 30 choose rounders; 30 choose swimming; 20 choose athletics.

2 180 boys choose from five summer sports options.
60 choose cricket; 45 choose athletics; 30 choose swimming; 30 choose tennis; 15 choose golf.

3 240 people travelled from London to the South Coast to visit a maritime museum which has a special exhibition.
80 travelled by train; 60 travelled by bus; 40 travelled by coach; 30 travelled by car; 30 travelled by motor cycle.

4 A large restaurant employs 108 people. They all have to work late on one night a week except Sunday, when the restaurant is closed.
12 work on Monday night; 12 work on Tuesday night; 12 work on Wednesday night; 18 work on Thursday night; 27 work on Friday night; 27 work on Saturday night.

5 720 people from six different cities attend a business meeting:
120 from Manchester; 120 from Liverpool; 80 from Birmingham; 80 from Leeds; 80 from Newcastle; 240 from London.

6 144 teenagers from London were asked which football team they supported. 36 supported Spurs; 24 supported Arsenal; 18 supported Chelsea; 18 supported West Ham; 16 supported Millwall; 16 supported Fulham; 16 supported QPR.

7 A small country town spends its revenue from local taxes as follows:
$\frac{1}{5}$ General public services; $\frac{1}{5}$ Education; $\frac{1}{6}$ Police;
$\frac{1}{8}$ Road maintenance; $\frac{1}{8}$ Public transport;
$\frac{1}{10}$ Landscape improvement; $\frac{1}{12}$ Investments.

8 The Wonderful Western coach company operates coaches from London to Bristol, Exeter, Plymouth, Newport, Cardiff and Swansea.
One year its bookings were as follows:
$\frac{7}{20}$ to Bristol; $\frac{1}{5}$ to Cardiff; $\frac{3}{20}$ to Swansea; $\frac{1}{8}$ to Newport;
$\frac{1}{8}$ to Plymouth; $\frac{1}{20}$ to Exeter.

19.6 Constructing and interpreting conversion graphs

Example 1

After a great recovery of the pound each £1 buys eight French Francs (FF).

a Copy and complete this table.

£10	£20	£30	£40	£50	£60	£70	£80	£90
FF80	FF160							

b Show the information on a conversion graph.

c Use your graph to find:

(i) how many French Francs you can buy for £35

(ii) how much it costs to buy FF440.

d Show that the gradient of your graph gives the Francs equivalent of one pound.

a

£	10	20	30	40	50	60	70	80	90	100
FF	80	160	240	320	400	480	560	640	720	800

b

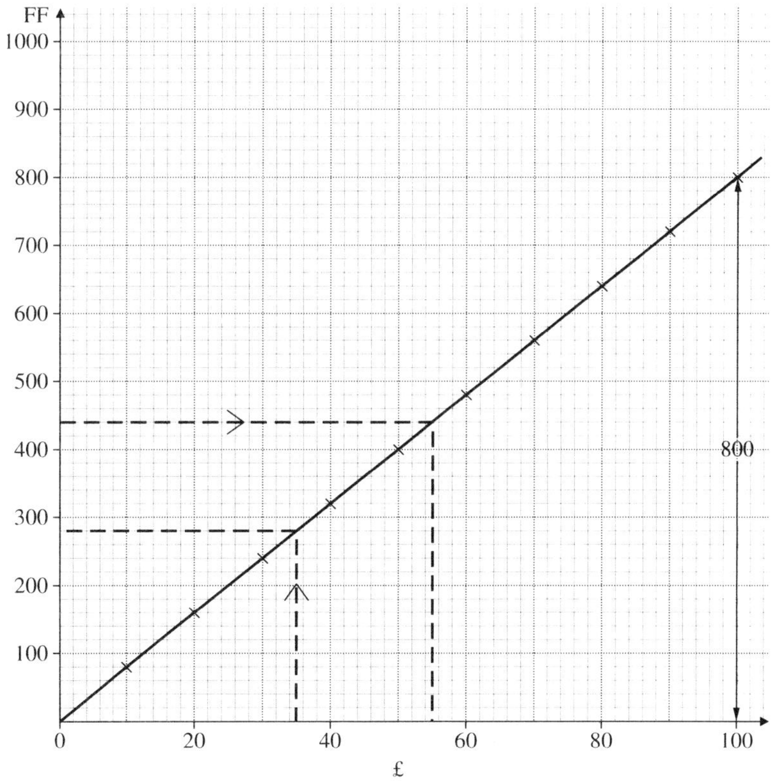

c (i) FF280 (ii) £55

d The gradient of the graph = 800 ÷ 100 = 8, i.e. there are FF8 to the £1

Exercise 19.13

1 When you go to Belgium, each £1 buys 50 Belgian Francs (BF).
a Copy and complete this table.

£	£10	£20	£30	£40	£50	£60	£70	£80	£90
BF	500	1000	1500						

b Show the information on a conversion graph.
(Horizontal scale 2 cm = £10, vertical scale 2 cm = BF500)

c Use your graph to find how many Belgian Francs can be bought for:
(i) £15 (ii) £25 (iii) £45 (iv) £75

d Use your graph to find how much it costs to buy:
(i) BF1750 (ii) BF2750 (iii) BF3250 (iv) BF4250

2 When you go to Spain, each £1 buys 200 Pesetas (Ptas).
a Copy and complete this table.

£	10	20	30	40	50	60	70	80	90
Ptas	2000	4000	6000						

b Show the information on a conversion graph.
(Horizontal scale 2 cm = £10, vertical scale 2 cm = 200 Ptas)

c Use your graph to find how many Pesetas can be bought for:
(i) £35 (ii) £45 (iii) £5 (iv) £75 (v) £15

d Use your graph to find how much it costs to buy:
(i) Ptas 11 000 (iii) Ptas 17 000
(ii) Ptas 13 000 (iv) Ptas 5000

3 The table shows conversions from pounds to United States dollars.

£	12	20	28	36	40	48	60	80	88
$	18	30	42	54	60	72	90	120	132

a Show the information on a conversion graph.
(Horizontal scale 2 cm = £10, vertical scale 2 cm = $20)

b Use your graph to find how many dollars can be bought for:
(i) £16 (ii) £44 (iii) £52 (iv) £68 (v) £72

c Use your graph to find how much it costs to buy:
(i) $36 (ii) $48 (iii) $84 (iv) $114 (v) $126

d Find the dollar equivalent of £1 from the gradient of your graph.

Example 2

Temperatures can be approximately converted from Celsius to Fahrenheit by:
'Double the Celsius figure and then add 30'

a Copy and complete this table.

Temperature °C	0	10	20	30	40
Temperature °F					

b Show the information on a conversion graph.

c Use your graph to find:
(i) The Fahrenheit equivalent of 15 °C
(ii) The Celsius equivalent of 80 °F

a

Temperature °C	0	10	20	30	40
Temperature °F	30	50	70	90	110

b

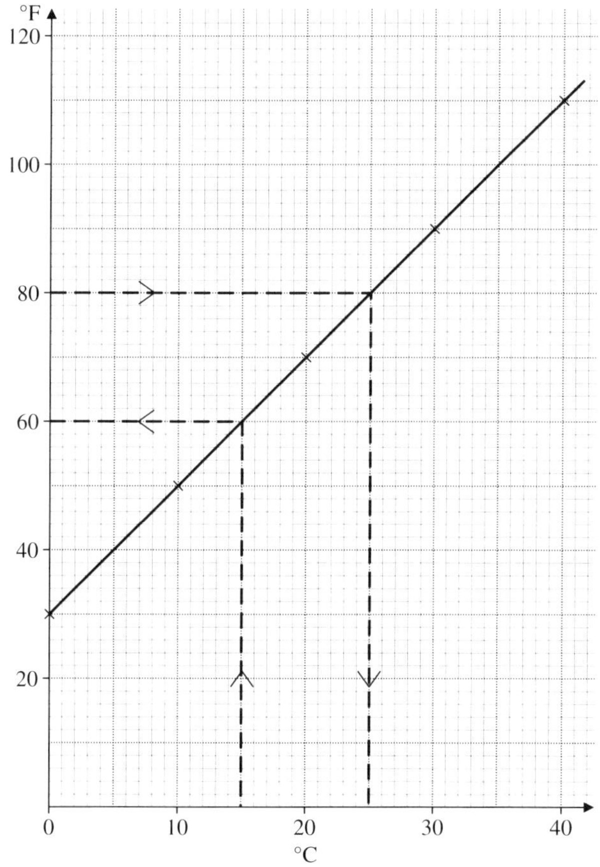

c (i) 15 °C = 60 °F (ii) 80 °F = 25 °C
(See the broken lines on the graph.)

Exercise 19.14

1 a The table shows conversions from miles to kilometres.
Copy and complete the table.

Miles	0	1	2	3	4	5	6	7	8	9
Kilometres	0	1.6	3.2	4.8						

b Show the information on a conversion graph.
(Horizontal scale 2 cm = 1 mile, vertical scale 2 cm = 2 km)

c Use your graph to find the number of kilometres which are equivalent to:
(i) 2.5 miles (ii) 4.5 miles (iii) 5.5 miles (iv) 8.5 miles

d Use your graph to find the number of miles which are equivalent to:
(i) 2.4 km (ii) 5.6 km (iii) 10.4 km (iv) 12 km

2 a The table shows the conversions from acres to hectares.
Copy and complete the table.

Acres	0	1	2	3	4	5	6
Hectares	0	0.4	0.8	1.2			

Acres	7	9	11	13	15	18
Hectares						

b Show the information on a conversion graph.
(Horizontal scale 2 cm = 2 acres, vertical scale 2 cm = 1 hectare)

c Use your graph to find the number of hectares which are equivalent to:
(i) 8 acres (ii) 12 acres (iii) 17 acres

d Use your graph to find the number of acres which are equivalent to:
(i) 4 hectares (ii) 5.6 hectares (iii) 6.4 hectares

3 a The table below shows conversions from square miles to kilometres.
Copy and complete the table.

Square miles	0	0.4	1.6	2.0	2.4	3.6	4.0	4.4	5.6	6.0
Square kilometres	0	1.0	4.0	5.0						

b Show the information on a conversion graph.
(Horizontal scale 2 cm = 1 square mile, vertical scale 2 cm = 2 square kilometres)

c Use your graph to find the number of square kilometres which are equivalent to:
(i) 4.8 square miles (ii) 3.2 square miles (iii) 0.8 square miles

d Use your graph to find the number of square miles which are equivalent to:
(i) 3 square km (ii) 7 square km (iii) 13 square km

e Find the number of square kilometres to one square mile from the gradient of your graph.

4 a The table shows the conversions from inches to centimetres. Copy and complete the table.

Inches	0	8	12	20	28	32
Centimetres	0			50		

Inches	40	48	52	60	68	72
Centimetres	100			150		

b Show the information on a conversion graph.
(Horizontal scale 2 cm = 10 inches, vertical scale 2 cm = 20 cm)

c Use your graph to find the number of centimetres which are equivalent to:
(i) 24 inches (ii) 56 inches (iii) 64 inches

d Use your graph to find the number of inches which are equivalent to:
(i) 40 cm (ii) 90 cm (iii) 110 cm

e Find the number of centimetres to one inch from the gradient of your graph.

5 a The table shows the conversions from pounds to kilograms. Copy and complete the table.

Pounds	0	6	12	18
Kilograms	0	2.7		

b Show the information on a conversion graph.
(Horizontal scale 2 cm = 2 pounds,
vertical scale 2 cm = 1 kilogram)

c Use your graph to find the number of kilograms which are equivalent to:
(i) 8 pounds (ii) 10 pounds (iii) 14 pounds

d Use your graph to find the number of pounds which are equivalent to:
(i) 7.2 kg (ii) 1.8 kg (iii) 0.9 kg

e Find the number of kilograms to one pound from the gradient of your graph.

19.7 Constructing and interpreting frequency diagrams and choosing class intervals for a continuous variable

Example 1

Look at the frequency diagram below.

a Write down the class intervals and state the frequency for each.

b Find the total number of children.

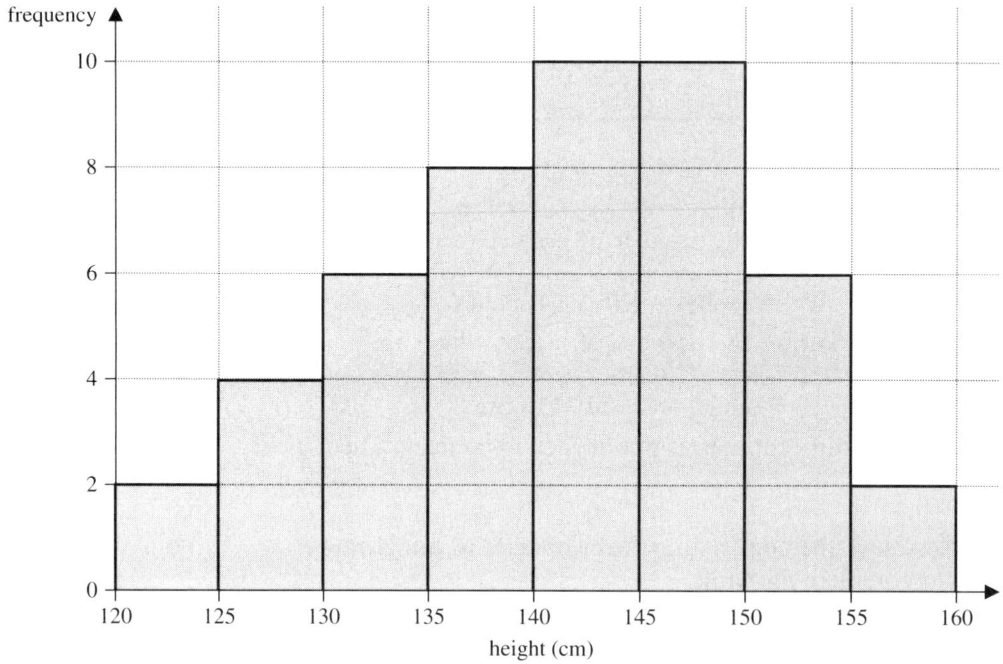

a

Class interval	Frequency	Class interval	Frequency
$120 \leqslant h < 125$	2	$140 \leqslant h < 145$	10
$125 \leqslant h < 130$	4	$145 \leqslant h < 150$	10
$130 \leqslant h < 135$	6	$150 \leqslant h < 155$	6
$135 \leqslant h < 140$	8	$150 \leqslant h < 160$	2

(Note that $120 \leqslant h < 125$ means that the height can be anything from 120 cm to 125 cm except 125 cm itself, which is in the next interval.)

b The total number of children $= 2 + 4 + 6 + 8 + 10 + 10 + 6 + 2 = 48$

Exercise 19.15

1 The frequency diagram shows the weights of the children in Class 5A.

a Write down the class intervals and state the frequency for each.

b Find the total number of children in the class.

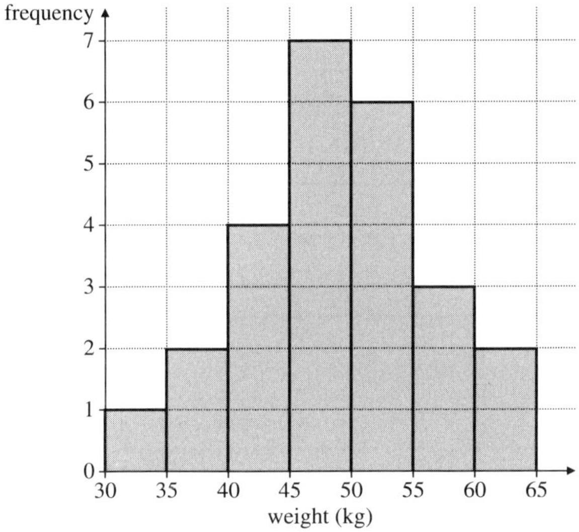

2 The frequency diagram below shows the temperature for the days of a certain month.

a Write down the class intervals and state the frequency for each.

b How many days did this month have?

c What month was this one likely to have been?

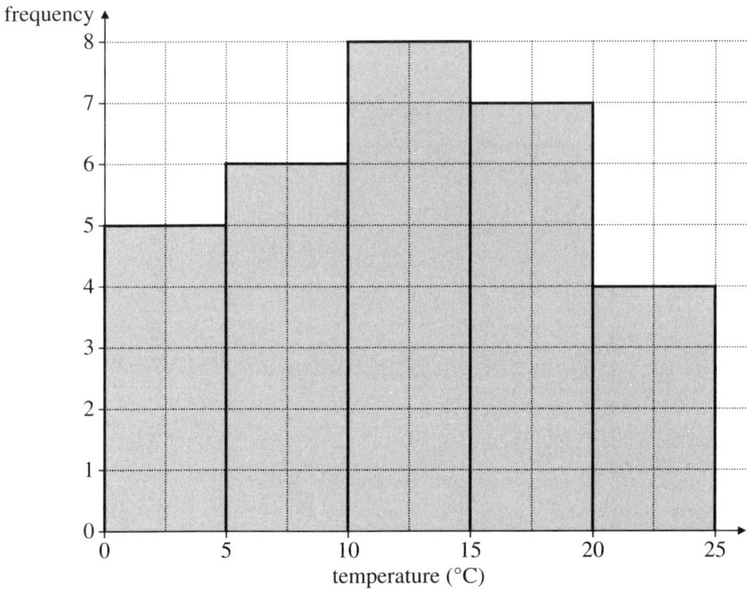

Example 2

The heights of 20 children to the nearest centimetre are:
127 cm, 128 cm, 128 cm, 129 cm, 130 cm, 134 cm, 135 cm, 137 cm, 139 cm, 140 cm, 140 cm, 142 cm, 144 cm, 145 cm, 148 cm, 149 cm, 150 cm, 154 cm, 155 cm, 157 cm

a Construct a frequency table using suitable class intervals of (i) 10 cm and (ii) 5 cm.

b Show the information on frequency diagrams.

a As the heights are measured to the nearest whole centimetre, a height of 129.49 cm would be recorded as 129 cm, whereas a height of 129.50 cm would be recorded as 130 cm. The boundary between these two class intervals is therefore 129.5 cm.

(i)

Class interval	119.5–129.5 cm	129.5–139.5 cm	139.5–149.5 cm	149.5–159.5 cm
Frequency	4	5	7	4

(ii)

Class interval	124.5–129.5 cm	129.5–134.5 cm	134.5–139.5 cm	139.5–144.5 cm	144.5–149.5 cm	149.5–154.5 cm	154.5–159.5 cm
Frequency	4	2	3	4	3	2	2

b

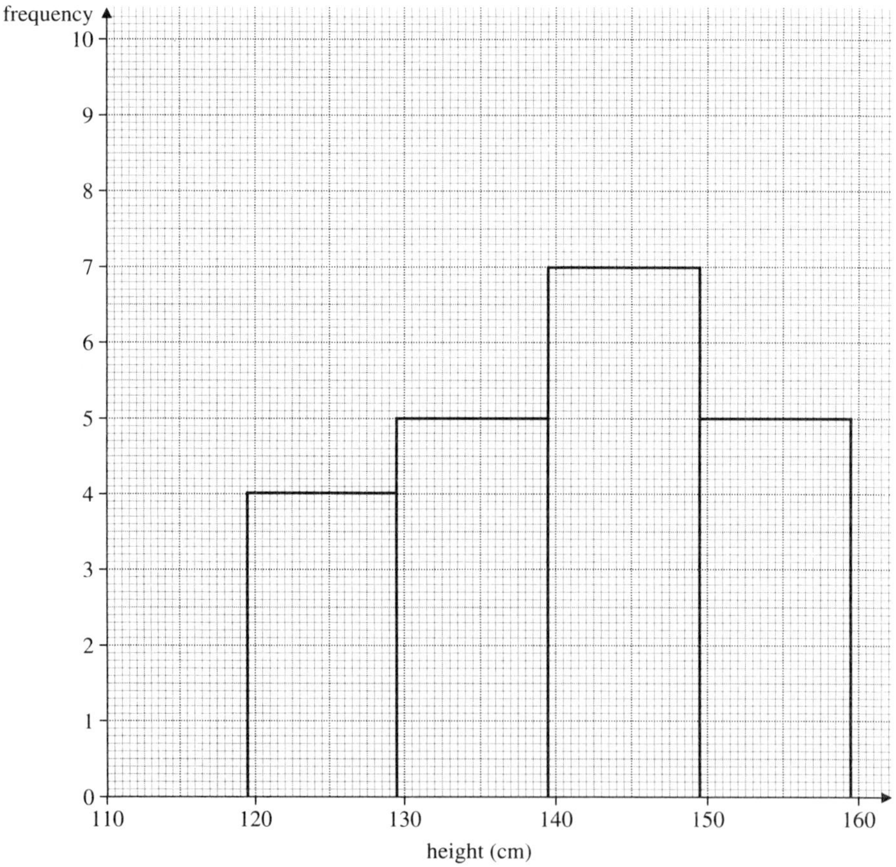

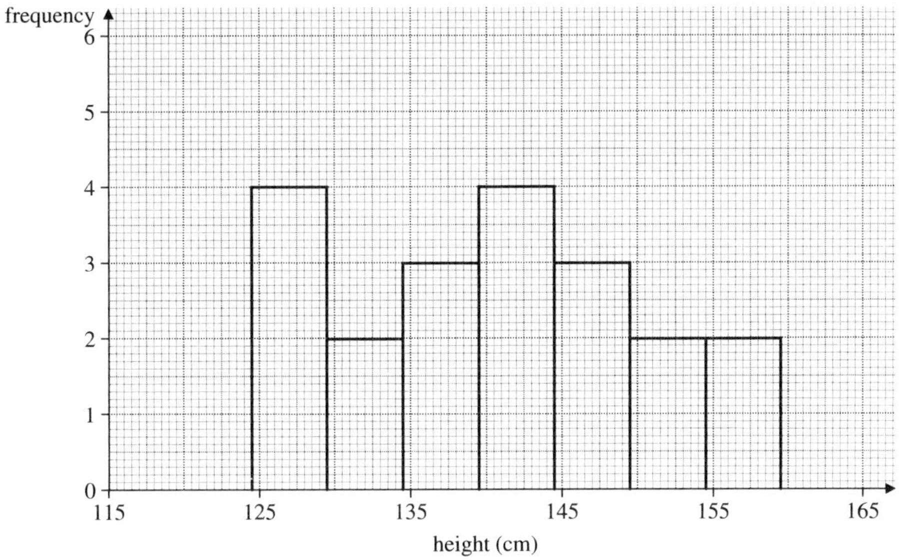

Exercise 19.16

1 The weights of 20 children are measured to the nearest kilogram are:

41 kg, 43 kg, 46 kg, 47 kg, 48 kg, 49 kg, 51 kg, 52 kg, 52 kg, 53 kg, 53 kg, 54 kg, 55 kg, 56 kg, 57 kg, 58 kg, 59 kg, 61 kg, 63 kg, 64 kg

a Construct a frequency table using suitable class intervals of (i) 10 kg (ii) 5 kg

b Show the information on frequency diagrams.

2 The lengths of 25 leaves from a certain tree measured to the nearest millimetre are:

42 mm, 46 mm, 48 mm, 49 mm, 51 mm, 52 mm, 52 mm, 53 mm, 54 mm, 55 mm, 56 mm, 56 mm, 57 mm, 57 mm, 58 mm, 62 mm, 63 mm, 64 mm, 65 mm, 66 mm, 68 mm, 69 mm, 72 mm, 74 mm, 76 mm

a Construct a frequency table using suitable class intervals of (i) 10 mm (ii) 5 mm

b Show the information on frequency diagrams.

3 All 30 children in Class 5B measured their best time for the 400 m race. The times to the nearest second are:

54 s, 57 s, 58 s, 61 s, 62 s, 62 s, 63 s, 64 s, 65 s, 66 s, 66 s, 67 s, 68 s, 69 s, 70 s, 71 s, 72 s, 72 s, 73 s, 74 s, 74 s, 75 s, 76 s, 77 s, 78 s, 79 s, 81 s, 83 s, 84 s, 86 s

a Construct a frequency table using suitable class intervals of (i) 10 s and (ii) 5 s.

b Show the information on frequency diagrams.

19.8 Creating scatter diagrams for discrete and continuous variables and having a basic understanding of correlation

In general terms it is probably true to say that the taller you are the heavier you will be. We say that there is a 'correlation' between height and weight.
When this is true it will be shown by the pattern of points on a 'scatter' graph.

Example 1

The height and weight of each of ten pupils are shown below.
Draw a scatter graph plotting weight against height and say whether or not there is any correlation between them.

	Height	Weight		Height	Weight
Jodie	140 cm	50 kg	Tania	125 cm	44 kg
Leroy	160 cm	56kg	Raj	155 cm	54 kg
Azil	150 cm	54 kg	Khalid	145 cm	50 kg
Anna	120 cm	42 kg	Imogen	130 cm	46 kg
David	135 cm	48kg	Justin	165 cm	58 kg

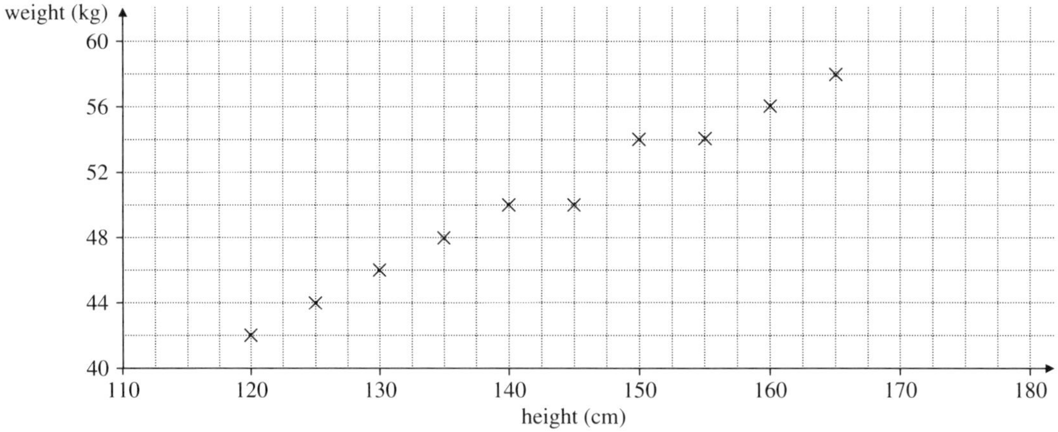

The general shape of this scatter graph is a pattern of points sloping upwards to the right. In other words: the greater the height the heavier the weight.
This suggests that there is a 'positive' correlation between a person's height and weight.

Example 2

Andy wanted to know whether there was any correlation between a person's age and the amount of pocket money he or she received. He collected data from ten pupils.
Show the data as a scatter graph and say whether you think there is a correlation.

	Age	Pocket money
Jodie	14	£2.50
Leroy	16	£4.00
Azil	15	£3.50
Anna	12	£2.00
David	14	£3.00
Tania	13	£2.50
Raj	16	£4.50
Khalid	15	£3.00
Imogen	13	£3.00
Justin	16	£5.00

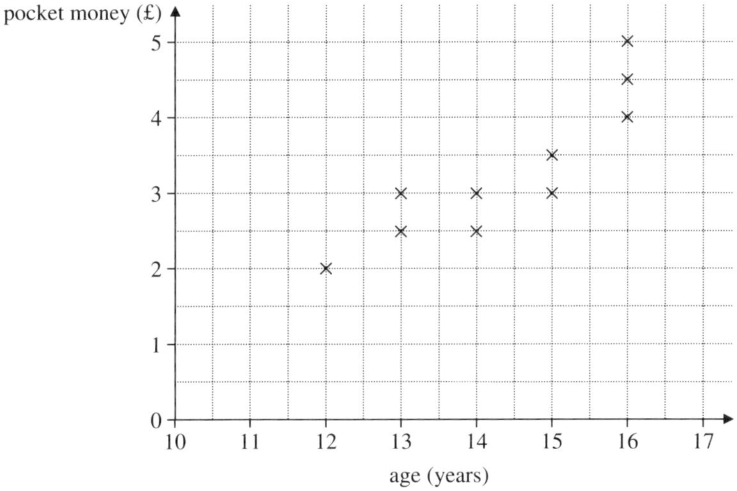

Again the general shape of this scatter graph shows a trend upwards to the right. In other words: the greater the age, the more pocket money.
This suggests that there is a 'positive' correlation between age and pocket money.

Exercise 19.17

For each question draw a scatter graph on a copy of the grid illustrated and say whether or not there is any correlation.
(Your teacher has permission to photocopy these grids.)

1 The table shows the height and age of the ten children in Example 1.

	Age	Height
Jodie	14	140 cm
Leroy	16	160 cm
Azil	15	150 cm
Anna	12	120 cm
David	14	135 cm
Tania	13	125 cm
Raj	16	155 cm
Khalid	15	145 cm
Imogen	13	130 cm
Justin	16	165 cm

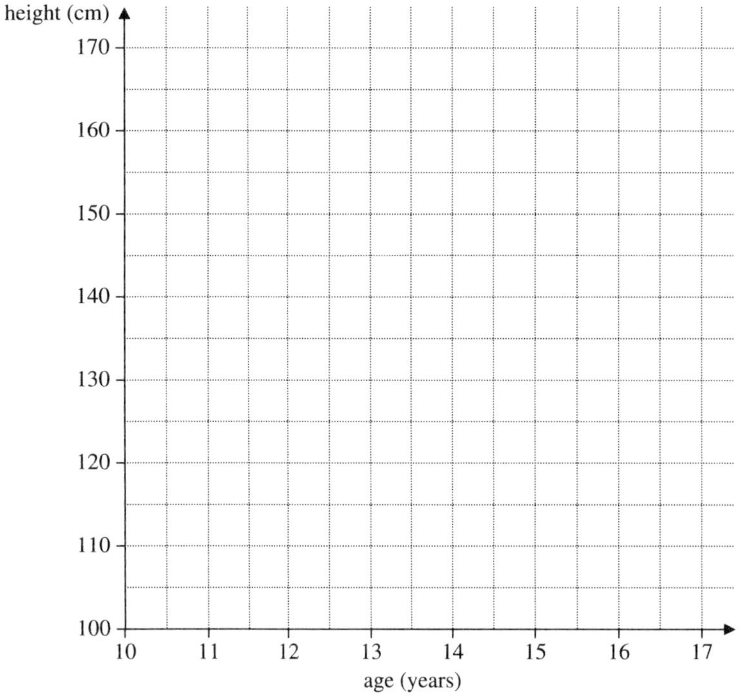

2 The children in Class 3A took tests in Mathematics and Science.

	Mathematics mark (%)	Science mark (%)
Jack	60	50
Pranav	25	35
Asif	55	65
David	90	70
Harry	35	40
Rani	15	25
Josie	85	80
Ruth	30	50
Becki	70	75
Shona	45	50

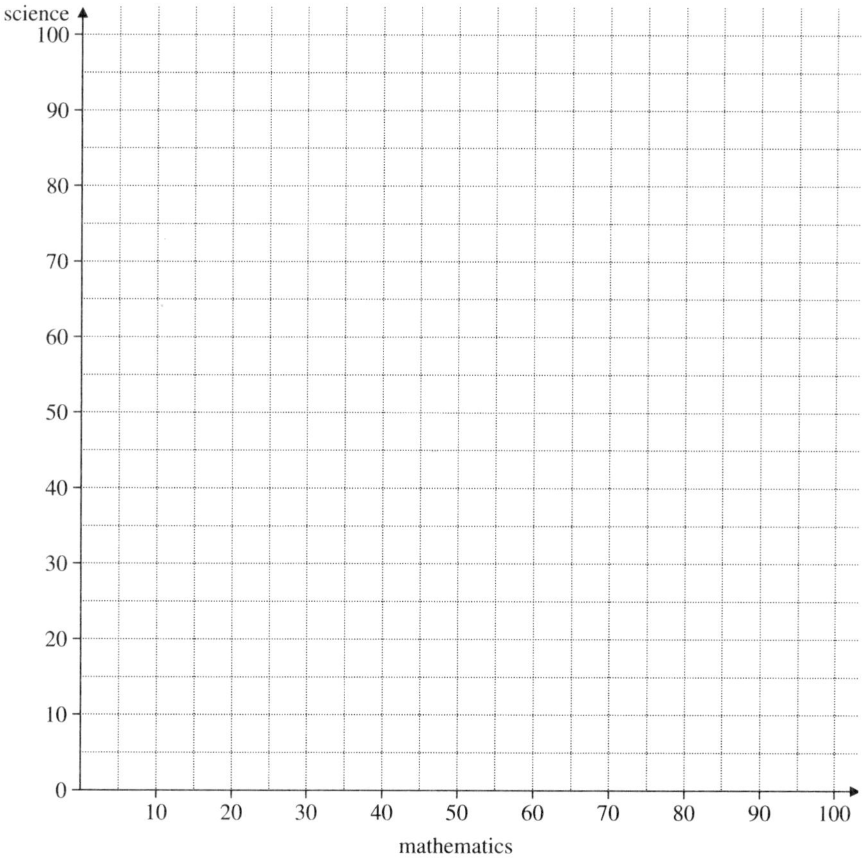

3 The table shows the height of each of the children in Class 3B and their shoe size.

	Shoe size	Height
Jack	7	160 cm
Jotinder	3	150 cm
Gareth	8	170 cm
Fred	5	140 cm
Sunil	4	130 cm
Anita	2	140 cm
Nicola	5	150 cm
Larissa	3	140 cm
Natalie	6	160 cm
Shani	1	130 cm

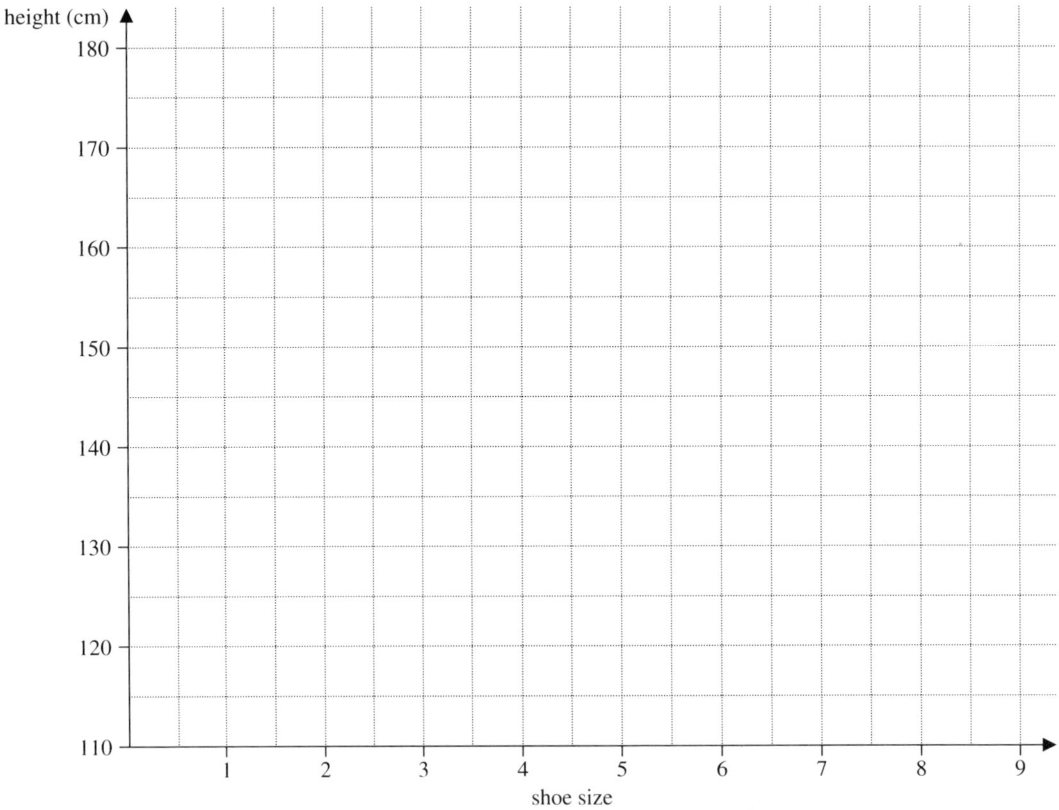

Example 3

The table below shows the petrol used, in miles per gallon at different speeds.
Find whether a correlation exists between fuel consumption and speed.

Speed	30 mph	35 mph	40 mph	45 mph	50 mph	55 mph	60 mph	65 mph	70 mph	75 mph
Petrol used (miles per gallon)	58	60	57	54	49	45	41	37	32	25

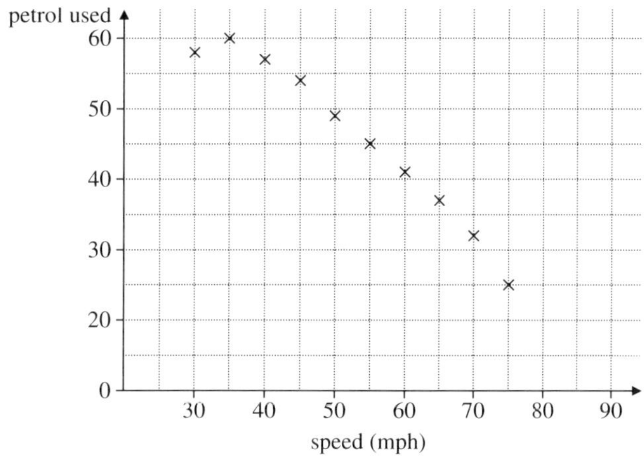

The scatter graph shows a pattern of points sloping downwards to the right. In other words: the greater the speed, the fewer miles per gallon.
This suggests that there is a 'negative' correlation between the speed of the car and the petrol used, in miles per gallon.

Example 4

Use the data below to find whether there is a correlation between the purchase cost of a car and its efficiency in terms of average fuel consumption.

Car	Cost	Miles per gallon
Mazda MR	£12 000	24
Volvo SXI	£10 000	27
Rover SE	£17 000	30
Mondeo L	£10 000	32
Cavalier LX	£9000	36
Fiesta XRS	£10 000	40
Peugot GRDT	£16 000	45
Orion D	£12 000	48
Fiat XR	£14 000	55
Citroen D	£7000	60

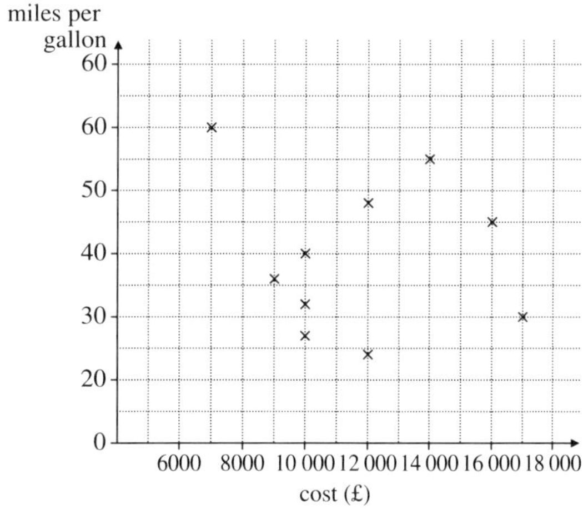

On this scatter graph there are no obvious trends upwards or downwards to the right.
It appears that there is no correlation between price and fuel consumption.

Exercise 19.18

For each question draw a scatter graph on a copy of the grid illustrated and comment on the correlation if there is any.

1 The table shows the best times in the 100 m and 400 m races for each of the ten boys in Class 4B.

	100 m race	400 m race
Marcus	13.5 s	62 s
Theo	14.5 s	65 s
Julian	12.5 s	61 s
Wayne	13.5 s	64 s
Martin	12.0 s	62 s
Waseem	14.5 s	66 s
Leroy	14.0 s	63 s
Richard	12.5 s	63 s
Raj	15.0 s	67 s
Sanjay	13.0 s	62 s

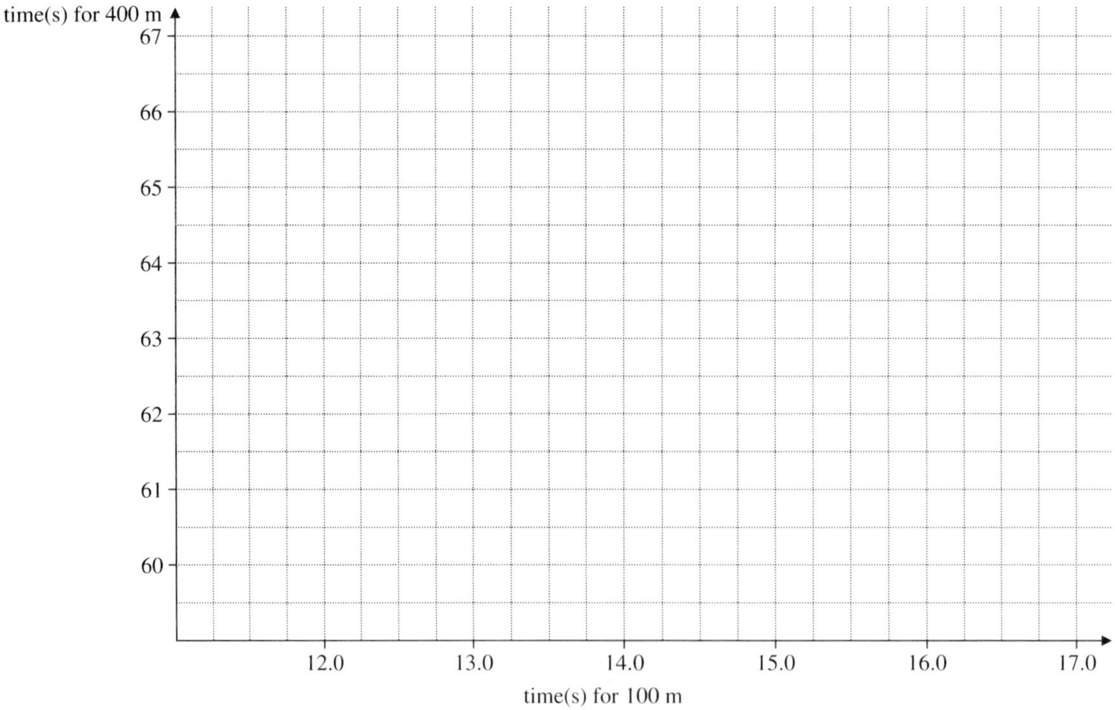

2 The table shows the best performance at high jump and long jump for each of the girls in Class 4A.

	High jump	Long jump
Polly	120 cm	410 cm
Govinder	150 cm	460 cm
Carly	130 cm	450 cm
Jodie	100 cm	400 cm
Lucy	140 cm	440 cm
Kate	125 cm	430 cm
Daisy	110 cm	430 cm
Samiya	135 cm	440 cm
Liz	115 cm	420 cm
Yogita	105cm	390 cm

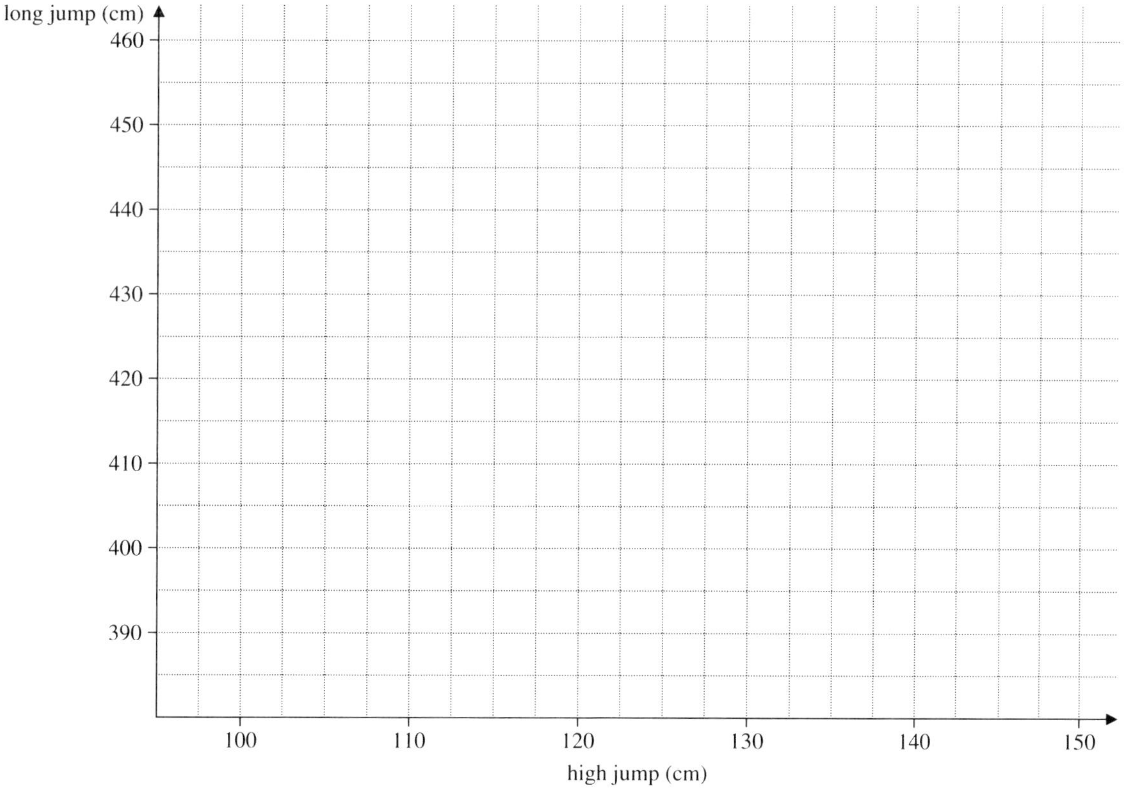

3 The table shows the mean length of the cones from a certain kind of tree together with the latitude of the forest where the measurements were estimated.

Latitude of forest	Mean length of cones
48 °N	12 cm
46 °N	13 cm
47 °N	11 cm
48 °N	11 cm
49 °N	13 cm
50 °N	12 cm
51 °N	9 cm
52 °N	10 cm
53 °N	8 cm
54 °N	9 cm

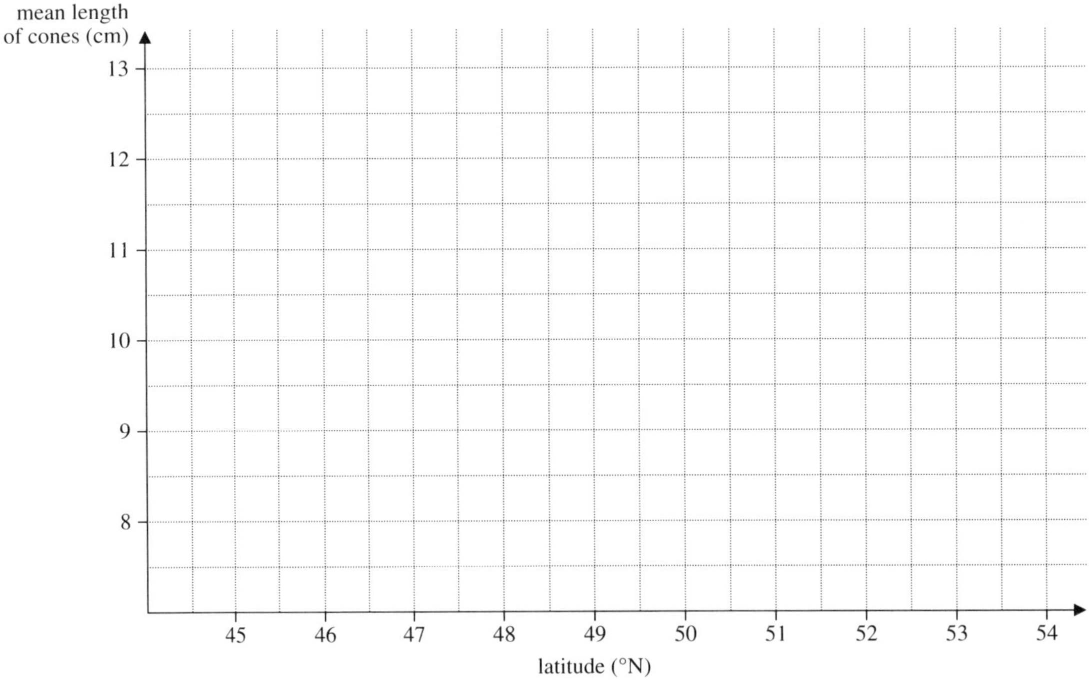

4 The table shows the train fare for certain distances.
Copy and complete the table before drawing your scatter graph.

Distance	Fare	Fare per mile
500 miles	£60.00	$6000 \div 500 = 12\,\text{p}$
450 miles	£54.00	
400 miles	£54.00	
350 miles	£45.50	
300 miles	£40.50	
250 miles	£36.25	
200 miles	£29.00	
150 miles	£20.25	
100 miles	£14.00	

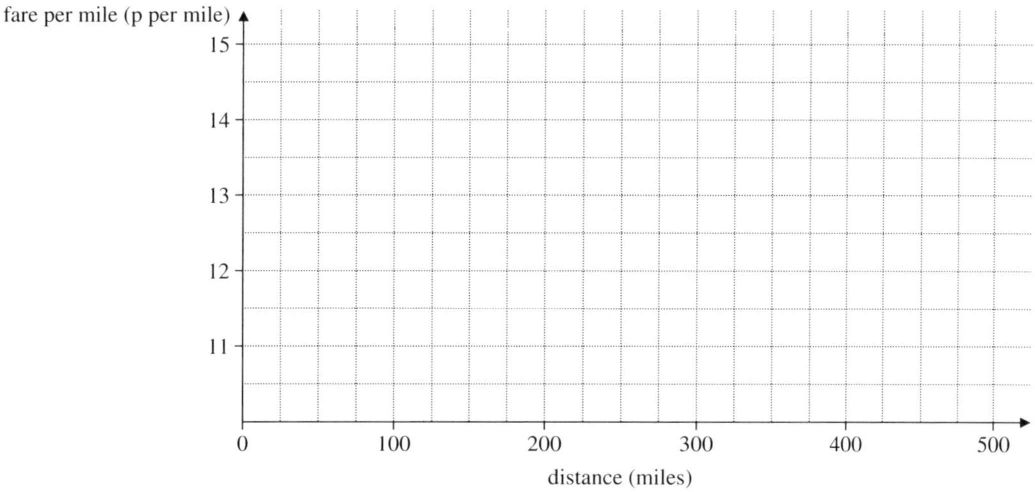

20 Median, mode, mean and range of a set of data

20.1 Understanding and using the median and mode in everyday contexts

The **median** of a set of data is the middle term when the data is arranged in ascending or descending order.

Example 1

Look back at this data from Example 2, Section 19.1.
What is the median birthday for the fifteen children?

Anne: 12 July
Emma: 9 June
Miranda: 15 October
Zorba: 12 March
Paul: 12 December
Adam: 15 May
Fran: 14 July
Razi: 17 December
Azil: 15 November
Satwinder: 1 February
David: 7 January
George: 12 January
Salik: 9 April
Melissa: 4 July
Yusuf: 14 January

Their birthdays in chronological order are:

7 January
12 January
14 January
1 February
12 March
9 April
15 May
9 June ← Median date, in the middle of list
4 July
12 July
14 July
15 October
15 November
12 December
17 December

The median birthday is 9 June because this is the one in the middle of the list.

Example 2

Kirsty bought five similar bags of potatoes.
The number of potatoes in each bag was:
9, 11, 8 10 and 12
What is the median number per bag?

Arrange the numbers in ascending order:
8, 9, 10, 11, 12
↑
10 is the median number because it occurs in the middle of the list.

Exercise 20.1

Find the median.

1 John counted the number of matches in each of five boxes. The numbers were:
45, 47, 41, 43 and 42

2 One week the number of children present in Class 1B was:
Monday 17, Tuesday 16, Wednesday 19, Thursday 15, Friday 18

3 One week Jason's school bus took these times from his home to his school.
Monday 35 minutes, Tuesday 32 minutes, Wednesday 30 minutes, Thursday 38 minutes, Friday 37 minutes

4 One week the thermometer in Janet's garden showed the temperatures at midday:
Sunday 8 °C, Monday 11 °C, Tuesday 13 °C, Wednesday 9 °C, Thursday 10 °C, Friday 12 °C, Saturday 14 °C

5 One week a cafe sold these numbers of lunches:
Sunday 7, Monday 5, Tuesday 4, Wednesday 8, Thursday 13, Friday 6, Saturday 19

6 A vending machine sells cups of coffee, and for one week the sales were as follows:
Sunday 16, Monday 24, Tuesday 17, Wednesday 31, Thursday 25, Friday 33, Saturday 45

7 Andrew recorded the number of hours of sunshine for one week as follows:
Sunday 2, Monday 3, Tuesday 8, Wednesday 10, Thursday 5, Friday 1, Saturday 0

8 At Susan's school there is a small class with only nine children. Their heights are:
120, 116, 117, 113, 122, 114, 121, 124 and 112 cm

9 The weights of the children in Question **8** are:
35, 32, 33, 31, 38, 39, 30, 40 and 29 kg

Example 3

Soraya bought eight punnets of strawberries at various times.
The prices she paid were: 36 p, 43 p, 22 p, 29 p, 34 p, 39 p, 49 p and 25 p
Find the median price.

Arraange the prices in ascending order:
22 p, 25 p, 29 p, 34 p, 36 p, 39 p, 43p, 49 p
↑ ↑
As the number of prices is even there are two middle terms.
We take the median price as the value half way between the middle terms, namely 35 p.

Exercise 20.2

Find the median in Questions **1–4**.

1 Jayesh delivers newspapers on six days of the week.
One week his times to complete his round in minutes were:
32, 36, 38, 37, 34, 33

2 Kelly ran in the 400 metres race for her school on six occasions.
Her times in seconds were: 72, 69, 66, 64, 67, 70

3 A helicopter flies between Penzance and the Isles of Scilly.
One day eight flights were made and the numbers of passengers were:
21, 15, 9, 18, 14, 25, 11, 17

4 One day the roadway over Tower Bridge in London had to be raised eight times.
The delay times to traffic in minutes were:
12, 18, 21, 23, 13, 9, 11, 17

5 A train runs from London to Plymouth. The number of passengers on board for each stage of the journey is shown below.
Find the median number for **a** first class passengers and **b** standard class passengers.

	First class	Standard class
London (Paddington) to Reading	50	250
Reading to Newbury	55	230
Newbury to Westbury	45	190
Westbury to Taunton	30	210
Taunton to Exeter	40	240
Exeter to Newton Abbott	25	260
Newton Abbott to Totnes	20	220
Totnes to Plymouth	15	180

The **mode** of a set of data is the item or term which occurs the most frequently.
It is the most common term or item.

Example 4

Look at this data from Example 1, Section 19.1.

Anne: tea	George: tea	Azil: hot milk
Adam: Horlicks	Miranda: coffee	Melissa: Bournvita
David: coffee	Razi: chocolate	Paul: coffee
Emma: coffee	Salik: hot milk	Satwinder: chocolate
Fran: hot milk	Zorba: coffee	Yusuf: coffee

Coffee is the modal drink because this was the most popular drink.

Example 5

What is the mode of the birth dates in Example 1 in this section?

The most common birth date was the 12th (four children) so this is the mode of the birth dates.

Example 6

What is the mode of the birthday months in Example 1 in this section?

Three children were born in January and three were born in July, so for this case January and July are joint modes.

Example 20.3

1 Look at this data from Exercise 19.1. Twenty children in London were asked which football team they supported.
What is the mode of the football teams?

George: Arsenal	Luke: Spurs
Peter: West Ham	Steven: Chelsea
William: Spurs	Melanie: Spurs
David: Arsenal	Anna: Queen's Park Rangers
Joanne: Chelsea	Sunil: Arsenal
Robert: Arsenal	Sanjay: Chelsea
Andrew: Spurs	Rahul: West Ham
Richard: Chelsea	Manjula: Spurs
Kelly: Queen's Park Rangers	Daniel: Spurs
Sophie: Arsenal	Thomas: West Ham

3 Look at this data from Exercise 19.1
All twenty children in Class 2A bought a can of pop during one morning break.
What is the modal flavour of the drinks?

Saraya: apple
Ranjit: lime
Abbie: lemon
Jennifer: raspberry
Candace: orange
Shani: lemon
Polly: lime
Tanya: cola
Natalie: lemon
Rebecca: raspberry
Asiti: orange
Dean: lime
David: orange
George: apple
Joshua: lemon
Tyrone: raspberry
Peter: orange
Patrick: lime
Andrew: cola
James: orange

3 Look at this data from Exercise 19.1.
All fifteen children in Class 2B bought a packet of crisps during one morning break.
What is the modal flavour of the crisps?

Kate: salt and vinegar
Carl: prawn cocktail
Lucy: plain
Amy: plain
William: salt and vinegar
Michael: plain
Aisha: salt and vinegar
Robert: salt and vinegar
Suzanne: cheese and onion
Chantelle: cheese and onion
David: Worcester sauce
Herjinder: salt and vinegar
Laura: Worcester sauce
Pritesh: cheese and onion
George: cheese and onion

4 Look at this data from Exercise 19.1.
All twenty-five girls in year 11 were asked to choose their summer sports option.
What is the mode of the sports options?

Nicola: swimming
Annie: rounders
Sian: athletics
Marie: tennis
Nina: golf
Kirsty: athletics
Amanda: swimming
Frances: tennis
Rani: athletics
Janis: swimming
Jaswinder: tennis
Vicky: rounders
Becky: tennis
Drashma: athletics
Laura: swimming
Joanne: rounders
Mandeep: tennis
Natalie: golf
Kayleigh: swimming
Jodie: golf
Deborah: rounders
Samantha: tennis
Katie: athletics
Shani: athletics
Rosie: tennis

5 A school athletics club has twenty members.
Each member's best time for the 200 m race is:

Aisha	28 seconds	Bobby	30 seconds
Becky	29 seconds	Linford	28 seconds
Kayleigh	27 seconds	George	27 seconds
Emma	29 seconds	Tyrone	29 seconds
Laura	28 seconds	Julian	27 seconds
Nita	31 seconds	Daniel	28 seconds
Rani	28 seconds	Wayne	31 seconds
Suzanne	31 seconds	Sandeep	29 seconds
Jamilla	29 seconds	Marcus	28 seconds
Candace	30 seconds	Amrit	30 seconds

What is the modal time?

6 Fifteen children went strawberry picking.
The amounts they each picked were:

Marissa	17 kg	Gaynor	17 kg	Jack	18 kg
Pritti	18 kg	Camilla	15 kg	James	19 kg
Joanna	15 kg	Lucy	18 kg	Saleem	17 kg
Martha	16 kg	Paul	17 kg	David	18 kg
Rosie	18 kg	Jerome	19 kg	Liam	16 kg

What is the modal weight picked?

Example 7

Look at this data from Example 3, Section 19.1.
What is the modal height group?

Height	Tally	Frequency
110–119 cm	|||	3
120–129 cm	||||	4
130–139 cm	|	1
140–149 cm	~~||||~~	5
150–159 cm	||	2
	Total	15

There are five children with heights in the 140–149 cm range, so this is the modal height group.

Exercise 20.4

1 Look at this data from Exercise 19.3.
It shows the numbers of caravans parked on a holiday site over a four-week period.
What is the modal group for the numbers of caravans?

Number of caravans	0–9	10–19	20–29	30–39
Frequency	5	8	9	6

2 Look at this data from Exercise 19.3.
It shows the best performances of members of class 5A at high jump.
What is the modal group for height clearances?

Height cleared	100–109 cm	110–119 cm	120–129 cm	130–139 cm	140–149 cm	150–159 cm	160–169 cm
Frequency	2	3	4	5	3	2	1

3 Look at this data from Exercise 19.3.
It shows the times members in class 4C take to travel to school.

Time	0–9 min	10–19 min	20–29 min	30–39 min	40–49 min
Frequency	6	7	5	4	3

What is the modal time group?

4 Twenty dogs are staying at a boarding kennel.
The weights of the dogs in kilograms are:
24, 51, 45, 35, 27, 42, 10, 37, 21, 32, 48, 18, 31, 29, 39, 53, 47, 28, 36, 15
Copy and complete the tally chart.
Find the modal weight group.

Weight	Tally	Frequency
10–19 kg		
20–29 kg		
30–39 kg		
40–49 kg		
50–59 kg		

5 There are fifteen classes at Willow Lane School.
The numbers of children in each class are:
25, 15, 30, 27, 8, 17, 32, 14, 24, 9, 31, 12, 28, 9, 19
Copy and complete the tally chart.
Find the modal group for the class sizes.

Number in class	Tally	Frequency
0–9		
10–19		
20–29		
30–39		

6 **a** Write down the number of days in each month of an ordinary year.
What is the modal number of days in a month?

b Write down the first letter of the name of each month.
Which of these letters is the mode?

20.2 Understanding, calculating and using the mean and range of a set of data

Example 1

Find the mean of: 1, 5, 9, 17

To find the mean of a set of numbers, add up the numbers and divide the results by the number of numbers in the set.

The mean of 1, 5, 9 and 17 is $(1 + 5 + 9 + 17) \div 4 = 8$

Example 2

Find the mean score of a batsman who made 71, 36, 105, 41 and 22 in five completed innings.

The mean score is: $(71 + 36 + 105 + 41 + 22) \div 5 = 55$

Exercise 20.5

Find the mean.

1 Sherene's school bus takes the following journey times during a certain week:
Monday 28 minutes Tuesday 27 minutes
Wednesday 26 minutes Thursday 34 minutes
Friday 30 minutes

2 In Paul's class the attendance figures during a certain week were:
Monday 13 Tuesday 14 Wednesday 14
Thursday 17 Friday 17

3 James put four light bulbs into a chandelier and he found that they lasted for:
75 days 80 days 72 days 65 days

4 Kirsty bought four bars of soap in a packet and their weights were:
125 grams 123 grams 119 grams 129 grams

5 During a certain week the numbers of hours of sunshine in London were:
Monday 7.5 hours Tuesday 5.5 hours Wednesday 6 hours
Thursday 9.5 hours Friday 10.5 hours Saturday 10 hours
Sunday 7 hours

6 A newspaper seller works on six evenings each week.
One week his sales were:
Monday 120 Tuesday 129 Wednesday 132
Thursday 140 Friday 152 Saturday 203

7 One week Govinder worked on a fruit picking farm.
His daily takings were as follows:
Monday £28.40 Tuesday £24.30 Wednesday £25.30
Thursday £29.20 Friday £27.90 Saturday £30.50

8 During the summer examinations Carly achieved the following marks:
English 45% Geography 70% Art 65%
Mathematics 54% French 50% Craft 53%
History 63% Science 52%

9 Amarjit performed for her school athletics team on 10 occasions.
Her high jump clearances were:
115 121 118 117 116 122 120 119 123 and 122 cm

The **mean** gives an indication of the 'average' item in a set of data. It does no give any indication of whether the data is closely spaced or well spread out.

The **range** of the data is the difference between the largest and smallest item.

Example 3

Compare the performances of two hockey players who score the following numbers of goals in their matches.
a 4, 0, 0, 1, 3, 2, 0, 1, 1, 5, 1, 0. **b** 2, 1, 2, 2, 1, 1, 2, 1.

a Mean score = total ÷ 12 = 18 ÷ 12 = 1.5, Range = 5 − 0 = 5
b Mean score = total ÷ 8 = 12 ÷ 8 = 1.5, Range = 2 − 1 = 1

Both teams have a mean score of 1.5 goals, but their ranges are 5 and 1 respectively.
This shows that the first team has a very varying record while the second team's is very consistent.

Exercise 20.6

1 Sally buys two packets of torch bulbs made by two different manufacturers. Each packet contains four bulbs.
She timed how long the bulbs lasted:

Manufacturer A	Manufacturer B
6.5 hours	5.0 hours
7.5 hours	5.5 hours
6.0 hours	8.5 hours
8.0 hours	9.0 hours

Find the mean and range for the bulbs from **a** Manufacturer A **b** Manufacturer B.
Comment on any differences between the two manufacturer's bulbs.

2 Two village teams meet in a local football final after a four-round contest.
Their match attendances are:

	Team A	Team B
First round	1370	560
Second round	1410	1040
Third round	1840	2130
Semi-final	2360	3250

Find the mean and range for **a** Team A and **b** Team B.
Comment on any differences between the two teams' attendances.

3 A ferry sails seven times daily across a tidal estuary.
The crossing times for each day of a certain weekend are:
Saturday 13, 15, 13, 16, 16, 13 and 12 minutes
Sunday 9, 12, 11, 14, 17, 17 and 18 minutes
Find the mean and range for the times for **a** Saturday and **b** Sunday.
Comment on any differences between the times on the two days.

4 The table shows how Edward, Drashma and Helena perform in their summer examinations.

	Edward	Helena	Drashma
English	42	78	55
History	35	76	61
Geography	36	81	60
French	32	80	59
Mathematics	95	40	54
Science	85	39	58
Art	70	40	62
Craft	77	38	63

Find the mean and range for **a** Edward **b** Helena **c** Drashma
Comment on any differences between their marks.

Example 4

The mean of five numbers is 20.
If four of the numbers are 17, 9, 24 and 29, what is the fifth number?

If the mean of the five numbers is 20, the total is $20 \times 5 = 100$
The sum of the four given numbers is $17 + 19 + 24 + 29 = 89$
Therefore the fifth number is $100 - 89 = 11$

Exercise 20.7

1 The mean height of five boys is 149 cm.
If the heights of four of them are 146, 142, 149 and 151 cm, what is the height of the fifth?

2 The mean weight of five girls is 40 kg.
If the weights of four of them are 37, 41, 42 and 36 kg, what is the weight of the fifth?

3 One week (Monday to Saturday) a certain train suffered a mean lateness of two minutes.
For Monday to Friday the latenesses were:
2, 1, 3, 0 and 1 minute
By how many minutes was it late on Saturday?

4 One week (Monday to Saturday) Mr. Jones found that his mean time for driving to work was 45 minutes. For Monday to Friday his times were:
52, 48, 50, 47 and 49 minutes.
What was his time on Saturday?

5 One week a cafe served an average of 145 meals a day.
For Monday to Saturday the numbers were:
150, 154, 140, 135, 146 and 151
How many meals were served on Sunday?

6 One week, from Monday to Thursday, Andrew found that he waited a mean time of 7 minutes for his school bus. On Friday, there was a new driver and the mean time for all five days became 9 minutes.
For how long did he wait on Friday?

7 One week from Monday to Thursday, Susan found that her mean time for cycling to school was 18 minutes. On Friday, however, it was very wet and windy, and her mean time for all five days was 21 minutes.
How long did it take her on Friday?

8 One week, from Monday to Saturday, Rajinder found the mean temperature at midday to be 9.5 °C. On Sunday it was rather cold and the mean temperature for all seven days was 9.0 °C.
What was the temperature on Sunday?

9 A small class of 10 children had an English test. The average mark for nine of them was 52%.
The tenth pupil was new to school and did not know all of the work, so the average mark for all 10 pupils was 50%.
What mark did the new pupil get?

10 A ferry sails ten times daily. One day the mean time for nine sailings was 21 minutes. Fot the tenth sailing, however, the weather was very foggy and the overall mean time was 23 minutes.
What was the time for the tenth sailing?

21 Understanding, estimating and finding probabilities

21.1 Giving and justifying subjective estimates of probabilities

Probability tells us how likely an event is to occur.
It does not tell us what is going to happen.

Example 1

Say whether the occurrence of each of the following events is very likely, likely, not sure, unlikely or very unlikely.

a You will not eat any food at all tomorrow.
b Manchester United will score at least one goal in their next match.
c It will rain at least once in the next three days.
d It will not rain at all in the next three days.
e When you toss a coin the result will be a head.

a Very unlikely. (Everyone normally eats at least once a day.)
b Likely. (A good team scores at least one goal a match.)
c Very likely. (Most parts of the country have rain at least once every three days.)
d Unlikely. (Unless there is a heatwave it normally rains at least once in every three days.)
e Not sure. (It is equally likely to give a head or a tail.)

Exercise 21.1

Say whether the occurrence of each of the following events is very likely, likely, not sure, unlikely or very unlikely.

1 If I toss two coins together, the result will be two heads.

2 There will be snow during February next year.

3 My school bus will be a few minutes late tomorrow morning.

4 If I roll a dice, the score will be an even number.

5 If I leave my car lights on for an hour, my car will still start easily.

6 If I drive to work, all three traffic lights on the way will be showing green.

7 Two trains which are travelling in opposite directions both arrive at and depart from the same station at the same time.

8 If I throw a dice, the score will not be a six.

9 If I go jogging regularly, I will not put on weight.

10 Every day next April will be warm.

11 If I climb up a mountain, I will find that it gradually gets colder.

12 If I roll two dice, the score will be 12.

21.2 Understanding and using the probability scale from 0 to 1

Probabilities are usually measured on a scale from 0 to 1.
We say that an event which definitely will not occur has a probability of 0.
We say that an event which certainly will occur has a probability of 1.

0	$\frac{1}{2}$ or 0.5	1
(Impossible)	(Head or tail with a coin)	(Certain)

Example 1

Use the numbers 0, $\frac{1}{4}$, $\frac{1}{2}$, $\frac{3}{4}$ or 1 to show what you think the probability of each of these events is.

a When a coin is tossed the result is a head.
b When you pick a card from a pack of 52 it will be a red card.
c When you pick a card from a pack of 52 it will be a heart.
d It will snow for a whole week somewhere in England next July.
e You will not live until your 150th birthday.
f You are likely to own a car before you are 40.

a $\frac{1}{2}$ (or 0.5) A head or tail is equally likely
b $\frac{1}{2}$ (or 0.5) A red or black card is equally likely
c $\frac{1}{4}$ (or 0.25) $\frac{1}{4}$ of the pack are hearts (13 out of 52)
d 0 It has never been known to snow for a whole week in July.
e 1 No one has ever lived to such an age.
f $\frac{3}{4}$ (or 0.75) Most families own at least one car.

Exercise 21.2

1 Write down four events that are certain to occur (i.e. with a probability of 1).

2 Write down four events that are most unlikely or impossible (i.e. with a probability of 0).

3 Write down four events that have a probability of $\frac{1}{2}$ (or 0.5).

Exercise 21.3

Use the numbers 0, $\frac{1}{4}$, $\frac{1}{2}$, $\frac{3}{4}$ or 1 to show what you think the probability of each of these events is.

1 Passengers are told to stand upstairs on a bus.

2 I will use more petrol if I drive my car faster.

3 A car is to be built from steel which cannot rust.

4 In winter weather the ice will form on top of the pond first.

5 If I am stopped at a level crossing, a train will pass from left to right.

6 Peter will pass a two-paper mathematics test if he needs to get an average of 50% but has only 20% on the first paper.

7 It will snow somewhere in England next winter.

8 If I throw a dice, the score will be an even number.

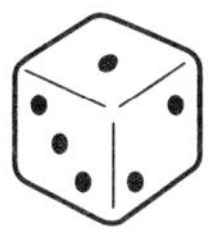

9 If I spin an octagonal spinner the score will be a square number.

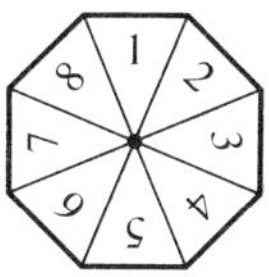

10 I will remove a blue pen from a case which contains 6 blue pens and 2 red ones.

11 The first pupil to get off a school bus at its last stop will be a boy if 4 boys and 12 girls are riding on it.

12 The first pupil from Class 2B to walk into the classroom will be a boy if there are 9 boys and 3 girls in the class.

21.3 Listing all the possible outcomes of an event

When a dice is rolled there are six possible outcomes, a score of 1, 2, 3, 4, 5 or 6.
Three of these are even numbers (2, 4, 6), so the probability of getting an even number is 3 out of 6, i.e. $\frac{3}{6}$ or $\frac{1}{2}$.
Two of these are multiples of 3 (3 and 6), so the probability of getting a multiple of 3 is 2 out of 6, i.e. $\frac{2}{6}$ or $\frac{1}{3}$.
Four of these are greater than 2 (3, 4, 5 and 6) so the probability of getting a number greater than 2 is 4 out of 6, i.e. $\frac{4}{6}$ or $\frac{2}{3}$.
If we can list all the possible outcomes of an event, and they are all equally likely, it is easy to calculate the probability of a particular outcome.

Example 1

a What is the probability of picking a picture card from the set of hearts in a pack of cards?
b What is the probability of picking a multiple of 4 from the set of spades in a pack of cards?

a There are 13 hearts and of these there are 3 picture cards: the King, the Queen and the Jack.
Therefore the probability of picking a picture card is $\frac{3}{13}$.
b There are 13 spades and of these two are multiples of 5: the 5 and the 10.
Therefore the probability of picking a multiple of 5 is $\frac{2}{13}$.

Exercise 21.4

1 The numbers from 1–10 are each printed on a counter and the 10 counters are placed in a bag.
If one counter is picked at random, what is the probability that the number on it is:
a a odd number **b** an even number **c** a multiple of 3
d a multiple of 4 **e** a multiple of 5 **f** a multiple of 6?

2 The numbers from 1–15 are each printed on a counter and the 15 counters are placed in a bag. If one counter is picked at random, what is the probability that the number on it is:
a a odd number **b** an even number **c** a multiple of 3
d a multiple of 4 **e** a multiple of 5 **f** a multiple of 6?

3 Twenty-four counters are placed in a bag. Two of them are red, three of them are orange, four of them are yellow, six of them are green and nine of them are blue.
If one counter is picked at random, what is the probability that it is:
a red **b** orange **c** yellow **d** green **e** blue?

4 Twenty counters are placed in a bag. Eight of them are red, six of them are yellow, four of them are green, and two of them are blue. If one counter is picked at random, what is the probability that it is:

a red **b** yellow **c** green **d** blue
e any colour except blue **f** any colour except green
g any colour except yellow **h** any colour except red?

5 If the spinner illustrated is spun, what is the probability that the score is:

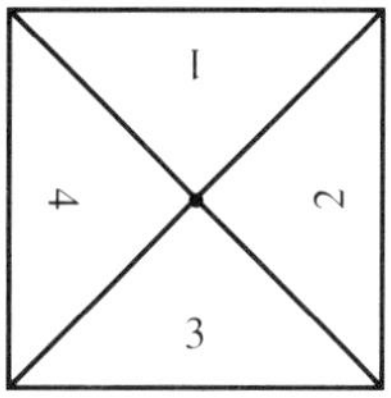

a an odd number **b** an even number
c a number greater than 2 **d** a number less than 2
e a number less than 3 **f** a number greater than 3?

6 If the spinner illustrated is spun, what is the probability that the score is:

a an odd number
b an even number
c a multiple of 3
d a multiple of 4
e a number greater than 4
f a number less than 4
g a number less than 5
h a number greater than 5?

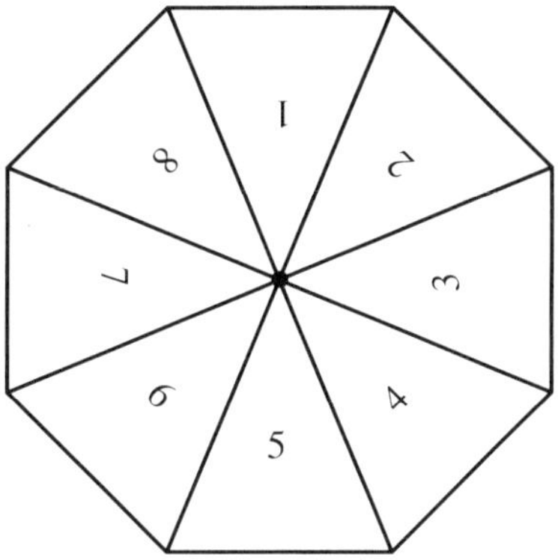

7 The letters of the five cities below are printed on cards. The cards for each city are put into a separate bag.

DUNDEE HEREFORD LEICESTER
EXETER ABERDEEN

What is the probability that a card with an E printed on it would be drawn at random from:

a the DUNDEE bag **b** the EXETER bag
c the HEREFORD bag **d** the ABERDEEN bag
e the LEICESTER bag?

8 The letters of the six names below are printed on cards. The cards for each name are put into a separate bag

JOANNA PATRICIA ALEXANDRA AMANDA
SAMANTHA BARBARA What is the probability that a card with an A printed on it would be drawn at random from:

a the JOANNA bag **b** the AMANDA bag
c the PATRICIA bag **d** the SAMANTHA bag
e the ALEXANDRA bag **f** the BARBARA bag?

Example 2

What are the possible outcomes when two coins are tossed together?

There are four possible outcomes: HH, HT, TH and TT

Example 3

What are the possible outcomes when a coin is tossed and a dice is rolled at the same time?

Coin	H	H	H	H	H	H	T	T	T	T	T	T
Dice	1	2	3	4	5	6	1	2	3	4	5	6

There are 12 possible outcomes.

Example 4

What are the possible outcomes when a red dice and a blue dice are rolled together?
What is the probability that the score is seven?

Red dice	Blue dice	Red dice	Blue dice	Red dice	Blue dice
1	1	1	2	1	3
2	1	2	2	2	3
3	1	3	2	3	3
4	1	4	2	4	3
5	1	5	2	5	3
6	1	6	2	6	3
Red dice	**Blue dice**	**Red dice**	**Blue dice**	**Red dice**	**Blue dice**
1	4	1	5	1	6
2	4	2	5	2	6
3	4	3	5	3	6
4	4	4	5	4	6
5	4	5	5	5	6
6	4	6	5	6	6

There are 36 possible outcomes.
There are six ways of scoring seven. Each one is shaded on the table above.
Therefore the probability of scoring seven $= \frac{6}{36} = \frac{1}{6}$.

Exercise 21.5

1 Three coins are tossed together.
List the possible outcomes.
What is the probability of:
a three heads **b** three tails **c** two heads and a tail **d** one head and two tails?

2 Fred has four counters, two red ones and two blue ones.
If he lays them out in a line, how many patterns can he make?
What is the probability of making an arrangement which has:
a two reds together **b** two blues together **c** a symmetrical pattern?

3 The two spinners illustrated are spun together.
List all the possible score outcomes.
Copy and complete this table.

Score	Number of ways
2	
3	
4	
5	
6	

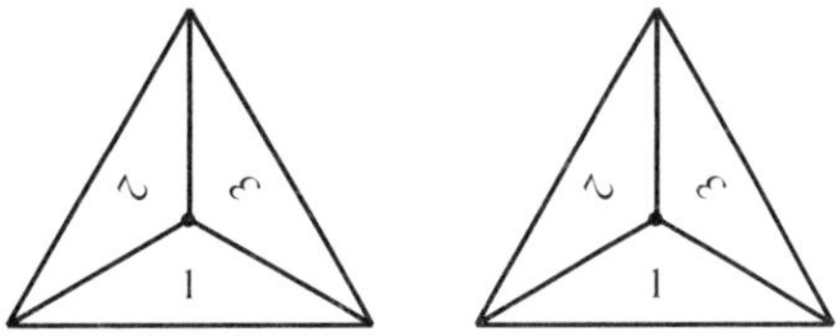

Find the probability of obtaining each of the scores.

4 The two spinners illustrated are spun together.
List all the possible score outcomes.
Copy and complete this table.

Score	Number of ways
2	
3	
4	
5	
6	
7	
8	
9	

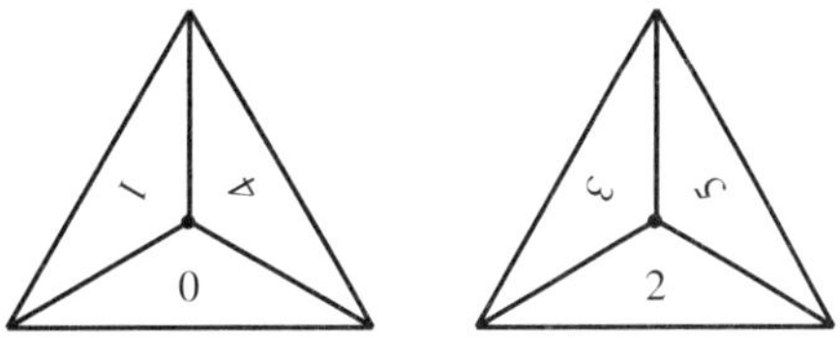

Find the probability of obtaining each of the scores.

5 These are the colours for the four teams who reached the semi-finals of the FA Cup during a certain year:
Chelsea (blue) Manchester United (red)
Everton (blue) Nottingham Forest (red)
List all the possible pairings for the FA Cup final.
If all outcomes are equally probable, find the probability that in the final:
a two red teams will meet **b** two blue teams will meet
c a red and a blue team will meet **d** there will be colour clash.

6 Near the end of a season a football team has only two matches left. List all nine possible outcomes for these last two matches.
Three points are awarded for a win, one for a draw and none for a loss.
Copy and complete this table.

Outcome	Number of points

Use your table to complete the table below.

Number of points	Number of possible ways
0	
1	
2	
3	
4	
5	
6	

If a team needs 4 or more points to win the championship, and all outcomes are equally probable, what is the probability that they will win it?

7 In a town there is a market day daily from Monday to Friday, but James is told that he can only trade on three of the days.
How many different sets of three different days are possible?
If the probability of all of the above outcomes is the same, what is the probability that James will trade on three consecutive days?

8 At a seaside resort Celia is told that she can only sell ice cream on any of the two days from Friday to Monday over the Easter holiday weekend.
How many different pairs of two different days are possible?
If the probability of all the above outcomes is the same, what is the probability that Celia will sell ice cream on two consecutive days?

21.4 Understanding that different outcomes may result from repeating an experiment

When a six-sided dice is rolled, each of the numbers 1, 2, 3, 4, 5 and 6 are equally likely results. (Assuming that the dice is unbiased.) Over a large number of rolls each of the numbers is likely to occur with a probability of $\frac{1}{6}$ (or 0.1666). For any particular set of rolls, however, the number of times that a given number occurs will not necessarily be exactly $\frac{1}{6}$ of the total.

Example 1

A dice is rolled 60, 120, 180 and 240 times and the results are:

(i)	Number	1	2	3	4	5	6	
	Frequency	8	12	10	11	10	9	(60 rolls in all)
(ii)	Number	1	2	3	4	5	6	
	Frequency	24	16	22	20	18	20	(120 rolls in all)
(iii)	Number	1	2	3	4	5	6	
	Frequency	30	33	24	36	27	30	(180 rolls in all)
(iv)	Number	1	2	3	4	5	6	
	Frequency	38	39	44	33	45	41	(240 rolls in all)

a Find for what fraction of the total number of rolls each number occurred.

b Total the results to find the number of times that each number occurred in the 600 rolls altogether.
From what fraction of the total does each number occur now?

a	(i)	Number	1	2	3	4	5	6
		Fraction	$\frac{2}{15}$	$\frac{1}{5}$	$\frac{1}{6}$	$\frac{11}{60}$	$\frac{1}{6}$	$\frac{3}{20}$
	(ii)	Number	1	2	3	4	5	6
		Fraction	$\frac{1}{5}$	$\frac{2}{15}$	$\frac{11}{60}$	$\frac{1}{6}$	$\frac{3}{20}$	$\frac{1}{6}$
	(iii)	Number	1	2	3	4	5	6
		Fraction	$\frac{1}{6}$	$\frac{11}{60}$	$\frac{2}{15}$	$\frac{1}{5}$	$\frac{3}{20}$	$\frac{1}{6}$
	(iv)	Number	1	2	3	4	5	6
		Fraction	$\frac{19}{120}$	$\frac{13}{80}$	$\frac{11}{60}$	$\frac{11}{80}$	$\frac{3}{16}$	$\frac{41}{240}$

b	Number	1	2	3	4	5	6
	Frequency	100	100	100	100	100	100
	Fraction	$\frac{1}{6}$	$\frac{1}{6}$	$\frac{1}{6}$	$\frac{1}{6}$	$\frac{1}{6}$	$\frac{1}{6}$

Note that with a larger number of rolls the fractions are likely to get closer to $\frac{1}{6}$, but unlike the example above they will not necessarily be exactly $\frac{1}{6}$.

Exercise 21.6

1 a Toss a coin 20, 40, 60 and 80 times. Copy and complete these tables.

(i)

Outcome	Head	Tail
Frequency		

(20 tosses)

(ii)

Outcome	Head	Tail
Frequency		

(40 tosses)

(iii)

Outcome	Head	Tail
Frequency		

(60 tosses)

(iv)

Outcome	Head	Tail
Frequency		

(80 tosses)

b Now total the results.
Copy and complete this table.

Outcome	Head	Tail
Frequency		
Fraction		

(200 tosses)

c Are both your fractions in part **b** close to $\frac{1}{2}$?

d If five groups do this experiment you can make a table for the results of 1000 tosses. This should give results very close to $\frac{1}{2}$.

2 a Write all the letters of the alphabet except Z on cards. Shuffle the pack of cards.
Pick a card (returning it to the pack each time) 20, 40, 60 and 80 times.
Copy and complete these tables.

(i)

Outcome	Vowel	Consonant
Frequency		

(20 cards picked)

(ii)

Outcome	Vowel	Consonant
Frequency		

(40 cards picked)

(iii)

Outcome	Vowel	Consonant
Frequency		

(60 cards picked)

(iv)

Outcome	Vowel	Consonant
Frequency		

(80 cards picked)

b Now total the results.
Copy and complete this table:

Outcome	Vowel	Consonant
Frequency		
Fraction		

(200 cards picked)

c Are your fractions in part **b** close to $\frac{1}{5}$ and $\frac{4}{5}$?

3 a Shuffle a pack of playing cards. Pick a card (returning it to the pack each time) 20, 40, 60 and 80 times.
Copy and complete these tables:

	Outcome	Diamond	Heart	Club	Spade	
(i)	Outcome	Diamond	Heart	Club	Spade	
	Frequency					(20 cards picked)
(ii)	Outcome	Diamond	Heart	Club	Spade	
	Frequency					(40 cards picked)
(iii)	Outcome	Diamond	Heart	Club	Spade	
	Frequency					(60 cards picked)
(iv)	Outcome	Diamond	Heart	Club	Spade	
	Frequency					(80 cards picked)

b Now total the results.
Copy and complete this table:

Outcome	Diamond	Heart	Club	Spade
Frequency				
Fraction				

(200 cards picked)

c Are your fractions in part **b** close to $\frac{1}{4}$?

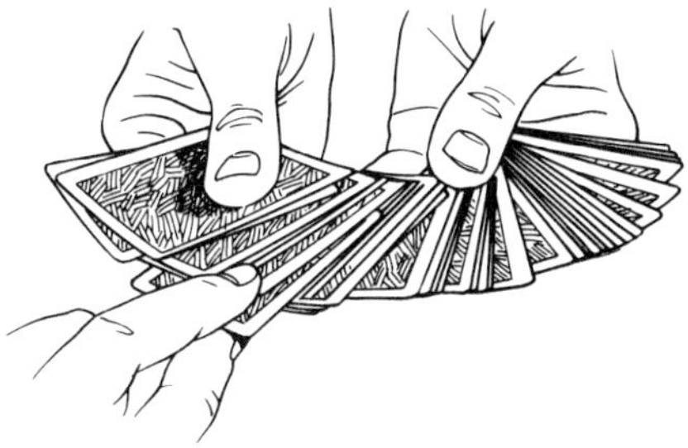

21.5 Recognising situations where estimates of probability can be based on equally likely events and others where estimates must be based on statistical evidence

In Section 21.4 the possible outcomes from throwing an unbiased dice are equally likely.
But the probability that the next vehicle to come along a road might be a lorry is certainly not equally likely at all times or on all roads. It will depend on the time of day, where the road leads to, and possibly many other things.
The example below assumes that the events are equally likely.

Example 1

A bag contains 5 red balls, 3 green balls and 2 white balls. One ball is drawn at random from it.
What is the probability that the colour of the selected ball is **a** red, **b** green, **c** white?

a Each of the balls (but not each of the colours) is equally likely to be withdrawn, therefore the probability is:

$$\frac{\text{number of reds}}{\text{total}} = \frac{5}{10} = \frac{1}{2} \quad \text{(or 0.5)}$$

b Probability is: $\dfrac{\text{number of greens}}{\text{total number}} = \dfrac{3}{10}$ (or 0.3)

c Probability is: $\dfrac{\text{number of whites}}{\text{total number}} = \dfrac{2}{10} = \dfrac{1}{5}$ (or 0.2)

Exercise 21.7

1 60 people are travelling on a bus. 27 are sitting downstairs, 8 are standing downstairs, 10 are sitting upstairs and smoking, and 15 are sitting upstairs but are not smoking. What is the probability that the first passenger to get off the bus was:
a sitting downstairs **b** standing downstairs
c a smoker from upstairs **d** a non-smoker from upstairs?

2 A carpenter wishes to drill a hole whose diameter may be anything between 4 mm and 8 mm inclusive. He has a box of 12 drill bits and their diameters are:
2 mm, 2.5 mm, 4 mm, 4.5 mm, 6 mm, 7.5 mm, 8 mm, 9 mm
12 mm, 12.5 mm, 13.5 mm, 15 mm

If he picks a drill bit out of the box at random, what is the probability that it will be:
a of the correct size **b** too small **c** too big?

3 The table shows the shoe sizes of the children in Class 4C.

Dean 5	Paul 4	Suzanne 5
William 3	David 3	Deena 3
Liam 4	Kelly 2	Anne 5
Michael 5	Emma 5	Laura 2
James 5	Nita 4	Nicola 4
Sunil 5	Candace 3	Yogita 3
Sanjay 3	Shani 5	

If one of the children is selected at random, what is the probability that he or she has shoe size:
a 2 **b** 3 **c** 4 **d** 5?

4 In Natalie's house the immersion heater is left on 24 hours a day. It is turned off for a total of $4\frac{1}{2}$ hours a day because of the thermostat setting.
If a red light shows when the power is on, what is the probability that this light will be **a** off, and **b** on, if it is looked at at random?

5 The diagram shows the cross-section of the Severn Estuary. The Grubby near the Welsh bank is only visible for 8 hours a day and the English Stones for only $7\frac{1}{2}$ hours a day.
If someone turns up at random, what is the probability that he or she will be able to see **a** The Grubby and **b** the English Stones?

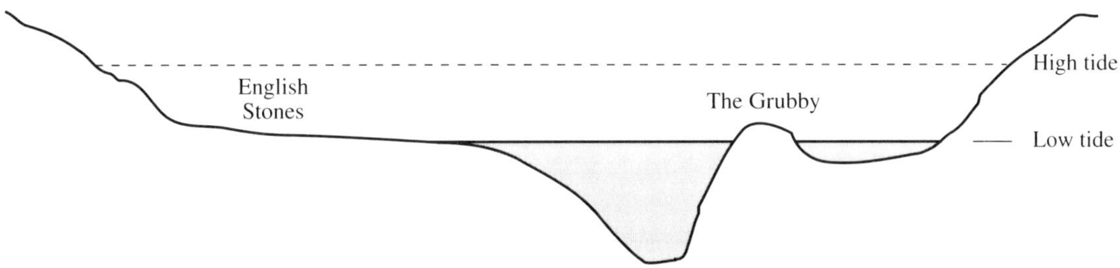

6 Near to my house there is a level crossing and 120 trains pass it between 6 a.m. and 10 p.m. The gates have to be closed for two minutes for each train to pass.
Find the probability that I will find the gates closed if I drive down the road between the above times.

7 My garden has an area of $200\,\text{m}^2$. It consists of a lawn of area $168\,\text{m}^2$ and a pond of area $32\,\text{m}^2$. If a wild duck lands in my garden at random, what is the probability that it will land **a** on the grass, and **b** on the water?

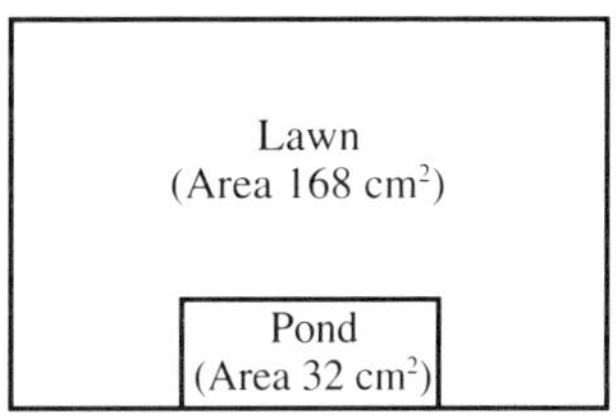

8 The surface area of the Earth is 500 000 000 km^2; 350 000 000 km^2 of which is ocean and 150 000 000 km^2 of which is land.
If a meteorite hits the Earth at random, what is the probability that it will come down on:
a water **b** land
c anywhere in the tropical zone if the area of the zone is 220 000 000 km^2?

If we ask what is the probability that a particular light bulb is faulty or not, we might assume that the answer is $\frac{1}{2}$ because the light bulb can only be faulty or not faulty. However, the two outcomes do not have to be equally likely. It would be reasonable to assume that most light bulbs are not faulty.

Example 2

It was suspected that a large consignment of light bulbs contained too many defectives.
The number of defectives in each of 12 packets are given in the table.
Estimate the probability of a given light bulb being defective.

Number of light bulbs in packet	6	6	6	8	8	8	10	10	10	12	12	12
Number of defective bulbs	0	0	1	1	1	0	2	0	1	1	1	1

The total number of light bulbs $= 6 + 6 + 6 + 8 + 8 + 8 + 10 + 10 + 10 + 12 + 12 + 12 = 108$
The total number of defective bulbs $= 0 + 0 + 1 + 1 + 1 + 0 + 2 + 0 + 1 + 1 + 1 + 1 = 9$
Therefore the estimated probability of a light bulb being defective is: $\frac{9}{108} = \frac{1}{12}$ (or 0.083)

Exercise 21.8

1 Estimate the probability that a new-born lamb will be a black one from this evidence:

Farm	High Lane	Manor Row	Hill Top	Dale Head	Riverside
Number of sheep	40	50	45	60	36
Number of black sheep	12	15	18	20	12

2 Estimate the probability that a show jumping horse will hit a fence from this evidence from six trials:

Number of fences on trial course	10	12	9	8	9	12
Number of fences hit	1	2	2	0	3	4

3 Ms Khan's train is due to arrive at the station near her work at 8.30 a.m. on Monday to Friday.
The table shows the actual arrival times over a four-week period.

Monday	8.30	Monday	8.30	Monday	8.30	Monday	8.30
Tuesday	8.25	Tuesday	8.29	Tuesday	8.28	Tuesday	8.29
Wednesday	8.30	Wednesday	8.30	Wednesday	8.36	Wednesday	8.30
Thursday	8.27	Thursday	8.34	Thursday	8.24	Thursday	8.32
Friday	8.40	Friday	8.30	Friday	8.30	Friday	8.30

Estimate the probability that her train will arrive:
a on time **b** late **c** early.

4 A football team played in 40 league matches during a season.
The overall record was:

HOME			
Played	Won	Drawn	Lost
20	14	5	1

AWAY			
Played	Won	Drawn	Lost
20	6	9	5

Estimate the probability that the team will:
a win at home **b** draw at home **c** lose at home
d win away **c** draw away **f** lose away
g win any match **h** draw any match **i** lose any match

21.6 Knowing that if each of n events is equally likely, the probability of one occurring is $\frac{1}{n}$

An unbiased dice has six faces and each of the numbers is equally likely to occur when the dice is rolled. The probability of getting any of the numbers is therefore $\frac{1}{6}$.

If you pick a telephone number at random on any page in the phone book, the probability that the last digit will be a 0 is $\frac{1}{10}$. There are 10 possible digits, 0, 1, 2, 3, 4, 5, 6, 7, 8, 9 and 0, and each one is equally likely.
However if you look at the first digit of any six-digit telephone number you will find that the different digits are not equally likely. For example, in some areas most six-digit numbers to start with a 2.

A similar situation can be found with car registration numbers. In the number

D 570 PNM

the very last letter is equally likely to be any of the letters used.
The first letter, which shows the delivery date, is not.
There will clearly be more cars with the letters for the more recent times of delivery.

Example 1

At Skylines Airport aeroplanes are sent out on one of eight directions given by the bearings:

045° 135° 225° 315° 090° 180° 270° 360°

If the traffic controller is equally likely to call any of the bearings overall, what is the probability that for the bearing:

a the first digit is 2? **b** the last digit is 5?
c the middle digit is 2? **d** the middle digit is 5?

a The probability $= \frac{2}{8} = \frac{1}{4}$
(Each of the four possible numbers for the first digit is equally likely.)

b The probability $= \frac{4}{8} = \frac{1}{2}$
(Each of the two possible numbers for the last digit is equally likely.)

c The probability $= \frac{1}{8}$
(Each of the eight possible numbers for the middle digit is equally likely.)

d The probability $= 0$
(5 is not one of the numbers used for the middle digit.)

Exercise 21.9

1 PAT writes each of the letters of her name on separate cards and places the three cards in a bag.
If she then withdraws a single card, what is the probability that it has:
a P **b** A **c** T written on it?

2 ADA writes each of the letters of her name on separate cards and places the three cards in a bag.
If she then withdraws a single card, what is the probability that it has:
a A **b** D written on it?

3 PETE writes the four letters of his name on separate cards and places the four cards in a bag.
If he then withdraws a single card, what is the probability that it has:
a P **b** T **c** E written on it?

4 LILLIAN writes the four letters of her shortened name LILL on separate cards and places the four cards in a bag.
If she then withdraws a single card, what is the probability that:

a it has L written on it
b it has I written on it?

5 At Cloudways Airport aeroplanes are sent out on one of 12 directions given by the bearings:

030° 120° 210° 300° 060° 150° 240° 330° 090° 180° 270° 360°

If the traffic controller is equally likely to call any of the bearings overall, what is the probability that for the bearing:

a the first digit is 2
b the first digit is 3
c the middle digit is 5
d the middle digit is 9
e the middle digit is 3
f the middle digit is 6
g the last digit is 0?

6 A small railway company has 80 locomotives. The passenger train locomotives are numbered 100 to 159 and the freight train locomotives are numbered 200 to 219.

If a train comes along what is the probability that:

a the last digit of the number is 9
b the last digit of the number is 5
c the first digit of the number is 1
d the first digit of the number is 2
e the middle digit of the number is 4
f the middle digit of the number is 2
g the middle digit of the number is 1
h the middle digit of the number is 0?

7 A small island has only 200 telephone subscribers. Numbers 100 to 149 are in West Area, 200 to 299 are in Mid Area, and 300–349 are in East Area.

If a number is chosen at random from the directory, what is the probability that:

a the last digit is 9
b the last digit is 5
c the first digit is 2
d the first digit is 3
e the middle digit is 9
f the middle digit is 1?

21.7 Identifying all the outcomes when dealing with two combined events which are independent, using diagrammatic, tabular and other forms

Example 1

Two coins are tossed.
Show the possible outcomes **a** as a table, **b** on a grid and **c** as a tree diagram.

a

		Second coin	
		Head	Tail
First coin	Head	HH	HT
	Tail	TH	TT

b

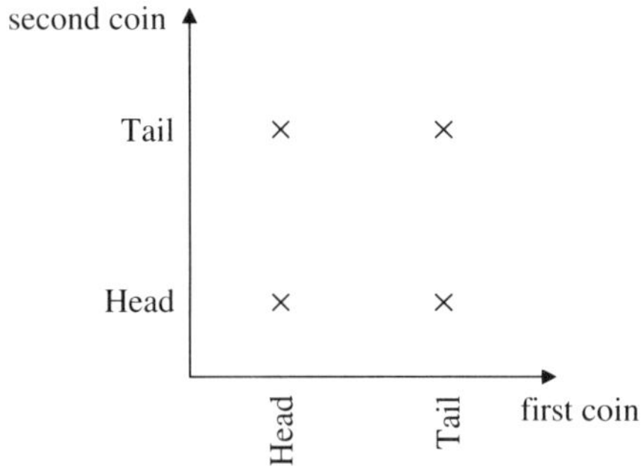

c

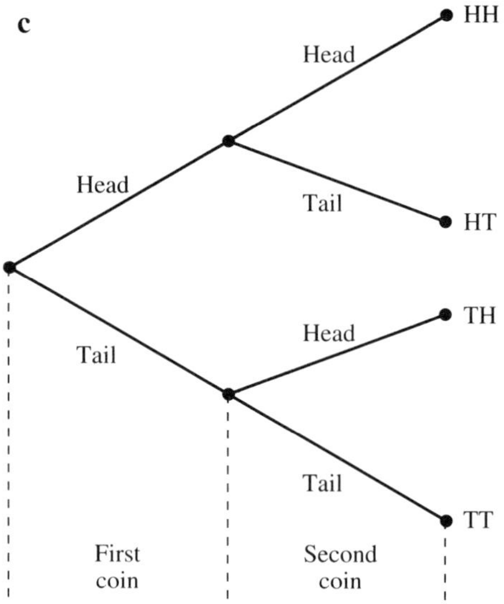

Example 2

A coin is tossed and a dice is rolled.
Show the possible outcomes **a** as a table, **b** on a grid, and **c** as a tree diagram.

a

		Dice					
		One	Two	Three	Four	Five	Six
Coin	Head	H, 1	H, 2	H, 3	H, 4	H, 5	H, 6
	Tail	T, 1	T, 2	T, 3	T, 4	T, 5	T, 6

b

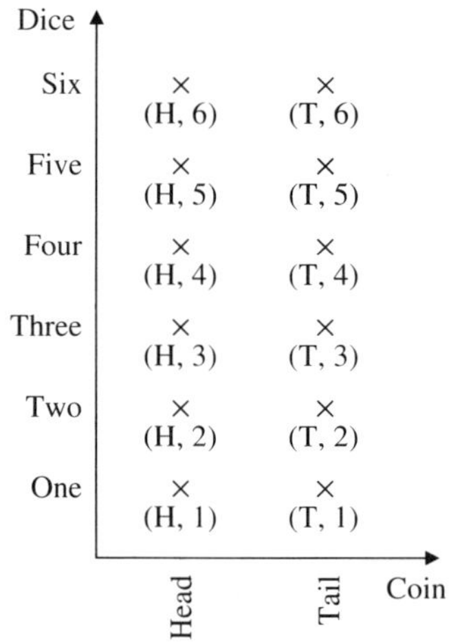

c

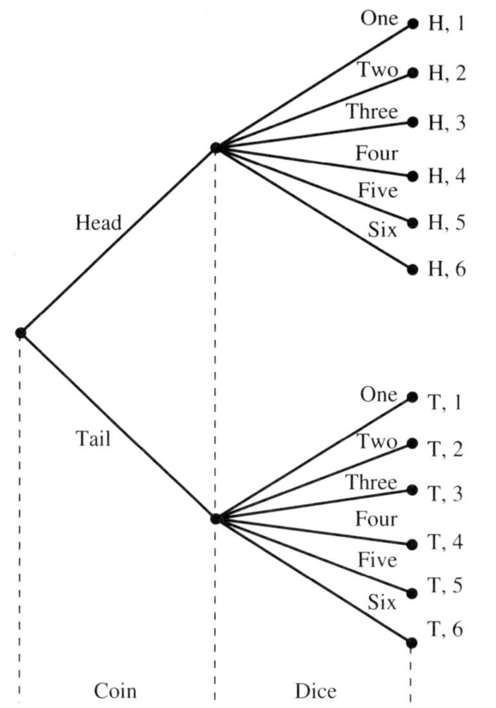

Exercise 21.10

1 Two dice are rolled.

a Copy and complete this outcome table.

Second dice

First dice	1	2	3	4	5	6
1	1, 1	1, 2				
2						
3						
4						
5						
6					6, 5	6, 6

b Show the outcomes on a grid.

c For how many outcomes is the score:

(i) an even number

(ii) a 'double' (both dice showing the same face)

(iii) a multiple of 3

(iv) a multiple of 5?

2 Two tetrahedral dice are thrown.

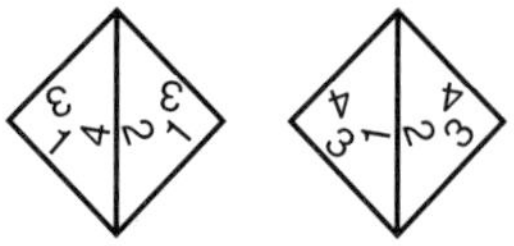

a Show the outcomes on a table similar to that in Question **1**.
b Show the outcomes on a grid.
c For how many outcomes is the score:
(i) an odd number (ii) a 'double' (iii) a multiple of 3 (iv) a multiple of 4?

3 An ordinary dice and a tetrahedral dice are thrown together.

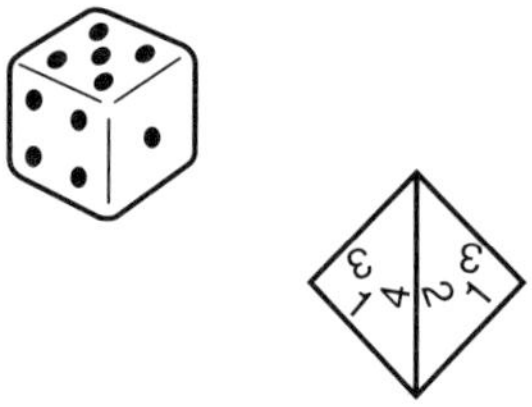

a Show the outcomes on a table similar to that in Question **1**.
b Show the outcomes on a grid.
c For how many outcomes is the score:
(i) a double (ii) a square number (iii) a prime number?

4 Two arrows are shot at the archery target shown.
(If the arrow misses the target the score is zero.)
a Copy and complete this table.

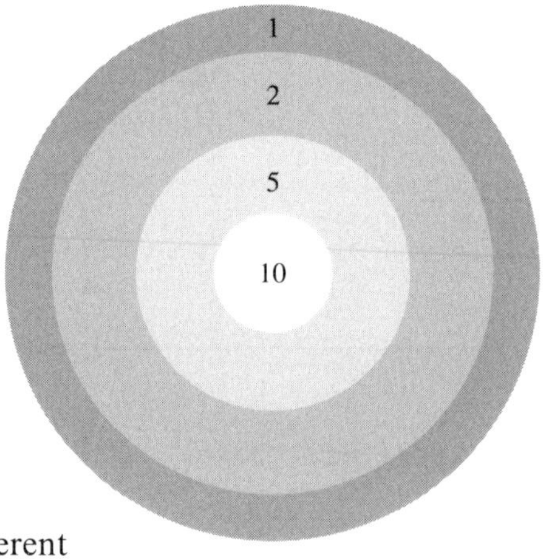

Second arrow

First arrow	0	1	2	5	10
0					
1			1, 2		
2					
5		5, 1		5, 5	
10					

b Show the outcomes on a grid.
c Only two of the possible scores can arise by three different methods.
Which two scores are these?

5 Two playing cards are picked at random from a shuffled pack.
a Copy and complete the table below:

Second card

	Diamond	Heart	Club	Spade
Diamond	DD	DH		
Heart				
Club				
Spade				

b Show the outcomes on a grid.
c For how many outcomes are the two cards:
(i) from the same suit
(ii) both from a red suit
(iii) both from a black suit?

6 Three coins are tossed in succession.

a Copy and complete this tree diagram to show the possible outcomes.

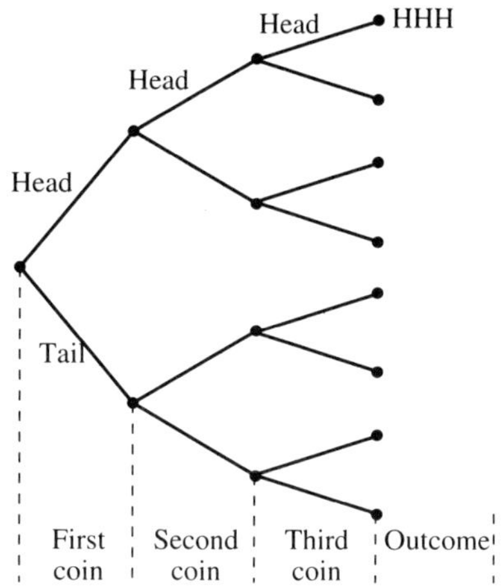

b How many outcomes result in:

(i) two heads and a tail
(ii) two tails and a head
(iii) two heads in succession
(iv) two tails in succession
(v) a symmetric sequence, e.g. HTH?

7 a Jodie has three days off school for a half-term holiday. Copy and complete the tree diagram to show the possible weather outcomes.

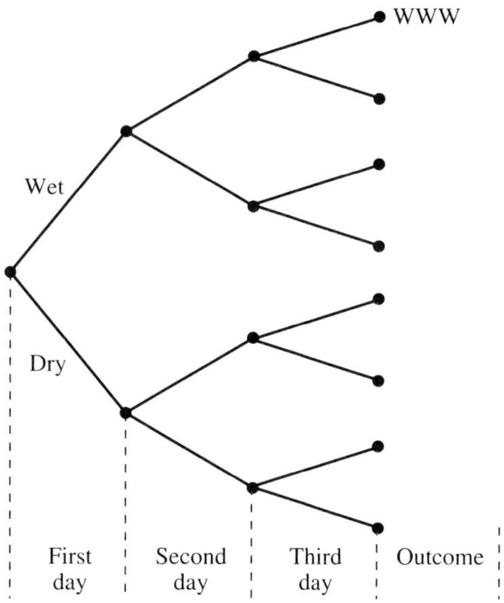

b How many outcomes have:

(i) successive wet days
(ii) successive dry days
(iii) alternating weather from one day to the next?

8 A bag contains red and blue counters. Three counters are withdrawn in succession.

a Copy and complete the tree diagram to show the outcomes.

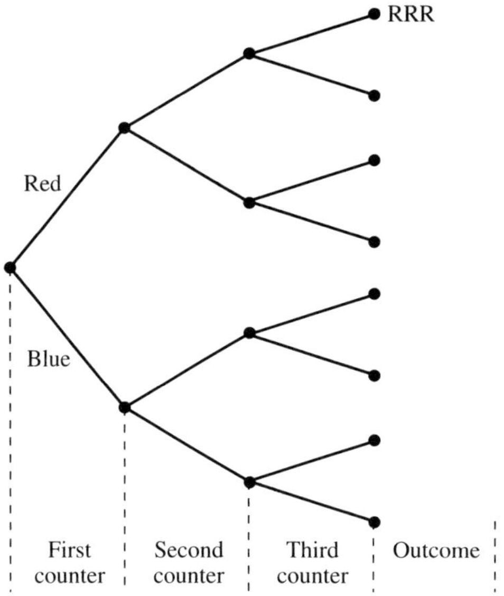

b If the order does not matter, how many different outcomes are there?

9 A bag contains red, white and blue counters and two counters are withdrawn in succession.

a Copy and complete the tree diagram to show the outcomes.

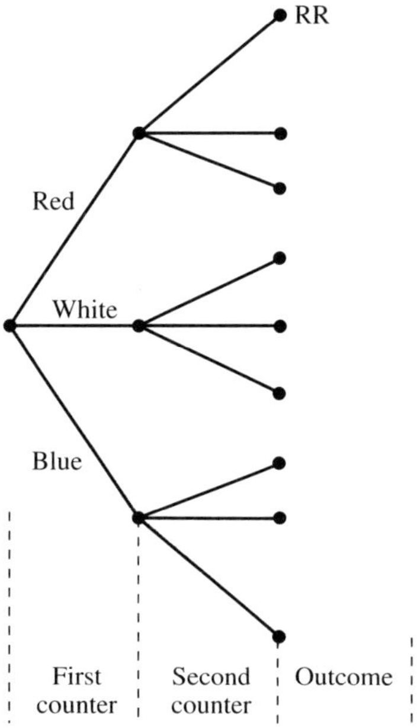

b If the order does not matter, how many different outcomes are there?

21.8 Appreciating that the total sum of the probabilities of mutually exclusive events is one and that the probability of something happening is one minus the probability of it not happening

Example 1

A bag contains 20 beads: 10 red, 5 yellow, 3 green and 2 blue. If one is picked at random, what is the probability that it is:
a red **b** yellow **c** green **d** blue?

a Probability is: $\frac{\text{number of reds}}{\text{total number}} = \frac{10}{20} = \frac{1}{2}$

b Probability is: $\frac{\text{number of yellows}}{\text{total number}} = \frac{5}{20} = \frac{1}{4}$

c Probability is: $\frac{\text{number of greens}}{\text{total number}} = \frac{3}{20}$

d Probability is: $\frac{\text{number of blues}}{\text{total number}} = \frac{2}{20} = \frac{1}{10}$

Note that the sum of the four 'mutually exclusive' probabilities is one.

$\frac{1}{2} + \frac{1}{4} + \frac{3}{20} + \frac{1}{10} = 1.$

Exercise 21.11

1 There are 60 counters in a bag. 20 are red, 18 are yellow, 16 are green and 6 are blue.
If one is withdrawn at random, what is the probability that its colour is:
a red **b** yellow **c** green **d** blue?

2 There are 24 dresses on a rack in a clothes shop. 9 of them are white, 8 of them are yellow, 4 of them are pink and 3 of them are crimson.
If one is picked from the rack at random, what is the probability that it is:
a white **b** yellow **c** pink **d** crimson?

3 Manjula's science teacher has 36 metal cylinders in a box. 18 are steel, 10 are aluminium, 6 are copper and 2 are brass.
If Manjula picks one at random for an experiment, what is the probability that it is made of:
a steel **b** aluminium **c** copper **d** brass?

4 There are 40 coins in a cash bag. 4 are ten-pence pieces, 12 are five-pence pieces, 16 are two-pence pieces and 8 are one-pence pieces. If one coin is removed from the bag at random, what is the probability that its value is:
a 10 p **b** 5 p **c** 2 p **d** 1 p?

5 LILLIAN writes the seven letters of her name on a card and then shuffles the cards.
If she picks one card at random, what is the probability that the letter on it is:
a L **b** I **c** A **d** N?

6 MINNIE writes the six letters of her name on cards and then shuffles the cards.
If she picks one card at random, what is the probability that the letter on it is:
a I **b** N **c** M **d** E?

Example 2

A bag contains 20 marbles: 15 red and 5 yellow.
If one is picked at random, what is the probability that it is:
a red **b** not a red one **c** yellow **d** not a yellow one?

a Probability is: $\dfrac{\text{number of reds}}{\text{total number}} = \dfrac{15}{20} = \dfrac{3}{4}$

b Probability is: $1 - \dfrac{3}{4} = \dfrac{1}{4}$ (i.e. one minus the probability from part **a**)

c Probability is: $\dfrac{\text{number of yellows}}{\text{total number}} = \dfrac{5}{20} = \dfrac{1}{4}$

d Probability is: $1 - \dfrac{1}{4} = \dfrac{3}{4}$ (i.e. one minus the probability from part **c**)

Exercise 21.12

1 A fete is being held in a large hall which has 9 doors. By mistake the caretaker has left 3 of them locked.
If someone tries a door at random, what is the probability that it will be **a** locked, **b** unlocked?

2 Marcus has 10 pairs of shoes and 2 of the pairs are brown. If he puts on pair at random what is the probability that he will be wearing **a** brown shoes, **b** a pair that are not brown?

3 A matchbox contains 48 matchsticks but 12 of them are used ones. If one is taken out of the box at random, what is the probability that it is **a** a used one, **b** not a used one?

4 A baker has 90 loaves in his oven and 15 of them are brown. What is the probability that the first loaf he takes out is **a** a brown one, **b** not a brown one?

5 Deepal's science teacher has 24 electrical meters on his laboratory shelf and 9 of them are voltmeters. If Deepal picks one at random for an experiment, what is the probability that he will pick **a** a voltmeter, **b** not a voltmeter?

6 AMANDA writes the six letters of her name on cards and shuffles the cards.
If she picks one card at random, what is the probability that the letter on it is **a** A, **b** not an A?

Example 3

A bag contains 10 counters: 6 red, 3 yellow and 1 blue.
If one counter is taken out at random, what is the probability that it is:

a a red counter **b** not a red counter
c a yellow counter **d** not a yellow counter
e a blue counter **f** not a blue counter?

Show that the answers to parts **a**, **c** and **e** total 1.

a Probability is: $\dfrac{\text{number of reds}}{\text{total number}} = \dfrac{6}{10} = \dfrac{3}{5}$

b Probability is: $1 - \dfrac{3}{5} = \dfrac{2}{5}$ (i.e. one minus the probability in part **a**)

c Probability is: $\dfrac{\text{number of yellows}}{\text{total number}} = \dfrac{3}{10}$

d Probability is: $1 - \dfrac{3}{10} = \dfrac{7}{10}$ (i.e. one minus the probability in part **c**)

e Probability is: $\dfrac{\text{number of blues}}{\text{total number}} = \dfrac{1}{10}$

f Probability is: $1 - \dfrac{1}{10} = \dfrac{9}{10}$ (i.e. one minus the probability in part **e**)

The probability of either a red, a yellow or a blue is

$$\frac{3}{5} + \frac{3}{10} + \frac{1}{10} = 1$$

You are certain to get one of these colours.
The three events are mutually exclusive.

Exercise 21.13

1 A packet contains 30 plastic cups: 12 white, 10 yellow and 8 pink.
If one cup is taken out at random, what is the probability that it is:
a a white cup **b** not a white cup **c** a yellow cup
d not a yellow cup **e** a pink cup **f** not a pink cup?

2 A large service station has 20 fuel pumps: 12 dispense unleaded petrol, 6 dispense leaded petrol, and 2 dispense diesel.
If a motorist stops at any pump, what is the probability that:
a it dispenses unleaded petrol
b it does not dispense unleaded petrol
c it dispenses leaded petrol
d it does not dispense leaded petrol
e it dispenses diesel
f it does not dispense diesel?

3 There are 24 newspapers in a pile in the newsagent's shop: 10 copies of the Daily Gossip, 8 copies of the Daily Looker and 6 copies of the Daily Viewer.
If one is picked out at random, what is the probability that:
a it is a copy of the Daily Gossip
b it is not a copy of the Daily Gossip
c it is a copy of the Daily Looker
d it is not a copy of the Daily Looker
e it is a copy of the Daily Viewer
f it is not a copy of the Daily Viewer?

4 A children's book has 60 pages. 24 of them are written pages, 21 of them are picture pages and 15 of them are blank pages for drawing on.
If any page at random is looked at, what is the probability that it is:
a a written page
b not a written page
c a picture page
d not a picture page
e a blank page for drawing on
f not a blank page for drawing on?

5 There are 6 kittens in a litter. 3 of them are black, 2 of them are ginger and one of them is white. What is the probability that the first one to be found a home is:
a a black one **b** not a black one **c** a ginger one
d not a ginger one **e** a white one **f** not a white one?

6 William studies mathematics, physics and chemistry for his A-Level exams and he has 15 textbooks altogether. 5 of them are on mathematics, 6 of them are on physics and 4 of them are on chemistry.
If one of them is picked off a shelf at random, what is the probability that it is:
a a mathematics book **b** not a mathematics book
c a physics book **d** not a physics book
e a chemistry book **f** not a chemistry book?

7 BARBARA writes the seven letters of her name on cards and shuffles the cards.
If she picks a card at random what is the probability that the letter on it is:
a an A **b** not an A **c** a B
d not a B **e** an R **f** not an R?

8 A set of traffic lights had display times:

Red	48 seconds	Red and Amber	12 seconds
Green	30 seconds	Amber	18 seconds

a Find the total time for the sequence.

If I approach these lights in my car, what is the probability that:
b the red display is showing
c the red display is not showing
d the red and amber display is showing
e the red and amber display is not showing
f the green display is showing
g the green display is not showing
h the amber display is showing
i the amber display is not showing?

SECTION E: INVESTIGATIONS

Chapter 22 contains 50 starting points for investigations. Chapter 23 has further examples of longer investigations.
In each of the investigations in this section:
(i) State how you are going to investigate the problem.
(ii) If your method does not work, explain why and try another method
(iii) Comment on observations obtained from any tables, diagrams or formulae.
(iv) Remember to check your statements before drawing conclusions.
(v) Try to develop the task on your own.
(vi) Use short sentences.

22 Fifty starting points for investigations

1 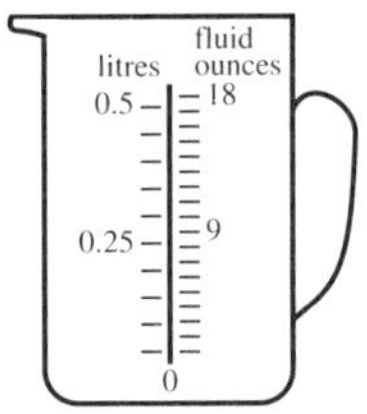

How many fluid ounces are there in one pint?

I buy a five-litre can of oil.
How many pints are there in that?
Show how you get your answer.

2 Usha says, 'When I square an odd number and then subtract 1, my answer can always be divided by 4 exactly'.
Test what Usha says using **four** different odd numbers:
What do you think?

3

Row 1	1	2	3	4	5
Row 2	6	7	8	9	10
Row 3	11	12	13	14	15
Row 4	16	17	18	19	20

(i) Write down two numbers from the table that are:
a prime numbers **b** square numbers
c multiples of 6 **d** factors of 36

(ii) Find the total of the numbers in Row 1.
Find the total of the numbers in Row 2.
What is the difference between the two totals?

Repeat this for Rows 2 and 3.
Is the difference between the two totals the same as before?
Explain why the difference between the total of Row 4 and the total of Row 3 will again be 25.

Add another row to the table.
Is the difference between the totals of two consecutive rows always 25?

4 Look at the pattern on the right which is made using consecutive numbers. What is the next one in this sequence?

$1 + 2 = 3$
$2 + 3 = 5$
$3 + 4 = 7$
$4 + 5 = 9$ etc.

What can you say about the results?

Investigate other patterns made using consecutive numbers.
Some possible starting points are:

$1 + 2 = 3$	$1 + 2 + 3 = 6$	$1 + 2 + 3 = 6$	$1 + 2 + 3 = 6$
$3 + 4 =$	$2 + 3 + 4 =$	$3 + 4 + 5 =$	$4 + 5 + 6 =$
$5 + 6 =$	$3 + 4 + 5 =$	$5 + 6 + 7 =$	$7 + 8 + 9 =$

5 Investigate patterns using:
a 2 or 3 consecutive even numbers.
b 2 or 3 consecutive odd numbers.

6 Write down all the multiples of 6 which are less than 100. Look at the numbers which are either one more or one less than a particular multiple of 6.
Do you agree that at least one of these numbers is a prime number?
Investigate for which multiples of 6 both numbers are prime numbers.
If both numbers are not prime, what can you say about the factors of the other number?

7 The factors of 6 are 1, 2, 3 and 6.
The sum of the first three factors is the number itself $\quad 1 + 2 + 3 = 6$
6 is called a 'happy number'.
Which other numbers less than 100 are also 'happy numbers'?

8 The 'digit' sums of 17, 27, 37 and 47 are shown on the right.
Which numbers less than 100 have a digit sum of:
a 1 **b** 2 **c** 3 **d** 9?

$17 \rightarrow 1 + 7 = 8$
$27 \rightarrow 2 + 7 = 9$
$37 \rightarrow 3 + 7 = 10 \rightarrow 1 + 0 \rightarrow 1$
$47 \rightarrow 4 + 7 = 11 \rightarrow 1 + 1 \rightarrow 2$

What can you say about all the numbers whose digit sum is 9?

9

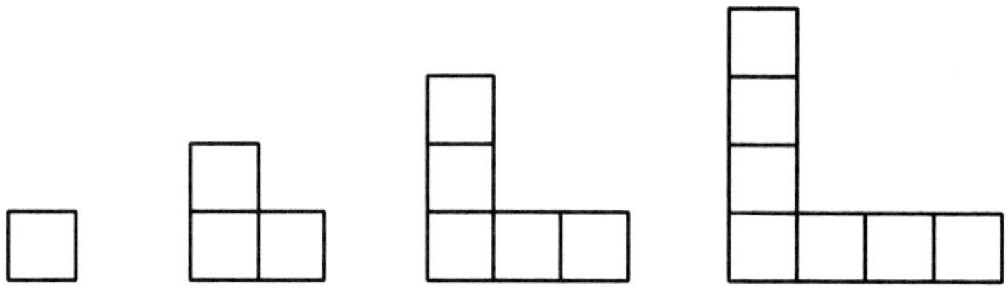

Write down the number of squares in each of the four L shapes.
Draw the next L shape in the sequence.
How many squares are there this time?
What can you say about these numbers?

How many squares will there be in the sixth L shape?
Explain your answer by looking at the pattern in the numbers.

How many squares will there be in the tenth L shape?

Can you predict how many squares there will be in the 100th L shape?

10 Complete the following:

$$1 = 1$$
$$1 + 2 + 1 = 4$$
$$1 + 2 + 3 + 2 + 1 =$$
$$1 + 2 + 3 + 4 + 3 + 2 + 1 =$$

What can you say about each of the totals?
What do you think that $1 + 2 + 3 + 4 + (5) + 4 + 3 + 2 + 1$ will be?
What is the connection between the total and the middle number in this set?

Try this with the next total
$1 + 2 + 3 + 4 + 5 + (6) + 5 + 4 + 3 + 2 + 1$.
What set of numbers like this will have a total of 100?

11 Square numbers can be formed by adding together consecutive odd numbers starting at 1.

$$1 = 1^2$$
$$1 + 3 = 4 = 2^2$$
$$1 + 3 + 5 = 9 = 3^2$$
$$1 + 3 + 5 + 7 = 16 = 4^2$$

The first five square numbers are shown on the right.
Write down the next five square numbers.

What can you say about the difference between any two consecutive square numbers?
(For example $4^2 - 3^2 = 7$)
Do you think this is always true?

Can you find consecutive square numbers whose difference is also a square number?
(For example $5^2 - 4^2 = 9 = 3^2$)

12 1, 3, 5, 7, 9, 11 are the first six odd numbers.
Write down the next two odd numbers.
Complete the following:

$1 \times 5 = 5$	$3^2 = 9$	$3^2 - 1 \times 5 = 4$
$3 \times 7 = 21$	$5^2 = 25$	$5^2 - 3 \times 7 =$
$5 \times 9 = 45$	$7^2 =$	$7^2 - 5 \times 9 =$
7×11	$9^2 =$	$9^2 - 7 \times 11 =$

What can you say about each of your answers in the third column? Do you think that this is true for any three consecutive odd numbers?

Try 23, 25, 27, i.e. $23 \times 27 =$, $25^2 =$, $25^2 - 23 \times 27 =$
Now try another set of three consecutive odd numbers..

13 Triangle numbers are formed by adding together consecutive whole numbers starting at 1. The first five triangle numbers are shown on the right.

$$1$$
$$1 + 2 = 3$$
$$1 + 2 + 3 = 6$$
$$1 + 2 + 3 + 4 = 10$$
$$1 + 2 + 3 + 4 + 5 = 15$$

Write down the next five triangle numbers.

What can you say about the sum of any two consecutive triangle numbers? Do you think this is always true?

14 1, 3, 6, 10, 15 are the first five triangle numbers.
Write down the next five triangle numbers.
Now double each of the triangle numbers.
Write the double of each triangle number as the product of two consecutive numbers.
What do you notice about your results?
Do you think this is always true?

15 1, 8, 27, 64, 125 are the first five cube numbers.
$1^3 = 1$, $2^3 = 8$, $3^3 = 27$, $4^3 = 64$, $5^3 = 125$.
Write down the next five cube numbers.
Find the sum of consecutive cubes starting at 1.
(For example $1^3 + 2^3 + 3^3 = 1 + 8 + 27 = 36 = 6^2$)

Is it true that the sum of consecutive cubes starting at 1 is always a square number? What do you notice about the resulting square number in each case?

16 What is the largest number you can make using each of the numbers 1, 2, 3 and 4 exactly once?
Try 4×321; 43×21; etc.

What is the largest number you can make using each of 1, 2, 3, 4, 5, 6 exactly once?

17 Using four 4s, the operations $+$, $-$, $\times$, $\div$, brackets and $\sqrt{\ }$ it is possible to make many numbers.
For example: $4+4+4+4=16$, $(4+4+4)\div 4=3$,
$(44-4)\div 4=10$
Investigate which numbers between 1 and 100 you can make using exactly four 4s.

18 The first 10 numbers in the Fibonacci Sequence are:
1, 1, 2, 3, 5, 8, 13, 21, 34, 55.
Write down the next five numbers in this sequence.

Find the ratios of each pair of consecutive terms in the Fibonacci Sequence,
i.e. $\frac{1}{1}$, $\frac{1}{2}$, $\frac{2}{3}$, $\frac{3}{5}$, etc.
What do you notice about these ratios?
Repeat this for $\frac{1}{1}$, $\frac{2}{1}$, $\frac{3}{2}$, $\frac{5}{3}$, etc.
What can you say about the ratios this time?

Find the difference between each ratio in the second list and the corresponding ratio in the first set.
What can you say about these differences?

19 Starting with $X_1 = 1$, use your calculator to find the value of $\dfrac{1}{(1+X_1)}$. Call this X_2.

Take the result X_2 as the next value of X and find $\dfrac{1}{(1+X)}$ again. Call this X_3.

Take X_3 as the next value of X and find $\dfrac{1}{(1+X)}$ again. Call this X_4.
Repeat this process to find X_5, X_6, X_7, X_8, X_9, and X_{10}.
What do you notice about the value of X_n as n increases?

20 Starting with $X_1 = 1$, use your calculator to find the value of $\sqrt{6+X_1}$. Call this X_2.
Take the result X_2 as the next value of X and find $\sqrt{6+X}$ again. Call this X_3.
Take X_3 as the next value of X and find $\sqrt{6+X}$ again. Call this X_4.
Repeat this process to find X_5, X_6, X_7, X_8, X_9 and X_{10}.
What do you notice about the value of X_n as n increases?

Repeat this process starting with $X_1 = 1$ and substituting:
a $\sqrt{2+X}$ instead of $\sqrt{6+X}$.
b $\sqrt{12+X}$ instead of $\sqrt{6+X}$.

21 How many squares can you find in this diagram?
How many of these are:
a 1 × 1 **b** 2 × 2 **c** 3 × 3?

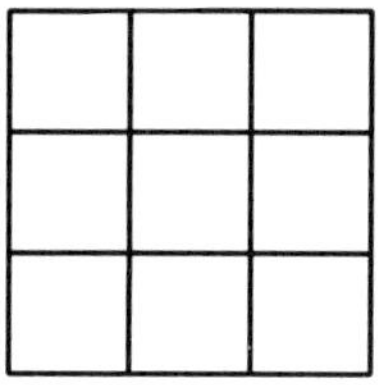

Investigate the number of squares of each size in each of these:

(i)  (ii) 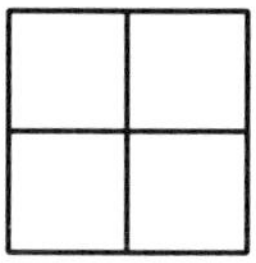 (iii) 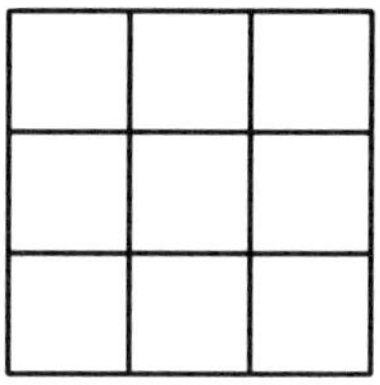 (iv) 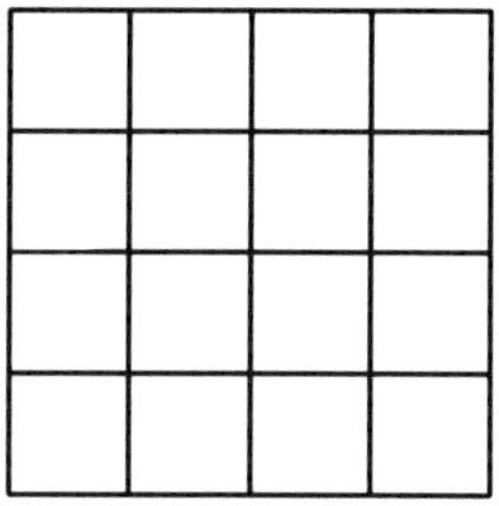

Can you predict how many squares of each size there will be in (v)?

(v) 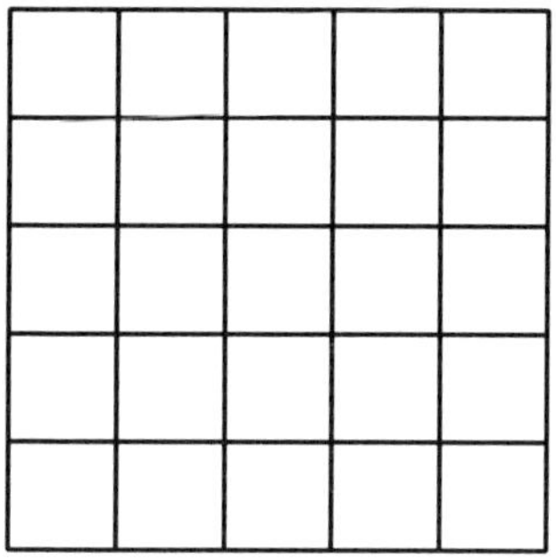

Do the numbers of squares of a particular size in each of diagrams (i)–(v) form a pattern?

22 How many triangles can you find in the diagram?

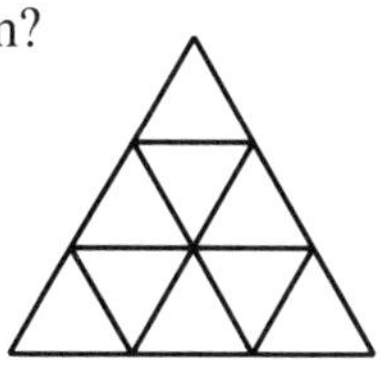

Investigate the patterns in the number of triangles of each size in:

(i)  (ii) 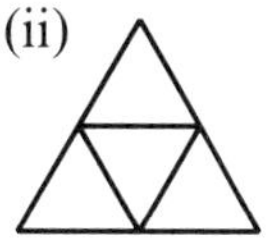 (iii) 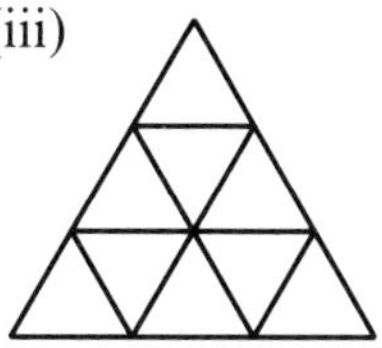 (iv) 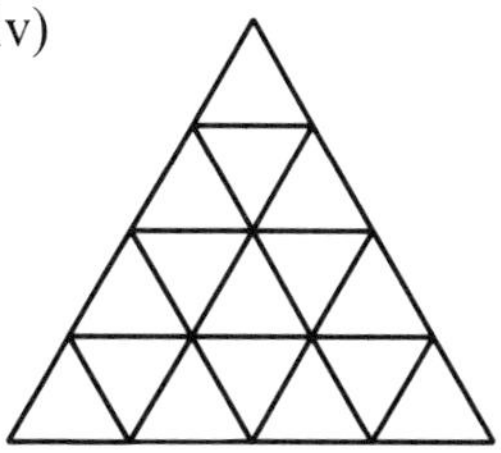

23 How many different shapes can you make using exactly 4 squares, where each square must have a complete edge in common with the adjacent square?
One of these shapes is shown.

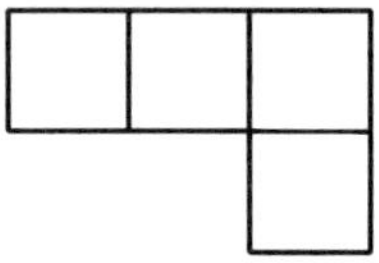

Investigate the number of different shapes that you can make with each of:

a  **b** 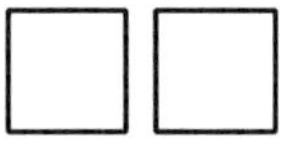 **c** 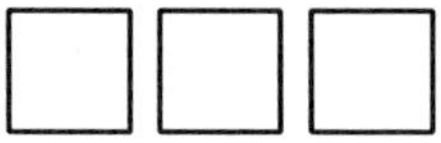

d 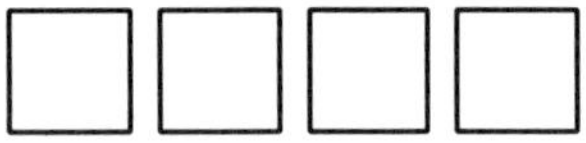 **e** 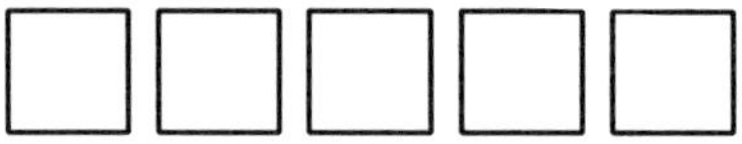

For each shape you find, show any lines of symmetry and indicate the order of rotational symmetry.
Note: Reflections or rotations of the same shape should not be counted as different, i.e.

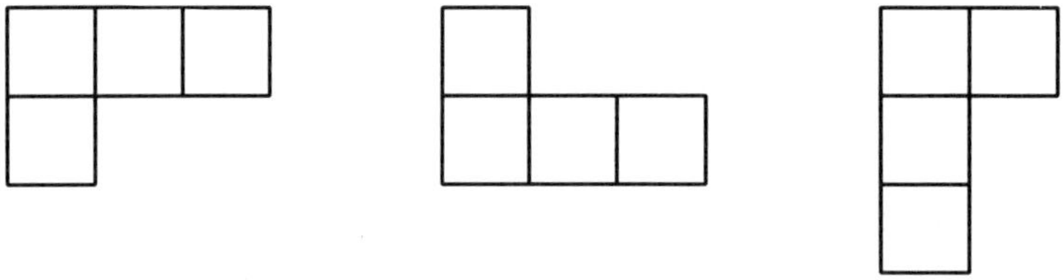

are all the same as each other.

24 How many different shapes can you make using exactly four equilateral triangles, where each triangle must have a complete edge in common with the adjacent triangle?
One of these shapes is shown. What are the names of each of these shapes?

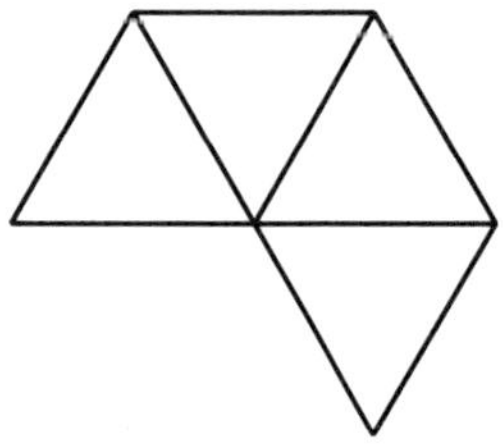

Investigate the number of different shapes that you can make with **a** one triangle, **b** two triangles, **c** three triangles, **d** five triangles, **e** six triangles.
For each shape show any lines of symmetry and indicate the order of rotational symmetry if any.

25 What different shapes can you make using:
a two equilateral triangles
b two isosceles triangles (acute, obtuse)
c two right-angled isosceles triangles
d two right-angled scalene triangles
e two scalene triangles (acute, obtuse)?
In each case name the shape you have made.

Investigate the shapes you can make using three triangles.

26 How many different shapes can you make using exactly four cubes, where each cube has a complete face in common with the adjacent cube?

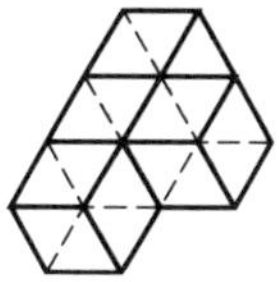

Make a sketch of each shape on isometric graph paper.

Investigate the number of different shapes you can make with **a** 1 cube, **b** 2 cubes, **c** 3 cubes and **d** 5 cubes.

27 How many different shapes can you make using exactly three tetrahedrons, where each has a complete face in common with the adjacent one?

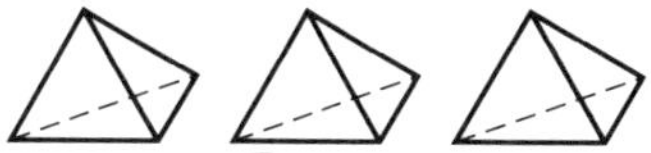

Investigate the number of different shapes you can make with four tetrahedrons.

28 These four squares are made using 12 matchsticks.
Show how, by taking away 2 matches, you can leave:
a 3 squares **b** 2 squares **c** 1 square.

How could you make three squares using all 12 matches?

29 This rectangle can be divided into two identical squares using a single line; which is shown dashed.
Investigate other pairs of identical shapes that can be formed using a single line.
Name each of the shapes you make.

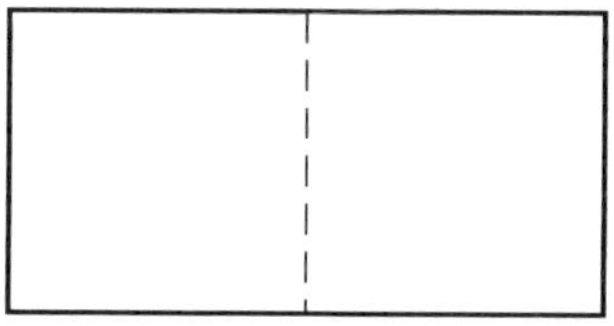

30 This rectangle can be divided into four quarters using two lines which are shown dashed.
Investigate other ways of dividing the rectangle into four quarters using just two lines.
Name the four shapes you can make in each case.

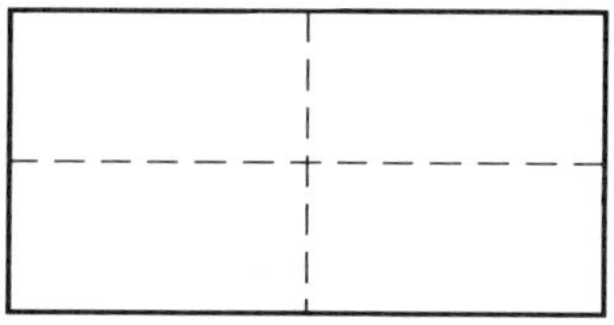

31 This rectangle can be divided into three thirds using two lines which are shown dotted.
Investigate other ways of dividing the rectangle into three thirds using just two lines.
Name the three shapes you make in each case.

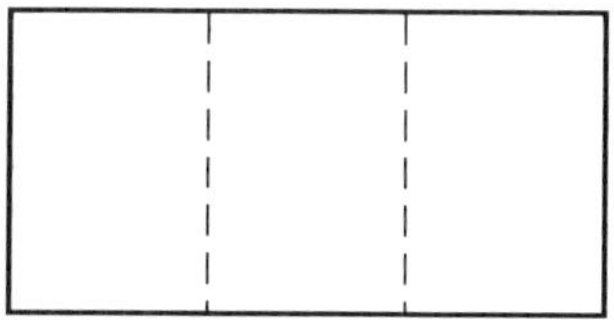

32 A design for a flag can be made by shading half the rectangles black in the diagram and leaving the other half white.
How many different designs can you make by shading half of the four rectangles?

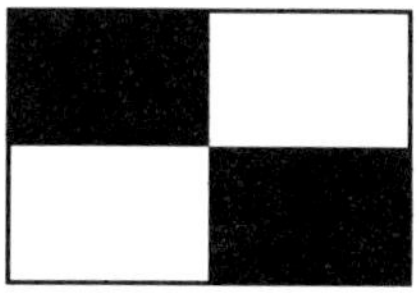

33 Investigate possible designs for a flag made by shading half of the six squares.
Which of your designs has a line or lines of symmetry?
Which of your designs has rotational symmetry?

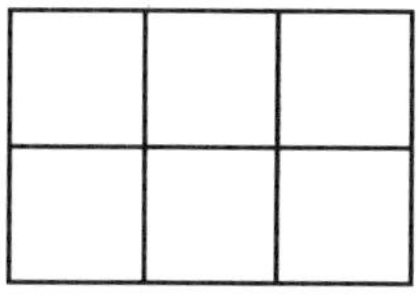

34 You will need some dotty paper (or squared paper).
Mark out nine dots like this:
Join three of the dots to form a triangle.
The triangle shown is an isosceles triangle. It has two sides the same length.
How many different triangles can you make by joining just three dots?

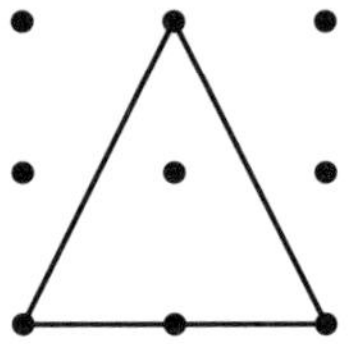

Draw each triangle and name it, i.e. isosceles, scalene, acute, obtuse or right-angled.

35 A right-angled scalene triangle can be made on a 9-pin grid as shown.
In how many different positions can you draw the same triangle on this grid?

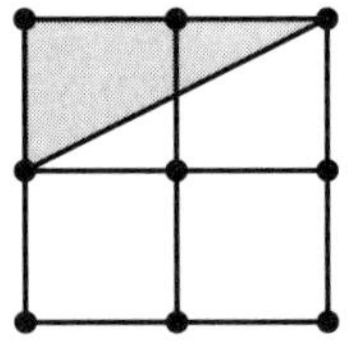

36 Two of the possible different right-angled scalene triangles are shown on the 9-pin grid.
One of these could be obtained from the other by reflecting it in the diagonal line which is shown dashed.
Find other pairs of triangles that can be obtained from each other by a reflection in a line.

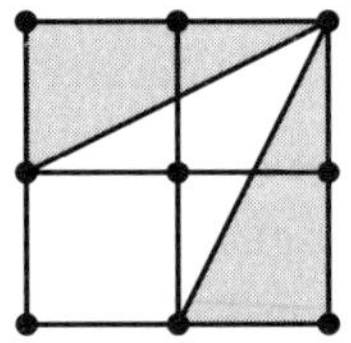

37 Another two of the possible different scalene triangles are shown on the 9-point grid.
One of these could be obtained from the other by a rotation of 90° about the centre point.
Find other pairs of triangles that can be obtained from each other by a rotation about the centre point.

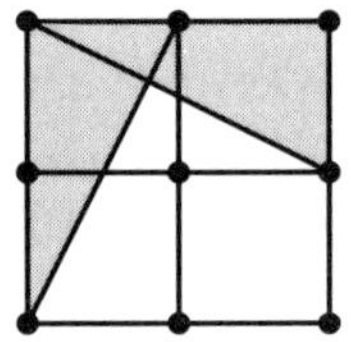

38 Investigate ways of mapping the shaded square onto one of the other squares by using:
a a reflection
b a rotation

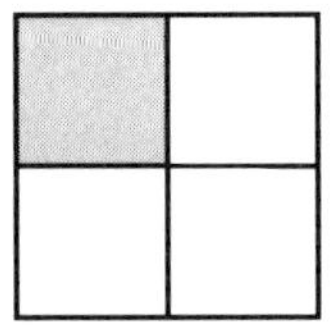

39 Look at these ways of drawing four lines.
There are no points of intersection if the lines are all parallel.

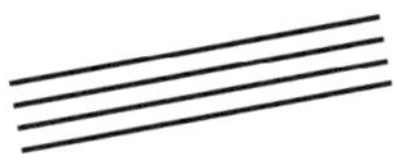

There is 1 point of intersection if the lines all go through the same point.
What is the maximum possible number of points of intersection when four lines are drawn?

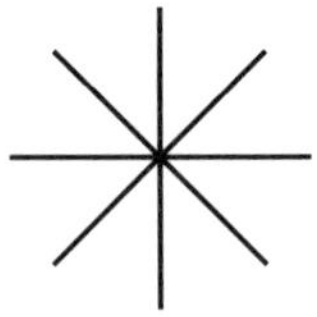

What other numbers of points of intersection are possible?
Investigate the possible number of points of intersection using 1, 2, 3, 5 or 6 lines.

40 Investigate the maximum possible number of possible points of intersection for each of 1, 2, 3, 4, 5 or 6 lines.

41 Find how many matchsticks you need to make each of the shapes in the following set:

(i)  (ii)  (iii) 

(iv)  (v) 

Predict how many matchsticks you will need for the next shape. Can you generalise for the number of matchsticks needed to make the *n*th shape in this set?

Investigate the number of rectangles (excluding squares) that you can find in each shape.

42 Find how many straws (and connections) you will need to make each shape in the following set.

(i) 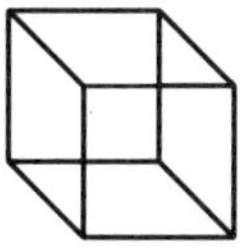 (ii) 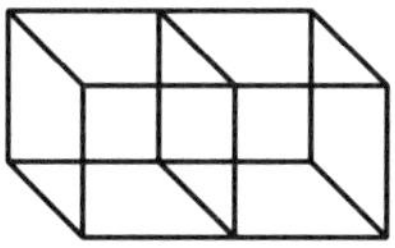 (iii) 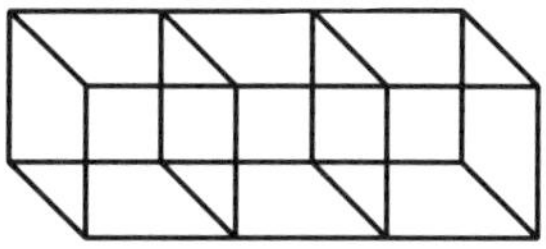

(iv) 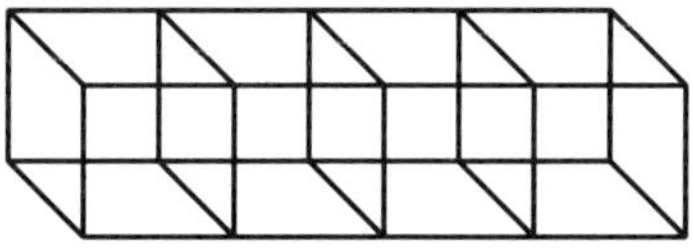

Predict how many straws you will need for the next shape. Can you generalise for the number of straws needed to make the *n*th shape in this set?

43 Find how many straws you will need to make each of the shapes in the following set.

(i) 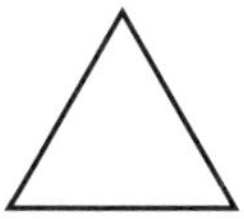 (ii) 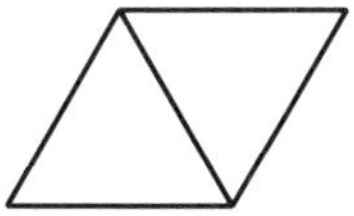 (iii) 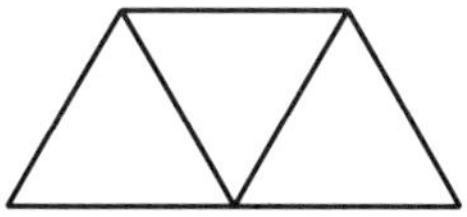

(iv) 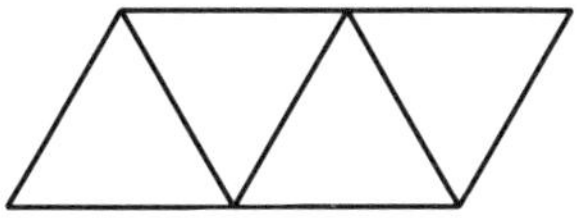 (v) 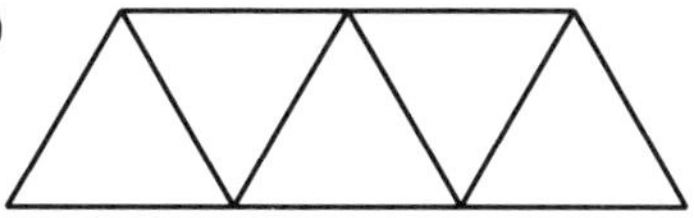

Predict how many straws you will need for the next shape. Can you generalise for the number of straws needed to make the *n*th shape in this set?

Investigate the number of parallelograms that you can find in each shape. (*Hint:* you may need to look at the odd and even terms separately.)

44

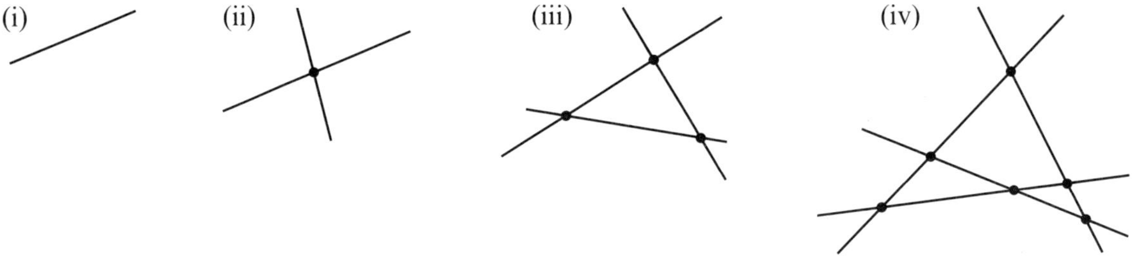

Write down the number of points of intersection formed by the lines in each diagram.
What do you notice about this sequence of numbers?
Predict the next number in this sequence.

Now draw six lines and find the number of points of intersection.
Does this number agree with your prediction?

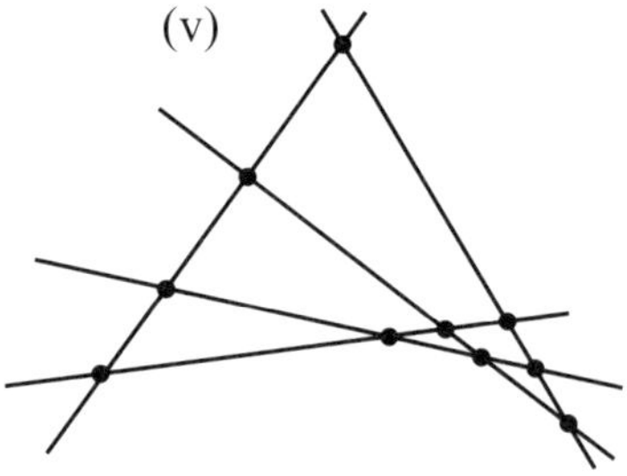

45 Look again at the diagrams in Question **44**.
In the first diagram the line divides the page into two regions.
In the second diagram the lines divide the page into four regions.
Write down the number of regions in each of the next three diagrams.
What do you notice about this sequence of numbers?

Predict the next number in this sequence.
Check your answer with the sixth diagram you drew in Question **44**.

46 Take a long strip of paper and fold it in half.
Unfold the paper and you should see two smaller rectangles and one fold line.

Refold the paper and fold it in half again.
Unfold the paper and you should see four smaller rectangles and three fold lines.

Repeat this to find the number of rectangles and fold lines for 3, 4, 5 and 6 folds.
Record your results in a table.

Folds	Rectangles	Fold lines
1	2	1
2	4	3
3		
4		
5		
6		

What do you notice about **a** the number of rectangles, **b** the number of fold lines?

Predict the next number in each sequence.
Check your answers by folding the paper once more.

47 In an old Chinese puzzle called the Tower of Hanoi, a pile of discs of decreasing size and placed on a peg.
They have to be moved on to another peg in the least number of possible moves, where only one disc may be moved at any time and no larger disc may be placed on a smaller disc.

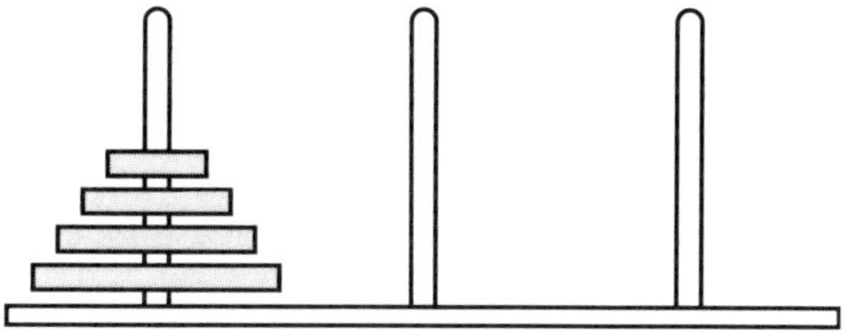

With two discs the least number of moves is 3.

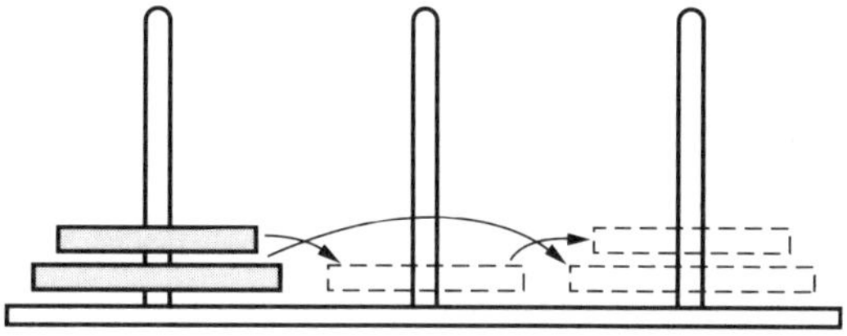

Find the least number of moves needed for
a 3 discs **b** 4 discs **c** 5 discs **d** 6 discs
and complete this table.

Number of discs	Least number of moves
1	1
2	3
3	
4	
5	
6	

What do you notice above the numbers in the number of moves column?

Predict the least number of moves for 7 discs.
Check your answer by doing it practically.
(You could use different-sized coins for your discs: 1 p, 2 p, 5 p, 10 p, 20 p, 50 p, £1.)

48 In this investigation you need two sets of counters of different colours. The counters have to swop places but you may only move a counter into an adjacent space or jump one space over a different coloured counter.
For two counters three moves are needed.

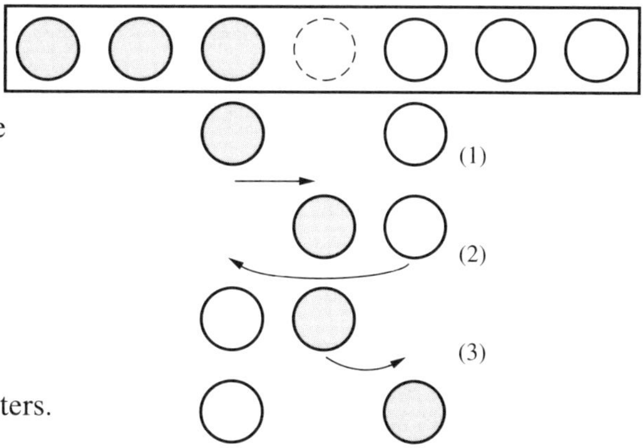

Find the least number of moves needed with
a 4 counters, **b** 6 counters, **c** 8 counters.
What do you notice about this sequence of numbers?

Predict the least number of moves for 10 counters.
Check your answer by doing it practically.

49 Joseph likes to be able to give the exact amount of money when he goes to the cinema so he doesn't have to wait for change.
On Tuesday he went out with £5 made up of 3×1 p, 1×2 p, 3×5 p, 1×10 p, 1×20 p, 3×50 p and $3 \times$ £1 coins.

(i) What coins would he use to pay for his bus fare of 58 p?

(ii) In what ways could he then pay for his cinema ticket costing £1.20?
(*Hint:* there are four different ways possible.)

(iii) If instead of going to the cinema he went to his local shop, could he buy **any** article up to £5 in cost without needing any change?
Explain your answer.

(iv) Another day Joseph had only one coin of each value up to 50 p, i.e. 1 p, 2 p, 5 p, 10 p, 20 p and 50 p.
What value articles could he buy this time without needing any change?
Explain your answer.
Why can't he buy any article costing between 39 p and 49 p?

50 A dice has three faces marked with a 1, two faces marked with a 2 and one face marked with a 3.
Which number are you most likely to get when you roll this dice?
Which number are you least likely to get?

If you rolled this dice 600 times, about how many times would you expect to get **a** 1, **b** 2, **c** 3?
A fairground stall uses this dice with the following rules:

Each go costs 1 p.
If you get a 1 you lose your money.
If you get a 2 you get your money back.
If you get a 3 you get 3 p back.

Are you likely to lose money or make money at this stall?
Explain your answer.

23 Examples of longer investigations

1 Moving squares

a Three movable squares, numbered 1, 2 and 3, are placed in a tray. The tray could contain four squares but one space is deliberately left empty.
Movement of the squares is possible by clockwise or anti-clockwise movement of the numbered squares.
For the clockwise case, the first two movements are illustrated.

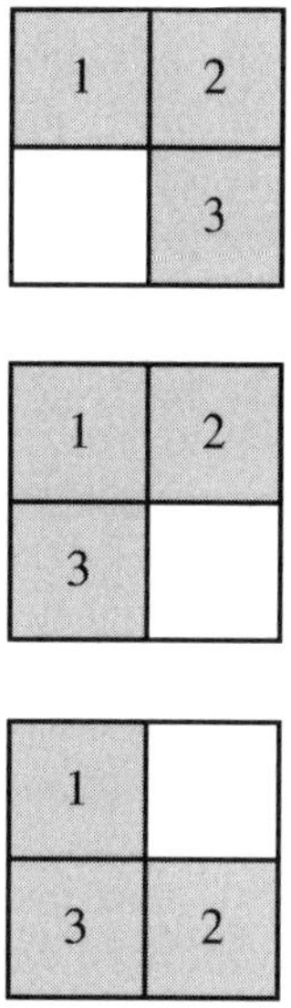

Twelve successive movements are required before the starting pattern reappears.

Copy and complete this 12-move sequence.

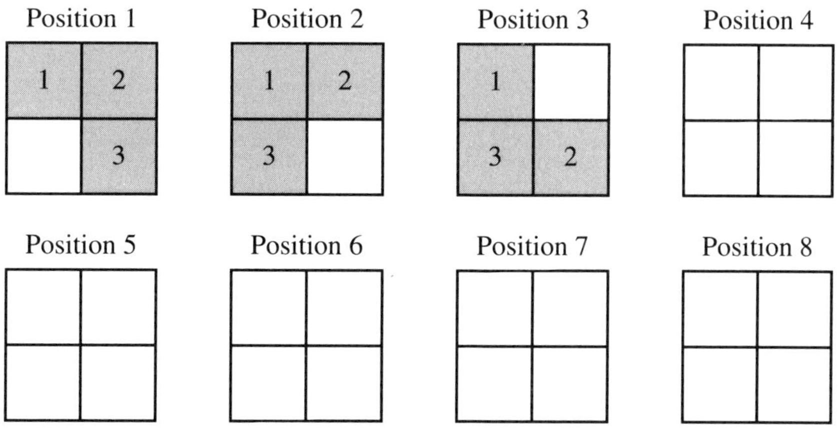

Position 9

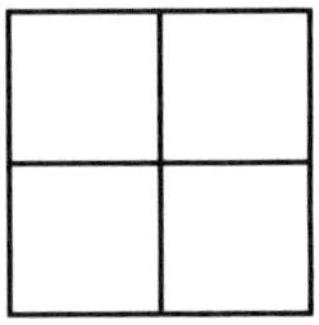

Position 10

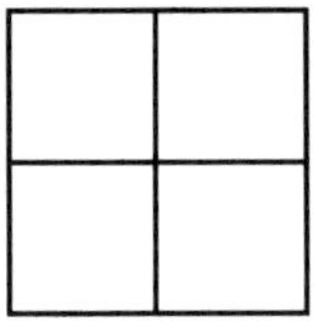

Position 11

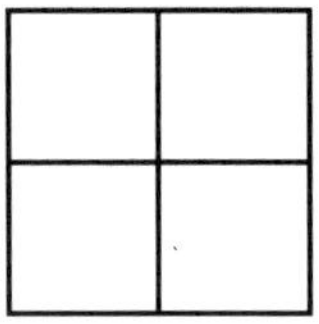

Position 12

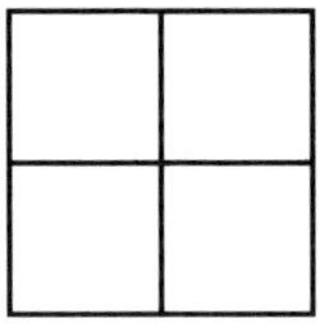

Position 1

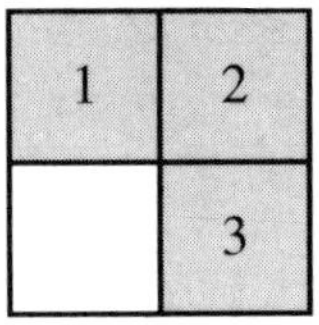

Draw position 1 beside position 4.
Position 4 can be described as a transformation of position 1.
What is this transformation?

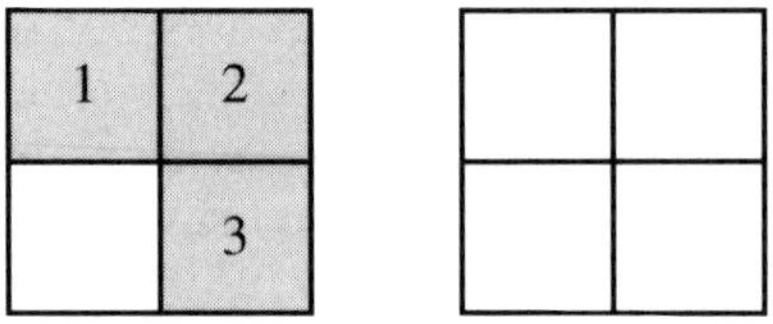

Draw position 1 beside position 7.
Position 7 can be described as a transformation of position 1.
What is this transformation?

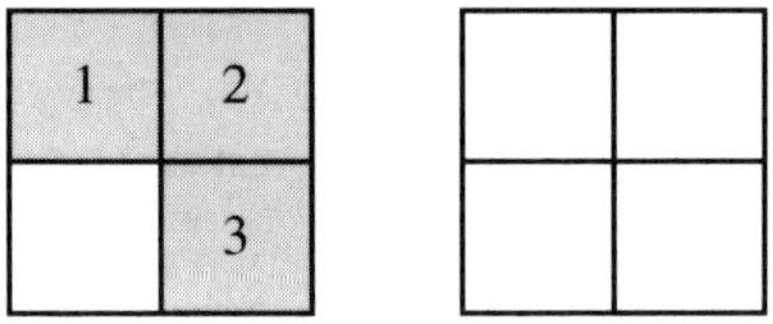

Draw position 1 beside position 10.
Position 10 can be described as a transformation of position 1.
What is this transformation?

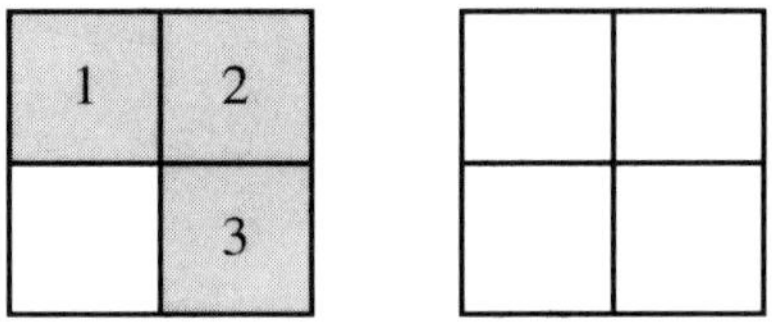

b Repeat part **a** for anticlockwise movement of the numbered squares.

c Arrange the numbers 1, 2, 3, 4, 5, 6, 7 and 8, in order, on a grid of nine squares.
You should have one square without a number on it.
By moving one number at a time into the empty square, can you rearrange the eight numbers so that they are now in the reverse order?

2 *Wall tiling*

When tiling a wall you use plastic spacers between the tiles.
There are three different types:

a corner spacer ┌
a T-shaped spacer ┬
a cross-shape spacer +

A 2 × 2 tiled square uses:

4 corner spacers
4 T-shaped spacers
1 cross-shaped spacer

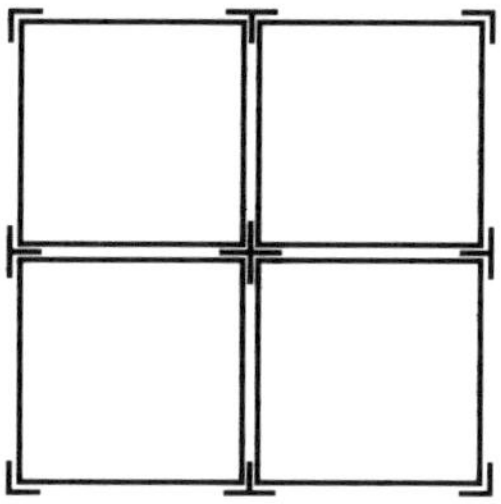

a Investigate the number of different spacers needed for different sized squares.

b What if the shape was a rectangle or a shape of your own choice?

3 *Diagonals and polygons*

A diagonal is a line joining two vertices of a polygon but which does not form one of the sides of the polygon.
For instance, in a regular pentagon (five sides) there are five diagonals.

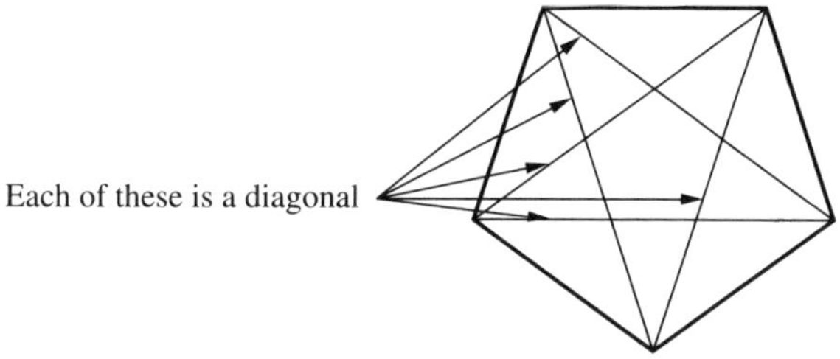

a Show that there are nine diagonals in regular hexagon (six sides).

b How many diagonals are there in
(i) an equilateral triangle?
(ii) a square?
(iii) a regular octagon (eight sides)?

c Investigate the relationship between the number of vertices of a regular polygon and the number of its diagonals.
Record your results with observations.

In particular, obtain the number of diagonals for a regular polygon with:
a 10 sides **b** 20 sides **c** 43 sides

d Investigate the number of regions formed in a given polygon when the diagonals are drawn.
What shape are the regions and how many regions are there?

e Which polygons and their diagonals can be drawn without taking your pen off the page and without going over any line more than once?

4 Borders and tiles

A border of tiles is formed by having an extra set of tiles in a different colour. For instance, starting with a row of 5 plain tiles, you need the 16 border tiles that are shown shaded.

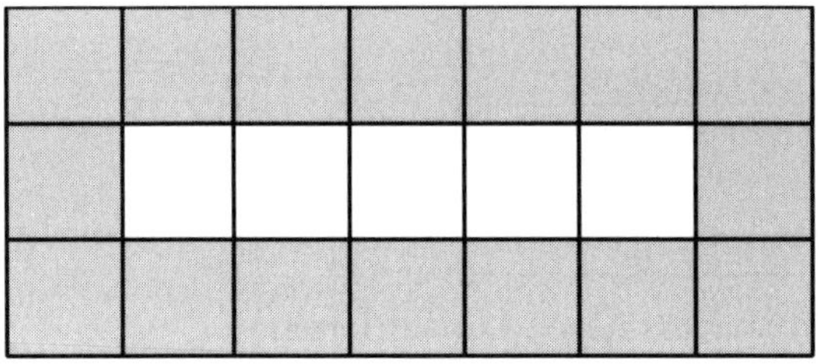

a Show that there are 12 border tiles if you start with a row of 3 plain tiles.

b How many border tiles are there if you start with a **row** of:
(i) 4 plain tiles (ii) 6 plain tiles (iii) 9 plain tiles?

c Investigate the relationship between the number of border tiles and the number of plain tiles in the row you start with.
Record your results with observations.
In particular, obtain the number of border tiles if you start with a row of:
(i) 10 plain tiles (ii) 20 plain tiles (iii) 43 plain tiles?

d What happens if you start with a rectangle rather than a row of plain tiles?

(i)
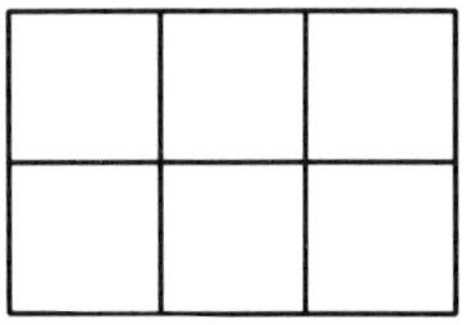
(ii)
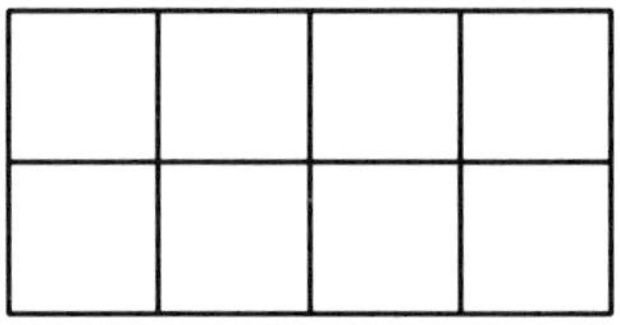
(iii)
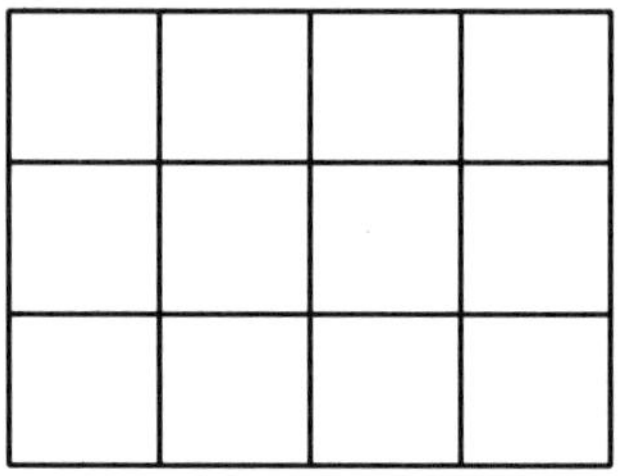

Can you find the relationship between the number of border tiles and the number of plain tiles?

5 Pick's theorem – dotty areas

Investigate the link between the area of a shape drawn on dotty paper and the number of dots using the following approach.
Count the dots in each of the shapes.

(i)
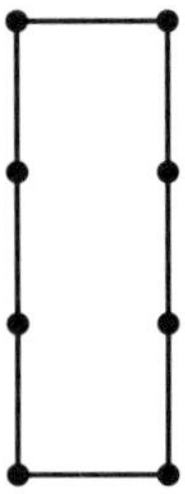
(ii)
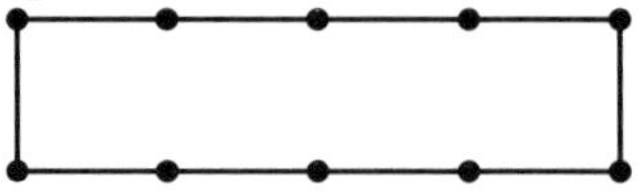
(iii)
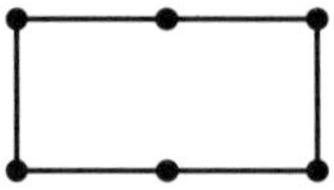

(iv) 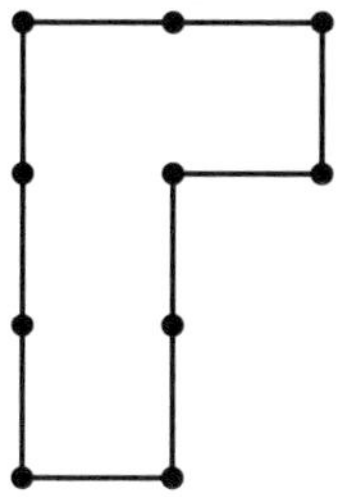

(v) 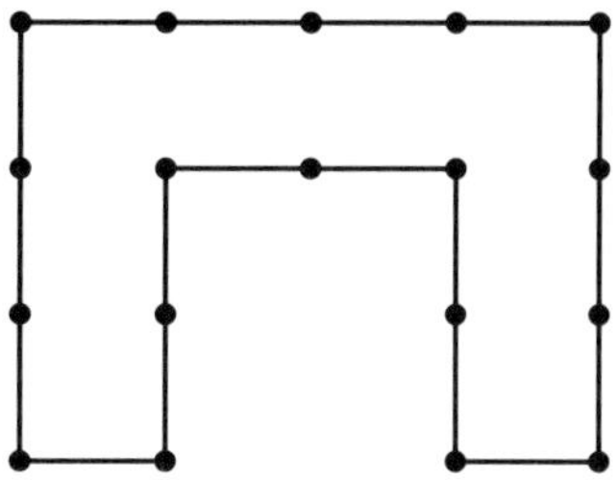

(vi) 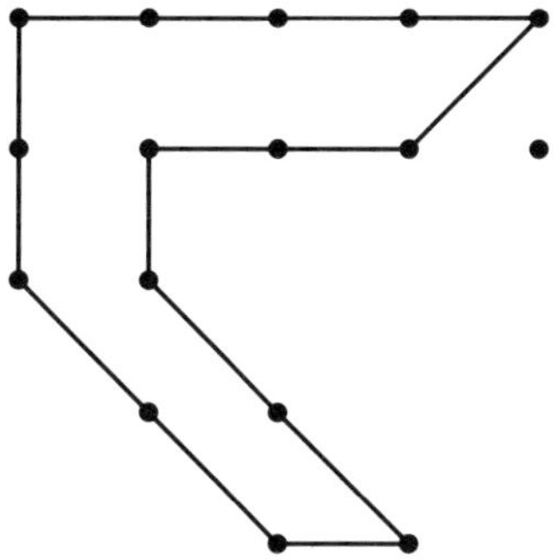

(vii) 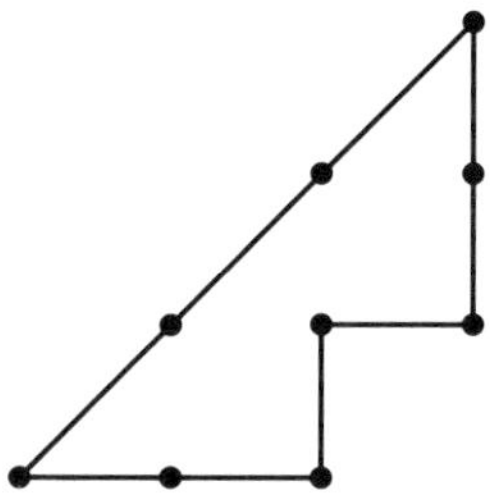

(viii) 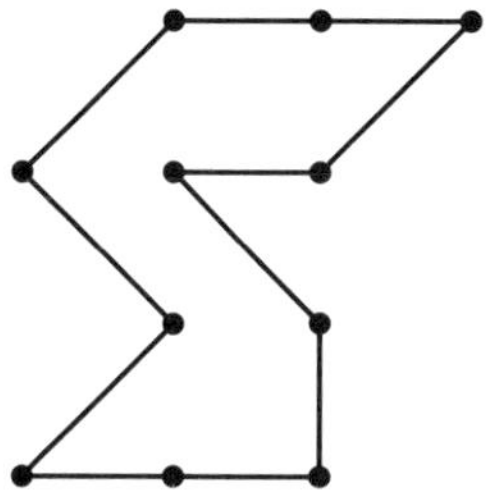

Record your results in a table like this:

	Area	Dots	Dots/2
(i)	3	8	4
(ii)	4	10	
(iii)			

a Can you find a connection?

b Now look at these shapes.
How many dots inside? How many dots on the edge?
What do you notice?

(i) 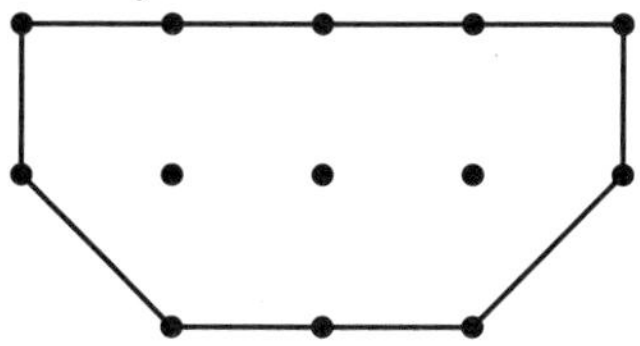

(ii) 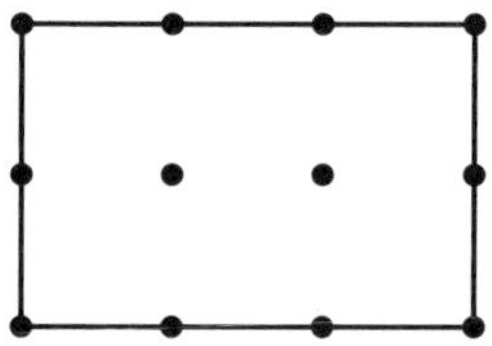

(iii) 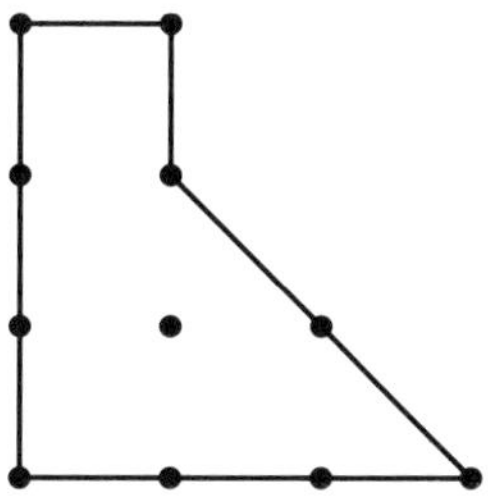

(iv) 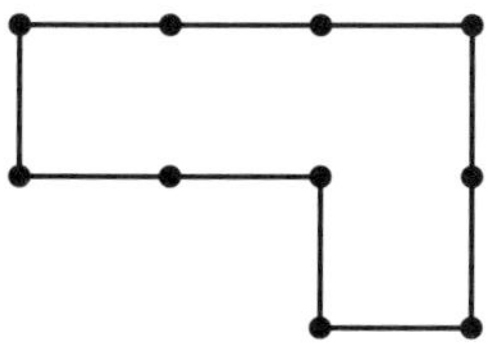

(Continued over the page)

(v)

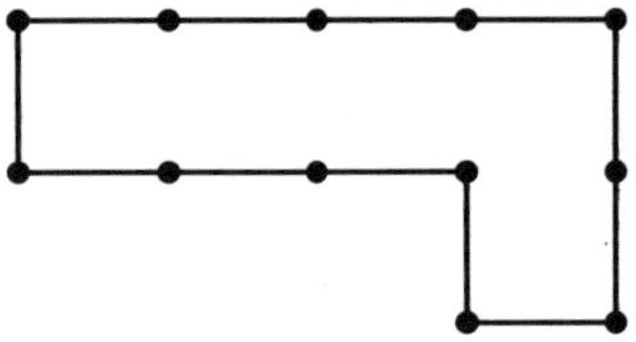

(vi)

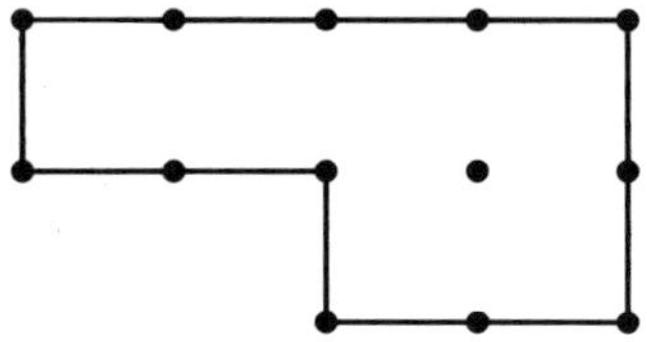

(vii)

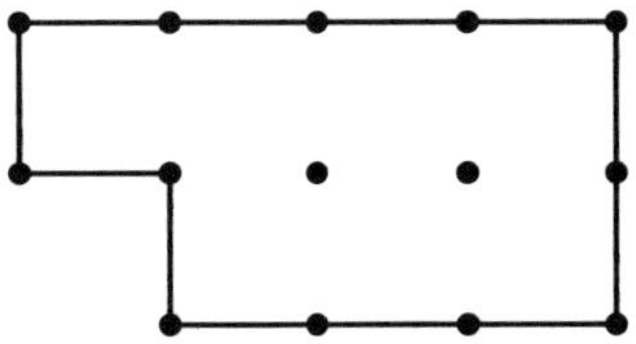

(viii)

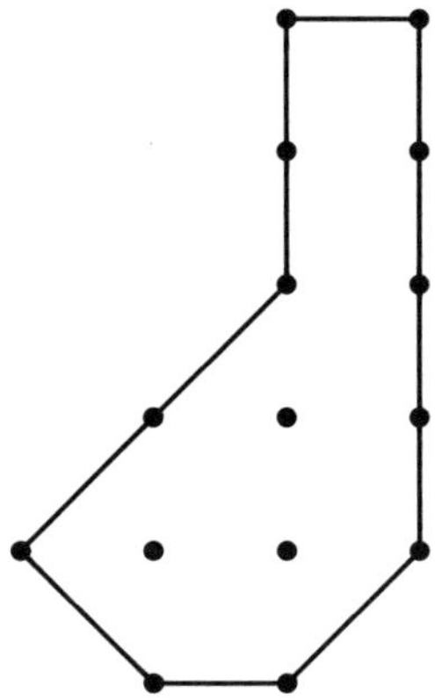

(ix)

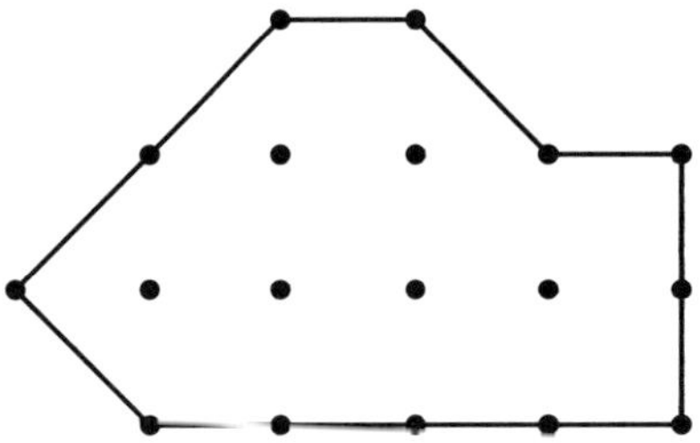

(x)

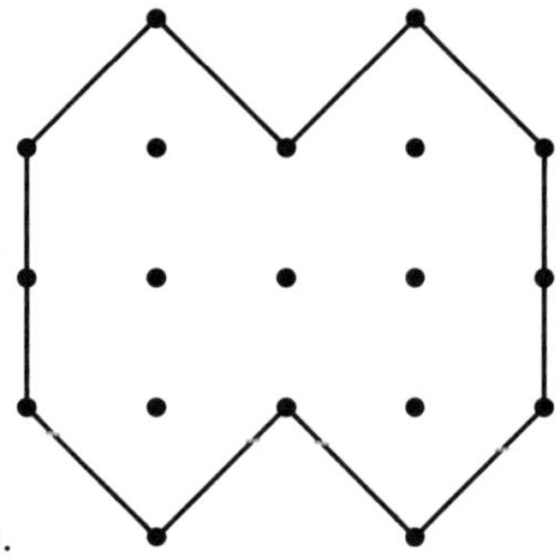

Make a table for: Area, Dots inside, Dots outside.
Can you find a connection?

c Now try your own shapes.
Does your rule still work?

6 *Planning a holiday*

You are planning a two-week camping holiday in Northern France for two people, using bicycles.
You will be staying on official campsites and cooking your own food.

a What do you need to think about to work out an estimate for the total cost of the holiday?

b Where would you need to look to find out some of the information you required?

ANSWERS

page 1 ***Exercise 1.1***

1 **a** twenty five **b** twenty nine **c** thirty two **d** thirty seven
2 **a** forty three **b** fifty six **c** seventy four **d** ninety one
3 **a** one hundred and twenty **b** one hundred and thirty **c** one hundred and thirty six **d** two hundred and twenty
4 **a** four hundred **b** four hundred and five **c** nine hundred and five **d** nine hundred
5 One hundred and twenty three metres
6 Three hundred and ninety four metres
7 Three hundred and thirty six kilometres
8 Six hundred and forty kilometres

page 2 ***Exercise 1.2***

1 18 **2** 13 **3** 28 **4** 34 **5** 92 **6** 90 **7** 379
8 524 **9** 795 **10** 825 **11** 838 km **12** 687 days **13** 905 t m

page 2 ***Exercise 1.3***

1 3 **2** 7 **3** 6 **4** 90 **5** 700 **6** 600 **7** 2 **8** 90 **9** 50 **10** 300

page 3 ***Exercise 1.4***

1 540, 504, 450, 405
2 210, 201, 120, 102
3 321, 312, 231, 213, 132, 123
4 987, 978, 897, 879, 798, 789
5 330, 320, 303, 302, 230, 220, 203, 202
6 122, 123, 132, 133
7 588, 589, 598, 599
8 933, 934, 943, 944
9 112, 121, 122, 211, 212, 221
10 556, 565, 566, 655, 656, 665

page 3 ***Exercise 1.5***

1 743, 734, 473, 437, 374, 347
2 952, 925, 592, 529, 295, 259
3 864, 846, 684, 648, 486, 468
4 531, 513, 351, 315, 153, 135
5 720, 702, 270, 207
6 960, 906, 690, 609
7 552, 525, 255
8 884, 848, 488
9 933, 393, 339
10 660, 606
11 368, 386, 638, 683, 836, 863
12 457, 475, 547, 574, 745, 754
13 246, 264, 426, 462, 624, 642
14 169, 196, 619, 916, 961
15 308, 380, 803, 830
16 405, 450, 504, 540
17 377, 737, 773
18 226, 262, 622
19 118, 181, 811
20 909, 990

page 4 ***Exercise 1.6***

1 20 **2** 80 **3** 250 **4** 430 **5** 700
6 990 **7** 400 **8** 700 **9** 600 **10** 900
11 300 **12** 200 **13** 150 mm **14** 270 mm **15** 600 mm
16 900 mm **17** 90 mm **18** 40 mm **19** 800 cm **20** 300 cm
21 600 cm **22** 20 kg **23** 160 cm **24** **a** 90 km **b** 900 km
25 **a** $21 \times 10 = 210$ **b** $35 \times 10 = 350$ **c** $90 \times 10 = 900$
d $8 \times 100 = 800$ **e** $2 \times 100 = 200$ **f** $7 \times 100 = 700$
g $5 \times 100 = 500$ **h** $9 \times 100 = 900$ **i** $4 \times 100 = 400$
j $36 \times 10 = 360$ **k** $45 \times 10 = 450$ **l** $40 \times 10 = 400$
m $80 \times 10 = 800$ **n** $100 \times 10 = 1000$ **o** $75 \times 10 = 750$
p $42 \times 10 = 420$ **q** $10 \times 10 = 100$ **r** $5 \times 100 = 500$
s $6 \times 100 = 600$ **t** $10 \times 100 = 1000$

page 5 ***Exercise 1.7***

1 **a** nine **b** ninety **c** nine hundred **d** nine thousand
2 **a** four **b** forty **c** four hundred **d** four thousand
3 **a** 2 **b** 20 **c** 200 **d** 2000
4 **a** 6 **b** 60 **c** 600 **d** 6000
5 **a** 8 **b** 80 **c** 800
6 **a** 3 **b** 30 **c** 300
7 **a** 2 **b** 20 **c** 200
8 **a** 9 **b** 90 **c** 900
9 **a** 7 **b** 70
10 **a** 5 **b** 50
11 **a** 3 **b** 30
12 **a** 8 **b** 80

page 6 ***Exercise 1.8***

1 30
2 80
3 **a** 20 **b** 200
4 **a** 70 **b** 700
5 90
6 50
7 **a** 300 **b** 30
8 20

page 7 ***Exercise 1.9***

1 500 g
2 20
3 **a** 400 g **b** 40 g
4 **a** 200 **b** 20
5 50

page 8 ***Exercise 1.10***

1 **a** 7 cm 8 mm **b** 7.8 cm **c** 78 mm
2 **a** 3 cm 6 mm **b** 3.6 cm **c** 36 mm
3 **a** 2 cm 4 mm **b** 2.4 cm **c** 24 mm
4 **a** 5 cm 4 mm **b** 5.4 cm **c** 54 mm
5 **a** 6 cm 2 mm **b** 6.2 cm **c** 62 mm

page 9 ***Exercise 1.11***

1 6.2 cm, 62 mm
2 8.6cm, 86 mm
3 3.7 cm, 37 mm
4 14.3 cm, 143 mm
5 16.2 cm, 162 mm
6 18 cm 5 mm, 185 mm
7 13 cm 1 mm, 131 mm
8 12 cm, 7 mm, 127 mm
9 25 cm 2 mm, 252 mm
10 24 cm 3 mm, 24.3 cm
11 28 cm 4 mm, 28.4 cm
12 22 cm 5 mm, 22.5 cm

page 10 ***Exercise 1.12***

1 1.36 m, 136 cm
2 1.28 m, 128 cm
3 1.16 m, 116 cm
4 1.08 m, 108cm
5 1.06 m, 106 cm

page 10 ***Exercise 1.13***

1 1 m 38 cm, 138 cm
2 1 m 29 cm, 129 cm
3 1 m 30 cm, 130 cm
4 1 m 7 cm, 107 cm
5 1 m 5 cm, 105 cm

page 10 ***Exercise 1.14***

1 1 m 34 cm, 1.34 m
2 1 m 25 cm, 1.25 m
3 1 m 13 cm, 1.13 m
4 1 m 9 cm, 1.09 m
5 1 m 1 cm, 1.01 m

page 11 ***Exercise 1.15***

1 1.55 m, 155 cm
2 1.10 m, 110 cm
3 2 m 15 cm, 215 cm
4 1 m 5 cm, 105 cm
5 1 m 15 cm, 1.15 m
6 3 m 50 cm, 3.50 m

page 11 ***Exercise 1.16***

1 6 tens 4 units 2 tenths 5 hundredths 7 thousandths
2 9 tens 2 units 3 tenths 4 hundredths 6 thousandths
3 2 tens 5 units 7 tenths 8 hundredths 3 thousandths
4 3 tens 5 units 2 tenths 9 hundredths
5 5 units 6 tenths 4 hundredths 7 thousandths
6 3 units 8 tenths 5 hundredths 2 thousandths
7 9 tens 6 units 1 tenth
8 9 units 7 tenths

page 12 ***Exercise 1.17***

1 0.65 0.58 0.46 0.37
2 0.549 0.458 0.367 0.276
3 0.823 0.732 0.641 0.55
4 0.54 0.458 0.36 0.279
5 0.721 0.6 0.54 0.48
6 0.4 0.36 0.2 0.173

page 12 ***Exercise 1.18***

1 5.234 5.243 5.324 5.342 5.423 5.432
2 6.245 6.354 6.435 6.453 6.534 6.543
3 7.456 7.465 7.546 7.564 7.645 7.654
4 7.089 7.098 7.809 7.89 7.908 7.98
5 1.002 1.02 1.022 1.2 1.202 1.22

page 13 ***Exercise 2.1***

1 32 **2** 35 **3** 63 **4** 24 **5** 42 p **6** 24 p **7** 45 p **8** 20 cm

page 15 ***Exercise 2.2***

1 8 **2** 9 **3** 9 **4** 4 **5** 9 cm **6** 8 cm

page 15 ***Exercise 2.3***

1 57 **2** 38 **3** 19 **4** 99 **5** 37 **6** 69 **7** 58
8 84 **9** 46 p **10** 37 p **11** 73 p **12** 27 cm **13** 67 cm **14** 32 minutes

page 16 ***Exercise 2.4***

1 32 **2** 91 **3** 72 **4** 68 **5** 45 **6** 53
7 39 **8** 27 **9** 17 years **10** 16 years **11** 55 cm **12** 26 p

page 17 ***Exercise 2.5***

1 9 **2** 7 **3** 8 **4** 9 **5** 12 **6** 11
7 15 **8** 13 **9** 12 **10** 14 **11** 12 **12** 17

page 18 **Exercise 2.6**

1 a 78 **b** 79 **c** 78 **b** **2 a** 87 **b** 86 **c** 86 **a** **3 a** 65 **b** 65 **c** 64 **c**
4 a 679 **b** 679 **c** 678 **c** **5 a** 967 **b** 968 **c** 967 **b** **6 a** 882 **b** 892 **c** 892 **a**
7 94 kg **8** 193 cm **9** 212 km **10** 827 kg

page 19 **Exercise 2.7**

1 a 24 **b** 23 **c** 24 **b** **2 a** 45 **b** 46 **c** 46 **a** **3 a** 38 **b** 39 **c** 39 **a**
4 a 324 **b** 325 **c** 324 **b** **5 a** 566 **b** 567 **c** 566 **b** **6 a** 285 **b** 275 **c** 275 **a**
7 a 563 **b** 563 **c** 573 **c** 8 a 376 **b** 386 **c** 376 **b** **9 a** 84 **b** 85 **c** 85 **a**
10 48 km **11 a** 131 cm **b** 119 cm **c** 12 cm **a** **12** 37 kg, 29 kg, 8 kg, 59 kg, 22 kg, 30 kg

page 20 **Exercise 2.8**

1 a 68 **b** 66 **c** 68 **b** **2 a** 98 **b** 96 **c** 96 **a** **3 a** 84 **b** 84 **c** 85 **c**
4 a 78 **b** 80 **c** 78 **b** **5 a** 90 **b** 91 **c** 90 **b** **6 a** 95 **b** 96 **c** 96 **a**
7 a 112 **b** 112 **c** 116 **c** **8 a** 108 **b** 104 **c** 108 **b**
9 a 78 km **b** 130 km **c** 156 km **d** 182 km **10** 57 kg, 95 kg

page 21 **Exercise 2.9**

1 a 13 **b** 14 **c** 13 **b** **2 a** 17 **b** 18 **c** 18 **a**
3 a 13 **b** 12 **c** 12 **a** **4 a** 13 **b** 13 **c** 11 **c**
5 a 15 **b** 14 **c** 15 **b** **6 a** 17 **b** 16 **c** 16 **a**
7 a 24 **b** 25 **c** 25 **a** **8 a** 15 **b** 15 **c** 16 **c**
9 17 kg **10** 12

page 22 **Exercise 2.10**

1 a £16.20 **b** £3.80 **2 a** £15.40 **b** £4.60 **3 a** £7.20 **b** £2.80 **4 a** £5.30 **b** £4.70
5 a 2.8 m **b** 3 m **c** 20 cm or 0.2 m **6 a** 2.5 cm **b** 3 m **c** 50 cm

page 23 **Exercise 2.11**

1 a £8.60 **b** £1.40 **2 a** £5.90 **b** £4.10 **3 a** £3.90 **b** £1.10 **4 a** £4.40 **b** £0.60
5 a 2.70 m **b** 3 m **c** 30 cm **6 a** 1.50 m **b** 2 m **c** 50 cm

page 25 **Exercise 2.12**

1 5, 2 **2** 13, 1 **3** 4, 2 m **4** 8, 2 p **5** 16, 4 **6** 15, 2

page 26 **Exercise 2.13**

1 a 676 **b** 868 **c** 949 **d** 768 **e** 656 **f** 558 **g** 636 **h** 357
2 a 728 **b** 912 **c** 714 **d** 414 **e** 658 **f** 817 **g** 855 **h** 966

page 26 **Exercise 2.14**

1 a 936 **b** 936 **c** 946 **c** **2 a** 884 **b** 864 **c** 864 **a**

page 27 **Exercise 2.15**

1 a 4032 **b** 4032 **c** 4048 **c** **2 a** 3276 **b** 3264 **c** 3264 **a** **3 a** 5396 **b** 5376 **c** 5376 **a**

page 27 *Exercise 2.16*

1 450 **2** 180 **3** **a** 364 km **b** 494 km **4** 900 kg

page 28 *Exercise 2.17*

1 **a** 23 **b** 24 **c** 23 **b** **2** **a** 25 **b** 26 **c** 26 **a**
3 **a** 32 **b** 32 **c** 33 **c** **4** **a** 43 **b** 42 **c** 42 **a**
5 **a** 52 **b** 53 **c** 53 **a** **6** **a** 32 **b** 33 **c** 32 **b**

page 28 *Exercise 2.18*

1 12 **2** 35 **3** 16 seconds **4** 12 **5** 30

page 28 *Exercise 2.19*

1 4800 **2** 1200 **3** 1200 **4** 5600 **5** 6000 **6** 15 000 **7** 24 000 **8** 36 000

page 29 *Exercise 2.20*

1 7 **2** 8 **3** 6 **4** 4 **5** 7 **6** 4
7 3 **8** 6 **9** 3 **10** 9 **11** 6 **12** 9

page 30 *Exercise 3.1*

1 69 (40 + 30) **2** 129 (90 + 40) **3** 71 (50 + 20) **4** 109 (70 + 40)
5 94 (40 + 50) **6** 95 (70 + 20) **7** 503 (300 + 200) **8** 602 (200 + 400)
9 470 (200 + 300) **10** 286 (100 + 200) **11** 536 (300 + 200) **12** 625 (500 + 100)
13 22 (50 − 30) **14** 45 (80 − 40) **15** 37 (90 − 60) **16** 39 (80 − 40)
17 18 (80 − 60) **18** 187 (300 − 100) **19** 274 (600 − 300) **20** 291 (500 − 200)
21 389 (700 − 300) **22** 494 (600 − 100) **23** 434 (700 − 300) **24** 130 (500 − 400)

page 31 *Exercise 3.2*

1 7 **2** 10 **3** 5 **4** 8 **5** 4 **6** 10
7 8 **8** 5 **9** 16 **10** 20 **11** 6 **12** 5

page 32 *Exercise 3.3*

1 13 **2** 16 **3** 20 **4** 32 **5** 26 **6** 22
7 39 **8** 45 **9** 56 **10** 84 **11** 33 **12** 59

page 32 *Exercise 3.4*

1 **a** 8 p **b** 47 p **c** 55 p **d** 16 p **e** 31 p **f** 70 p **g** 23 p **h** 62 p
2 **a** 8 p **b** 49 p **c** 57 p **d** 16 p **e** 33 p **f** 74 p **g** 25 p **h** 66 p
3 **a** $1\frac{2}{5}$ cm **b** 8 cm **c** 4 cm **d** 11 cm **e** 6 cm **f** 13 cm **g** 3 cm **h** 10 cm

page 33 *Exercise 3.5*

1 **a** 7 **b** 2 **2** **a** 5 **b** 10 **3** **a** 3 **b** 10 **4** **a** 7 **b** 20 **5** **a** 4 **b** 3

page 34 **Exercise 3.6**

1 a 5 **b** 6 **2 a** 5 **b** 8 **3 a** 5 **b** 9
4 a 3 **b** 2 **5 a** 39 **b** 10 **c** 1 **6 a** 4 **b** 1

page 35 **Exercise 3.7**

1 a 48.8 **b** 48.7 **c** 48.7 **a** **2 a** 58.5 **b** 58.6 **c** 58.5 **b**
3 a 73.5 **b** 73.6 **c** 73.5 **b** **4 a** 66.4 **b** 66.5 **c** 66.5 **a**
5 a 81.6 **b** 82.6 **c** 81.6 **b** **6 a** 23.5 **b** 23.5 **c** 24.5 **c**

page 35 **Exercise 3.8**

1 a 19.15 **b** 19.25 **c** 19.15 **b** **2 a** 23.38 **b** 23.48 **c** 23.48 **a**
3 a 64.4 **b** 64.4 **c** 63.4 **c** **4 a** 43.5 **b** 42.5 **c** 42.5 **a**
5 a 43.54 **b** 43.64 **c** 43.64 **a** **6 a** 4.61 **b** 4.51 **c** 4.61 **b**

page 36 **Exercise 3.9**

1 a 52.8 **b** 52.8 **c** 52.7 **c** **2 a** 22.4 **b** 23.4 **c** 23.4 **a**
3 a 6.9 **b** 6.8 **c** 6.9 **b** **4 a** 9960 **b** 9968 **c** 9968 **a**
5 a 70 **b** 70 **c** 60 **c** **6 a** 12 **b** 15 **c** 12 **b**

page 36 **Exercise 3.10**

1 a 33 **b** 32 **c** 33 **b** **2 a** 26.5 **b** 27.5 **c** 27.5 **a**
3 a 51.8 **b** 52.8 **c** 52.8 **a** **4 a** 1.55 **b** 1.65 **c** 1.55 **b**
5 a 37.5 **b** 37.5 **c** 37.6 **c** **6 a** 62.5 **b** 67.5 **c** 67.5 **a**

page 37 **Exercise 3.11**

1 a £7.85 **b** £2.15 **2 a** £5.75 **b** £4.25 **3 a** £4.55 **b** £5.45
4 a 7.65 m **b** 2.35 m **5 a** 3.71 m **b** 1.29 m

page 38 **Exercise 3.12**

1 a £23.75 **b** £26.25 **2 a** £18.65 **b** £31.35 **3 a** £19.35 **b** £30.65
4 a £30.25 **b** £19.75 **5 a** 5.2 m **b** 80 cm **6 a** 2.8 m **b** 3.2 m

page 39 **Exercise 3.13**

1 £3.29 per kg; £3.28 per kg; the second **2** £1.94 per kg; £1.90 per kg; the second
3 £1.32 per kg; £1.33 per kg; the first **4** £1.45 per kg; £1.44 per kg; the second
5 21p each; 22 p each; the first

page 39 **Exercise 3.14**

1 £1.21 per kg; £1.22 per kg the first **2** £1.10 per kg; £1.08 per kg the second
3 85 p per kg; 84 p per kg the second
4 a 68 p per litre **b** 69 p per litre **c** 70 p per litre **a**
5 a 95 p per litre **b** 92 p per litre **c** 90 p per litre **c**

page 40 ***Exercise 3.15***

1 59 **2** 36 **3** 53 **4** 77 **5** 42 **6** 43 **7** 34 **8** 56
9 31 **10** 38 **11** 45 **12** 26 **13** 33 **14** 36 m **15** 29 **16** 17 m

page 41 ***Exercise 3.16***

1 27 m **2** 25 m **3** 28 m **4** 24 cm **5** 47 cm
6 53 cm **7** 48 mm **8** 78 mm **9** 97 mm

page 42 ***Exercise 3.17***

1 3260 **2** 7370 **3** 3770 **4** 5940 **5** 7240 **6** 2480
7 3980 **8** 7210 **9** 8410 **10** 6710 **11** 3800 **12** 8600

page 43 ***Exercise 3.18***

1 380 **2** 360 **3** 960 **4** 480 **5** 850 **6** 760
7 710 **8** 500 **9** 300 **10** 900 **11** 400 **12** 800

page 43 ***Exercise 3.19***

1 6400 **2** 9400 **3** 9300 **4** 6400 **5** 4600 **6** 9800
7 4100 **8** 6000 **9** 5000 **10** 7000 **11** 5600 **12** 4200

page 43 ***Exercise 3.20***

1 27.4 **2** 51.3 **3** 34.6 **4** 53.1 **5** 46.1 **6** 79.1
7 37.9 **8** 56.9 **9** 43.0 **10** 55.0 **11** 35.0 **12** 43.0

page 43 ***Exercise 3.21***

1 65 **2** 83 **3** 33 **4** 57 **5** 67 **6** 76
7 50 **8** 60 **9** 90 **10** 71 **11** 31 **12** 20

page 44 ***Exercise 3.22***

1 0.33 **2** 0.75 **3** 0.58 **4** 0.11 **5** 0.094 **6** 0.074 **7** 0.056 **8** 0.069

page 44 ***Exercise 3.23***

1 5.25 **2** 7.84 **3** 3.68 **4** 6.92 **5** 4.63 **6** 7.47
7 8.21 **8** 5.71 **9** 7.90 **10** 7.50 **11** 8.60 **12** 1.70

page 45 ***Exercise 3.24***

1 4.7 **2** 7.5 **3** 6.4 **4.** 1.7 **5** 2.9 **6** 6.1
7 4.1 **8** 7.0 **9** 3.0 **10** 6.0 **11** 4.0 **12** 4.9

page 45 ***Exercise 3.25***

1 4.9 **2** 2.4 **3** 3.6 **4** 7.6 **5** 6.4 **6** 5.8
7 5.0 **8** 9.0 **9** 9.3 **10** 3.1 **11** 2.1 **12** 6.0

page 45 ***Exercise 3.26***

		4dp	3dp	2dp	1dp
1	3.138 27	3.1383	3.138	3.14	3.1
2	5.247 39	5.2474	5.247	5.25	5.2
3	6.426 18	6.4262	6.426	6.43	6.4
4	17.915 83	17.9158	17.916	17.92	17.9
5	19.807 52	19.8075	19.808	19.81	19.8
6	12.476 32	12.4763	12.476	12.48	12.5

page 45 ***Exercise 3.27***

1 0.24 **2** 0.43 **3** 0.55 **4** 0.38 **5** 0.29 **6** 0.64
7 0.03 **8** 0.09 **9** 0.06 **10** 0.08 **11** 0.05 **12** 0.07

page 46 ***Exercise 3.28***

1 1598 $(50 \times 30 = 1500)$ **2** 3698 $(90 \times 40 = 3600)$ **3** 3588 $(90 \times 40 = 3600)$
4 3478 $(70 \times 50 = 3500)$ **5** 6014 $(200 \times 30 = 6000)$ **6** 5093 $(500 \times 10 = 5000)$
7 6006 $(200 \times 30 = 6000)$ **8** 5966 $(300 \times 20 = 6000)$ **9** 8136 $(1000 \times 8 = 8000)$
10 40 071 $(2000 \times 20 = 40000)$ **11** 92 597 $(3000 \times 30 = 90000)$ **12** 74 594 $(2000 \times 40 = 80000)$

page 47 ***Exercise 3.29***

1 17.5 $(900 \div 50 = 18)$ **2** 9.5 $(600 \div 60 = 10)$ **3** 13.5 $(700 \div 50 = 14)$
4 21 $(800 \div 40 = 20)$ **5** 10.5 $(600 \div 60 = 10)$ **6** 5.5 $(2000 \div 400 = 5)$
7 3.5 $(2000 \div 500 = 4)$ **8** 5.5 $(3000 \div 500 = 6)$

page 47 ***Exercise 3.30***

1 13.5 $(60 \times 20 \div 80 = 15)$ **2** 50.75 $(90 \times 30 \div 50 = 54)$ **3** 109.25 $(70 \times 60 \div 40 = 105)$
4 38 $(60 \times 60 \div 90 = 40)$ **5** 88 $(60 \times 80 \div 50 = 96)$ **6** 77 $(90 \times 40 \div 50 = 72)$

page 49 ***Exercise 4.1***

1 a 324 km **b** 216 km **c** 162 km **d** 108 km **e** 81 km **f** 72 km
2 a 450 m **b** 300 m **c** 225 m **d** 180 m **e** 150 m **f** 100 m **g** 90 m
3 a 60 **b** 40 **c** 30 **d** 24 **e** 20 **f** 15 **g** 12

page 50 ***Exercise 4.2***

1 a 432 km **b** 336 km **c** 144 km **d** 384 km **e** 288 km **f** 192 km **g** 400 km **h** 360 km
i 320 km **j** 420 km **k** 300 km **l** 180 km
2 a 480 g **b** 420 g **c** 300 g **d** 450 g **e** 405 g **f** 360 g

page 51 ***Exercise 4.3***

1 a 91% **b** 73% **c** 69% **d** 83% **e** 99%
f 77% **g** 39% **h** 21% **i** 11% **j** 7%
2 a $\frac{97}{100}$ **b** $\frac{51}{100}$ **c** $\frac{49}{100}$ **d** $\frac{89}{100}$ **e** $\frac{93}{100}$
f $\frac{61}{100}$ **g** $\frac{27}{100}$ **h** $\frac{33}{100}$ **i** $\frac{17}{100}$ **j** $\frac{3}{100}$
3 a $\frac{57}{100}$ **b** 57% **c** $\frac{43}{100}$ **d** 43%
4 a $\frac{59}{100}$ **b** 59% **c** $\frac{41}{100}$ **d** 41%
5 a $\frac{63}{100}$ **b** 63% **c** $\frac{37}{100}$ **d** 37%

page 52 ***Exercise 4.4***

1 6 **2** 24 **3** 40 **4** 16 **5** 42 **6** 8 **7** 20, 12, 8 **8** 36, 18, 6

page 53 ***Exercise 4.5***

1 £16 **2** £18 **3** £54 **4** £28 **5** 24 m **6** 12 m **7** 6 m **8** 12 m

page 54 ***Exercise 4.6***

1 £1.60 **2** £1.80 **3** £2.80 **4** £5.40 **5** 2.5 m **6** 6.3 m **7** 2.4 m **8** 2.4 m

page 54 ***Exercise 4.7***

1 80% **2** 60% **3** 75% **4** 20% **5** 50%, 30%, 20%

page 55 ***Exercise 4.8***

1	**a** £44	**b** £45	**c** £44	**b**	**2**	**a** £35	**b** £35	**c** £36	**a**
3	**a** £56	**b** £55	**c** £55	**a**	**4**	**a** £42	**b** £44	**c** £44	**a**
5	**a** £84	**b** £81	**c** £84	**b**	**6**	**a** £9.75	**b** £9.45	**c** £9.45	**a**
7	**a** 155 m	**b** 154 m	**c** 154 m	**a**	**8**	**a** 34 m	**b** 34 m	**c** 36 m	**c**
9	**a** 26 m	**b** 28 m	**c** 28 m	**a**	**10**	**a** 6.25 m	**b** 6.75 m	**c** 6.25 m	**b**
11	**a** 8.25 m	**b** 8.24 m	**c** 8.25 m	**b**	**12**	**a** 9.35 m	**b** 9.45 m	**c** 9.45 m	**a**

page 56 ***Exercise 4.9***

1	96	84	60		
2	270	60	30		
3	192	160	48	80	
4	60	45	12	3	
5	27	45	60	30	18

page 56 ***Exercise 4.10***

1	**a** £93	**b** £91	**c** £91	**a**	**2**	**a** £68	**b** £68	**c** £63	**c**
3	**a** £96	**b** £99	**c** £99	**a**	**4**	**a** £12.24	**b** £12.24	**c** £12.21	**c**
5	**a** £15.36	**b** £15.96	**c** £15.36	**b**	**6**	**a** £14.40	**b** £15.40	**c** £14.40	**b**
7	**a** 54 cm	**b** 56 cm	**c** 54 cm	**b**	**8**	**a** 84 cm	**b** 84 cm	**c** 85 cm	**c**
9	**a** 95 cm	**b** 96 cm	**c** 96 cm	**a**	**10**	**a** 2.8 cm	**b** 2.4 cm	**c** 2.4 cm	**a**
11	**a** 3.6 cm	**b** 3.5 cm	**c** 3.6 cm	**b**	**12**	**a** 6.5 cm	**b** 6.3 cm	**c** 6.3 cm	**a**

page 57 ***Exercise 4.11***

1	40	60	25		
2	30	36	9		
3	195	60	45		
4	20	16	8	6	
5	60	70	65	55	
6	18	21	9	12	15
7	28	20	16	12	4

page 58 ***Exercise 4.12***

1 $\frac{1}{3}=\frac{4}{12}=\frac{5}{15}=\frac{6}{18}=\frac{7}{21}=\frac{9}{27}$ **2** $\frac{1}{5}=\frac{4}{20}=\frac{6}{30}=\frac{7}{35}=\frac{8}{40}=\frac{10}{50}$ **3** $\frac{1}{6}=\frac{2}{12}=\frac{5}{30}=\frac{6}{36}=\frac{10}{60}=\frac{12}{72}$

4 $\frac{1}{8}=\frac{3}{24}=\frac{4}{32}=\frac{5}{40}=\frac{7}{56}=\frac{10}{80}$ **5** $\frac{1}{9}=\frac{3}{27}=\frac{5}{45}=\frac{7}{63}=\frac{8}{72}=\frac{10}{90}$ **6** $\frac{1}{10}=\frac{3}{30}=\frac{4}{40}=\frac{7}{70}=\frac{8}{80}=\frac{10}{100}$

7 $\frac{3}{4}=\frac{9}{12}=\frac{12}{16}=\frac{15}{20}=\frac{24}{32}=\frac{30}{40}$ **8** $\frac{2}{5}=\frac{4}{10}=\frac{8}{20}=\frac{12}{30}=\frac{18}{45}=\frac{24}{60}$ **9** $\frac{3}{5}=\frac{6}{10}=\frac{18}{30}=\frac{21}{35}=\frac{24}{40}=\frac{30}{50}$

10 $\frac{4}{5}=\frac{8}{10}=\frac{16}{20}=\frac{24}{30}=\frac{36}{45}=\frac{40}{50}$ **11** $\frac{5}{6}=\frac{20}{24}=\frac{25}{30}=\frac{30}{36}=\frac{45}{54}=\frac{50}{60}$ **12** $\frac{3}{8}=\frac{9}{24}=\frac{15}{40}=\frac{18}{48}=\frac{27}{72}=\frac{36}{96}$

13 $\frac{1}{3}=\frac{2}{6}=\frac{3}{9}=\frac{5}{15}=\frac{9}{27}=\frac{12}{36}$ **14** $\frac{1}{5}=\frac{2}{10}=\frac{3}{15}=\frac{5}{25}=\frac{9}{45}=\frac{12}{60}$ **15** $\frac{1}{6}=\frac{3}{18}=\frac{4}{24}=\frac{7}{42}=\frac{8}{48}=\frac{9}{54}$

16 $\frac{1}{8}=\frac{2}{16}=\frac{6}{48}=\frac{8}{64}=\frac{9}{72}=\frac{12}{96}$ **17** $\frac{4}{5}=\frac{12}{15}=\frac{20}{25}=\frac{28}{35}=\frac{32}{40}=\frac{48}{60}$ **18** $\frac{5}{6}=\frac{10}{12}=\frac{15}{18}=\frac{35}{42}=\frac{40}{48}=\frac{60}{72}$

page 59 ***Exercise 4.13***

1 $\frac{8}{10}=\frac{4}{5}$ **2** $\frac{6}{10}=\frac{3}{5}$ **3** $\frac{4}{10}=\frac{2}{5}$ **4** $\frac{2}{10}=\frac{1}{5}$

5 $\frac{2}{12}=\frac{1}{6}$ **6** $\frac{6}{8}=\frac{3}{4}$ **7** $\frac{9}{15}=\frac{3}{5}$ **8** $\frac{6}{15}=\frac{2}{5}$

9 $\frac{35}{50}=\frac{7}{10}$ **10** $\frac{15}{20}=\frac{3}{4}$ **11** $\frac{5}{20}=\frac{1}{4}$ **12** $\frac{5}{25}=\frac{1}{5}$

13 $\frac{18}{24}=\frac{3}{4}$ **14** $\frac{16}{40}=\frac{2}{5}$ **15** $\frac{32}{40}=\frac{4}{5}$ **16** $\frac{16}{24}=\frac{2}{3}$

page 59 ***Exercise 4.14***

1 $\frac{10}{16}=\frac{5}{8}$ **2** $\frac{6}{16}=\frac{3}{8}$ **3** $\frac{14}{16}=\frac{7}{8}$ **4** $\frac{2}{16}=\frac{1}{8}$

5 $\frac{9}{30}=\frac{3}{10}$ **6** $\frac{9}{12}=\frac{3}{4}$ **7** $\frac{12}{28}=\frac{3}{7}$ **8** $\frac{20}{28}=\frac{5}{7}$

9 $\frac{5}{15}=\frac{1}{3}$ **10** $\frac{5}{10}=\frac{1}{2}$ **11** $\frac{5}{20}=\frac{1}{4}$ **12** $\frac{15}{20}=\frac{3}{4}$

13 $\frac{24}{40}=\frac{3}{5}$ **14** $\frac{8}{40}=\frac{1}{5}$ **15** $\frac{8}{16}=\frac{1}{2}$ **16** $\frac{8}{80}=\frac{1}{10}$

page 60 ***Exercise 4.15***

1 a $\frac{3}{5}$ **b** $\frac{3}{5}$ **c** $\frac{5}{8}$ **c** **2 a** $\frac{3}{4}$ **b** $\frac{3}{4}$ **c** $\frac{2}{3}$ **c**

3 a $\frac{4}{5}$ **b** $\frac{4}{5}$ **c** $\frac{5}{6}$ **c** **4 a** $\frac{2}{5}$ **b** $\frac{1}{2}$ **c** $\frac{2}{5}$ **b**

5 a $\frac{1}{4}$ **b** $\frac{3}{8}$ **c** $\frac{1}{4}$ **b**

page 60 ***Exercise 4.16***

1 a 0.84 **b** 0.85 **c** 0.84 **b** **2 a** 0.68 **b** 0.65 **c** 0.65 **a**

3 a 0.6 **b** 0.6 **c** 0.625 **c** **4 a** 0.44 **b** 0.45 **c** 0.45 **a**

5 a 0.16 **b** 0.175 **c** 0.175 **a**

page 61 ***Exercise 4.17***

1 $\frac{4}{20}+\frac{3}{20}=\frac{7}{20}$ **2** $\frac{4}{20}+\frac{7}{20}=\frac{11}{20}$ **3** $\frac{8}{20}+\frac{1}{20}=\frac{9}{20}$ **4** $\frac{8}{20}+\frac{3}{20}=\frac{11}{20}$ **5** $\frac{12}{20}+\frac{1}{20}=\frac{13}{20}$ **6** $\frac{12}{20}+\frac{7}{20}=\frac{19}{20}$

page 61 ***Exercise 4.18***

1 $\frac{11}{15}$ **2** $\frac{14}{15}$ **3** $\frac{7}{15}$ **4** $\frac{4}{15}$ **5** $\frac{8}{15}$ **6** $\frac{14}{15}$ **7** $\frac{13}{15}$ **8** $\frac{13}{15}$ **9** $\frac{8}{9}$ **10** $\frac{7}{9}$

11 $\frac{9}{20}$ **12** $\frac{19}{20}$ **13** $\frac{7}{20}$ **14** $\frac{13}{20}$ **15** $\frac{17}{20}$ **16** $\frac{19}{20}$ **17** $\frac{7}{12}$ **18** $\frac{11}{12}$ **19** $\frac{7}{8}$ **20** $\frac{5}{8}$

page 62 ***Exercise 4.19***

1 $\frac{16}{20}-\frac{3}{20}=\frac{13}{20}$ **2** $\frac{16}{20}-\frac{7}{20}=\frac{9}{20}$ **3** $\frac{12}{20}-\frac{1}{20}=\frac{11}{20}$ **4** $\frac{12}{20}-\frac{3}{20}=\frac{9}{20}$ **5** $\frac{8}{20}-\frac{1}{20}=\frac{7}{20}$ **6** $\frac{8}{20}-\frac{7}{20}=\frac{1}{20}$

page 61 *Exercise 4.20*

1 $\frac{11}{20}$ **2** $\frac{7}{20}$ **3** $\frac{1}{20}$ **4** $\frac{9}{20}$ **5** $\frac{1}{10}$ **6** $\frac{7}{10}$ **7** $\frac{3}{10}$ **8** $\frac{1}{10}$ **9** $\frac{5}{8}$ **10** $\frac{1}{8}$
11 $\frac{1}{6}$ **12** $\frac{5}{12}$ **13** $\frac{5}{9}$ **14** $\frac{2}{9}$ **15** $\frac{1}{9}$ **16** $\frac{1}{8}$ **17** $\frac{3}{8}$ **18** $\frac{1}{12}$ **19** $\frac{5}{12}$ **20** $\frac{4}{15}$

page 62 *Exercise 4.21*

1 $\frac{14}{20}=\frac{7}{10}$ **2** $\frac{18}{20}=\frac{9}{10}$ **3** $\frac{6}{20}=\frac{3}{10}$ **4** $\frac{18}{20}=\frac{9}{10}$ **5** $\frac{8}{10}=\frac{4}{5}$
6 $\frac{6}{10}=\frac{3}{5}$ **7** $\frac{4}{6}=\frac{2}{3}$ **8** $\frac{12}{15}=\frac{4}{5}$ **9** $\frac{9}{12}=\frac{3}{4}$ **10** $\frac{3}{12}=\frac{1}{4}$
11 $\frac{9}{12}=\frac{3}{4}$ **12** $\frac{3}{18}=\frac{1}{6}$ **13** $\frac{8}{12}=\frac{2}{3}$ **14** $\frac{4}{12}=\frac{1}{3}$ **15** $\frac{15}{20}=\frac{3}{4}$
16 $\frac{5}{20}=\frac{1}{4}$ **17** $\frac{15}{20}=\frac{3}{4}$ **18** $\frac{15}{20}=\frac{3}{4}$ **19** $\frac{10}{15}=\frac{2}{3}$ **20** $\frac{5}{15}=\frac{1}{3}$

page 63 *Exercise 4.22*

1 $\frac{6}{20}=\frac{3}{10}$ **2** $\frac{2}{20}=\frac{1}{10}$ **3** $\frac{14}{20}=\frac{7}{10}$ **4** $\frac{2}{20}=\frac{1}{10}$ **5** $\frac{2}{18}=\frac{1}{9}$
6 $\frac{14}{18}=\frac{7}{9}$ **7** $\frac{6}{15}=\frac{2}{5}$ **8** $\frac{3}{15}=\frac{1}{5}$ **9** $\frac{9}{12}=\frac{3}{4}$ **10** $\frac{3}{12}=\frac{1}{4}$
11 $\frac{3}{6}=\frac{1}{2}$ **12** $\frac{3}{18}=\frac{1}{6}$ **13** $\frac{4}{12}=\frac{1}{3}$ **14** $\frac{8}{12}=\frac{2}{3}$ **15** $\frac{5}{20}=\frac{1}{4}$
16 $\frac{15}{20}=\frac{3}{4}$ **17** $\frac{5}{20}=\frac{1}{4}$ **18** $\frac{5}{20}=\frac{1}{4}$ **19** $\frac{5}{15}=\frac{1}{3}$ **20** $\frac{10}{15}=\frac{2}{3}$

page 63 *Exercise 4.23*

1 $\frac{33}{40}$ **2** $\frac{17}{40}$ **3** $\frac{9}{40}$ **4** $\frac{19}{40}$ **5** $\frac{5}{24}$
6 $\frac{11}{24}$ **7** $\frac{17}{24}$ **8** $\frac{23}{24}$ **9** $\frac{13}{24}$ **10** $\frac{19}{24}$
11 $\frac{7}{24}$ **12** $\frac{8}{30}=\frac{4}{15}$ **13** $\frac{28}{30}=\frac{14}{15}$ **14** $\frac{22}{30}=\frac{11}{15}$ **15** $\frac{2}{30}=\frac{1}{15}$
16 $\frac{3}{30}=\frac{1}{10}$ **17** $\frac{3}{30}=\frac{1}{10}$ **18** $\frac{13}{20}$ **19** $\frac{3}{20}$ **20** $\frac{17}{24}$

page 64 *Exercise 4.24*

1 **a** 95% **b** 96% **c** 96% **a**
2 **a** 45% **b** 48% **c** 45% **b**
3 **a** 80% **b** 84% **c** 84% **a**
4 **a** 24% **b** 25% **c** 25% **a**
5 **a** 15% **b** 16% **c** 15% **b**

page 64 *Exercise 4.25*

1 $\frac{3}{4}$ **2** $\frac{5}{6}$ **3** $\frac{5}{8}$ **4** $\frac{3}{8}$ **5** $\frac{3}{5}$ **6** $\frac{2}{5}$ **7** $\frac{2}{3}$ **8** $\frac{5}{6}$
9 $\frac{5}{8}$ **10** $\frac{3}{8}$ **11** $\frac{9}{10}$ **12** $\frac{7}{10}$ **13** $\frac{1}{4}$ **14** $\frac{1}{3}$ **15** $\frac{1}{2}$ **16** $\frac{1}{6}$
17 $\frac{5}{6}$ **18** $\frac{3}{10}$ **19** $\frac{1}{4}$ **20** $\frac{1}{5}$ **21** $\frac{2}{5}$ **22** $\frac{3}{5}$ **23** $\frac{4}{5}$ **24** $\frac{3}{4}$

page 65 *Exercise 4.26*

1 $\frac{10}{25}$ $\frac{12}{30}$ $\frac{14}{35}$ $\frac{16}{40}$ $\frac{18}{45}$
2 $\frac{15}{25}$ $\frac{18}{30}$ $\frac{21}{35}$ $\frac{24}{40}$ $\frac{27}{45}$
3 $\frac{25}{40}$ $\frac{30}{48}$ $\frac{35}{56}$ $\frac{40}{64}$ $\frac{45}{72}$
4 $\frac{35}{40}$ $\frac{42}{48}$ $\frac{49}{56}$ $\frac{56}{64}$ $\frac{63}{72}$
5 $\frac{5}{30}$ $\frac{6}{36}$ $\frac{7}{42}$ $\frac{8}{48}$ $\frac{9}{54}$
6 $\frac{5}{40}$ $\frac{6}{48}$ $\frac{7}{56}$ $\frac{8}{64}$ $\frac{9}{72}$

page 65 *Exercise 4.27*

1 $\frac{24}{30}$ $\frac{20}{25}$ $\frac{16}{20}$ $\frac{12}{15}$ $\frac{8}{10}$ $\frac{4}{5}$
2 $\frac{30}{36}$ $\frac{25}{30}$ $\frac{20}{24}$ $\frac{15}{18}$ $\frac{10}{12}$ $\frac{5}{6}$
3 $\frac{54}{60}$ $\frac{45}{50}$ $\frac{36}{40}$ $\frac{27}{30}$ $\frac{18}{20}$ $\frac{9}{10}$
4 $\frac{18}{120}$ $\frac{15}{100}$ $\frac{12}{80}$ $\frac{9}{60}$ $\frac{6}{40}$ $\frac{3}{20}$
5 $\frac{6}{30}$ $\frac{5}{25}$ $\frac{4}{20}$ $\frac{3}{15}$ $\frac{2}{10}$ $\frac{1}{5}$
6 $\frac{6}{12}$ $\frac{5}{10}$ $\frac{4}{8}$ $\frac{3}{6}$ $\frac{2}{4}$ $\frac{1}{2}$

page 66 **Exercise 4.28**

1 $\frac{4}{5}$ $\frac{12}{15}$ $\frac{16}{20}$	**2** $\frac{7}{10}$ $\frac{21}{30}$ $\frac{28}{40}$	**3** $\frac{5}{8}$ $\frac{15}{24}$ $\frac{20}{32}$	**4** $\frac{5}{12}$ $\frac{15}{36}$ $\frac{20}{48}$	**5** $\frac{2}{5}$ $\frac{4}{10}$ $\frac{8}{20}$	**6** $\frac{4}{7}$ $\frac{8}{14}$ $\frac{16}{28}$
7 $\frac{3}{10}$ $\frac{6}{20}$ $\frac{9}{30}$	**8** $\frac{3}{8}$ $\frac{6}{16}$ $\frac{9}{24}$	**9** $\frac{7}{20}$ $\frac{14}{40}$ $\frac{21}{60}$	**10** $\frac{1}{7}$ $\frac{2}{14}$ $\frac{3}{21}$	**11** $\frac{2}{7}$ $\frac{4}{14}$ $\frac{6}{21}$	**12** $\frac{2}{9}$ $\frac{4}{18}$ $\frac{6}{27}$

page 67 **Exercise 4.29**

1 $\frac{9}{12}$ $\frac{12}{20}$	**2** $\frac{12}{15}$ $\frac{16}{20}$	**3** $\frac{3}{9}$ $\frac{5}{15}$ $\frac{7}{21}$ $\frac{9}{27}$

page 68 **Exercise 4.30**

1 24	**2** 25 m	**3** 65 kg	**4** 20 km	**5** 42
6 45 minutes	**7** £264	**8** £207	**9** £43 200	**10** 168 cm

page 69 **Exercise 4.31**

1 24	**2** 32	**3** 54 seconds	**4** 100	**5** 220 cm
6 75	**7** 12 hours	**8** 42 minutes	**9** £68	**10** £5100

page 70 **Exercise 4.32**

1 0.45	**2** 0.15	**3** 0.95	**4** 0.65	**5** 0.52	**6** 0.64
7 0.48	**8** 0.84	**9** 0.12	**10** 0.28	**11** 0.175	**12** 0.725
13 0.25	**14** 0.18	**15** 0.34	**16** 0.78	**17** 0.98	**18** 0.74
19 0.2	**20** 0.05	**21** 0.04	**22** 0.075	**23** 0.048	**24** 0.072

page 71 **Exercise 4.33**

1 $\frac{26}{100}=\frac{13}{50}$	**2** $\frac{46}{100}=\frac{23}{50}$	**3** $\frac{22}{100}=\frac{11}{50}$	**4** $\frac{86}{100}=\frac{43}{50}$	**5** $\frac{42}{100}=\frac{21}{50}$	**6** $\frac{66}{100}=\frac{33}{50}$
7 $\frac{44}{100}=\frac{11}{25}$	**8** $\frac{32}{100}=\frac{8}{25}$	**9** $\frac{85}{100}=\frac{17}{20}$	**10** $\frac{55}{100}=\frac{11}{20}$	**11** $\frac{325}{1000}=\frac{13}{40}$	**12** $\frac{675}{1000}=\frac{27}{40}$
13 $\frac{6}{10}=\frac{3}{5}$	**14** $\frac{7}{10}$	**15** $\frac{1}{10}$	**16** $\frac{1}{100}$	**17** $\frac{7}{100}$	**18** $\frac{9}{100}$
19 $\frac{2}{100}=\frac{1}{50}$	**20** $\frac{8}{100}=\frac{2}{25}$	**21** $\frac{25}{1000}=\frac{1}{40}$	**22** $\frac{35}{1000}=\frac{7}{200}$	**23** $\frac{65}{1000}=\frac{13}{200}$	**24** $\frac{55}{1000}=\frac{11}{200}$

page 71 **Exercise 4.34**

1 25%	**2** 29%	**3** 23%	**4** 36%	**5** 32%	**6** 42%
7 85%	**8** 18%	**9** 15%	**10** 12%	**11** 4%	**12** 5%
13 8%	**14** 7%	**15** 40%	**16** 90%	**17** 20%	**18** 10%
19 76.5%	**20** 16.5%	**21** 24.5%	**22** 94.5%	**23** 64.5%	**24** 1.5%

page 72 **Exercise 4.35**

1 0.67	**2** 0.53	**3** 0.49	**4** 0.76	**5** 0.84	**6** 0.16
7 0.30	**8** 0.60	**9** 0.80	**10** 0.09	**11** 0.06	**12** 0.02
13 0.735	**14** 0.935	**15** 0.075	**16** 0.025	**17** 0.035	**18** 0.095

page 72 **Exercise 4.36**

1	**a** 20%	**b** 20%	**c** 25%	**c**	**2**	**a** 30%	**b** 35%	**c** 30%	**b**
3	**a** 10%	**b** 15%	**c** 10%	**b**	**4**	**a** 40%	**b** 35%	**c** 35%	**a**
5	**a** 80%	**b** 75%	**c** 75%	**a**	**6**	**a** 70%	**b** 70%	**c** 75%	**c**
7	**a** 17.5%	**b** 12.5%	**c** 12.5%	**a**	**8**	**a** 37.5%	**b** 32.5%	**c** 37.5%	**b**

page 73 *Exercise 4.37*

1 20% **2** 30% **3** 80% **4** 8% **5** 15%
6 35% **7** 20% **8** 15% **9** 25% **10** 30%

page 74 *Exercise 4.38*

1 '$\frac{1}{4}$ off' = 25%; 5%
2 '$\frac{1}{5}$ off' = 20%; 5%
3 '$\frac{1}{25}$ off' = 4%; 1%
4 '$\frac{1}{5}$ off' = 20%, so 25% is 5% better
5 '$\frac{1}{4}$ off' = 25%, so 30% is 5% better
6 '$\frac{1}{25}$ off' = 4%, so 6% is 2% better

page 74 *Exercise 4.39*

1 £80 £96 second by £16
2 £40 £30 first by £10
3 £12 £15 second by £3
4 £60 £48 first by £12
5 £40 £50 second by £10

page 76 *Exercise 5.1*

1 30° C 20° C 10° C −30° C −40° C −50° C
2 50° C 30° C 10° C −20° C −40° C −50° C
3 50° C 40° C 10° C −20° C −30° C −50° C
4 40° C 30° C 20° C −10° C −30° C −50° C
5 50° C 40° C 30° C −10° C −20° C −50° C
6 −50° C −40° C −30° C 10° C 20° C 40° C
7 −50° C −40° C −10° C 20° C 30° C 50° C
8 −50° C −30° C −20° C 10° C 20° C 40° C
9 −50° C −30° C −10° C 20° C 40° C 50° C
10 −40° C −30° C −20° C 10° C 20° C 50° C

page 76 *Exercise 5.2*

1 £300(c) £200 £150 £50(c) £100(o) −£150 £200(o) −£250
2 £300 £200 £100(c) £50(c) £150(o) £200(o) −£250 −£300
3 £300 £250(c) £200(c) £150 £50(o) −£100 −£200 £300(o)
4 £300(c) £250(c) £100 £50 −£100 £150(o) −£200 £250(o)
5 £300 £150(c) £100 £50(c) £100(o) −£150 −£200 £250(o)

page 78 *Exercise 5.3*

1 10° C 11° C 5° C 3° C 4° C 6° C 7° C
2 −2° C −3° C −2° C −4° C −3° C −2° C −4° C −3° C −1° C −1° C −5° C −6° C −5° C −4° C
3 £225 £350 £250 −£100 −£150 −£75 −£175 −£150

page 79 *Exercise 5.4*

1 3° C 4° C 8° C 2° C 1° C 3° C 2° C 2° C 1° C 2° C 2° C 1° C 5° C 6° C
2 £25 £45 −£30 −£30 £45 £15 −£15 −£15

page 80 *Exercise 5.5*

1 8° C 10° C 9° C 11° C 10° C 7° C 7° C 4° C 8° C 6° C 8° C 9° C 7° C 4° C
2 £150 £200 £350 £300 £50 £25 £75 £25

page 81 *Exercise 5.6*

1
12° C 10° C 2° C
13° C 11° C 2° C
11° C 9° C 2° C
10° C 6° C 4° C
12° C 6° C 6° C
12° C 8° C 4° C
12° C 8° C 4° C

2
£50 £60 £110
£25 £50 £75
£35 £75 £110
£50 £40 £90
£15 £55 £70

page 82 **Exercise 5.7**

	a	b	c	d	e
1	−9° C	−15° C	−18° C	−12° C	−21° C
2	−£45	−£75	−£105	−£60	−£90
3	−12 m	−28 m	−84 m	−116 m	−396 m
4	−9° C	−4° C	−12° C	−3° C	−6° C
5	−£30	−£40	−£15	−£20	−£24

page 84 **Exercise 6.1**

1 16 **2** 64 **3** 225 **4** 625 **5** 1600 **6** 3600
7 3375 **8** 27 000 **9** 16 **10** 256 **11** 625 **12** 1296
13 15 625 **14** 1 000 000 **15** 16 384 **16** 78 125 **17** 256 **18** 6561

page 85 **Exercise 6.2**

1 2^2 **2** 5^2 **3** 6^2 **4** 7^2 **5** 11^2 **6** 12^2
7 20^2 **8** 30^2 **9** 50^2 **10** 3^3 **11** 6^3 **12** 7^3
13 12^3 **14** 20^3 **15** 3^5 **16** 5^5

page 85 **Exercise 6.3**

1 $64 = 2^6 = 4^3 = 8^2$
2 $256 = 2^8 = 4^4 = 16^2$
3 $512 = 2^9 = 8^3$
4 $1024 = 2^{10} = 4^5 = 32^2$
5 $4096 = 2^{12} = 4^6 = 8^4 = 16^3 = 64^2$
6 $729 = 3^6 = 9^3 = 27^2$
7 $6561 = 3^8 = 9^4 = 81^2$
8 $19\,683 = 3^9 = 27^3$
9 $1296 = 6^4 = 36^2$
10 $2\,401 = 7^4 = 49^2$
11 $15\,625 = 5^6 = 25^3 = 125^2$
12 $10\,000 = 10^4 = 100^2$
13 $1\,000\,000 = 10^6 = 100^3 = 1000^2$
14 $1\,000\,000\,000 = 10^9 = 1000^3$

page 86 **Exercise 7.1**

1 100 g **2** 120 g **3** 12 kg **4** 180 **5** 136

page 87 **Exercise 7.2**

1 4 **2** 35 **3** 25 **4** 13 **5** 16

page 87 **Exercise 7.3**

	a	b	c	d	e
1	2 kg	5 kg	7 kg	3.5 kg	0.5 kg
2	5 g	4 g	3.75 g	2.25 g	0.75 g
3	25 g	15 g	35 g	22.5 g	12.5 g
4	0.3 g	0.2 g	0.24 g	0.16 g	0.18 g

page 88 **Exercise 7.4**

1 2 : 10 3 : 15 5 : 25 7 : 35
2 2 : 16 3 : 24 5 : 40 7 : 56
3 4 : 12 6 : 18 8 : 24 9 : 27
4 4 : 36 6 : 54 8 : 72 9 : 81
5 4 : 40 6 : 60 8 : 80 9 : 90
6 2 : 100 3 : 150 5 : 250 7 : 350
7 2 : 8 3 : 12 5 : 20 7 : 28
8 4 : 20 6 : 30 8 : 40 9 : 45
9 10 : 80 4 : 32 6 : 48 8 : 64
10 4 : 48 6 : 72 10 : 120 8 : 96
11 4 : 80 6 : 120 8 : 160 9 : 180
12 4 : 200 6 : 300 8 : 400 10 : 500

page 90 **Exercise 7.5**

1 a 2.5 cm 25 000 cm 250 m **b** 2 cm 20 000 cm 200 m **c** 4 cm 40 000 cm 400 m
d 5 cm 50 000 cm 500 m **e** 1.5 cm 15 000 cm 150 m **f** 3.5 cm 35 000 cm 350 m
g 3 cm 30 000 cm 300 m
2 a 6 cm 30 000 cm 300 m **b** 10 cm 50 000 cm 500 m **c** 7 cm 35 000 cm 350 m
d 8 cm 40 000 cm 400 m **e** 8 cm 40 000 cm 400 m **f** 7 cm 35 000 cm 350 m
g 4 cm 20 000 cm 200 m

page 91 **Exercise 7.6**

1 20 cm **2** 50 cm **3** 16 cm; 12 cm
4 25 cm; 15 cm **5** 20 cm by 16 cm by 12 cm

page 92 **Exercise 7.7**

1 2000 cm = 20 m **2** 1200 cm = 12 m **3** 21 m by 12 m
4 1 m by 2 m **5** 1.5 m by 1.2 m by 0.75 m

page 92 **Exercise 7.8**

1 1 : 6 **2** 1 : 2 **3** 1 : 16 **4** 1 : 200 **5** 1 : 2000

page 93 **Exercise 7.9**

1 a 17 cm **b** 170 cm **2 a** 3 m **b** 27 m **3 a** 7 cm **b** 21 cm
4 a 40 m **b** 280 m **5 a** 3 atmospheres **b** 9 atmospheres **6 a** 6 days **b** 12 days
7 a 20 volts **b** 220 volts **8 a** 1400 **b** 19 600

page 95 **Exercise 7.10**

1 a 550 m **b** 770 m **c** 990 m **d** 1650 m **e** 2200 m **f** 2310 m **g** 2750 m
2 a 30 p **b** 45 p **c** 75 p **d** £1.35 **e** £1.65 **f** £1.95 **g** £2.40
3 a £2.25 **b** £3.75 **c** £5.25 **d** £9.00 **e** £11.25 **f** £13.50 **g** £15.00
4 a 100 g **b** 250 g **c** 300 g **d** 400 g **e** 450 g **f** 500 g **g** 600 g
5 a 30 m **b** 75 m **c** 36 m **d** 54 m **e** 120 m **f** 135 m **g** 162 m **h** 198 m
6 a 4 min **b** 6 min **c** 12 min **d** 18 min **e** 20 min **f** 24 min **g** 28 min
7 a 2 **b** 3 **c** 5 **d** 7 **e** 9 **f** 12 **g** 14 **h** 20
8 a 3 km **b** 4.5 km **c** 7.5 km **d** 10.5 km **e** 16.5 km **f** 21 km **g** 27 km **h** 30 km
9 a 2 **b** 3 **c** 5 **d** 6 **e** 9 **f** 10 **g** 12
10 a £1.20 **b** £2.00 **c** £2.80 **d** £4.80 **e** £6.00 **f** £7.20 **g** £8.00

page 96 **Exercise 7.11**

1 15 : 18 = 5 : 6 **2** 12 : 21 = 4 : 7 **3** 12 : 16 = 3 : 4 **4** 16 : 28 = 4 : 7 **5** 20 : 25 = 4 : 5 **6** 25 : 30 = 5 : 6
7 5 : 15 = 1 : 3 **8** 12 : 18 = 2 : 3 **9** 24 : 30 = 4 : 5 **10** 16 : 24 = 2 : 3 **11** 32 : 40 = 4 : 5 **12** 36 : 45 = 4 : 5

page 97 **Exercise 7.12**

1 £20, £30 **2** 100 kg, 350 kg **3** 150 g, 200 g
4 75 cm, 200 cm **5** 80 millilitres, 100 millilitres **6** 2.8 m, 7.7 m
7 25, 30 **8** 25, 45 **9** 105 g, 315 g
10 60 millilitres, 120 millilitres **11** 150 kg, 750 kg **12** 24 millilitres, 96 millilitres

page 99 ***Exercise 8.1***

1	1	3	5	7	9	11	13	15	17	19
2	2	4	6	8	10	12	14	16	18	20
3	3	6	9	12	15	18	21	24	27	30
4	4	8	12	16	20	24	28	32	36	40
5	5	10	15	20	25	30	35	40	45	50
6	10	20	30	40	50	60	70	80	90	100
7	8	16	24	32	40	48	56	64	72	80
8	6	12	18	24	30	36	42	48	54	60
9	9	18	27	36	45	54	63	72	81	90
10	7	14	21	28	35	42	49	56	63	70
11	1	4	9	16	25	36	49	64		
12	1	3	6	10	15	21	28	36		

page 100 ***Exercise 8.2***

1 a

+	2	3	4	5	6	7	8
1	3	4	5	6	7	8	9
2	4	5	6	7	8	9	10
3	5	6	7	8	9	10	11
4	6	7	8	9	10	11	12
5	7	8	9	10	11	12	13
6	8	9	10	11	12	13	14
7	9	10	11	12	13	14	15

b Odd numbers starting at 3
c Yes. The numbers on either side of the diagonal are the same.

2 a

+	1	2	3	4	5	6	7
2	3	4	5	6	7	8	9
3	4	5	6	7	8	9	10
4	5	6	7	8	9	10	11
5	6	7	8	9	10	11	12
6	7	8	9	10	11	12	13
7	8	9	10	11	12	13	14
8	9	10	11	12	13	14	15

b Odd numbers starting at 3
c Yes

3 a

+	1	2	3	4	5	6	7
3	4	5	6	7	8	9	10
4	5	6	7	8	9	10	11
5	6	7	8	9	10	11	12
6	7	8	9	10	11	12	13
7	8	9	10	11	12	13	14
8	9	10	11	12	13	14	15
9	10	11	12	13	14	15	16

b Even numbers starting at 4
c Yes

4 a

+	2	3	4	5	6	7	8
2	4	5	6	7	8	9	10
3	5	6	7	8	9	10	11
4	6	7	8	9	10	11	12
5	7	8	9	10	11	12	13
6	8	9	10	11	12	13	14
7	9	10	11	12	13	14	15
8	10	11	12	13	14	15	16

b Even numbers starting at 4
c Yes

page 102 ***Exercise 8.3***

1 a

×	1	2	4	8	16	32
1	1	2	4	8	16	32
2	2	4	8	16	32	64
4	4	8	16	32	64	128
8	8	16	32	64	128	256
16	16	32	64	128	256	512
32	32	64	128	256	512	1024

b All powers of 4
c Yes

2 a

×	1	3	9	27	81
1	1	3	9	27	81
3	3	9	27	81	243
9	9	27	81	243	729
28	27	81	243	729	2187
81	81	243	729	2187	6561

b All powers of 9
c Yes

3 a

×	1	4	16	64
1	1	4	16	64
4	4	16	64	256
16	16	64	256	1024
64	64	256	1024	4096

b All powers of 16
c Yes

4 a

×	1	2	4	8	16	32
2	2	4	8	16	32	64
4	4	8	16	32	64	128
8	8	16	32	64	128	256
16	16	32	64	128	256	512
32	32	64	128	256	512	1024
64	64	128	256	512	1024	2048

b All odd powers of 2
c Yes

5 a

×	1	3	9	27
3	3	9	27	81
9	9	27	81	243
27	27	81	243	729
81	81	243	729	2187

b All odd powers of 3
c Yes

6 a

×	3	5	7	9
1	3	5	7	7
2	6	10	14	18
3	9	15	21	27
4	12	20	28	36

b Alternate triangle numbers starting with 3
c No

page 102 ***Exercise 8.4***

1 1, 2, 3, 6
2 1, 2, 5, 10
3 1, 2, 4, 8
4 1, 2, 7, 14
5 1, 2, 11, 22
6 1, 2, 13, 26
7 1, 3, 5, 15
8 1, 3, 9, 27
9 1, 3, 11, 33
10 1, 5, 7, 35
11 1, 5, 11, 55
12 1, 3, 9
13 1, 2, 4
14 1, 5, 25
15 1, 7, 49

page 103 ***Exercise 8.5***

1 a

×	20	10	5
1	20	10	5
2	50	20	10
4	80	40	20

b 20
c 1, 2, 4, 5, 10, 20

2 a

×	32	16	8
1	32	16	8
2	64	32	16
4	128	64	32

b 32
c 1, 2, 4, 8, 16, 32

3 a

×	44	22	11
1	44	22	11
2	88	44	22
4	176	88	44

b 44
c 1, 2, 4, 11, 22, 44

4 a

×	66	33	22	11
1	66	33	22	11
2	132	66	44	22
3	198	99	66	33
6	396	198	132	66

b 66
c 1, 2, 3, 6, 11, 22, 33, 66

5 a

×	40	20	10	8
1	40	20	10	8
2	80	40	20	16
4	160	80	40	32
5	200	100	50	40

b 40
c 1, 2, 4, 5, 8, 10, 20, 40

6 a

×	56	28	14	8
1	56	28	14	8
2	128	56	28	16
4	256	128	56	32
7	392	196	98	56

b 56
c 1, 2, 4, 7, 8, 14, 28, 56

7 a

×	100	50	25	20	10
1	100	50	25	20	10
2	200	100	50	40	20
4	400	200	100	80	40
5	500	250	125	100	50
10	1000	500	250	200	100

b 100
c 1, 2, 4, 5, 10, 20, 25, 50, 100

8 a

×	81	27	9
1	81	27	9
3	243	81	27
9	729	243	81

b 81
c 1, 3, 9, 27, 81

9 a

×	64	32	16	8
1	64	32	16	8
2	128	64	32	16
4	256	128	64	32
8	512	256	128	64

b 64
c 1, 2, 4, 8, 16, 32, 64

page 104 ***Exercise 8.6***

1

a	b	c
4	13	16
5	16	20
6	19	24
7	22	28

2

a	b	c
7	16	22
9	20	28
11	24	34
13	28	40

3

a	b	c
16	25	40
25	36	60
36	49	84
49	64	112

page 106 ***Exercise 8.7***

1 41, 61, 85, 113
2 50, 72, 98, 128
3 21, 28, 36, 45

page 107 ***Exercise 8.8***

1 22 44 88
2 30 60 120
3 50 100 200
4 64 128 256
5 150 300 600
6 250 500 1000

page 107 ***Exercise 8.9***

1 40 20 10
2 120 60 30
3 200 100 50
4 212 106 53
5 340 170 85
6 420 210 105

page 108 ***Exercise 8.10***

1 8 kg 16 kg 32 kg
2 $12\,\text{km h}^{-1}$ $24\,\text{km h}^{-1}$
3 £14 £28 £56
4 120 km 60 km 30 km
5 36 18 9

page 108 ***Exercise 8.11***

1 6, 12, 24, 48, 96, 192, 384, 768
2 10, 20, 40, 80, 160, 320, 640
3 14, 28, 56, 112, 224, 448, 896, 1792
4 22, 44, 88, 176, 352, 704
5 26, 52, 104, 208, 412, 832, 1664, 3328

page 109 ***Exercise 8.12***

1 $3 \times 2 \times 2$
2 $7 \times 2 \times 2$
3 $5 \times 2 \times 2$
4 $11 \times 2 \times 2$
5 $13 \times 2 \times 2$
6 $17 \times 2 \times 2$
7 $19 \times 2 \times 2 \times 2$
8 $2 \times 2 \times 2 \times 2$
9 $3 \times 2 \times 2 \times 2 \times 2$
10 $7 \times 2 \times 2 \times 2 \times 2$
11 $11 \times 2 \times 2 \times 2 \times 2$
12 $5 \times 2 \times 2 \times 2 \times 2$

page 109 ***Exercise 8.13***

1 60 300
2 200 1000
3 80 400
4 800 4000
5 40 200
6 55 cm 275 cm

page 110 ***Exercise 8.14***

1 30 6
2 50 10
3 160 32
4 320 64
5 180 km 36 km
6 300 60

page 111 ***Exercise 8.15***

1 **a** 5, 9, 17, 33, 65, 129 **b** 7, 13, 25, 49, 97, 193 **c** 9, 17, 33, 65, 129 **d** 11, 21, 41, 81, 161
e 13, 25, 49, 97, 193 **f** 15, 29, 57, 113 **g** 17, 33, 65, 129
2 **a** 3, 5 and 9 **b** 4 and 7

page 112 ***Exercise 8.16***

1 Divide by 6; 210
2 Divide by 8; 360
3 Divide by 12; 336
4 400
5 672
6 120
7 Multiply by 4; 14
8 Multiply by 5; 17
9 Multiply by 20; 18
10 Multiply by 75; 12
11 60
12 9

page 113 ***Exercise 8.17***

1 **a** 32 400 **b** yes **c** yes
2 **a** 12 000 **b** yes **c** yes
3 **a** 21 000 **b** yes **c** yes
4 **a** 1012 **b** yes **c** yes
5 **a** 576 **b** yes **c** yes

page 114 ***Exercise 8.18***

1 **a** 40
2 **a** 135
3 **a** 56
4 56

page 115 ***Exercise 8.19***

1 **a** 125 cm^3 **b** 216 cm^3 **c** 512 cm^3 **d** 1000 cm^3 **e** 729 cm^3 **f** 343 cm^3
2 **a** 1331 **b** 2197 **c** 9261 **d** 29 791 **e** 421 875
3 **a** 3.375 **b** 15.625 **c** 13.824 **d** 32.768 **e** 0.027 **f** 0.001

page 115 ***Exercise 8.20***

1 5
2 8
3 7
4 9
5 2
6 1
7 40
8 70
9 50
10 80
11 90
12 100

page 115 ***Exercise 8.21***

1 24
2 25
3 22
4 28
5 26
6 27
7 32
8 35
9 39
10 45
11 37
12 2.3
13 2.9
14 1.7
15 1.9
16 1.5
17 3.6
18 4.1
19 3.8
20 4.9
21 0.9
22 0.6
23 0.7
24 0.8
25 0.4
26 0.3
27 0.2
28 0.1

page 116 ***Exercise 8.22***

1 36 cm
2 34 cm
3 1.2 m
4 1.3 m
5 65 m
6 200 m
7 55 m
8 23 m

page 117 **Exercise 8.23**

1 11 prime; 12 rectangular; 13 prime; 14 rectangular; 15 rectangular; 16 square; 17 prime; 18 rectangular; 19 prime; 20 rectangular
r = rectangular, p = prime, s = square
2 21r, 22r, 23p, 24r, 25s, 26r, 27r, 28r, 29p, 30r
3 41p, 42r, 43p, 44r, 45r, 46r, 47p, 48r, 49s, 50r
4 61p, 62r, 63r, 64s, 65r, 66r, 67p, 68r, 69r, 70r

page 119 **Exercise 8.24**

1 a 20 **b** 40 **c** 60 **d** 80 **2 a** 17 **b** 19 **c** 18 **d** 22 **3 a** 1.6 **b** 1.9 **c** 1.7 **d** 2.1

page 118 **Exercise 8.25**

1 12 cm **2** 3.5 cm **3** 1.7 cm **4** 1.3 m

page 119 **Exercise 9.1**

1 10, 13, 16, 19, 22
2 13, 17, 21, 25, 29
3 8, 10, 12, 14, 16; even numbers
4 12, 15, 18, 21, 24; multiples of 3

page 120 **Exercise 9.2**

1 27, 81, 243, 729, 2187; powers of 3
2 64, 256, 1024, 4096, 16 384; powers of 4
3 1000, 10 000, 100 000, 1 000 000; powers of 10

page 121 **Exercise 9.3**

1 20, 30, 42, 56, 72 **2** 30, 45, 63, 84, 108 **3** 16, 25, 36, 49, 64; square numbers

page 121 **Exercise 9.4**

1 18, 29, 47, 76, 123 **2** 21, 34, 55, 89, 144 **3** 15, 24, 39, 63, 102

page 122 **Exercise 9.5**

1 2, 3, 5, 7, 11, 13, 17, 19, 23, 29, 31, 37, 41, 43, 47, 53, 59, 61, 67, 71, 73, 79, 83, 89, 97; 25
2 All the prime numbers except 2 and 3 are either one more or one less than a multiple of 6.
3 All the prime numbers except 2 are either one more or one less than a multiple of 4.
4 12, 15, 21, 27, 33, 39, 51, 57, 63, 69, 81, 93 (all are multiples opf 3)
5 9, 21, 27, 33, 51, 57, 63, 69, 87 (all are multiples of 3)
6 101, 103, 107, 109, 113, 127, 131, 137, 139, 149, 151, 157, 163, 167, 173, 179, 181, 191, 193, 197, 199 (there are 21)

page 124 **Exercise 9.6**

1 a 12, 16, 20, 24 **b** 13, 17, 21, 25
2 a 9, 12, 15, 18 **b** 10, 13, 16, 19
3 a 4, 5, 6, 7 **b** 6, 7, 8, 9 **c** 9, 11, 13, 15
4 a 4, 5, 6, 7 **b** 6, 10, 15, 21 **c** 15, 21, 28, 36 **d** 24, 35, 48, 63
5 a 4, 5, 6, 7, 8, 9, 10; 20; 50 **b** 3, 4, 5, 6, 7, 8, 9, 19; 49; 99 **c** 7, 9, 11, 13, 15, 17, 19; 39; 99
6 b 25, 36, 49, 64 **c** 36, 49, 64, 81 **d** 55, 91, 140, 204

page 126 Exercise 9.7

1 6 8 9 11; 2 5 8 9
2 20 40 50 70; 30 50 60 65
3 1 3 4 6; 6 9 11 13
4 3 8 13 28; 35 45 50 70

page 128 Exercise 9.8

1 6 12 24 42 48 72; 3 5 6 9 10 15
2 60 105 120 150 180 300; 2 3 5 6 9 15
3 1 3 4 6 8 15; 6 15 21 27 30 108

page 130 Exercise 9.9

1 6 10 12 16; 2 5 7 8
2 1 5 7 13; 3 6 7 10
3 5 14 17 26; 2 3 6 7

page 131 Exercise 9.10

1 **a** $\times 2$; 12, 14, 16 **b** $\times 2$ then $+1$; 13, 15, 17
c $\times 2$ then $+2$; 14, 18, 22 **d** $\times 2$ then -1; 11, 15, 19
2 $\times 2$ then $+5$; 13 cm, 15 cm, 17 cm
3 $\times 2$ then $+10$; 18 m, 20 m, 22 m
4 $\times 5$ then $+10$; 30° C, 35° C, 40° C

page 133 Exercise 9.11

1 10, 12, 14, 20, $2n$; $n \rightarrow 2n$
2 30, 35, 50, 100; $5n$; $n \rightarrow 5n$
3 20, 30, 40, 60, 150, 250, $10n$; $n \rightarrow 10n$

page 134 Exercise 9.12

1 (i) 24 p (ii) 36 p (iii) 48 p (iv) 60 p (v) 96 p (vi) 120 p (vii) $12a$ p (viii) $12b$ p (ix) $12c$ p (x) $12d$ p
2 (i) 40 p (ii) 60 p (iii) 80 p (iv) £1.20 (v) £2.00 (vi) £2.40 (vii) $20a$ p (viii) $20b$ p (ix) $20c$ p (x) $20d$ p
3 (i) £1.50 (ii) £2.00 (iii) £3.00 (iv) £4.00 (v) £4.50 (vi) £5.00 (vii) $50p$ p (vii) $50q$ p (ix) $50r$ p (x) $50s$ p

page 136 Exercise 9.13

1 16, 25, 31, 46, 61, 91, $3n+1$; $n \rightarrow 3n+1$
2 19, 28, 31, 40, 79, 124, $3n+4$; $n \rightarrow 3n+4$

page 137 Exercise 9.14

1 (i) 26 (ii) 29 (iii) 32 (iv) 35 (v) 44 (vi) $3a+20$ (vii) $3b+20$
2 (i) £37 (ii) £43 (iii) £55 (iv) £73 (v) £85 (vi) £$6a+25$ (vii) £$6b+25$
3 (i) 150 cm (ii) 200 cm (iii) 225 cm (iv) 350 cm (v) $(25p+100)$ cm (vi) $(25q+100)$ cm

page 139 Exercise 9.15

1 21, 24, 27, 30, 36, 42, $3(n+2)$; $n \rightarrow 3(n+2)$
2 (i) 50° F (ii) 60° F (iii) 70° F (iv) 90° F (v) 40° F (vi) $2(p+15)$° F (vii) $2(q+15)$° F (viii) $2(r+15)$° F (ix) $2(s+15)$° F
3 (i) 50° F (ii) 60° F (iii) 70° F (iv) 90° F (v) 40° F (vi) $2(p+15)$° F (vii) $2(q+15)$° F (viii) $2(r+15)$° F (ix) $2(s+15)$° F
4 (i) 14 cm (ii) 16 cm (iii) 18 cm (iv) 15 cm (v) 13 cm (vi) 17 cm (vii) $2(p+5)$ cm (viii) $2(q+5)$ cm (ix) $2(r+5)$ cm (x) $2(s+5)$ cm

page 141 Exercise 10.1

1 **a** 72 p **b** £1.20 **c** £1.92
2 **a** £1.20 **b** £1.60 **c** £2.80
3 **a** £1.75 **b** £2.10 **c** £2.80
4 **a** £3.00 **b** £4.20 **c** £4.80
5 **a** £1.00 **b** £1.50 **c** £2.50

page 142 **Exercise 10.2**

	a	b	c	d	e	f
1	17 cm	19 cm	21 cm	25 cm	35 cm	
2	26 cm	32 cm	44 cm	50 cm	62 cm	
3	3 lb	5 lb	13 lb	17 lb	19 lb	25 lb
4	150	200	300	600	700	1000
5	940	1340	1740	2140	2940	

page 143 **Exercise 10.3**

	a	b	c	d	e	f	g
1	13	18	28	38	4	6	8
2	7	17	22	37	3	6	7
3	9	11	15	17	1	7	8
4	10	12	0	16	3	5	6

page 144 **Exercise 10.4**

1 15 **2** 16 **3** 9 **4** 15 **5** 6 **6** 8 **7** 4 **8** 8

page 144 **Exercise 10.5**

1 26 cm **2** 35 cm **3** 32 cm **4** 56 mm **5** 7 m

page 145 **Exercise 10.6**

1 12 cm **2** 24 cm **3** 27 cm **4** 63 mm **5** 7.5 m **6** 4.5 m

page 145 **Exercise 10.7**

1 20 cm **2** 25 cm **3** 37 mm **4** 57 mm **5** 8 m

page 146 **Exercise 10.8**

1 20 cm **2** 24 cm **3** 32 cm **4** 80 mm **5** 14 m **6** 10 m

page 146 **Exercise 10.9**

1 26 cm **2** 26 cm **3** 38 cm **4** 46 cm **5** 100 mm **6** 92 mm

page 147 **Exercise 10.10**

1 27 cm^2 **2** 48 cm^2 **3** 96 cm^2 **4** 105 cm^2 **5** 9 m^2 **6** 3 m^2

page 147 **Exercise 10.11**

1 24 cm^2 **2** 63 cm^2 **3** 7 m^2 **4** 51 m^2 **5** 840 mm^2 **6** 720 mm^2

page 148 **Exercise 10.12**

1 49 cm^2 **2** 36 cm^2 **3** 16 cm^2 **4** 2.25 m^2 **5** 6.25 m^2 **6** 1.44 m^2

page 148 **Exercise 10.13**

1 $20\,cm^2$ **2** $20\,cm^2$ **3** $64\,cm^2$ **4** $250\,mm^2$ **5** $96\,mm^2$ **6** $2\,m^2$

page 149 **Exercise 10.14**

1 $27\,cm^2$ **2** $50\,cm^2$ **3** $280\,cm^2$ **4** $400\,mm^2$ **5** $960\,mm^2$ **6** $14\,m^2$ **7** $15\,m^2$ **8** $25\,m^2$

page 149 **Exercise 10.15**

1 $720\,cm^3$ **2** $840\,cm^3$ **3** $7200\,cm^3$ **4** $10\,800\,mm^3$ **5** $57\,600\,mm^3$ **6** $135\,m^3$ **7** $21\,m^3$ **8** $330\,m^3$

page 150 **Exercise 10.16**

1 $x = 5$ **2** $x = 12$ **3** $x = 18$ **4** $y = 35$ **5** $y = 15$ **6** $y = 55$
7 $t = 40$ **8** $t = 13$ **9** $t = 11$ **10** $m = 25$ **11** $m = 40$ **12** $n = 23$
13 $n = 40$ **14** $x = 3$ **15** $x = 2$ **16** $y = 6$

page 151 **Exercise 10.17**

1 $x = 15$ **2** $x = 16$ **3** $x = 17$ **4** $y = 14$ **5** $y = 27$ **6** $t = 13$
7 $t = 17$ **8** $m = 15$ **9** $m = 16$ **10** $n = 24$ **11** $p = 11$ **12** $p = 12$
13 $x = 24$ **14** $x = 36$ **15** $y = 42$ **16** $y = 15$ **17** $t = 72$ **18** $t = 60$
19 $m = 200$ **20** $m = 90$ **21** $n = 50$ **22** $n = 190$ **23** $f = 1000$ **24** $f = 637$

page 152 **Exercise 10.18**

1 $x = 3$ **2** $x = 6$ **3** $x = 9$ **4** $x = 4$ **5** $y = 6$
6 $m = 8$ **7** $m = 11$ **8** $m = 10$ **9** $m = 10$ **10** $y = 2$
11 $y = 3$ **12** $x = 4$ **13** $x = 28$ **14** $y = 60$ **15** $y = 18$
16 $t = 20$ **17** $m = 36$ **18** $n = 36$ **19** $n = 42$ **20** $p = 20$

page 152 **Exercise 10.19**

1 8 **2** 5 **3** 36 **4** 60 km **5** 80 km **6** 6 **7** 8 **8** 6 **9** 150 m

page 154 **Exercise 10.20**

1 $x = 4$ **2** $x = 3$ **3** $x = 2$ **4** $t = 4$ **5** $x = 5$ **6** $x = 13$
7 $b = 3$ **8** $b = 2$ **9** $m = 3$ **10** $x = 8$ **11** $x = 6$ **12** $y = 5$
13 $n = 42$ **14** $n = 34$ **15** $t = 3$ **16** $q = 2$ **17** $q = 2$ **18** $a = 3$
19 $c = 4$ **20** $x = 2$ **21** $x = 2$ **22** $t = 3$ **23** $t = 3$ **24** $m = 2$

page 155 **Exercise 10.21**

1 $x = 4.9$ **2** $x = 4.7$ **3** $y = 5.3$ **4** $y = 3.9$ **5** $q = 1.4$ **6** $q = 1.7$ **7** $m = 4.1$ **8** $m = 2.8$

page 155 **Exercise 10.22**

1 $x = 3.7$ **2** $x = 4.1$ **3** $y = 2.5$ **4** $y = 1.3$

page 156 *Exercise 10.23*

1 a 4, 4, 4, 4; 25, 29, 33, 37, 41
b 5, 5, 5, 5; 27, 32, 37, 42, 47
c 6, 6, 6, 6; 31, 37, 43, 49, 55
d 4, 4, 4, 4; 23, 27, 31, 35, 39
e 0.5, 0.5, 0.5, 0.5; 4.5, 5, 5.5, 6, 6.5
f 1.5, 1.5, 1.5, 1.5; 8.5, 10, 11.5, 13, 14.5
g 4.5, 4.5, 4.5, 4.5; 22.5, 27, 31.5, 36, 40.5
h 2.5, 2.5, 2.5, 2.5; 14.5, 17, 19.5, 22, 24.5

2 a 1, 2, 3, 4; 16, 22, 29, 37, 46
b 1, 2, 3, 4; 18, 24, 31, 39, 48
c 2, 3, 4, 5; 23, 30, 38, 47, 57
d 3, 4, 5, 6; 26, 34, 43, 53, 64
e 4, 5, 6, 7; 32, 41, 51, 62, 74
f 1, 1, 1, 1; 5.5, 6.5, 7.5, 8.5, 9.5

3 a 1, 4, 9, 16; 56, 92, 141, 205, 286
b 0, 1, 4, 9; 34, 59, 95, 144, 208
c 4, 9, 16, 25; 96, 145, 209, 290, 390
d 1, 3, 6, 10; 36, 57, 85, 121, 166
e 1, 3, 6, 10; 35, 56, 84, 120, 165
f 3, 6, 10, 15; 66, 94, 130, 175, 230

page 156 *Exercise 10.24*

1 A, C, E, G, I, K, M, O, Q, S, U, W, Y
2 A, E, I, M, Q, U, Y
3 A, F, K, P, U, Z
4 Z, X, V, T, R, P, N, L, J, H, F, D, B
5 Z, W, T, Q, N, K, H, E, B
6 Z, V, R, N, J, F, B

page 158 *Exercise 10.25*

1 4; 24, 28, 32, 36, 40; $n \to 4n$
2 4; 25, 29, 33, 37, 41; $n \to 4n + 1$
3 5; 30, 35, 40, 45, 50; $n \to 5n$
4 5; 31, 36, 41, 46, 51; $n \to 5n + 1$
5 6; 37, 43, 49, 55, 61; $n \to 6n + 1$
6 6; 33, 39, 45, 51, 57; $n \to 6n - 3$

page 158 *Exercise 10.26*

1 38, 51, 66, 83, 102; $n \to n^2 + 2$
2 39, 52, 67, 84, 103; $n \to n^2 + 3$
3 45, 58, 73, 90, 109; $n \to n^2 + 9$
4 34, 47, 62, 79, 98; $n \to n^2 - 2$

page 159 *Exercise 10.27*

1 3, 4, 5, 6, 7, 8
2 5, 6, 7, 8, 9, 10
3 2, 3, 4, 5, 6, 7
4 1, 2, 3, 4, 5, 6
5 −1, 0, 1, 2, 3, 4
6 −2, −1, 0, 1, 2, 3
7 −4, −3, −2, −1, 0, 1
8 −3, −2, −1, 0, 1, 2

page 160 *Exercise 10.28*

1 6, 5, 4, 3, 2, 1
2 5, 4, 3, 2, 1, 0
3 3, 2, 1, 0, −1, −2
4 4, 3, 2, 1, 0, −1
5 1, 0, −1, −2, −3, −4

page 162 *Exercise 10.29*

1

$x =$	−4	−3	−2	−1	0	1	2	3	4
$x^2 =$	16	9	4	1	0	1	4	9	16
$y = x^2 + 2 =$	18	11	6	3	2	3	6	11	18

2

$x =$	−4	−3	−2	−1	0	1	2	3	4
$x^2 =$	16	9	4	1	0	1	4	9	16
$y = x^2 + 3 =$	19	12	7	4	3	4	7	12	19

3

$x =$	-4	-3	-2	-1	0	1	2	3	4
$x^2 =$	16	9	4	1	0	1	4	9	16
$y = x^2 - 1 =$	15	8	3	0	-1	0	3	8	15

page 164 ***Exercise 10.30***

1 The points (0, 0), (20, 100), (40, 200), (60, 300), (80, 400) all lie on a straight line. 50, 150, 250, 350 metres.
2 The points (0, 0), (5, 10), (10, 20), (15, 30), (20, 40), (25, 50), (30, 60) all lie on a straight line. 5, 15, 25, 45 metres.
3 The points (0, 0), (2, 10), (4, 20), (6, 30), (8, 40), (10, 50) all lie on a straight line. 5, 15, 25, 35, 45 metres.
4 The points (0, 0), (10, 2), (20, 4), (30, 6), (40, 8), (50, 10), (60, 12) all lie on a straight line. 1, 3, 9, 11 km.

page 167 ***Exercise 11.1***

1 a 17 cm **b** 19 cm **c** 12 cm **d** 24 cm **e** 30 cm **f** 24 cm **g** 40 cm
2 a 36 cm **b** 34 cm **c** 30 cm **d** 18 cm **e** 24 cm **f** 16 cm **g** 50 cm
3 a 16 cm **b** 12 cm **c** 20 cm **d** 12 cm

page 169 ***Exercise 11.3***

1 4 cm **2** 3 cm **3** 5.5 cm **4** 4.3 cm

page 170 ***Exercise 11.4***

1 4 cm, 7 cm, 8 cm **2** 3 cm, 4 cm, 5 cm **3** 5 cm, 5 cm, 6 cm **4** 7 cm, 7 cm, 10 cm **5** 5 cm, 12 cm, 13 cm

page 171 ***Exercise 11.5***

1 2 cm, 4 cm, 5 cm, 5 cm **2** 3 cm, 3 cm, 5 cm, 7 cm **3** 2 cm, 7 cm, 6 cm, 8 cm **4** 5 cm, 5 cm, 5 cm, 5 cm

page 177 ***Exercise 12.1***

1 a A′(13, 1), B′(13, 5), C′(16, 1) **b** L′(13, 6), M′(13, 13), N′(17, 6) **c** P′(13, 14), Q′(13, 17), R′(17, 14)
2 a A′(13, 5), B′(16, 5), C′(16, 1) **b** L′(13, 9), M′(17, 9), N′(17, 6) **c** P′(13, 14), Q′(20, 14), R′(20, 10)

page 179 ***Exercise 12.2***

1 a A′(13, 1), B′(17, 1), C′(17, 3), D′(13, 6) **b** P′(13, 7), Q′(16, 7), R′(16, 10), S′(13, 14)
2 a A′(13, 1), B′(17, 1), C′(17, 6), D′(13, 3) **b** W′(13, 7), X′(16, 7), Y′(16, 14), Z′(13, 10)
3 a A′(13, 1), B′(13, 8), C′(16, 8), D′(20, 1) **b** E′(13, 10), F′(20, 10), G′(20, 15), H′(16, 15)

page 179 ***Exercise 12.3***

1 a A′(1, 17), B′(9, 17), C′(5, 10) **b** L′(9, 14), M′(15, 14), N′(12, 10) **c** P′(16, 13), Q′(24, 13), R′(20, 10)
2 a A′(1, 10), B′(9, 10), C′(5, 17) **b** P′(10, 10), Q′(16, 10), R′(13, 14) **c** X′(17, 10), Y′(25, 10), Z′(21, 13)

page 180 ***Exercise 12.4***

1 Rotational symmetry, order 4
2 Rotational symmetry, order 3
3 Rotational symmetry, order 2
4 Rotational symmetry, order 2
5 Rotational symmetry, order 2
6 Rotational symmetry, order 3
7 Rotational symmetry, order 2
8 Rotational symmetry, order 2
9 Rotational symmetry, order 4
10 Rotational symmetry, order 2

page 184 ***Exercise 12.7***

1 2 lines of symmetry; order 2
2 2 lines of symmetry; order 2
3 2 lines of symmetry; order 2
4 3 lines of symmetry; order 3
5 4 lines of symmetry; order 4
6 6 lines of symmetry; order 6

page 185 ***Exercise 12.8***

1 2 planes of symmetry; order 2
2 1 plane of symmetry; order 2
3 3 planes of symmetry; order 2 in each case
4 5 planes of symmetry; order 2, order 2, order 4

page 187 ***Exercise 12.9***

1 Rhombus
2 Square, parallelogram
3 Rectangle, kite, 2 parallelograms
4 2 Trapeziums, 2 irregular quadrilaterals
5 Parallelogram
6 Rectangle, parallelogram, square

page 188 ***Exercise 12.10***

1 2 squares, rectangle, parallelogram, kite, trapezium, irregular quadrilateral
2 2 squares, rectangle, 2 parallelograms, 2 kites, trapezium, 2 irregular quadrilaterals
3 3 squares, rectangle, 2 parallelograms, 2 trapeziums, isosceles trapezium, 3 kites, 4 irregular quadrilaterals

page 189 ***Exercise 12.11***

1 Trapezium
2 Rectangle
3 Isosceles trapezium
4 Trapezium
5 Kite
6 Rectangle

page 191 ***Exercise 13.1***

1 90° **2** 180° **3** 45° **4** 270° **5** 135° **6** 180° **7** 135° **8** 270°

page 191 ***Exercise 13.2***

1 90° **2** 180° **3** 60° **4** 60° **5** 120°

page 192 ***Exercise 13.3***

1 60°, acute
2 30°, acute
3 90°, right
4 120°, obtuse
5 180°, straight
6 90°, right
7 120°, obtuse
8 150°, obtuse
9 180°, straight

page 193 ***Exercise 13.4***

1 80° **2** 70° **3** 31° **4** 45° **5** 51.5° **6** 27.5°
7 160° **8** 140° **9** 90° **10** 70° **11** 99.5° **12** 44.5°

page 193 **Exercise 13.5**

1 $a = A\hat{X}Y$, $b = B\hat{X}Y$,
2 $a = X\hat{Y}Z$, $b = Y\hat{X}Z$, $c = X\hat{Z}Y$, $d = X\hat{Z}W$
3 $a = R\hat{P}Q$, $b = P\hat{Q}R$, $c = P\hat{R}S$, $d = P\hat{S}R$
4 $a = P\hat{S}T$, $b = Q\hat{P}T$, $c = R\hat{Q}T$, $d = R\hat{T}S$
(Consult your teacher for other possible answers).

page 195 **Exercise 13.6**

1 **a** 120° **b** 30° **c** 20° **d** 15° **e** 105° **f** 115° **g** 150° **h** 160°
i 60° **j** 75° **k** 65° **l** 165°
2 **a** 31° **b** 74° **c** 4° **d** 81° **e** 134° **f** 148° **g** 106° **h** 149°
i 176° **j** 46° **k** 32° **l** 99°

page 196 **Exercise 13.7**

1 33° **2** 55° **3** 112° **4** 133°

page 198 **Exercise 13.9**

1 90° **2** 90° **3** 100° **4** 120° **5** 30° **6** 40°

page 199 **Exercise 13.10**

1 65° **2** 120° **3** 40° **4** 80° **5** 70° **6** 120°

page 199 **Exercise 13.11**

1 60°, 120° **2** 36°, 144° **3** 15°, 165° **4** 72°, 108°

page 200 **Exercise 13.12**

1 40° **2** 35° **3** 78° **4** 54° **5** 30°, 60° **6** 9°, 81°

page 201 **Exercise 13.13**

1 90° **2** 160° **3** 160° **4** 70° **5** 120° **6** 90°

page 202 **Exercise 13.14**

1 $a = 105°$, adjacent to 75°; $b = 75°$, vertically opposite 75°; $c = 105°$, adjacent to 75°
$d = 75°$, corresponding to 75°; $e = 105°$, adjacent to d; $f = 75°$, vertically opposite d
$g = 105°$, adjacent to d
2 $x = 65°$, adjacent to 115°; $y = 115°$, vertically opposite to 115°; $z = 65°$, adjacent to 115°
$t = 115°$, corresponding to 115°; $u = 65°$, adjacent to t; $v = 115°$, vertically opposite t
$w = 65°$, adjacent to t

3	$a = 125°$	$b = 55°$	$c = 125°$	$d = 55°$	$e = 125°$	$m = 70°$	$n = 110°$	$p = 70°$	$q = 110°$	$r = 70°$
4	$a = 80°$	$b = 100°$	$c = 80°$	$d = 100°$	$e = 80°$	$u = 50°$	$v = 130°$	$w = 50°$	$x = 130°$	$y = 50°$
5	$a = 155°$	$b = 25°$	$c = 25°$	$d = 155°$	$e = 25°$	$m = 150°$	$n = 30°$	$p = 30°$	$q = 150°$	$r = 30°$
6	$a = 160°$	$b = 20°$	$c = 20°$	$d = 160°$	$e = 20°$	$m = 35°$	$n = 145°$	$p = 145°$	$q = 35°$	$r = 145°$

page 204 **Exercise 13.15**

1 **a** 30° **b** 80° **c** 50° **d** 55° **e** 55° **f** 25°

2 **a** $\hat{A} = 12°$, $\hat{B} = 24°$, $\hat{C} = 144°$ **b** $\hat{A} = 12°$, $\hat{B} = 36°$, $\hat{C} = 132°$ **c** $\hat{A} = 20°$, $\hat{B} = 60°$, $\hat{C} = 100°$ **d** $\hat{A} = 20°$, $\hat{B} = 40°$, $\hat{C} = 120°$ **e** $\hat{A} = 30°$, $\hat{B} = 60°$, $\hat{C} = 90°$ **f** $\hat{A} = 18°$, $\hat{B} = 72°$, $\hat{C} = 90°$

3 **a** $\hat{P} = 90°$, $\hat{Q} = \hat{R} = 45°$ **b** $\hat{P} = 108°$, $\hat{Q} = \hat{R} = 36°$ **c** $\hat{P} = 120°$, $\hat{Q} = \hat{R} = 30°$ **d** $\hat{P} = 140°$, $\hat{Q} = \hat{R} = 20°$ **e** $\hat{P} = 36°$, $\hat{Q} = \hat{R} = 72°$ **f** $\hat{P} = 20°$, $\hat{Q} = \hat{R} = 80°$

4 **a** 64° **b** 58° **c** 49° **d** 77° **e** 51° **f** 9° **g** 28° **h** 21°

page 205 **Exercise 13.16**

1 $a = 61°$, $b = 56°$ **2** $m = 77°$, $n = 31°$ **3** $p = 24°$, $q = 71°$ **4** $x = 123°$, $y = 35°$ **5** $a = 60°$, $b = 120°$ **6** $m = 35°$, $n = 145°$

page 205 **Exercise 13.17**

1 $a = 50°$, $b = 80°$ **2** $m = 30°$, $n = 120°$ **3** $p = 40°$, $q = 100°$

page 206 **Exercise 13.18**

1 $a = 40°$, $b = 20°$, $c = 40°$, $d = 120°$ **2** $a = 115°$, $b = 30°$, $c = 35°$, $d = 115°$ **3** $x = 125°$, $y = 15°$, $z = 40°$, $t = 125°$ **4** $e = 15°$, $f = 20°$, $g = 145°$, $h = 145°$ **5** $a = 40°$, $b = 45°$, $c = 45°$, $d = 95°$ **6** $x = 15°$, $y = 5°$, $z = 5°$, $t = 160°$

page 207 **Exercise 13.19**

1 **a** 21 cm **b** 30 cm **c** 10.5 cm **d** 45° **e** 135°

2 **a** 21 cm **b** 24 cm **c** 6 cm **d** 30° **e** 150°

3 **a** 16 cm **b** 16 cm **c** 31 cm **d** 30° **e** 150°

4 **a** 5.6 cm **b** 8 cm **c** 4 cm **d** 45° **e** 105°

5 **a** 8 cm **b** 10 cm **c** 7 cm **d** 60° **e** 45° **f** 75°

6 **a** 8 cm **b** 8 cm **c** 6 cm **d** 6 cm **e** 10 cm **f** 10 cm

page 209 **Exercise 13.20**

1 3; 3 **2** 4; 4 **3** 5; 5 **4** 7; 7 **5** 8; 8

6 The number of sides, number of axes of symmetry and order of rotational symmetry are all the same for any given regular polygon.

page 210 **Exercise 13.21**

1 **a** 45° **b** 135° **2** **a** 24° **b** 156° **3** **a** 15° **b** 165° **4** **a** 10° **b** 170°

5 **a** 36° **b** 10 **6** **a** 20° **b** 18 **7** **a** 9° **b** 40 **8** **a** 4° **b** 90

page 211 **Exercise 13.22**

1 **a** 1260° **b** 140° **2** **a** 3240° **b** 162° **3** **a** 5040° **b** 168° **4** **a** 7740° **b** 172°

5 **a** 12 **b** 150° **6** **a** 60 **b** 174° **7** **a** 72 **b** 175° **8** **a** 3 **b** 60°

page 213 ***Exercise 14.1***

	Metres	Centimetres	Millimetres
1	5	500	5000
2	12	1200	12 000
3	25	2500	25 000
4	24	2400	24 000
5	16	1600	16 000
6	1	100	1000
7	6	600	6000
8	11	1100	11 000
9	26	2600	26 000
10	50	5000	50 000

page 214 ***Exercise 14.2***

	Litres	Millilitres	Kilograms	Grams
1	6	6000	6	6000
2	12	12 000	12	12 000
3	10	10 000	10	10 000
4	2	2000	2	2000
5	18	18 000	18	18 000
6	25	25 000	25	25 000
7	32	32 000	32	32 000
8	28	28 000	28	28 000
9	36	36 000	36	36 000
10	40	40 000	40	40 000
11	17	17 000	17	17 000
12	38	38 000	38	38 000
13	16	16 000	16	16 000
14	50	50 000	50	50 000

page 215 ***Exercise 14.3***

	Hours	Minutes	Seconds
1	5	300	18 000
2	11	660	39 600
3	10	600	36 000
4	17	1020	61 200
5	13	780	46 800
6	2	120	7200
7	16	960	57 600
8	40	2400	144 000

page 217 ***Exercise 14.4***

1 15 cm	**2** 10 cm	**3** 25 cm	**4** 40 cm	**5** 60 cm	**6** 35 cm
7 2 in	**8** 8 in	**9** 12 in	**10** 18 in	**11** 44 in	**12** 32 in

page 217 ***Exercise 14.5***

1 12 m	**2** 6 m	**3** 18 m	**4** 54 m	**5** 162 m	**6** 963 m
7 33 ft	**8** 99 ft	**9** 66 ft	**10** 231 ft	**11** 1056 ft	**12** 4422 ft

page 218 ***Exercise 14.6***

1 45 m	**2** 36 m	**3** 81 m	**4** 225 m	**5** 1584 m	**6** 810 m
7 11 yd	**8** 44 yd	**9** 396 yd	**10** 715 yd	**11** 1100 yd	

page 219 Exercise 14.7

1 632 km **2** 840 km **3** 312 km **4** 168 km
5 360 km **6** 208 km **7** 120 km **8** 80 km
9 495 miles **10** 355 miles **11** 215 miles **12** 305 miles
13 135 miles **14** 290 miles **15** 515 miles **16** 795 miles

page 219 Exercise 14.8

1 56 g **2** 280 g **3** 448 g **4** 560 g **5** 672 g
6 3 oz **7** 4 oz **8** 8 oz **9** 13 oz **10** 15 oz

page 220 Exercise 14.9

1 9 kg **2** 27 kg **3** 45 kg **4** 72 kg **5** 63 kg **6** 90 kg
7 33 lb **8** 55 lb **9** 165 lb **10** 77 lb **11** 143 lb **12** 176 lb

page 221 Exercise 14.10

1 0.48 litres, 0.9 litres, 1.5 litres, 2.4 litres
2 0.24 litres, 0.36 litres, 0.84 litres, 1.44 litres
3 1.4 pints, 2.1 pints, 3.5 pints, 4.9 pints
4 0.35 pints, 0.875 pints, 2.625 pints, 4.375 pints

page 211 Exercise 14.11

1 18 litres, 45 litres, 27 litres, 36 litres, 54 litres, 9 litres
2 3.3 gallons, 4.4 gallons, 7.7 gallons, 11 gallons, 13.2 gallons, 9.9 gallons
3 b 675 litres, **c** 720 litres, **e** 693 litres, **f** 648 litres, **g**, **h**, **b**, **i**, **e**, **d**, **c**, **f**, **a**,

page 222 Exercise 14.12

1 a 1400 cm **b** 14 000 mm **2 a** 1900 cm **b** 19 000 mm **3 a** 1200 cm **b** 12 000 mm
4 a 800 cm **b** 8000 mm **5 a** 332.8 cm **b** 3328 mm **6 a** 674.5 cm **b** 6745 mm
7 a 241.9 cm **b** 2419 mm **8 a** 863.4 cm **b** 8634 m **9 a** 456 cm **b** 4560 mm
10 a 795 cm **b** 7950 mm **11 a** 520 cm **b** 5200 mm **12 a** 840 cm **b** 8400 mm
13 a 81.9 cm **b** 819 mm **14 a** 27.3 cm **b** 273 mm **15 a** 5.3 cm **b** 53 mm
16 a 3.6 cm **b** 36 mm

page 222 Exercise 14.13

1 a 2840 mm **b** 2.84 m **2 a** 6590 mm **b** 6.59 m **3 a** 6700 mm **b** 6.7 m
4 a 9400 mm **b** 9.4 m **5 a** 56 000 mm **b** 56 m **6 a** 74 000 mm **b** 74 m
7 a 6345 mm **b** 6.345 m **8 a** 5286 mm **b** 5.286 m **9 a** 248 mm **b** 0.248 m
10 a 536 mm **b** 0.536 m **11 a** 725 mm **b** 0.725 m **12 a** 873 mm **b** 0.873 m
13 a 43 mm **b** 0.043 m **14 a** 57 mm **b** 0.057 m **15 a** 92 mm **b** 0.092 m
16 a 71 mm **b** 0.071 m

page 223 Exercise 14.14

1 a 961.5 cm **b** 9.615 m **2 a** 897.6 cm **b** 8.976 m **3 a** 354 cm **b** 3.54 m
4 a 627 cm **b** 6.27 m **5 a** 410 cm **b** 4.1 m **6 a** 670 cm **b** 6.7 m
7 a 900 cm **b** 9 m **8 a** 400 cm **b** 4 m **9 a** 7800 cm **b** 78 m
10 a 6500 cm **b** 65 m **11 a** 9200 cm **b** 92 m **12 a** 8300 cm **b** 83 m
13 a 72.5 cm **b** 0.725 m **14 a** 37.9 cm **b** 0.379 m **15 a** 9.6 cm **b** 0.096 m
16 a 7.8 cm **b** 0.078 m

page 223 **Exercise 14.15**

1 86 000 g	**2** 6254 g	**3** 9386 g	**4** 7521 g
5 3860 g	**6** 7520 g	**7** 4370 g	**8** 372 g
9 190 g	**10** 45 g	**11** 82 g	**12** 74 g
13 56 g	**14** 29 g	**15** 3.6 g	**16** 5.7 g

page 223 **Exercise 14.16**

1 25 kg	**2** 2.475 kg	**3** 5.286 kg	**4** 7.394 kg
5 8.39 kg	**6** 4.28 kg	**7** 2.95 kg	**8** 0.526 kg
9 0.17 kg	**10** 0.076 kg	**11** 0.059 kg	**12** 0.043 kg
13 0.028 kg	**14** 0.017 kg	**15** 0.0048 kg	**16** 0.0063 kg

page 224 **Exercise 14.17**

1 5000 ml	**2** 60 000 ml	**3** 6400 ml	**4** 3700 ml	**5** 600 ml
6 300 ml	**7** 50 ml	**8** 80 ml	**9** 3 ml	**10** 9 ml

page 224 **Exercise 14.18**

1 6 litres	**2** 30 litres	**3** 9.5 litres	**4** 1.4 litres	**5** 0.8 litres
6 0.5 litres	**7** 0.09 litres	**8** 0.04 litres	**9** 0.007 litres	**10** 0.005 litres

page 225 **Exercise 15.1**

1 98 m	**2** 324 m	**3** 220 m	**4** 350 m	**5** 810 m

page 226 **Exercise 15.2**

1 a 24 cm	**b** 36 cm	**c** 60 cm	**d** 96 mm	**e** 260 mm	**f** 360 mm
2 a 24 cm	**b** 18 cm	**c** 60 cm	**d** 135 mm	**e** 216 mm	**f** 288 mm
3 a 72 cm	**b** 108 cm	**c** 150 cm	**d** 252 mm	**e** 450 mm	**f** 360 mm

page 227 **Exercise 15.3**

1 c, 76 cm	**2 a**, 244 cm	**3 b**, 256 mm	**4 c**, 340 mm	**5** 20 cm	**6** 18 cm
7 24 cm	**8** 18 cm	**9** 26 cm	**10** 18 cm	**11** 26 cm	**12** 16 cm

page 230 **Exercise 15.4**

1 a 8 cm	**b** 12 cm	**c** 25 cm	**d** 45 cm	**e** 68 mm	**f** 87 mm
2 a 12 cm	**b** 17 cm	**c** 25 cm			
3 a 28 cm	**b** 39 cm	**c** 52 cm			
4 a 12 cm	**b** 9 cm	**c** 15 cm			
5 a 6 cm	**b** 9 cm	**c** 15 cm			

page 231 **Exercise 15.5**

1 1 cm and 8 cm; 2 cm and 7 cm; 3 cm and 6 cm; 4 cm and 5 cm
2 1 cm and 6 cm; 2 cm and 5 cm; 3 cm and 4 cm
3 1 cm and 7 cm; 2 cm and 6 cm; 3 cm and 5 cm; 4 cm and 4 cm
4 1 cm and 5 cm; 2 cm and 4 cm; 3 cm and 3 cm

page 232 ***Exercise 15.6***

1 1 cm, 6 cm, 6 cm; 3 cm, 5 cm, 5 cm; 5 cm, 4 cm, 4 cm
2 1 cm, 5 cm, 5 cm; 3 cm, 4 cm, 4 cm; 5 cm, 3 cm, 3 cm
3 2 cm, 8 cm, 8 cm; 4 cm, 7 cm, 7 cm; 6 cm, 6 cm, 6 cm; 8 cm, 5 cm, 5 cm
4 2 cm, 3 cm, 3 cm

page 233 ***Exercise 15.7***

1 18 cm^2 **2** 14 cm^2 **3** 6 cm^2 **4** 16 cm^2 **5** 25 cm^2
6 14 cm^2 **7** 10 cm^2 **8** 16 cm^2 **9** 14 cm^2 **10** 15 cm^2

page 235 ***Exercise 15.8***

1 15 cm^2 **2** 10 cm^2 **3** 18 cm^2 **4** 25 cm^2 **5** 15 cm^2
6 20 cm^2 **7** 20 cm^2 **8** 5 cm^2 **9** 8 cm^2 **10** 24.5 cm^2

page 237 ***Exercise 15.9***

1 8 cm^2 **2** 20.5 cm^2 **3** 26 cm^2 **4** 17 cm^2 **5** 45 cm^2 **6** 16 cm^2

page 239 ***Exercise 15.10***

1 24 cm^3 **2** 18 cm^3 **3** 36 cm^3 **4** 36 cm^3
5 4 cm^3 **6** 6 cm^3 **7** 7 cm^3 **8** 9 cm^3

page 241 ***Exercise 15.11***

1 30 cm^3 **2** 15 cm^3 **3** 15 cm^3 **4** 6 cm^3 **5** 52.5 cm^3

page 242 ***Exercise 15.12***

1 25 cm^2 **2** 144 mm^2 **3** 900 mm^2 **4** 1.96 m^2 **5** 5.76 m^2

page 242 ***Exercise 15.13***

1 6 cm **2** 15 cm **3** 70 mm **4** 1.3 m **5** 1.7 m

page 243 ***Exercise 15.14***

1 **a** 144 cm^2 **b** 144 cm^2 **c** 140 cm^2 **c**
2 **a** 2700 mm^2 **b** 2700 mm^2 **c** 2660 mm^2 **c**
3 **a** 4.2 m^2 **b** 4.8 m^2 **c** 4.8 m^2 **a**
4 **a** 5.6 m^2 **b** 5.4 m^2 **c** 5.6 m^2 **b**
5 **a** 13.5 m^2 **b** 16.5 m^2 **c** 16.5 m^2 **a**

page 244 ***Exercise 15.15***

1 225 cm^2 **2** 540 cm^2 **3** 640 mm^2 **4** 1680 mm^2 **5** 1.5 m^2
6 1.2 m^2 **7** 1.8 m^2 **8** 2 m^2 **9** 315 cm^2 **10** 1350 mm^2

page 245 ***Exercise 15.16***

1 5.4 m^2 **2** 13.5 m^2 **3** 25.2 m^2 **4** 90 cm^2
5 1100 cm^2 **6** 770 cm^2 **7** 420 cm^2 **8** 3150 mm^2

page 246 ***Exercise 15.17***

1 360 cm^2 **2** 660 cm^2 **3** 3780 mm^2 **4** 7560 mm^2 **5** 2.1 m^2 **6** 20.7 m^2

page 247 ***Exercise 15.18***

1 2850 m^2 **2** 13 500 m^2 **3** 1125 cm^2
4 6600 cm^2 **5** 608 cm^2 **6** 2960 m^2
7 a 28 m^2 **b** 332 m^2 **8** 650 cm^2 **9** 72 m^2
10 a 15 cm^2 **b** 5 cm^2 **c** 2.6 m^2 **d** 40.2 cm^2 **e** 40.6 cm^2

page 249 ***Exercise 15.19***

1 512 cm^3 **2** 1728 cm^3 **3** 125 000 mm^3 **4** 15 625 mm^3

page 249 ***Exercise 15.20***

1 3 cm **2** 4 cm **3** 10 cm **4** 20 mm **5** 30 mm

page 250 ***Exercise 15.21***

1 a 286 cm^3 **b** 288 cm^3 **c** 288 cm^3 **a**
2 a 980 mm^3 **b** 960 mm^3 **c** 960 mm^3 **a**
3 a 2.4 m^3 **b** 2.8 m^3 **c** 2.4 m^3 **b**
4 18 000 mm^3 **5** 2625 cm^3

page 251 ***Exercise 15.22***

1 135 cm^3 **2** 425 cm^3 **3** 1080 cm^3 **4** 2560 cm^3 **5** 750 cm^3

page 253 ***Exercise 16.1***

1 15.7 cm **2** 12.56 cm **3** 47.1 cm **4** 109.9 cm **5** 14.13 cm **6** 23.55 m
7 18.84 cm **8** 62.8 cm **9** 157 cm **10** 141.3 mm **11** 78.5 mm **12** 4.71 m

page 254 ***Exercise 16.2***

1 8 cm **2** 9 cm **3** 10 cm
4 65 cm **5** 75 cm **6** 55 cm
7 a 7 mm **b** 3.5 mm **8 a** 3 mm **b** 1.5 mm **9 a** 11 mm **b** 5.5 mm
10 a 160 mm **b** 80 mm **11 a** 180 mm **b** 90 mm **12 a** 210 mm **b** 105 mm

page 255 ***Exercise 16.3***

1 a 4.71 cm **b** 706.5 cm **2 a** 47.1 cm **b** 942 cm **3** 79.5 cm
4 a 3.925 cm **b** 60 **5 a** 125.6 cm **b** 75 **6 a** 1.884 m **b** 7500

page 256 ***Exercise 16.4***

1 400 m **2** 200 m **3** 150 m **4** 100 m

page 257 ***Exercise 16.5***

1 $78.5\,\text{cm}^2$ **2** $28.26\,\text{cm}^2$ **3** $706.5\,\text{cm}^2$ **4** $1256\,\text{cm}^2$
5 $5024\,\text{mm}^2$ **6** $3846.5\,\text{mm}^2$ **7** $11304\,\text{mm}^2$ **8** $6358.5\,\text{mm}^2$
9 $9498.5\,\text{m}^2$ **10** $94.985\,\text{m}^2$ **11** $19.625\,\text{m}^2$ **12** $3.14\,\text{m}^2$

page 258 ***Exercise 16.6***

1 a 25 mm **b** 50 mm **2 a** 30 mm **b** 60 mm **3 a** 50 mm **b** 100 mm
4 a 2 cm **b** 4 cm **5 a** 4.5 cm **b** 9 cm **6 a** 7.5 cm **b** 15 cm

page 259 ***Exercise 16.7***

1 a $100\,\text{cm}^2$ **b** $78.5\,\text{cm}^2$ **c** $21.5\,\text{cm}^2$
2 a $78.5\,\text{cm}^2$ **b** $49\,\text{cm}^2$ **c** $29.5\,\text{cm}^2$
3 a $153.86\,\text{cm}^2$ **b** $63\,\text{cm}^2$ **c** $90.86\,\text{cm}^2$
4 a $63\,\text{cm}^2$ **b** $38.465\,\text{cm}^2$ **c** $24.535\,\text{cm}^2$
5 $628\,\text{mm}^2$ **6** $199.04\,\text{cm}^2$
7 $606.5\,\text{cm}^2$ **8** $1721.5\,\text{cm}^2$
9 a $157\,\text{cm}^2$ **b** $628\,\text{cm}^2$ **c** $39.25\,\text{cm}^2$ **d** $2512\,\text{mm}^2$ **e** $5652\,\text{mm}^2$
10 a $49\,\text{cm}^2$ **b** $39.25\,\text{cm}^2$ **c** $9.75\,\text{cm}^2$
11 a $200\,\text{cm}^2$ **b** $157\,\text{cm}^2$ **c** $43\,\text{cm}^2$
12 $446.75\,\text{m}^2$

page 261 ***Exercise 16.8***

1 7.08 cm **2** 10.62 cm **3** 106.2 cm **4** 2.124 m **5** 0.354 m **6** 0.531 m

page 262 ***Exercise 16.9***

1 2.82 cm **2** 8.46 cm **3** 28.2 cm **4** 22.56 mm
5 39.48 mm **6** 0.705 m **7** 5.64 m **8** 14.1 cm

page 263 ***Exercise 17.1***

1 A1 – circle, A2 – square, A3 – triangle, A4 – star, A5 – rectangle, A6 – arrow
B1 – square, B2 – rectangle, B3 – circle, B4 – triangle, B5 – arrow, B6 – star
C1 – star, C2 – triangle, C3 – square, C4 – arrow, C5 – circle, C6 – rectangle
D1 – triangle, D2 – arrow, D3 – rectangle, D4 – circle, D5 – star, D6 – square
E1 – rectangle, E2 – star, E3 – arrow, E4 – square, E5 – triangle, E6 – circle
F1 – arrow, F2 – circle, F3 – star, F4 – rectangle, F5 – square, F6 – triangle

2 a A5 B1 C3 D2 E4 F6
b A2 B3 C1 D4 E6 F5
c A1 B2 C6 D5 E3 F4
d A3 B6 C4 D1 E5 F2
e A4 B5 C2 D6 E1 F3
f A6 B4 C5 D3 E2 F1

page 265 Exercise 17.2

1 (1, 2) (1, 3) (1, 4) (1, 8) (1, 10) (2, 3) (2, 4) (2, 5) (2, 8) (3, 1) (3, 2) (3, 3) (3, 6) (3, 7) (4, 3) (4, 4) (4, 5) (4, 8) (4, 9) (4, 0) (5, 1) (5, 2) (5, 7) (5, 8) (5, 0)

2 **a** (1, 2) (1, 10) (9, 10) (9, 2)
b (1, 12) (1, 18) (7, 18) (7, 12)
c (12, 2) (12, 12) (18, 12) (18, 2)
d (10, 14) (10, 18) (18, 18) (18, 14)

3 **a** (5, 1) (2, 5) (6, 8) (9, 4)
b (6, 10) (2, 13) (5, 17) (9, 14)
c (15, 1) (11,4) (14, 8) (18, 5)
d (14, 10) (11, 14) (15, 17) (18, 13)

page 268 Exercise 17.3

2 Hexagon
3 Octagon
4 4-pointed star

page 270 Exercise 17.4

2 **b** (i) 10 p (ii) 14 p (iii) 18 p
3 **b** (i) £6 (ii) £8 (iii) £12 (iv) £16
4 **b** (i) 6 cm (ii) 7 cm (iii) 9 cm

page 273 Exercise 17.5

1 (1, 2) (1, 3) (1, 4) (1, 5) (1, 6)
(2, 1) (2, 2) (2, 3) (2, 5) (2, 6)
(3, 0) (3, 1) (3, 2) (3, 3) (3, 4) (3, 6)
(4, 0) (4, 1) (4, 3) (4, 4) (4, 5)
(0, 1) (0, 2) (0, 3) (0, 5)

2 (1, −2) (1, −3) (1, −4) (1, −5) (1, −6)
(2, −1) (2, −2) (2, −3) (2, −5) (2, −6)
(3, −1) (3, −2) (3, −3) (3, −4) (3, −6)
(4, −1) (4, −3) (4, −4) (4, −5)
(0, −1) (0, −2) (0, −3) (0, −5)

3 (−1, 2) (−1, 3) (−1, 4) (−1, 5) (−1, 6)
(−2, 1) (−2, 2) (−2, 3) (−2, 5) (−2, 6)
(−3, 0) (−3, 1) (−3, 2) (−3, 3) (−3, 4) (−3, 6)
(−4, 0) (−4, 1) (−4, 3) (−4, 4) (−4, 5)

4 (−1, −2) (−1, −3) (−1, −4) (−1, −5) (−1, −6)
(−2, −1) (−2, −2) (−2, −3) (−2, −5) (−2, −6)
(−3, −1) (−3, −2) (−3, −3) (−3, −4) (−3, −6)
(−4, −1) (−4, −3) (−4, −4) (−4, −5)

page 275 Exercise 17.6

1 **a** (4, 4) (4, −4) **b** (−4, 4) (−4, −4)
2 **a** (4, 8) (8, 0) (4, −8) **b** (−4, 8) (−8, 0) (−4, −8)
3 **a** (0, 10) (4, 4) (10, 0) (4, −4) (0, −10) **b** (0, 10) (−4, 4) (−10, 0) (−4, −4) (0, −10)
4 **a** (0, 11) (4, 2) (10, −6) (0, −5) **b** (0, 11) (−4, 2) (−10, −6) (0, −5)

page 277 Exercise 17.7

1 A screwdriver
2 An arrow with head and tail
3 An octagon (eight sides)
4 An octagon (eight sides)

page 280 Exercise 17.8

1 **a** (16, 18) (16, −18) (−16, 18) (−16, −18)
b (14, 12) (14, −12) (−14, 12) (−14, −12)
c (12, 8) (12, −8) (−12, 8) (−12, −8)
d (10, 6) (10, −6) (−10, 6) (−10, −6)
e (6, 4) (6, −4) (−6, 4) (−6, −4)
f (2, 2) (2, −2) (−2, 2) (−2, −2)

2 **a** (2, 4) (10, 4) (10, 9) (2, 9)
b (12, 2) (18, 2) (18, 14) (12, 14)
c (4, −3) (8, −3) (11, −7) (1, −7)
d (9, −11) (13, −11) (18, −16) (4, −16)
e (−6, 4) (−4, 8) (−6, 12) (−8, 8)
f (−14, 1) (−10, 8) (−14, 15) (−18, 8)
g (−6, −2) (−2, −2) (−2, −6) (−6, −6)
h (−11, −3) (−6, −8) (−11, −13) (−16, −8)

3 Each shape is a rectangle.

4 **a** isosceles triangle
b equilateral triangle
c right-angled triangle
d right-angled triangle

page 282 ***Exercise 17.9***

1 **a** (1, 4) (1, 10) or (13, 4) (13, 10) or (4, 7) (10, 7)
b (1, −4) (1, −10) or (13, −4) (13, −10) or (4, −7) (10, −7)
c (−1, 4) (−1, 10) or (−13, 4) (−13, 10) or (−4, 7) (−10, 7)
d (−1, −4) (−1, −10) or (−13, −4) (−13, −10) or (−4, −7) (−10, −7)

2 **a** (4, 1) (10, 1) or (4, 13) (10, 13) or (7, 4) (7, 10)
b (4, −1) (10, −1) or (4, −13) (10, −13) or (7, −4) (7, −10)
c (−4, 1) (−10, 1) or (−4, 13) (−10, 13) or (−7, 4) (−7, 10)
d (−4, −1) (−10, −1) or (−4, −13) (−10, −13) or (−7, −4) (−7, −10)

3 **a** (7, 7) (14, 14) or (0, 7) (7, 0) or (14, 21) (21, 14)
b (7, −7) (14, −14) or (0, −7) (7, 0) or (14, −21) (21, −14)
c (−7, 7) (−14, 14) or (0, 7) (−7, 0) or (−14, 21) (−21, 14)
d (−7, −7) (−14, −14) or (0, −7) (−7, 0) or (−14, −21) (−21, −14)

page 284 ***Exercise 17.10***

1 **a** 0103 **b** 0104 **c** 0106 **d** 0203 **e** 0205 **f** 0206 **g** 0305 **h** 0306
i 0103 **j** 0105 **k** 0106 **l** 0204 **m** 0205 **n** 0304 **o** 0305 **p** 0306
q 0104 **r** 0105 **s** 0203 **t** 0204 **u** 0205 **v** 0304 **w** 0305 **x** 0307

2 **a** Lochans, Portpatrick, Stoney Kirk
b Castle Kennedy, Dunragit
c Glenluce, Milton, Auchenmalg
d Kirkolm, Cairnryan, Stranraer
e Penwhirn Reservoir, New Luce, Craig Fell
f Urrall Fell, Artfield Fell, Carscreugh
g Carlock Hill, Portencalzie
h Altimeg Hill, Glenwhilly
i Lochton, Benbroke Fell, Drumlanford

3 (i)
a Hunslet
b Potternewton
c Moortown
d Alwoodley
e Eccup Reservoir
f Harewood Bridge
g Burmantofts
h Gledhow
i Harewood
j Kearby
k Osmondthorpe
l Harehills
m Wike
n Kirby Overblow
o Stourton
p Oakwood
q Roundhay
r Netherby
s Halton
t Beechwood
u East Keswick

(ii)
a Crossgates
b Monkswood
c Bardsey
d Manston
e Seacroft
f East Rigton
g Sicklinghall
h Swillington
i Sandhills
j Thorner
k Collingham
l Garforth Bridge
m Linton
n Great Preston
o Garforth
p Barwick in Elmet
q Wetherby
r Kirk Deighton
s Leeds
t Woodhouse
u Chapel Allerton
v Weardley
w Dunkeswick
x Buttersyke Bar

page 288 **Exercise 17.11**

1 **a** Hardwick **b** Burntwood **c** Rugeley **d** Chelmsley Wood
e Fillongley **f** Bedworth **g** Hollywood **h** Mappleborough Green
i Alcester **j** Blackheath **k** Wordsley **l** Six Ashes
m Wydle Green **n** Fazeley **o** No Man's Heath **p** Solihull
q Honiley **r** Leamington Spa **s** Weat Heath **t** Bromsgrove
u Droitwich Spa **v** West Bromwich **w** Wolverhampton **x** Albrighton

2 **a** 10 km N **b** 20 km N **c** 30 km N **d** 10 km NE **e** 20 km NE **f** 30 km NE
g 10 km E **h** 20 km E **i** 30 km E **j** 10 km SE **k** 20 km SE **l** 30 km SE
m 10 km S **n** 20 km S **o** 30 km S **p** 10 km SW **q** 20 km SW **r** 30 km SW
s 10 km W **t** 20 km W **u** 30 km W **v** 10 km NW **w** 20 km NW **x** 30 km NW

page 291 **Exercise 17.12**

1 A 10 km, 030° B 20 km, 030° C 40 km, 030° D 20 km, 060° E 30 km, 060°
F 60 km, 060° G 20 km, 120° H 30 km, 120° I 50 km, 120° J 30 km, 150°
K 40 km, 150° L 60 km, 150° M 20 km, 210° N 30 km, 210° P 50 km, 210°
Q 30 km, 240° R 40 km, 240° S 60 km, 240° T 10 km, 300° U 20 km, 300°
V 50 km, 300° W 20 km, 330° X 30 km, 330° Y 50 km, 330°

page 294 **Exercise 17.14**

1 A 40 km, 030° B 32 km, 060° C 42 km, 045° D 48 km, 030° E 28 km, 045° F 48 km, 060°
G 40 km, 120° H 48 km, 150° I 35 km, 135° J 21 km, 135° K 24 km, 150° L 24 km, 120°
M 48 km, 210° N 48 km, 240° P 28 km, 225° Q 40 km, 240° R 57 km, 225° S 40 km, 210°
T 32 km, 330° U 24 km, 300° V 48 km, 330° W 49 km, 315° X 32 km, 300° Y 21 km, 315°

2 A (35 km E, 20 km N) B (28 km E, 28 km N) C (14 km E, 24 km N)
D (21 km E, 12 km N) E (30 km E, 30 km N) F (8 km E, 14 km N)
G (21 km E, −12 km, N) H (16 km E, −28 km N) I (42 km E, −24 km N)
J (35 km E, −35 km N) K (21 km E, −21 km N) L (20 km E, −35 km N)
M (−16 km E, −28 km N) N (−21 km E, −12 km N) P (−28 km E, −28 km N)
Q (−15 km E, −15 km N) R (−12 km E, −21 km N) S (−28 km E, −16 km N)
T (−20 km E, 35 km N) U (−25 km E, 25 km N) V (−35 km E, 20 km N)
W (−40 km E, 40 km N) X (−42 km E, −24 km N) Y (−12 km E, 21 km N)

3 **a** (i) (28,13) (ii) (26,56) (iii) (36,36) (iv) (13, −28)
(v) (17, −47) (vi) (24, −24) (vii) (−13, 28) (viii) (−36, 36)
b (i) 31m, 025° (ii) 50m, 020° (iii) 34m, 045° (iv) 31m, 115°
(v) 62m, 155° (vi) 31m, 295° (vii) 50m, 340° (viii) 34m, 315°

page 295 **Exercise 17.15**

1 10 km, 030° **2** 10 km, 045° **3** 10 km, 060° **4** 10 km, 090°
5 5 km, 120° **6** 5 km, 135° **7** 5 km, 150° **8** 5 km, 180°
9 10 km, 210° **10** 10 km, 225° **11** 10 km, 240° **12** 10 km, 270°

page 297 **Exercise 18.1**

1 A (3, 3), B (3, 9) C (9, 3)
2 **a** P (2, 2), Q (2, 4), R (4, 2) **b** P (4, 4), Q (4, 8), R (8, 4)
3 **a** P (3, 3), Q (3, 6), R (6, 3) **b** P (5, 5), Q (5, 10), R (10, 5)
4 **a** X (3, 6), Y (6, 6), Z (6, 3) **b** X (5, 10), Y (10, 10), Z (10, 5)
5 **a** X (2, 4), Y (4, 4), Z (4, 2) **b** X (4, 8), Y (8, 8), Z (8, 4)
6 **a** L (2, 6), M (6, 6), N (6, 2) **b** L (3, 9), M (9, 9), N (9, 3)

page 298 Exercise 18.2

1 A(0, 0), B(6, 4) C(8, 0)
2 a P(0, 0), Q(2, 4), R(8, 0) **b** P(0, 0), Q(3, 6), R(12, 0)
3 a L(0, 0), M(4, 4), N(6, 0) **b** L(0, 0), M(6, 6), N(9, 0) **c** L(0, 0), M(8, 8), N(12, 0)

page 299 Exercise 18.3

1 a A(4, 4), B(4, 8), C(8, 8), D(8, 4) **b** A(2, 2), V(2, 10), C(10, 10), D(10, 2)
2 A(2, 4), B(2, 8), C(10, 8), D(10, 4)
3 W(4, 2), X(4, 10), Y(8, 10), Z(8, 2)

page 300 Exercise 18.4

1 3 **2** 2 **3** 3 **4** 4 **5** 2 **6** 2

page 302 Exercise 19.1

1 Arsenal 5, Chelsea 4, QPR 2, Spurs 6, West Ham 3
2 Apple 2, cola 2, lemon 4, lime 4, orange 5, raspberry 3
3 Prawn 1, cheese 4, plain 3, salt and vinegar 5, Worcester sauce 2
4 Athletics 6, golf 3, rounders 4, swimming 5, tennis 7

page 304 Exercise 19.2

1 a January 2, February 1, March 2, April 2, May 1, June 2, July 1, August 2, September 2, October 1, November 2, December 2
b Monday 4, Tuesday 2, Wednesday 4, Thursday 2, Friday 2, Saturday 2, Sunday 4
2 a East 4, Midland 4, North 2, South 3, West 2
b Monday 2, Tuesday 2, Wednesday 3, Thursday 3, Friday 2, Saturday 3

page 306 Exercise 19.3

1 a 0–9 5, 10–19 8, 20–29 9, 30–39 6
2 a 100–109 cm 2, 110-119 cm 3, 120–129 cm 4, 130–139 cm 5, 140–149 cm 3, 150–159 cm 2, 160–169 cm 1
3 a 0–9 min 6, 10–19 min 7, 20–29 min 5, 30–39 min 4, 40–49 min 3

page 308 Exercise 19.4

1 a

Year	1984	1986	1988	1990	1992	1994	1996
Height (cm)	40	80	100	110	130	140	160

c (i) 60 cm (ii) 90 cm (iii) 120 cm (iv) 150 cm

2 a

Time	8 a.m.	9 a.m.	10 a.m.	11 a.m.	12 noon	1 p.m.	2 p.m.	3 p.m.	4 p.m.	5 p.m.
Length (m)	0	2	6	12	14	18	18	26	30	34

c (i) 4 m (ii) 16 m (iii) 22 m (iv) 28 m (v) 32 m
d Between 1 p.m. and 2 p.m. since the trench did not increase in length.

3 a

Time	8 a.m.	9 a.m.	10 a.m.	11 a.m.	12 noon	1 p.m.	2 p.m.	3 p.m.	4 p.m.	5 p.m.
Temperature (°C)	10	30	40	50	45	50	45	50	30	20

c (i) 20 °C (ii) 35 °C (iii) 45 °C (iv) 40 °C (v) 25 °C
d Either the immersion heater has a thermostat which cuts out at 50 °C and restarts at 45 °C or water is drawn off and then reheated.

4 **c** (i) 4 cm (ii) 16 cm (iii) 22 cm (iv) 28 cm d Between 1 p.m. and 3 p.m.
5 **c** (i) 10 $km\,h^{-1}$ (ii) 40 $km\,h^{-1}$ (iii) 70 $km\,h^{-1}$ (iv) 100 $km\,h^{-1}$ (v) 100 $km\,h^{-1}$ (vi) 70 $km\,h^{-1}$ (vii) 30 $km\,h^{-1}$
d 120 $km\,h^{-1}$ **e** 20 s
6 **b** (i) 10 m (ii) 45 m (iii) 40 m (iv) 20 m **c** (i) 30 m (ii) 130 m

page 314 ***Exercise 19.5***

1 **a** 8–9 a.m. **c** People going to work and people coming home from work.
2 **b** 1–2 p.m. **c** Normal meal times, breakfast, lunch and tea.
3 **a** 2–3 p.m. **c** People going home, temperature drops, nearer tea time.
4 **a** 1.5, 0.5, 0.4, 0.4, 0.6, 0.6, 0.3, 0.8, 1.0, 2.0, 2.3, 2.2, 1.8, 1.8 **c** 6–7 p.m. Most people cooking their evening meal.

page 318 ***Exercise 19.6***

1

First course	fish and chips	pie and chips	baked potato
Frequency	3	8	7

Second course	ice cream	fruit salad	sponge pudding
Frequency	7	6	5

Drink	tea	coffee	lemonade	orange squash
Frequency	5	2	4	7

2 **a**

Type of train	Local passenger	Intercity	Freight
Frequency	6	4	2

Direction	north	south	east
Frequency	4	5	3

b Local town south of junction (people going to work) **c** Major city north of junction (people going to work)

page 321 ***Exercise 19.7***

1 **a** 31, 32, 33, 34, 36, 36, 37, 37, 38, 39, 41, 41, 42, 43, 44, 46, 47, 48 kg
b

Weight	30–34 kg	35–39 kg	40–44 kg	45–49 kg
Frequency	4	6	5	3

2 **a** 6, 7, 8, 9, 11, 11, 12, 13, 14, 15, 16, 16, 17, 17, 18, 18, 19, 20, 21, 22
b

Time	5–9 min	10–14 min	15–19 min	20–24 min
Frequency	4	5	8	3

3 **a** 7, 1, 6, 3, 11, 1, 8, 10, 3, 6, 16, 0, 9, 0, 13 minutes
b 0, 0, 1, 1, 3, 3, 6, 6, 7, 8, 9, 10, 11, 13, 16 minutes late
c

Lateness	0–4 min	5–9 min	10–14 min	15–19 min
Frequency	6	5	3	1

4 a 95, 98, 102, 105, 108, 112, 115, 116, 118, 122, 125, 128, 132, 136, 143 cm

b

Height	90–99 cm	100–109 cm	110–119 cm	120–129 cm	130–139 cm	140–149 cm
Frequency	2	3	4	3	2	1

5 a 8, 13, 24, 27, 32, 35, 39, 42, 45, 48, 49, 51, 53, 54, 55, 56, 64, 65, 68, 72, 75, 81, 85, 90

b

Mark	0–9	10–19	20–29	30–39	40–49	50–59	60–69	70–79	80–89	90–99
Frequency	1	1	2	3	4	5	3	2	2	1

c 25, 28, 32, 35, 36, 38, 39, 42, 43, 44, 45, 48, 49, 51, 55, 56, 57, 59, 64, 67, 68, 71, 75, 80

d

Mark	0–9	10–19	20–29	30–39	40–49	50–59	60–69	70–79	80–89	90–99
Frequency	0	0	2	5	6	5	3	2	1	0

e The Maths marks are more spread out, ranging from 8 to 90. The English marks all lie between 25 and 80. There are no English marks in the 0–9, 10–19 or 90–99 ranges.

page 325 ***Exercise 19.8***

1 fish and chips 60°, pie and chips 160°, baked potato 140°, ice cream 140°, fruit salad 120°, sponge pudding 100° tea 100°, coffee 40°, lemonade 80°, orange squash 140°
2 local 180°, Intercity 120°, freight 60°, north 120°, south 150°, east 90°
3 (i) saloon 144°, hatchback 144°, estate 72°
(ii) blue 108°, green 72°, red 108°, white 72°
4 (i) 2-door 108°, 4-door 252°
(ii) 1 252°, 2 36°, 3 0°, 4 36°, 5 36°

page 326 ***Exercise 19.9***

1 30–34 kg 80°, 35–39 kg 120°, 40–44 kg 100°, 45–49 kg 60°
2 5–9 min 72°, 10–14 min 90°, 15–19 min 144°, 20–24 min 54°
3 0–4 144°, 5–9 96°, 10–14 96°, 15–19 24°
4 90–99 cm 48°, 100–109 cm 72°, 110–119 cm 96°, 120–129 cm 72°, 130–139 cm 48°, 140–149 cm 24°
5 (i) 0–9 15°, 10–19 15°, 20–29 30°, 30–39 45°, 40–49 60°, 50–59 75°, 60–69 45°, 70–79 30°, 80–89 30°, 90–99 15°
(ii) 0–9 0°, 10–19 0°, 20–29 30°, 30–39 75°, 40–49 90°, 50–59 75°, 60–69 45°, 70–79 30°, 80–89 15°, 90–99 0°
6 a 125–129 cm 18°, 130–134 cm 0°, 135–139 cm 54°, 140–144 cm 72°, 145–149 cm 72°, 150–154 cm 54°, 155–159 cm 72°, 160–164 cm 18°
b 35–39 kg 72°, 40–44 kg 90°, 45–49 kg 126°, 50–54 kg 72°

page 328 ***Exercise 19.10***

1 squash – 18, cross country – 20, hockey – 30, rugby – 40, football – 72
2 Parkstone – 30, Branksome – 45, Poole – 50, Bournemouth – 75, Southampton – 100
3 prawn cocktail – 15; 10%, salt and vinegar – 30; 20%, cheese and onion – 45; 30%, plain – 60; 40%
4 Wednesday – 8, 10%, Tuesday – 8, 10%, Monday – 12; 15%, Friday – 16, 20%, Thursday – 36, 45%

page 328 ***Exercise 19.11***

1 north – 144°, south – 90°, east – 72°, west – 54°
2 chunnel – 144°, air – 54°, boat – 126°, hovercraft – 36°
3 black – 144°, white – 90°, red – 54°, green – 36°, blue – 36°

page 330 ***Exercise 19.12***

1 tennis – 120°, rounders – 90°, swimming – 90°, athletics – 60°
2 cricket – 120°, athletics – 90°, swimming – 60°, tennis – 60°, golf – 30°
3 train – 120°, bus – 90°, coach – 60°, car – 45°, motor cycle – 45°
4 Monday – 40°, Tuesday – 40°, Wednesday – 40°, Thursday – 60°, Friday – 90°, Saturday – 90°
5 Manchester – 60°, Liverpool – 60°, Birmingham – 40°, Leeds – 40°, Newcastle – 40°, London – 120°
6 Spurs – 90°, Arsenal – 60°, Chelsea – 45°, West Ham – 45°, Millwall – 40°, Fulham – 40°, QPR – 40°
7 general – 72°, education – 72°, police – 60°, roads – 45°, transport – 45°, landscape – 36° investments – 30°
8 Bristol – 126°, Cardiff – 72°, Swansea – 54°, Newport – 45°, Plymouth – 45°, Exeter – 18°

page 332 ***Exercise 19.13***

1 a

£	10	20	30	40	50	60	70	80	90
BF	500	1000	1500	2000	2500	3000	3500	4000	4500

c (i) 750BF (ii) 1250BF (iii) 2250BF (iv) 3750BF
d (i) £35 (ii) £55 (iii) £65 (iv) £85

2 a

£	10	20	30	40	50	60	70	80	90
Ptas	2000	4000	6000	8000	10 000	12 000	14 000	16 000	18 000

c (i) 7000 Ptas (ii) 9000 Ptas (iii) 1000 Ptas (iv) 15000 Ptas (v) 3000 Ptas
d (i) £55 (ii) £65 (iii) £85 (iv) £25
3 b (i) \$24 (ii) \$66 (iii) \$78 (iv) \$102 (v) \$108
c (i) £24 (ii) £32 (iii) £56 (iv) £76 (v) £84
d \$1.5 = £1

page 334 ***Exercise 19.14***

1 a

Miles	0	1	2	3	4	5	6	7	8	9
km	0	1.6	3.2	4.8	6.4	8.0	9.6	11.2	12.8	14.4

c (i) 4 km (ii) 7.2 km (iii) 8.8 km (iv) 13.6 km
d (i) 1.5 miles (ii) 3.5 miles (iii) 6.5 miles (iv) 7.5 miles

2 a

Acres	0	1	2	3	4	5	6	7	9	11	13	15	18
ha	0	0.4	0.8	1.2	1.6	2.0	2.4	2.8	3.6	4.4	5.2	6.0	7.2

c (i) 3.2 hectares (ii) 4.8 hectares (iii) 6.8 hectares
d (i) 10 acres (ii) 14 acres (iii) 16 acres

3 a

SquareMiles	0	0.4	1.6	2.0	2.4	3.6	4.0	4.4	5.6	6.0
Square km	0	1.0	4.0	5.0	6.0	9.0	10.0	11.0	14.0	15.0

c (i) 12 sq km (ii) 8 sq km (iii) 2 sq km
d (i) 1.2 sq miles (ii) 2.8 sq miles (iii) 5.2 sq miles
e 2.5 sq miles

4 a

Inches	0	8	12	20	28	32	40	48	52	60	68	72
cm	0	20	30	50	70	80	100	120	130	150	170	180

c (i) 60 cm (ii) 140 cm (iii) 160 cm
d (i) 16 in (ii) 36 in (iii) 44 in
e 2.5 cm = 1 in

5 a

Pounds	0	6	12	18
Kilograms	0	2.7	5.4	8.1

c (i) 3.6 kg (ii) 4.5 kg (iii) 6.3 kg
d (i) 16 lb (ii) 4 lb (iii) 2 lb
e 0.45 kg = 1 lb

page 337 ***Exercise 19.15***

1 a $30 \leqslant w < 35$; 1, $35 \leqslant w < 40$; 2, $40 \leqslant w < 45$; 4, $45 \leqslant w < 50$; 7, $50 \leqslant w < 55$; 6, $55 \leqslant w < 60$; 3, $60 \leqslant w < 65$; 2
b 25 children
2 a $0 \leqslant t < 5$; 5, $5 \leqslant t < 10$; 6, $10 \leqslant t < 15$; 8, $15 \leqslant t < 20$; 7, $20 \leqslant t < 25$; 4
b 30 days **c** April or November

page 339 ***Exercise 19.16***

1 a (i)

39.5–49.5 kg	49.5–59.5 kg	59.5–69.5 kg
6	11	3

(ii)

39.5–44.5 kg	44.5–49.5 kg	49.5–54.5 kg	54.5–59.5 kg	59.5–64.5 kg
2	4	6	5	3

2 a (i)

39.5–49.5 mm	49.5–59.5 mm	59.5–69.5 mm	69.5–79.5 mm
4	11	7	3

(ii)

39.5–44.5 mm	44.5–49.5 mm	49.5–54.5 mm	54.5–59.5 mm	59.5–64.5 mm	64.5–69.5 mm	69.5–74.5 mm	74.5–70.5 mm
1	3	5	6	3	4	2	1

3 a (i)

49.5–59.5 mm	59.5–69.5 mm	69.5–79.5 mm	79.5–89.5 mm
3	11	12	4

(ii)

49.5–54.5 s	54.5–59.5 s	59.5–64.5 s	64.5–69.5 s	69.5–74.5 s	74.5–79.5 s	79.5–84.5 s	84.5–89.5 s
1	2	5	6	7	5	3	1

page 342 ***Exercise 19.17***

1 Yes; a positive correlation. **2** Yes; a positive correlation. **3** Yes; a positive correlation.

page 346 ***Exercise 19.18***

1 A positive correlation. **2** A positive correlation. **3** A negative correlation.
4 12 p, 13.5 p, 13 p, 13.5 p, 14.5 p, 14.5 p, 13.5 p, 14 p; a negative correlation.

page 351 ***Exercise 20.1***

1 43 **2** 17 **3** 35 min **4** 11 °C **5** 7 **6** 25 cm **7** 3 hours **8** 117 cm **9** 33 kg

page 352 **Exercise 20.2**

1 35 min **2** 68 s **3** 16 **4** 15 min **5 a** 35 **b** 225

page 353 **Exercise 20.3**

1 Spurs **2** Orange **3** Salt and vinegar **4** Tennis **5** 28 s **6** 18 kg

page 355 **Exercise 20.4**

1 20–29 **2** 130–139 cm **3** 10–19 mm
4 10–19 kg 3, 20–29 kg 5, 30–39 kg 6, 40–49 kg 4, 50–59 kg 2; 30–39 kg
5 0–9 3, 10–19 5, 20–29 4, 30–39 3; 10–19 **6 a** 31 **b** J

page 357 **Exercise 20.5**

1 29 min **2** 15 **3** 73 days
4 124 grams **5** 8 hours **6** 146 papers
7 £27.60 **8** 56.5% **9** 119.3 cm

page 358 **Exercise 20.6**

1 a 7 hours, 2 hours **b** 7 hours, 4 hours
The first set of bulbs appears to be more reliable.
2 a 1745, 990 **b** 1745, 2690
The second team have more varied attendances.
As these are increasing dramatically they could have started playing very well.
3 a 14 min, 4 min **b** 14 min, 9 min
The first set of times are more consistent.
On the Sunday the tide may have been more variable which could affect the crossing times.
4 a 59, 63 **b** 59, 43 **c** 59, 9
Edward has a wide range of work: some very good, some very weak. Drashma is very consistent.

page 360 **Exercise 20.7**

1 157 cm **2** 44 kg **3** 5 min **4** 24 min **5** 139
6 17 min **7** 33 min **8** 6 °C **9** 32 % **10** 41 min

page 361 **Exercise 21.1**

(Clearly these answers are subjective.)
1 Unlikely **2** Likely **3** Not sure **4** Not sure **5** Unlikely **6** Very unlikely
7 Very unlikely **8** Likely **9** Very likely **10** Very unlikely **11** Very likely **12** Very unlikely

page 363 **Exercise 21.3**

(Clearly these answers are subjective.)
1 0 **2** 1 **3** 0 **4** 1 **5** $\frac{1}{2}$ **6** $\frac{1}{4}$
7 $\frac{3}{4}$ **8** $\frac{1}{2}$ **9** $\frac{1}{4}$ **10** $\frac{3}{4}$ **11** $\frac{1}{4}$ **12** $\frac{3}{4}$

page 364 **Exercise 21.4**

1 **a** $\frac{5}{10}=\frac{1}{2}$ **b** $\frac{5}{10}=\frac{1}{2}$ **c** $\frac{3}{10}$ **d** $\frac{2}{10}=\frac{1}{5}$ **e** $\frac{2}{10}=\frac{1}{5}$ **f** $\frac{1}{10}$

2 **a** $\frac{8}{15}$ **b** $\frac{7}{15}$ **c** $\frac{5}{15}=\frac{1}{3}$ **d** $\frac{3}{15}=\frac{1}{5}$ **e** $\frac{3}{15}=\frac{1}{5}$ **f** $\frac{2}{15}$

3 **a** $\frac{2}{24}=\frac{1}{12}$ **b** $\frac{3}{24}=\frac{1}{8}$ **c** $\frac{4}{24}=\frac{1}{6}$ **d** $\frac{6}{24}=\frac{1}{4}$ **e** $\frac{9}{24}=\frac{3}{8}$

4 **a** $\frac{8}{20}=\frac{2}{5}$ **b** $\frac{6}{20}=\frac{3}{10}$ **c** $\frac{4}{20}=\frac{1}{5}$ **d** $\frac{2}{20}=\frac{1}{10}$ **e** $\frac{18}{20}=\frac{9}{10}$ **f** $\frac{16}{20}=\frac{4}{5}$ **g** $\frac{14}{20}=\frac{7}{10}$ **h** $\frac{12}{20}=\frac{3}{5}$

5 **a** $\frac{2}{4}=\frac{1}{2}$ **b** $\frac{2}{4}=\frac{1}{2}$ **c** $\frac{2}{4}=\frac{1}{2}$ **d** $\frac{1}{4}$ **e** $\frac{2}{4}=\frac{1}{2}$ **f** $\frac{1}{4}$

6 **a** $\frac{4}{8}=\frac{1}{2}$ **b** $\frac{4}{8}=\frac{1}{2}$ **c** $\frac{2}{8}=\frac{1}{4}$ **d** $\frac{2}{8}=\frac{1}{4}$ **e** $\frac{4}{8}=\frac{1}{2}$ **f** $\frac{3}{8}$ **g** $\frac{4}{8}=\frac{1}{2}$ **h** $\frac{3}{8}$

7 **a** $\frac{2}{6}=\frac{1}{3}$ **b** $\frac{3}{6}=\frac{1}{2}$ **c** $\frac{2}{8}=\frac{1}{4}$ **d** $\frac{3}{8}$ **e** $\frac{3}{9}=\frac{1}{3}$

8 **a** $\frac{2}{6}=\frac{1}{3}$ **b** $\frac{3}{6}=\frac{1}{2}$ **c** $\frac{2}{8}=\frac{1}{4}$ **d** $\frac{3}{8}$ **e** $\frac{3}{9}=\frac{1}{3}$ **f** $\frac{3}{7}$

page 367 **Exercise 21.5**

1 HHH, HHT, HTH, THH, HTT, THT, TTH, TTT
a $\frac{3}{6}=\frac{1}{2}$ **b** $\frac{3}{6}=\frac{1}{2}$ **c** $\frac{2}{6}=\frac{1}{3}$

2 6; RRBB, RBRB, RBBR, BRRB, BRBR, BBRR
a $\frac{1}{8}$ **b** $\frac{1}{8}$ **c** $\frac{3}{8}$ **d** $\frac{3}{8}$

3 2, 1; 3, 2; 4, 3; 5, 2; 6, 1; 2, $\frac{1}{9}$; 3, $\frac{2}{9}$; 4, $\frac{1}{3}$; 5, $\frac{2}{9}$; 6, $\frac{1}{9}$

4 2, 1; 3, 2; 4, 1; 5, 1; 6, 2; 7, 1; 8, 0; 9, 1; 2, $\frac{1}{9}$; 3, $\frac{2}{9}$; 4, $\frac{1}{9}$; 5, $\frac{1}{9}$; 6, $\frac{2}{9}$; 7, $\frac{1}{9}$; 8, $\frac{0}{9}$; 9, $\frac{1}{9}$

5 CE, CM, CN, EM, EN, MN; **a** $\frac{1}{6}$ **b** $\frac{1}{6}$ **c** $\frac{4}{6}=\frac{2}{3}$ **d** $\frac{2}{6}=\frac{1}{3}$

6

Outcome	Number of points
WW	6
WD	4
WL	3
DW	4
DD	2
DL	1
LW	3
LD	1
LL	0

Number of points	Number of possible ways
0	1
1	2
2	1
3	2
4	2
5	0
6	1

$\frac{3}{9}$ or $\frac{1}{3}$

7 10; $\frac{3}{10}$ **8** 6; $\frac{3}{6}=\frac{1}{2}$

page 370 ***Exercise 21.6***

1 c Both fractions in part **b** should be fairly close to $\frac{1}{2}$ or 0.5.

2 c The fractions in part **b** should be close to $\frac{1}{5}$ (0.2) and $\frac{4}{5}$ (0.8) respectively.

3 c The fractions in part **b** should each be close to $\frac{1}{4}$ or 0.25.

page 372 ***Exercise 21.7***

1 a $\frac{27}{60} = \frac{9}{20}$ **b** $\frac{8}{60} = \frac{2}{15}$ **c** $\frac{10}{60} = \frac{1}{6}$ **d** $\frac{15}{60} = \frac{1}{4}$

2 a $\frac{5}{12}$ **b** $\frac{2}{12} = \frac{1}{6}$ **c** $\frac{5}{12}$

3 a $\frac{2}{20} = \frac{1}{10}$ **b** $\frac{6}{20} = \frac{3}{10}$ **c** $\frac{4}{20} = \frac{1}{5}$ **d** $\frac{8}{20} = \frac{2}{5}$

4 a $\frac{4\frac{1}{2}}{24} = \frac{9}{48} = \frac{3}{16}$ **b** $\frac{19\frac{1}{2}}{24} = \frac{39}{48} = \frac{13}{16}$

5 a $\frac{8}{24} = \frac{1}{3}$ **b** $\frac{7\frac{1}{2}}{24} = \frac{15}{48} = \frac{5}{16}$

6 $\frac{240}{960} = \frac{1}{4}$

7 a $\frac{168}{200} = \frac{21}{25}$ **b** $\frac{32}{200} = \frac{4}{25}$

8 a $\frac{350}{500} = \frac{7}{10}$ **b** $\frac{150}{500} = \frac{3}{10}$ **c** $\frac{220}{500} = \frac{11}{25}$

page 374 ***Exercise 21.8***

1 $\frac{77}{231} = \frac{1}{3}$ **2** $\frac{12}{60} = \frac{1}{5}$

3 a $\frac{10}{20} = \frac{1}{2}$ **b** $\frac{4}{20} = \frac{1}{5}$ **c** $\frac{6}{20} = \frac{3}{10}$

4 a $\frac{14}{20} = \frac{7}{10}$ **b** $\frac{5}{20} = \frac{1}{4}$ **c** $\frac{1}{20}$ **d** $\frac{6}{20} = \frac{3}{10}$ **e** $\frac{9}{20}$ **f** $\frac{5}{20} = \frac{1}{4}$ **g** $\frac{20}{40} = \frac{1}{2}$ **h** $\frac{14}{40} = \frac{7}{20}$ **i** $\frac{6}{40} = \frac{3}{20}$

page 376 ***Exercise 21.9***

1 a $\frac{1}{3}$ **b** $\frac{1}{3}$ **c** $\frac{1}{3}$ **2 a** $\frac{2}{3}$ **b** $\frac{1}{3}$

3 a $\frac{1}{4}$ **b** $\frac{1}{4}$ **c** $\frac{2}{4} = \frac{1}{2}$ **4 a** $\frac{3}{4}$ **b** $\frac{1}{4}$

5 a $\frac{3}{12} = \frac{1}{4}$ **b** $\frac{3}{12} = \frac{1}{4}$ **c** $\frac{1}{12}$ **d** $\frac{1}{12}$ **e** $\frac{2}{12} = \frac{1}{6}$ **f** $\frac{2}{12} = \frac{1}{6}$ **g** $\frac{12}{12} = 1$

6 a $\frac{8}{80} = \frac{1}{10}$ **b** $\frac{8}{80} = \frac{1}{10}$ **c** $\frac{60}{80} = \frac{3}{4}$ **d** $\frac{20}{80} = \frac{1}{4}$ **e** $\frac{10}{80} = \frac{1}{8}$ **f** $\frac{10}{80} = \frac{1}{8}$ **g** $\frac{20}{80} = \frac{1}{4}$ **h** $\frac{20}{80} = \frac{1}{4}$

7 a $\frac{20}{200} = \frac{1}{10}$ **b** $\frac{20}{200} = \frac{1}{10}$ **c** $\frac{100}{200} = \frac{1}{2}$ **d** $\frac{50}{200} = \frac{1}{4}$ **e** $\frac{10}{200} = \frac{1}{20}$ **f** $\frac{30}{200} = \frac{3}{20}$

page 379 ***Exercise 21.10***

1 a Second dice

First dice	1	2	3	4	5	6
1	1, 1	1, 2	1, 3	1, 4	1, 5	1,6
2	2, 1	2, 2	2, 3	2, 4	2, 5	2, 6
3	3, 1	3, 2	3, 3	3, 4	3, 5	3, 6
4	4, 1	4, 2	4, 3	4, 4	4, 5	4, 6
5	5, 1	5, 2	5, 3	5, 4	5, 5	5, 6
6	6, 1	6, 2	6, 3	6, 4	6, 5	6, 6

c (i) 18 (ii) 6 (iii) 12 (iv) 7

2 b Second dice

First dice

	1	2	3	4
1	1, 1	1, 2	1, 3	1, 4
2	2, 1	2, 2	2, 3	2, 4
3	3, 1	3, 2	3, 3	3, 4
4	4, 1	4, 1	4, 3	4, 4

c (i) 8 (ii) 4 (iii) 5 (iv) 4

3 a Second dice

First dice

	1	2	3	4	5	6
1	1, 1	1, 2	1, 3	1, 4	1, 5	1, 6
2	2, 1	2, 2	2, 3	2, 4	2, 5	2, 6
3	3, 1	3, 2	3, 3	3, 4	3, 5	3, 6
4	4, 1	4, 2	4, 3	4, 4	4, 5	4, 6

c (i) 4 (ii) 5 (iii) 11

4 a Second arrow

First arrow

	0	1	2	5	10
0	0, 0	0, 1	0, 2	0, 5	0, 10
1	1, 0	1, 1	1, 2	1, 5	1, 10
2	2, 0	2, 1	2, 2	2, 5	2, 10
5	5, 0	5, 1	5, 2	5, 5	5, 10
10	10, 0	10, 1	10, 2	10, 5	10,10

c 2 – (0, 2), (1, 1), (2, 2); 10 – (10,0), (5, 5), (0, 10)

5 a Second card

	Diamond	Heart	Club	Spade
Diamond	DD	DH	DC	DS
Heart	HD	HH	HC	HS
Club	CD	CH	CC	CS
Spade	SD	SH	SC	SS

c (i) 4 (ii) 4 (iii) 4

6 b (i) 3 – HHT, HTH, THH (ii) 3 – TTH, THT, HTT
(iii) 3 – HHH, HHT, THH (iv) 3 – TTT, TTH, HTT
(v) 4 – HTH, THT, HHH, TTT

7 b (i) 3 – WWW, WWD, DWW (ii) 3 – DDD, DDW, WDD
(iii) 2 – WDW, DWD

8 b 4 – RRR, RRB, BBR, BBB

9 b 6 – RR, WW, BB, RB, RW, WB

page 383 **Exercise 21.11**

1 a $\frac{20}{60}$ or $\frac{1}{3}$ **b** $\frac{18}{60}$ or $\frac{3}{10}$ **c** $\frac{16}{60}$ or $\frac{4}{15}$ **d** $\frac{6}{60}$ or $\frac{1}{10}$

2 a $\frac{9}{24}$ or $\frac{3}{8}$ **b** $\frac{8}{24}$ or $\frac{1}{3}$ **c** $\frac{4}{24}$ or $\frac{1}{6}$ **d** $\frac{3}{24}$ or $\frac{1}{8}$

3 a $\frac{18}{36}$ or $\frac{1}{2}$ **b** $\frac{10}{36}$ or $\frac{5}{18}$ **c** $\frac{6}{36}$ or $\frac{1}{6}$ **d** $\frac{2}{36}$ or $\frac{1}{18}$

4 a $\frac{4}{40}$ or $\frac{1}{10}$ **b** $\frac{12}{40}$ or $\frac{3}{10}$ **c** $\frac{16}{40}$ or $\frac{2}{5}$ **d** $\frac{8}{40}$ or $\frac{1}{5}$

5 a $\frac{3}{7}$ **b** $\frac{2}{7}$ **c** $\frac{1}{7}$ **d** $\frac{1}{7}$

6 $\frac{2}{6}$ or $\frac{1}{3}$ **b** $\frac{2}{6}$ or $\frac{1}{3}$ **c** $\frac{1}{6}$ **d** $\frac{1}{6}$

page 384 **Exercise 21.12**

1 a $\frac{3}{9}$ or $\frac{1}{3}$ **b** $\frac{6}{9}$ or $\frac{2}{3}$

2 a $\frac{2}{10}$ or $\frac{1}{5}$ **b** $\frac{8}{10}$ or $\frac{4}{5}$

3 a $\frac{2}{48}$ or $\frac{1}{4}$ **b** $\frac{36}{48}$ or $\frac{3}{4}$

4 a $\frac{15}{90}$ or $\frac{1}{6}$ **b** $\frac{75}{90}$ or $\frac{5}{6}$

5 a $\frac{9}{24}$ or $\frac{3}{8}$ **b** $\frac{15}{24}$ or $\frac{5}{8}$

6 a $\frac{3}{6}$ or $\frac{1}{2}$ **b** $\frac{3}{6}$ or $\frac{1}{2}$

page 386 **Exercise 21.13**

1 a $\frac{12}{30}$ or $\frac{2}{5}$ **b** $\frac{18}{30}$ or $\frac{3}{5}$ **c** $\frac{10}{30}$ or $\frac{1}{3}$ **d** $\frac{20}{30}$ or $\frac{2}{3}$ **e** $\frac{8}{30}$ or $\frac{4}{15}$ **f** $\frac{22}{30}$ or $\frac{11}{15}$

2 a $\frac{12}{20}$ or $\frac{3}{5}$ **b** $\frac{8}{20}$ or $\frac{2}{5}$ **c** $\frac{6}{20}$ or $\frac{3}{10}$ **d** $\frac{14}{20}$ or $\frac{7}{10}$ **e** $\frac{2}{20}$ or $\frac{1}{10}$ **f** $\frac{18}{20}$ or $\frac{9}{10}$

3 a $\frac{10}{24}$ or $\frac{5}{12}$ **b** $\frac{14}{24}$ or $\frac{7}{12}$ **c** $\frac{8}{24}$ or $\frac{1}{3}$ **d** $\frac{16}{24}$ or $\frac{2}{3}$ **e** $\frac{6}{24}$ or $\frac{1}{4}$ **f** $\frac{18}{24}$ or $\frac{3}{4}$

4 a $\frac{24}{60}$ or $\frac{2}{5}$ **b** $\frac{36}{60}$ or $\frac{3}{5}$ **c** $\frac{21}{60}$ or $\frac{7}{20}$ **d** $\frac{39}{60}$ or $\frac{13}{20}$ **e** $\frac{15}{60}$ or $\frac{1}{4}$ **f** $\frac{45}{60}$ or $\frac{3}{4}$

5 a $\frac{3}{6}$ or $\frac{1}{2}$ **b** $\frac{3}{6}$ or $\frac{1}{2}$ **c** $\frac{2}{6}$ or $\frac{1}{3}$ **d** $\frac{4}{6}$ or $\frac{2}{3}$ **e** $\frac{1}{6}$ **f** $\frac{5}{6}$

6 a $\frac{5}{15}$ or $\frac{1}{3}$ **b** $\frac{10}{15}$ or $\frac{2}{3}$ **c** $\frac{6}{15}$ or $\frac{2}{5}$ **d** $\frac{9}{15}$ or $\frac{3}{5}$ **e** $\frac{4}{15}$ **f** $\frac{11}{15}$

7 a $\frac{3}{7}$ **b** $\frac{4}{7}$ **c** $\frac{2}{7}$ **d** $\frac{5}{7}$ **e** $\frac{2}{7}$ **f** $\frac{5}{7}$

8 a 108 seconds **b** $\frac{48}{108}$ or $\frac{4}{9}$ **c** $\frac{60}{108}$ or $\frac{5}{9}$ **d** $\frac{12}{108}$ or $\frac{1}{9}$ **e** $\frac{96}{108}$ or $\frac{8}{9}$ **f** $\frac{30}{108}$ or $\frac{5}{18}$
g $\frac{78}{108}$ or $\frac{13}{18}$ **h** $\frac{18}{108}$ or $\frac{1}{6}$ **i** $\frac{90}{108}$ or $\frac{5}{6}$

INDEX